Smart Sensors for Structural Health Monitoring

Smart Sensors for Structural Health Monitoring

Special Issue Editors

Simon Laflamme
Filippo Ubertini
Jian Li

MDPI • Basel • Beijing • Wuhan • Barcelona • Belgrade

MDPI

Special Issue Editors

Simon Laflamme
Iowa State University
USA

Filippo Ubertini
University of Perugia
Italy

Jian Li
The University of Kansas
USA

Editorial Office
MDPI
St. Alban-Anlage 66
4052 Basel, Switzerland

This is a reprint of articles from the Special Issue published online in the open access journal *Sensors* (ISSN 1424-8220) from 2018 to 2019 (available at: https://www.mdpi.com/journal/sensors/special_issues/Sensors_SHM).

For citation purposes, cite each article independently as indicated on the article page online and as indicated below:

LastName, A.A.; LastName, B.B.; LastName, C.C. Article Title. *Journal Name* **Year**, *Article Number*, Page Range.

ISBN 978-3-03921-758-8 (Pbk)
ISBN 978-3-03921-759-5 (PDF)

Contents

About the Special Issue Editors

Simon Laflamme is an Associate Professor in the Department of Civil, Construction, and Environmental Engineering at Iowa State University. He holds a Courtesy Appointment in the Department of Electrical and Computer Engineering, and is an Associate Director of the Center for Nondestructive Evaluation. Dr. Laflamme received his Ph.D in Structures and Materials (2011). His research interests include structural health monitoring, sensor development, structural control, and smart systems. Dr. Laflamme has been awarded the Early Career Engineering Research Faculty Award by Iowa State University (2016), holds four U.S. patents, and published a textbook on Structural Control.

Filippo Ubertini is a Full Professor of Structural Design at the Department of Civil and Environmental Engineering of University of Perugia where he teaches Advanced Structural Design and Earthquake Engineering and where he coordinates the International Ph.D Program in Civil and Environmental Engineering. He graduated cum laude in Civil Engineering at University of Perugia in 2005 and received his Ph.D in Civil Engineering from University of Pavia in 2009. He was visiting scholar at Columbia University in 2008. Author of more than 75 papers published in high impact international journals, his research is mainly focused on structural health monitoring, with emphasis on smart materials and applications to earthquake engineering and cultural heritage structures. He is a member of the editorial boards of: *Sensors, Advances in Civil Engineering, Shock and Vibration, Mathematical Problems in Engineering* and *Engineering Research Express*. Dr. Ubertini has been awarded by the Italian Association for Wind Engineering in 2010 and has received best papers awards in EVACES 2011 and IOMAC 2019.

Jian Li is an Associate Professor in the Department of Civil, Environmental and Architectural Engineering at the University of Kansas. He received Ph.D in 2013 from the University of Illinois at Urbana-Champaign, and MS in 2007 and BS in 2005 from Harbin Institute of Technology, all in Civil Engineering. His research focuses on both theoretical and experimental developments of advanced sensing and health monitoring techniques to improve the resiliency and sustainability of civil infrastructure under operational and extreme loading conditions. His specific research interests include vibration-based damage detection and model updating, wireless smart sensor networks, innovative sensing techniques, computer vision, data assimilation, uncertainty quantification, risk assessment and mitigation. Dr. Li was awarded the Takuji Kobori Prize by the International Association of Structural Control and Monitoring (IASCM) in 2018, and the Rising Stars in Structural Engineering Award from the Civil + Structural Engineer Magazine in 2019. He currently serves as the secretary and a member of the Board of Directors of the US-China Earthquake Engineering Foundation, as well as editorial board members of several international journals such as *Smart Structures and Systems, Engineering Research Express, Mathematical Problems in Engineering*.

sensors

MDPI

Article

An Optical Crack Growth Sensor Using the Digital Sampling Moiré Method

Xinxing Chen [1,*], Chih-chen Chang [1], Jiannan Xiang [2], Chaobo Zhang [1] and Ming Liu [2]

[1] Department of Civil and Environmental Engineering, Hong Kong University of Science and Technology, Hong Kong, China; cechang@ust.hk (C-c.C.); czhangbd@connect.ust.hk (C.Z.)
[2] Department of Electronic and Computer Engineering, Hong Kong University of Science and Technology, Hong Kong, China; jxiangad@connect.ust.hk (J.X.); eelium@ust.hk (M.L.)
* Correspondence: xchenak@connect.ust.hk; Tel.: +852-5989-3264

Received: 6 September 2018; Accepted: 9 October 2018; Published: 15 October 2018

check for
updates

Abstract: High-accuracy crack growth measurement is crucial for the health assessment of concrete structures. In this work, an optical crack growth sensor using the digital sampling moiré (DSM) method is developed for two-dimensional (2D) crack growth monitoring. The DSM method generates moiré fringes from a single image through digital image processing, and it measures 2D displacements using the phase difference of moiré fringes between motion. Compared with the previous sensors using traditional photogrammetric algorithms such as the normalized cross-correlation (NCC) method, this new DSM-based sensor has several advantages: First, it is of a higher sensitivity and lower computational cost; second, it requires no prior calibration to get accurate 2D displacements which can greatly simplify the practical application for multiple crack monitoring. In addition, it is more robust to the change of imaging distance, which is determined by the height difference between two sides of a concrete crack. These advantages break the limitation of the NCC method and broaden the applicability of the crack growth sensor. These advantages have been verified with one numerical simulation and two laboratory tests.

Keywords: optical crack growth sensor; digital sampling moiré; 2D crack growth; calibration; concrete crack

1. Introduction

Concrete deterioration is usually initiated with the appearance of surface cracks. Excessive crack propagation may result in possible dysfunction or even failure of concrete structures [1]. Therefore, crack growth monitoring is of great significance in the risk assessment of a variety of constructed facilities [2]. Currently, a wide range of sensors have been developed to quantitatively measure the crack growth for structural health monitoring purposes. For instance, an optical fiber sensor was imbedded into the structural components to detect and track the opening of a crack on a real bridge [3]. Mao et al. proposed a method for corrosion process monitoring by combining the fiber Bragg grating and Brillouin optical time domain analysis [4]. Li et al. integrated the acoustic emission technique and fiber optic sensing for concrete deterioration tracking [5]. The strain gauges and the linear variable differential transformer (LVDT) were applied to measure the change in the crack width of a concrete specimen in the splitting tension test [6]. The eddy current sensor and the string potentiometer were attached on an adobe house to track the crack extension induced by the vibratory compassion excitation and climatological effects at micro-meter level [7]. These sensors are either imbedded into the structures or mounted on the structural surface, and they are cabled sensors that need to be connected with wires for data transmission. However, cable installation and maintenance are expensive and time-consuming, and the installation of cabled sensors would influence the normal operations of the facilities [8,9].

In contrast, the wireless crack growth sensors transmit data without cables and they have been applied for crack width monitoring on various civil structures: Bennett et al. deployed the wireless sensors with 12 µm resolution for long-term crack monitoring in the Prague Metro and the London Underground [10]; Hoult et al. applied the wireless sensors with a resolution of 10 µm for long-term crack monitoring on bridges [8]; Zhou et al. used a high-accuracy wireless sensors with 1 µm resolution to monitor the dynamic crack width on an in-service reactor containment building during an every-ten years pressure test [9]; Hughi and Marzouk investigated the performance of low-cost piezo-ceramic sensors for crack monitoring [11]; Caizzone and DiGiampaolo developed a passive radio frequency identification (RFID) crack width sensor with submillimeter resolution based on electromagnetic coupling [12].

One problem of these sensors is that they can only measure displacements along the alignment direction of the sensors, while cracks may propagate in multiple directions. Therefore, it is necessary to develop a wireless sensor that can measure two-dimensional (2D) crack growth. Zhang et al. proposed a smart film crack sensor for monitoring the cracks' location, shape, and propagation [13]. The smart film was composed of enameled wires, and the crack monitoring was based on the proportional relationship between the crack width and ultimate strain of the broken wire. The image processing technique is another full-field method that can track multi-directional crack propagation. Shan et al. presented a stereovision-based method for crack width measurement [14]. Two cameras were used to track the 3D coordinates of two crack edges, and the minimum distance between two edges is regarded as the crack width. Another image processing method, the digital image correlation method, has been applied to assess cracks on fiber concrete beams [15], where a single camera was required that must be aligned well to ensure that the image plane is parallel with the interested region. A "stick and detect" crack growth sensor using the normalized cross-correlation (NCC) method as the image processing algorithm was developed for 2D crack monitoring [16]. This crack growth sensor was glued on one side of the crack and took images of the pattern attached on the other side of the crack. The height difference between the two sides of the crack is the main factor that influences the imaging distance, which is the distance between the sensor and the target. The NCC algorithm tracks the displacement of multiple points in a series of images according to their light intensity distribution [17]. Since the obtained displacement is in the image domain with the unit of pixels, a scale factor (pixel/mm) relating the physical domain and the image domain should be calibrated to calculate the displacement in the physical domain. One challenge is that the scale factor is sensitive to the imaging distance, which is the distance between the sensor and the target [18]. The imaging distances are not constant for all cracks due to their variant height differences between the two sides. Therefore, the scale factor obtained through the calibration process with a fixed imaging distance cannot be ubiquitously applied on uneven cracks with different imaging distances and heights. The mismatch of the scale factor between calibration and application may lead to measurement errors and affect the accuracy of the NCC method [18].

The moiré technique has been adopted for displacement measurement for a long time [19]. Traditionally, the geometric moiré method generates moiré fringes through the overlapping of two regular gratings [20]. The generated moiré fringes can amplify the grating's motion, so a displacement measurement with a high accuracy can be obtained [19]. A scanning moiré method was proposed to generate moiré fringes by sampling a grating pattern using a scanner composed by a set of regularly spaced dots or lines [21]. However, the sensitivity of the geometric and scanning moiré methods are limited as they only use the information of the moiré fringes' centerlines [22,23]. To achieve higher sensitivity, a temporal phase shifting method which makes use of the whole intensity profiles of the grating was developed [24–26]. A series of moiré fringes were produced by moving the grating with a transducer, and then the obtained moiré fringes were used to compute the phase distribution of the grating. At least three moved moiré fringes are required for the phase calculation of one single grating, therefore this method is not suitable for dynamic analysis. Recently, a digital sampling moiré (DSM) method was developed for high-sensitivity and dynamic displacement measurements [22]. In this method, a series of phase-shifted moiré fringes was generated from one image of the captured grating pattern through down-sampling and up-sampling. The discrete Fourier transform was then used to calculate the phase distribution of the recorded grating. In the final step, the phase difference between the two recorded gratings was used to compute the relative displacement. Results showed that the accuracy of the DSM method could approach 3.8 μm in a three-point bending test of a steel beam, which was equivalent to a 0.01 pixel in the image domain [22]. Comparing this with the NCC method adopted in the wireless crack sensor, one advantage of the DSM method is that the obtained results are in the phase domain, and the corresponding physical displacements can be calculated easily using the predefined pitch length. As a result, the moiré technique does not require any prior calibration to compute the scale factor which relates the image domain with the physical domain, and this technique is robust to the imaging distance since the predefined pitch length is not affected by the imaging distance.

In this study, an optical crack growth monitoring sensor incorporating the DSM algorithm is developed. This sensor is composed of an optical navigation sensor board (ADNS-3080, Avago technologies, San Jose, CA, USA), a processor (Arduino UNO), a wireless platform (XBee), and a battery. The crack displacements are computed based on a series of images taken by the ADNS-3080. The captured images are processed by the Arduino UNO using the DSM method to calculate the 2D translations of cracks. In the following sections, the development of this sensor is presented, and its performance is tested by the numerical simulation and laboratory tests.

2. Prototype and Hardware Components

Figure 1a shows a prototype of the optical crack growth sensor, it is composed of four parts: Arduino UNO, ADNS-3080, XBee, and battery. The Arduino UNO is a microprocessor that contains 32 kilobytes (KB) flash memory and 2 KB static random-access memory (SRAM) [27]. The Arduino UNO contains several types of interfaces such as an inter-integrated circuit bus (I2C), a Serial Peripheral Interface Bus (SPI), and a Tx/Rx serial port. The variety of interfaces makes the Arduino UNO adaptable and able to connect with different types of sensors and devices. ADNS-3080 is an optical sensor with a complementary metal-oxide-semiconductor (CMOS) camera which acquires images of the underneath surface illuminated by the embedded light-emitting diode (LED) [28]. It also includes a digital signal processor (DSP) and a serial port. XBee is a wireless communication board with a receiver's sensitivity as high as −92 dBm [29]. According to the IEEE802.15.4 standard, XBee has a longer communication range than Bluetooth and requires less power [30]. As shown in Figure 1b, the ADNS-3080 and the XBee interface with the Arduino UNO through the Tx/Rx serial port and the SPI port, respectively. The battery can provide power through Universal Serial Bus (USB).

For crack growth monitoring, the ADNS-3080 receives the command from the Arduino UNO to capture images. The captured images are then sent to the Arduino UNO and are processed with the DSM algorithm to calculate 2D displacements. After the calculation, the results are transmitted wirelessly through the XBee to a computer for further analysis. Similar to other crack growth sensors for long-term crack monitoring [8–10], this crack growth sensor should be mounted across the crack. As shown in Figure 2, the grating pattern was attached on side 1 of the crack, and it was recorded by the developed sensor fixed on side 2 of the crack. The crack's growth was measured by detecting the relative motion between the fixed support and the 2D grating pattern. In Figure 2b, *d* is the imaging distance between the sensor and the target pattern. Since the thickness of the epoxy glue used to fix the sensor is ignorable, *d* is mainly determined by the height difference between the two sides of the crack.

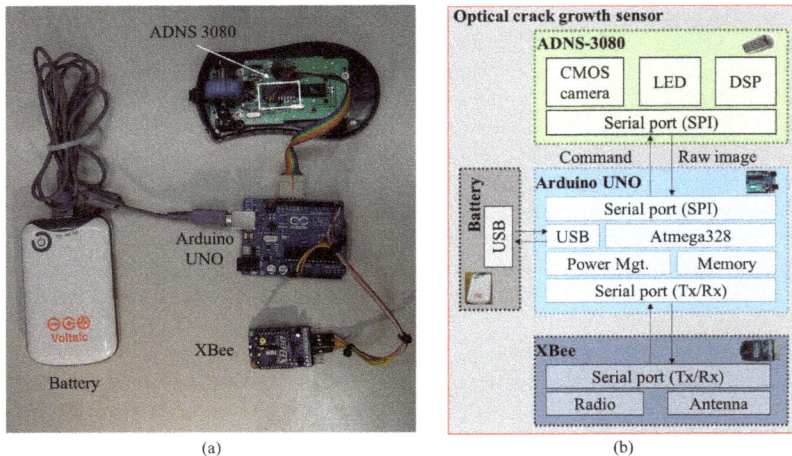

Figure 1. Optical crack growth sensor: (**a**) prototype; (**b**) schematic diagram.

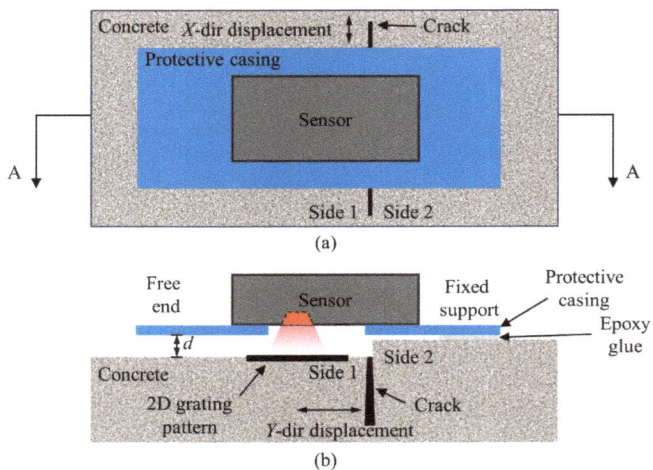

Figure 2. Alignment of the sensor: (**a**) top view; and (**b**) side view of section A-A.

3. The Digital Sampling Moiré (DSM) Method

The principle of the DSM method [22] is briefly summarized here. As shown in Figure 3, a 2D grating pattern with X- and Y-directional gratings is used to generate moiré fringes. The XY coordinates are attached on the 2D grating pattern, and their projection on the image plane in the uv image coordinates are represented as the xy coordinates. The physical lengths of one cycle of the corresponding grating (one white and one black bar) along the X and Y directions are denoted as P_X and P_Y, respectively, and their corresponding pitch lengths in the image domain are denoted as p_x and p_y, respectively. As shown in Figure 3b,c, the nearest integer values of pitch lengths of X- and Y-directional gratings in the image plane are s_x and s_y, respectively. To simplify the following derivation, P_X and P_Y are set to be identical as P, and likewise $p_x = p_y = p$, and $s_x = s_y = s$.

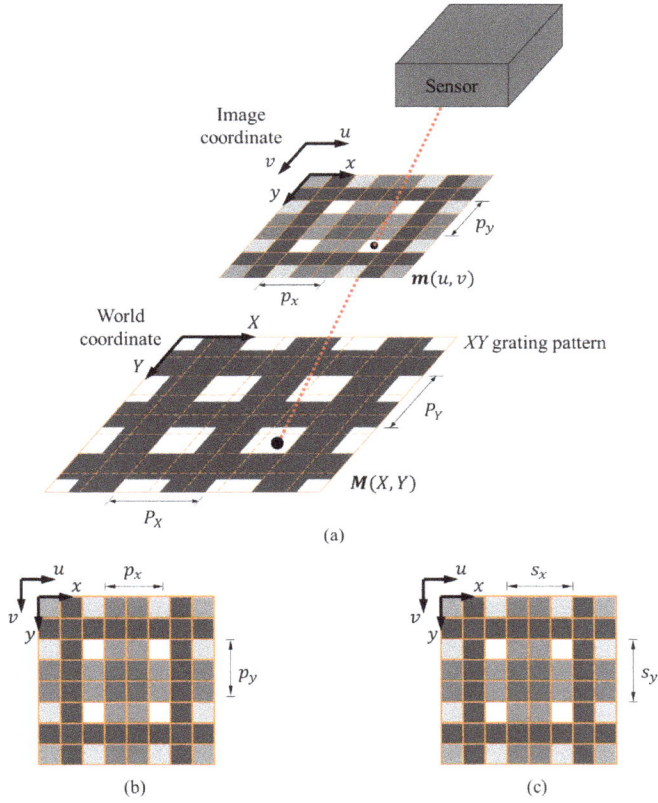

Figure 3. (**a**) The projection of the 2D grating pattern on the image plane; (**b**,**c**) the pitch length (p_x and p_y) and sampling pitch length (s_x and s_y) of the captured 2D grating pattern.

The DSM method computes the relative 2D displacement using a reference image I and a subsequent image I' acquired between the motion of the grating pattern (see Figure 4a,b). The first step is to extract the x- and y-directional gratings using $s \times 1$ pixel and $1 \times s$ pixel spatial average filters, respectively. The x-directional grating I_x and x'-directional gratings I'_x filtered from I and I' are shown in Figure 4c,d, respectively. The y-directional grating I_y and y'-directional gratings I'_y filtered from I and I' are shown in Figure 4e,f, respectively. The calculation process of X-directional displacement T_X is demonstrated in Figure 5. The extracted x-directional grating $I_x(u, v)$ and $I'_x(u, v)$ are shown in Figure 5a,b, respectively. Express $I_x(u, v)$ as

$$I_x(u, v) = a(u, v) \cos[\varphi_x(u, v)] + b(u, v), \tag{1}$$

where $a(u, v)$ is the amplitude of the grating intensity, and $b(u, v)$ represents the background intensity, and $\varphi_x(u, v)$ is the phase distribution of the x-directional grating.

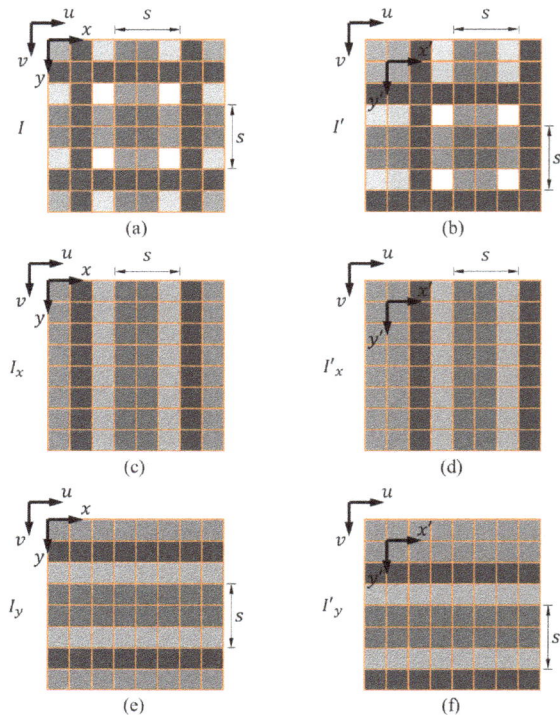

Figure 4. The captured 2D grating pattern in (**a**) the reference image I and (**b**) the subsequent image I'; the extracted (**c**) x-directional grating I_x and (**d**) x'-directional grating I'_x; the extracted (**e**) y-directional grating I_y and (**f**) y'-directional grating I'_y.

The DSM method generates moiré fringes through down-sampling and interpolation. Figure 5c,d illustrate the down-sampling of the gratings I_x and I'_x, in which every s pixels (in this illustration, $s = 3$ pixels) are chosen starting from the 1st, 2nd, and 3rd pixels to obtain a series of down-sampled gratings, respectively. Next, interpolation is performed to generate the corresponding moiré fringes shown in Figure 5e,f. The following equation presents the k-th moiré fringe $I_m(u, v; k)$

$$I_m(u, v; k) = a(u, v) \cos[\theta_x(u, v) + \frac{2\pi(k-1)}{s}] + b(u, v), \ (k = 1, \ldots, s) \tag{2}$$

where $\theta_x(u, v)$ is the phase distribution of the first moiré fringe, as shown in Figure 5g. Using the discrete Fourier transform, its value can be calculated as,

$$\theta_x(u, v) = -arctan \frac{\sum_{k=1}^{s} I_m(u, v; k) \times \sin(\frac{2\pi k}{s})}{\sum_{k=1}^{s} I_m(u, v; k) \times \cos(\frac{2\pi k}{s})}, \tag{3}$$

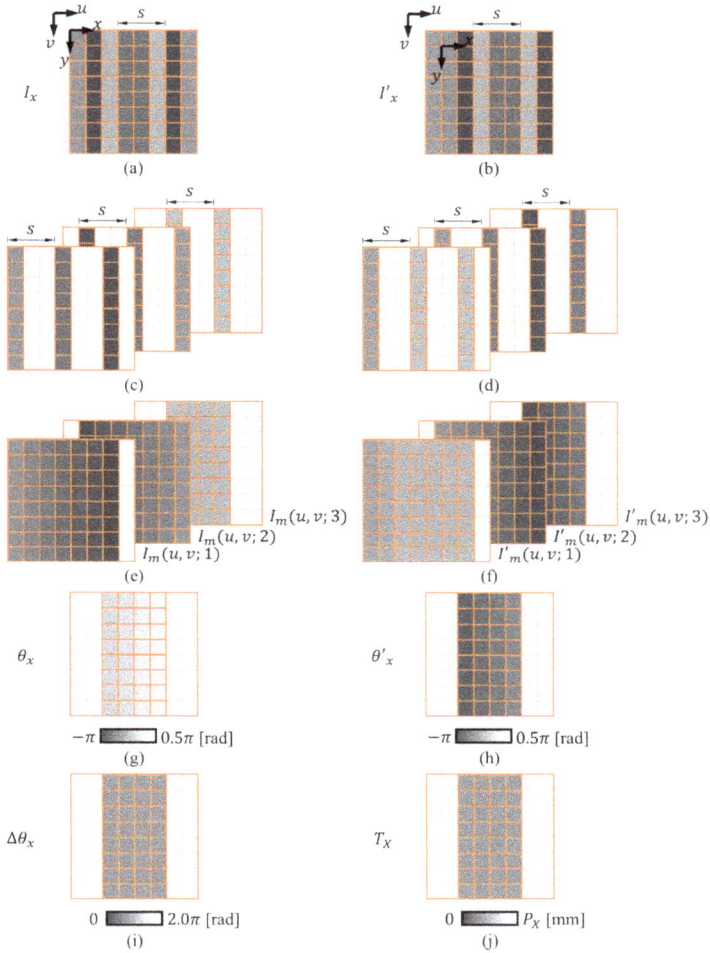

Figure 5. The calculation process of *x*-directional displacement T_X using the digital sampling moiré (DSM) method. (**a,b**) The filtered *x*-directional grating between motion; (**c,d**) corresponding down-sampled gratings; (**e,f**) corresponding moiré fringes; (**g,h**) phase distribution of corresponding moiré fringes; (**i**) the phase difference between the motion; (**j**) the obtained T_X.

Following the same procedure, the phase distribution of the moiré fringe after motion $\theta'_x(u,\ v)$ can be obtained by utilizing the moiré fringes $I'_m(u,\ v;\ k)$ (Figure 5h). Next, the phase difference $\Delta\theta_x$ (Figure 5i) can be calculated from the following equation:

$$\Delta\theta_x(u,\ v)\ =\ \theta_x(u,\ v)\ -\ \theta'_x(u,\ v). \qquad (4)$$

Finally, knowing the phase difference, the translation along *X*-direction, $T_X\ (u,\ v)$ (Figure 5j), can be calculated as follows

$$T_X(u,\ v)\ =\ \Delta\theta_x\ (u,\ v)\ \times\ \frac{P}{2\pi}, \qquad (5)$$

the translation along *Y*-direction $T_Y\ (u,\ v)$ can be obtained in the same way.

4. Performance Tests of the Optical Crack Growth Sensor

4.1. Simulation

In a previous study, the NCC method was embedded in an optical crack growth sensor to measure 2D crack displacements [16]. In this simulation, the DSM method was compared with the NCC method in terms of efficiency and accuracy. The NCC and DSM methods were coded in C language and uploaded to the Arduino UNO to process the same set of simulated images. The images were generated by MATLAB with a resolution of 24 × 24 pixels, corresponding to an area of 1.4 × 1.4 mm in the physical domain. The images were programmed to move at 10 μm/step for 10 steps. Different sets of simulated images with pitch lengths ranging from 3 to 7 pixels were analyzed in this simulation. Figure 6 shows the three representative 2D grating patterns with pitch lengths equal to 3, 5, and 7 pixels.

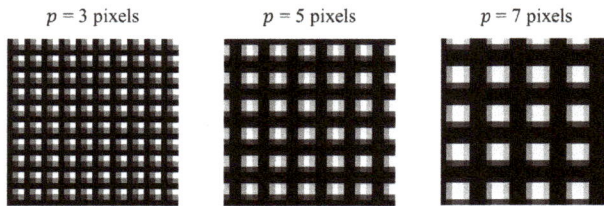

Figure 6. Simulation: the 2D grating patterns with different pitch lengths.

The performance of the NCC method is determined by the size of the region of interest (ROI) and the searching step for correlation calculation. A larger ROI produces more stable results and smaller searching step leads to higher sensitivity. On the other hand, the improved accuracy will result in a higher computational cost [16]. Following the previous study, the size of ROI was set to be 6 × 6 pixels and the searching step was equal to 0.2 pixel [16]. The simulated images were processed with the NCC algorithm embedded in the crack growth sensor, and the computational time for one output was 1.61 s. For the DSM method, its computational cost is related to the pitch length (p). When p ranges from 3 to 7 pixels, the corresponding computational cost ranges from 1.42 to 1.59 s per output, which is smaller than that of the NCC method.

Since the displacements obtained by the NCC method are in the image domain with pixel dimensions, a scale factor should be used to convert the displacements to the physical domain. Therefore, prior to the displacement measurement using the NCC method, calibration should be performed to obtain the scale factor [18]. Figure 7a,b show the measured displacements in pixel dimensions along the u- and v-direction, T_u and T_v obtained by the NCC method when $p = 3$ pixels, respectively. A line was fitted to the results using the least square method. The slopes of the fitted lines are the scale factors between the image coordinates and the physical coordinates. The average scale factor for two directions was found to be 16.2 pixels/mm.

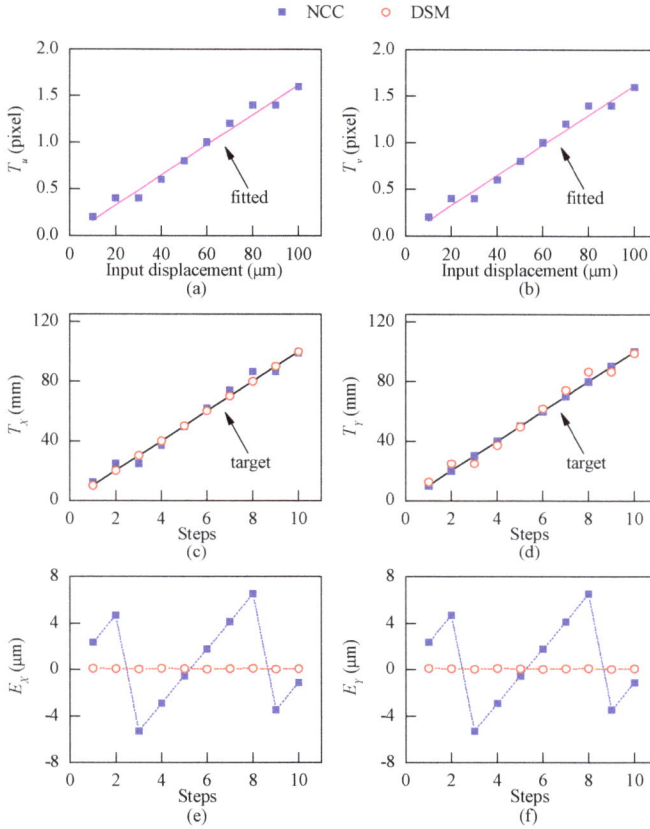

Figure 7. (**a**,**b**) Calibration for the normalized cross-correlation (NCC) method; (**c**,**d**) displacements obtained by the NCC and the DSM method; (**e**,**f**) displacement errors of two methods in the simulation.

In the DSM method, as shown in Equation (5), the phase difference of moiré fringes can be converted to the displacement in the physical domain with a predefined pitch length, hence no calibration is required. Figure 7c,d show the displacements in the physical domain obtained by the NCC method and the DSM method, respectively. It was observed that both methods can track the displacements along two directions. To compare the accuracy of the two methods, their errors were defined as the difference between the calculated and the actual displacements ($T_{calculate} - T_{true}$) and plotted in Figure 7e,f. The errors of the DSM method are significantly smaller than those of the NCC method. The zigzag pattern of the error distribution of the NCC method is due to the limitation of its resolution. The resolution of the NCC method is mainly determined by the searching step [16], which was 0.2 pixel in this case. As the calibrated scale factor is 16.2 pixel/mm, the resolution of the NCC method is 12.5 μm. The mean absolute error (MAE) and standard deviation (SD) are used here to quantify the error. For the DSM method, the MAE and SD are 0.04 μ*m* and 0.03 μm, respectively. In comparison, the MAE and SD of the NCC method are 3.41 μm and 3.90 μm, respectively. Hence, for $p = 3$ pixels, the accuracy of the DSM method is about 100 times higher than the NCC method.

Figure 8 shows the error analysis of the two methods when p changes from 3 to 7 pixels with an interval of 0.5 pixels. The MAE and SD of the NCC method range from 3 μm to 4 μm, while the corresponding values of the DSM method are smaller than 0.12 μm. This numerical simulation illustrates that the NCC method requires calibration to obtain the scale factor, which is not required by the DSM method. Moreover, the DSM method is of higher accuracy than the NCC method.

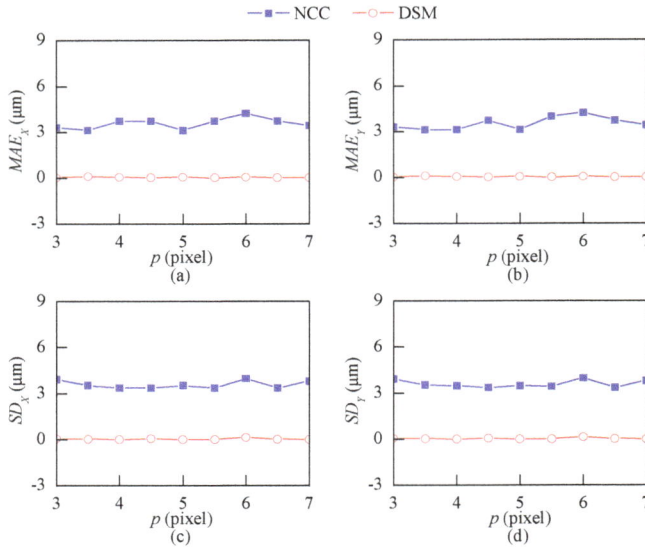

Figure 8. Error analysis for patterns with different pitch lengths. (**a**,**b**) The mean absolute error (MAE); (**c**,**d**) standard deviation (SD).

4.2. The XYZ-Table Test

In this section, to demonstrate the accuracy of the DSM method and its robustness to the change of imaging distance, an *XYZ*-table test was performed.

The setup of *XYZ*-table test is shown in Figure 9a. In brief, ADNS-3080 was mounted on a fixed over-hang platform as a cantilever support on one side, while the camera of the ADNS-3080 was targeted at the 2D grating pattern attached onto the *XYZ*-table. The images captured by the ADNS-3080 would be sent to Arduino UNO and XBee for analysis and wireless transmission. The *XYZ*-table can be manually controlled to translate along three directions with an accuracy of 1 μm. The relative 2D motion between the fixed over-hang platform and the *XYZ*-table can be adjusted by the *X*- and *Y*-directional bars to simulate 2D crack growth. The *Z*-directional bar controls the distance (*d*) from the optical crack growth sensor and the target, which was used to simulate the variation of *d* for monitoring cracks with different relative heights between two surfaces (Figure 2b).

In this test, similar to the previous part, the NCC method and the DSM method were used to measure the displacement of the grating pattern attached to the *XYZ*-table. As mentioned before, the NCC method requires calibration to get the scale factor relating the physical domain and the image domain. The calibration process for the NCC method was performed in a similar way as described in the previous study [16]. During the calibration, the sensor is usually kept very close to the target, while no friction was allowed between them (*d* = 0 mm). After the calibration, the crack growth sensor is installed on a real crack to measure crack displacements with the scale factor obtained in the calibration. As mentioned earlier, the scale factor is a function of the distance *d*. For a real case with a different height of the concrete surfaces on the two sides of an uneven crack (Figure 2b), it is likely that there would be a discrepancy between the calibration and the actual condition. Therefore, the inconsistency of the scale factor between the calibration and the real application could be a source of measurement errors for the NCC method. On the other hand, as shown in Equation (5), the DSM method obtains the displacement in the physical domain using the predefined value P that is independent of the distance d. Therefore, the accuracy of the DSM method is less sensitive to the distance d.

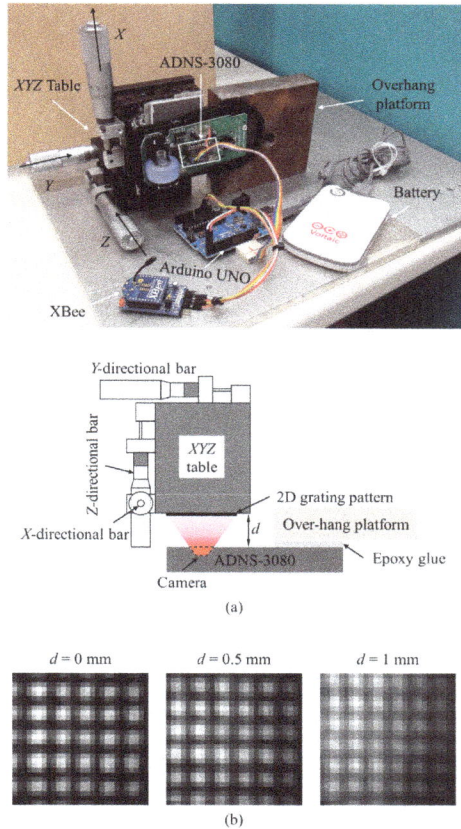

Figure 9. The XYZ-table test: (**a**) schematic of the setup; (**b**) the captured 2D grating patterns at different distances (d).

In this XYZ-table test, the robustness of the DSM and the NCC methods to the d were studied. As shown in Figure 9a, d can be controlled by rotating the Z-directional bar of the XYZ-table. In the test, d values were controlled to be 0 mm, 0.5 mm, and 1 mm, and the captured images of the 2D grating patterns for different d values are shown in Figure 9b. It is found that the captured 2D grating pattern contains more gratings with the increase of d. The enlarged field of view is due to the larger scale factor (mm/pixel) with the increase of d.

Calibration was performed for $d = 0$ mm by moving the 2D grating pattern along X- and Y-directions simultaneously with 10 μm per step for 10 steps in total. Linear fitting was used to get the scale factor based on the input displacements and the output T_u and T_v. The scale factor was calculated to be 16.7 pixel/mm. After calibration, for each d value ($d = 0$ mm, 0.5 mm, or 1.0 mm), the target pattern was controlled to translate in 2D with 10 μm per step along X- and Y-directions.

The captured grating images with the different d values were analyzed by both the DSM and the NCC methods, and the results are plotted in Figure 10. It shows that the DSM method tracks the displacements precisely for the three values of d. On the contrary, the errors of the NCC method increased as d changes from 0 mm to 1 mm. The increased errors mainly result from the change of scale factor when *d* alters. The MAE and SD of the measurement results with the two methods for each value of *d* are also shown in Figure 10. The error of the DSM method is not affected by *d* and its MAE and SD remain smaller than 2 μm and 1 μm, respectively. In contrast, for the NCC method, the MAE and SD of measurements increase from 3.37 μm to 8.76 μm and from 3.29 μm to 6.21 μm, respectively, as *d* changes from 0 mm to 1 mm.

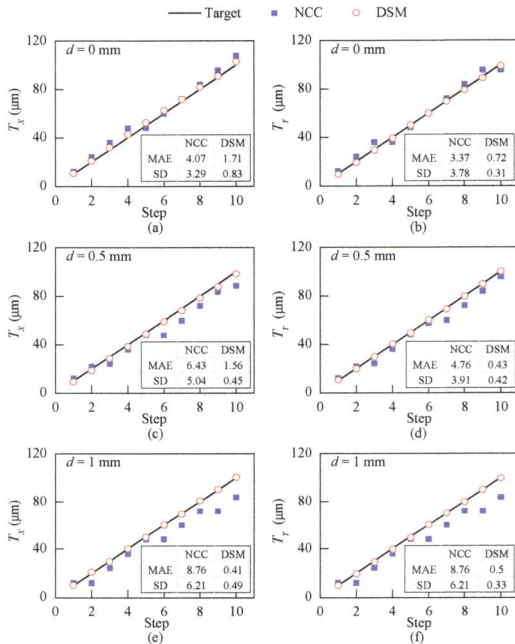

Figure 10. The *XYZ*-table test: the *X*- and *Y*-direction displacement measurement results when (**a**,**b**) *d* = 0 mm, (**c**,**d**) *d* = 0.5 mm and (**e**,**f**) *d* = 1.0 mm.

This laboratory *XYZ*-table test shows that the DSM method can achieve higher accuracy than the NCC method, while needs no prior calibration. Moreover, it is verified that the DSM method is more robust to the change of *d* (see Figure 2). The NCC method requires a calibrated scale factor (mm/pixel) to convert the 2D displacement from pixel domain (T_u and T_v) to the physical domain (T_X and T_Y). The results presented in Figure 10 show that the DSM method is more robust to the change of *d* than the NCC method. This is because the NCC requires a calibrated scale factor (mm/pixel) to convert the 2D displacement from pixel domain (T_u and T_v) to the physical domain (T_X and T_Y). This scale factor is affected by *d*, hence different *d* values would lead to different measurement results. On the other hand, the DSM method measures the physical displacement using a predefined pitch length *P* which is input to the algorithm as a constant. The robustness to imaging distance makes the DSM method especially suitable to measure the propagation of uneven cracks in which the distance between sensor and crack surface is difficult to control.

4.3. Concrete Crack Test

To evaluate the performance of the developed optical crack growth sensor for real crack monitoring, a reinforcement concrete (RC) structure test was conducted to measure the motions of two cracks with different *d* values: a flat crack (*d* = 0 mm) and an uneven crack (*d* > 0 mm). The cracks were located on two RC beam-column knee joints. Figure 11a shows the top view of the specimens.

The lengths of the beams and the columns for both specimens were 1800 mm. For Specimen 1, the beam and the column had the same square cross-section of 300 mm × 300 mm, while the corresponding cross-sections of Specimen 2 were 300 mm × 300 mm and 300 mm × 400 mm, respectively. A diagonally placed hydraulic actuator was connected to the beam and column tips of the specimens to produce harmonic motion. Specimen 1 and Specimen 2 were forced by the actuator to move with 3 mm and 30 mm amplitude, respectively, under the frequency of 0.01 Hz. Figure 11b,c show the location of the considered cracks on Specimen 1 and Specimen 2, respectively. The crack on Specimen 1 was flat, while the crack on Specimen 2 was uneven and its d value approached 1.5 mm. The optical crack growth sensor was attached across the cracks on two specimens to monitor their harmonic motions, as shown in Figure 11d,e. Apart from the optical sensor, an LVDT with 5 μm resolution was installed beneath the sensor to measure the one-dimensional displacement of the cracks, as shown in Figure 11d,e. The LVDT measured Y_1- and Y_2-directional motion for the 1st and the 2nd crack, respectively.

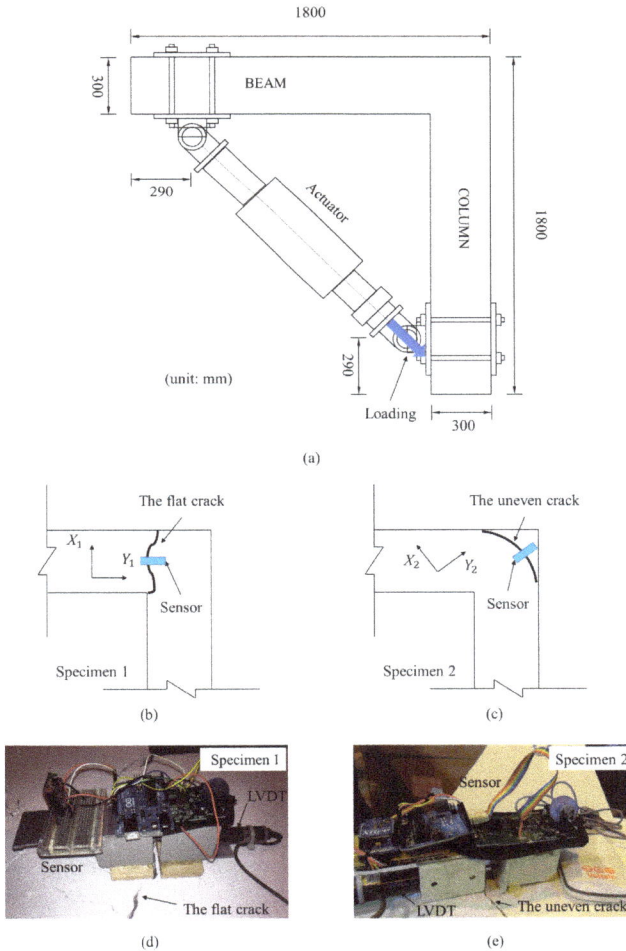

Figure 11. The beam-column knee joint test: (**a**) the vertical view of the RC specimens; schematics of (**b**) the 1st crack and (**c**) the 2nd crack; the alignment of the sensors on (**d**) the 1st crack and (**e**) the 2nd crack.

The results obtained by the LVDT and the optical sensor on two cracks are plotted in Figure 12. The NCC and DSM algorithms were integrated into the optical crack growth sensor, respectively, to compare with the LVDT. Figure 12a,b show the X_1- and Y_1-directional displacements of the 1st crack, in which the displacements obtained by the DSM and the NCC methods match well with each other. In Figure 12b, the Y_1-directional displacement measured by the LVDT shows a good agreement with the other two methods. However, Figure 12c,d show that for the second specimen the displacements obtained by the NCC method are significantly smaller than those measured by the DSM method. Besides, as shown in Figure 12d, the displacements measured by the LVDT agree well with those of the DSM method, indicating that only the DSM method can accurately track the propagation of cracks with uneven surfaces. The lower accuracy of the NCC method is due to the mismatch of the scale factor between the calibration and real application. As the calibration was performed for $d = 0$ mm, the uneven surfaces of the crack increase the d value and make the calibrated scale factor inaccurate for real application.

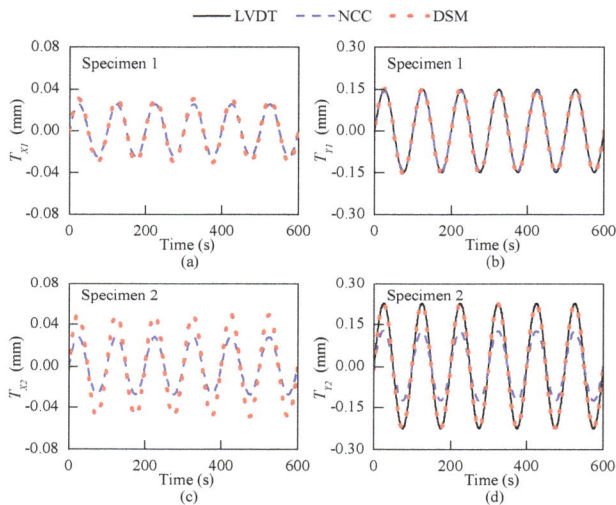

Figure 12. The beam-column knee joint test: displacement responses of cracks on two specimens along X- and Y-direction.

This laboratory test on real cracks demonstrates the advantages of the DSM method to measure the displacements of uneven cracks with high accuracy. This calibration-free and robust DSM-based crack growth sensor can play an important role in the crack growth monitoring on a variety of civil structures.

5. Discussion

In this work, a DSM-based optical crack growth sensor for the health assessment of concrete structures was developed by integrating Arduino UNO, XBee, and an optical navigation sensor board (ANDS-3080). Compared to the previous sensor which uses the NCC method as displacement measurement algorithm, this newly developed sensor adopts the DSM method and exhibits a few improvements. First, a calibration process is not required to get the accurate scale factor between the image domain and the physical domain; second, it can be applied on cracks with different d (the imaging distance between the target pattern and the sensor); third, a higher accuracy can be achieved with a lower computational cost.

Simulations and laboratory tests were performed to compare the NCC and DSM methods in multiple aspects. The simulation shows that the DSM method does not need calibration to obtain the scale factor of the imaging system, and it can achieve a higher sensitivity at a lower computational cost than the NCC method. The *XYZ*-table test demonstrates that the DSM-based sensor can achieve high-accuracy displacement measurements with MAE and SD smaller than 2 μm and 1 μm, respectively. Meanwhile, it also shows the robustness of the DSM method to the imaging distance *d*. The optical sensor was then used to track the 2D harmonic motion of two real cracks. The results show that the NCC method can measure accurate displacements only for the crack with a flat surface, of which the scale factor is consistent between calibration and the real test. On the other hand, the DSM-based sensor can detect precise crack motions for both flat and uneven cracks with different *d* values. The results exhibit the robustness of the proposed optical sensor to uneven cracks on real structures. These advantages make the DSM-based optical crack growth sensor a powerful tool for high-accuracy monitoring of 2D concrete crack growth.

While the experimental tests in this proof-of-concept study were performed in a laboratory under relatively stable temperature and humidity conditions, the proposed sensor exhibits some characteristics which might give it some edges for actual implementation. The operating temperature of the sensor's main components (ADNS-3080, Arduino UNO, XBee, and battery) is in the range of 0–40 °C, which seems to meet the demand of most structures under normal environmental conditions. In addition, this sensor has an embedded camera and a built-in LED as light source. In application, the sensor is attached closely to the concrete surface and the lighting for image acquisition is solely supplied by the built-in LED. Its fixed mechanical structure creates a stable illumination and image acquisition condition that minimizes the influence of ambient light. However, the sensor's components should be condensed, integrated, and weather-proofed before the sensor can be used for long-term monitoring under ambient conditions. Moreover, the temperature change would induce the deformation of the concrete and the grating pattern. To reduce the influence of this, the concrete deformation caused by thermal effects should be compensated for, and the grating pattern could be printed on a solid material such as a high-resistance magnet [30].

Author Contributions: X.C. conceived the presented idea, conducted the experiments and wrote the paper. C-c.C. supervised the progress of the study and helped to structure the paper. J.X. and M.L. contributed to the development of software. C.Z. helped in the hardware and experiments.

Funding: This study is partly supported by the Hong Kong Research Grants CERG 611112.

Conflicts of Interest: The authors declare no conflict of interest.

References

1. Mohammad, I.; Huang, H. Monitoring fatigue crack growth and opening using antenna sensors. *Smart Mater. Struct.* **2010**, *19*, 1–8. [CrossRef]
2. Hoult, N.A.; Dutton, M.; Hoag, A.; Take, W.A. Measuring crack movement in reinforced concrete using digital image correlation: Overview and application to shear slip measurements. *Proc. IEEE* **2016**, *104*, 1561–1574. [CrossRef]
3. Enckell, M.; Glisic, B.; Myrvoll, F.; Bergstrand, B. Evaluation of a large-scale bridge strain, temperature and crack monitoring with distributed fibre optic sensors. *J. Civ. Struct. Health Monit.* **2011**, *1*, 37–46. [CrossRef]
4. Mao, J.; Chen, J.; Cui, L.; Jin, W.; Xu, C.; He, Y. Monitoring the corrosion process of reinforced concrete using BOTDA and FBG sensors. *Sensors* **2015**, *15*, 8866–8883. [CrossRef] [PubMed]
5. Li, W.; Xu, C.; Ho, S.C.M.; Wang, B.; Song, G. Monitoring concrete deterioration due to reinforcement corrosion by integrating acoustic emission and FBG strain measurements. *Sensors* **2017**, *17*, 657. [CrossRef] [PubMed]
6. Rapoport, J.; Aldea, C.; Shah, S.P.; Ankenman, B.; Karr, A. Permeability of cracked steel fiber-reinforced concrete. *J. Mater. Civ. Eng.* **2002**, *14*, 355–358. [CrossRef]
7. Abeel, P.; Aimone-Martin, C.; Dowding, C. Autonomous crack measurement for comparison of vibratory compaction excitation and climatological effects. In Proceedings of the 7th European Workshop on Structural Health Monitoring (EWSHM), Nantes, France, 8–11 July 2014.

8. Hoult, N.A.; Fidler, P.R.; Hill, P.G.; Middleton, C.R. Long-term wireless structural health monitoring of the Ferriby Road Bridge. *J. Bridg. Eng.* **2010**, *15*, 153–159. [CrossRef]

9. Zhou, J.; Xu, Y.; Zhang, T. A wireless monitoring system for cracks on the surface of reactor containment buildings. *Sensors* **2016**, *16*, 883. [CrossRef] [PubMed]

10. Bennett, P.J.; Soga, K.; Wassell, I.; Fidler, P.; Abe, K.; Kobayashi, Y.; Vanicek, M. Wireless sensor networks for underground railway applications: Case studies in Prague and London. *Smart Struct. Syst.* **2010**, *6*, 619–639. [CrossRef]

11. Hughi, D.; Marzouk, H. Crack width monitoring system for reinforced concrete beams using piezo-ceramic sensors. *J. Civ. Struct. Health Monit.* **2015**, *5*, 57–66. [CrossRef]

12. Caizzone, S.; DiGiampaolo, E. Wireless passive RFID crack width sensor for structural health monitoring. *IEEE Sens. J.* **2015**, *15*, 6767–6774. [CrossRef]

13. Zhang, B.; Wang, S.; Li, X.; Zhang, X.; Yang, G.; Qiu, M. Crack width monitoring of concrete structures based on smart film. *Smart Mater. Struct.* **2014**, *23*, 1–15. [CrossRef]

14. Shan, B.; Zheng, S.; Ou, J. A stereovision-based crack width detection approach for concrete surface assessment. *KSCE J. Civ. Eng.* **2016**, *20*, 803–812. [CrossRef]

15. Hamrat, M.; Boulekbache, B.; Chemrouk, M.; Amziane, S. Flexural cracking behavior of normal strength, high strength and high strength fiber concrete beams, using digital image correlation technique. *Constr. Build. Mater.* **2016**, *106*, 678–692. [CrossRef]

16. Man, S.H.; Chang, C.C. Design and performance tests of a LED-based two-dimensional wireless crack propagation sensor. *Struct. Control Health Monit.* **2016**, *23*, 668–683. [CrossRef]

17. Yoo, J.C.; Han, T.H. Fast normalized cross-correlation. *Circuits Syst. Signal Process* **2009**, *28*, 819–843. [CrossRef]

18. Palacin, J.; Valganon, I.; Pernia, R. The optical mouse for indoor mobile robot odometry measurement. *Sens. Actuators A Phys.* **2006**, *126*, 141–147. [CrossRef]

19. Walker, C.A. *Handbook of Moiré Measurement*; CRC Press: Boca Raton, FL, USA, 2003.

20. Yokozeki, S.; Kusaka, Y.; Patorski, K. Geometric parameters of moiré fringes. *Appl. Opt.* **1976**, *15*, 2223–2227. [CrossRef] [PubMed]

21. Morimoto, Y.; Hayashi, T. Deformation Measurement during powder compaction by a scanning-moire Method. *Exp. Mech.* **1984**, *24*, 112–116. [CrossRef]

22. Ri, S.; Fujigaki, M.; Morimoto, Y. Sampling moiré method for accurate small deformation distribution measurement. *Exp. Mech.* **2010**, *50*, 501–508. [CrossRef]

23. Wang, Q.H.; Ri, S.; Tsuda, H.; Koyama, M.; Tsuzaki, K. Two-dimensional moiré phase analysis for accurate strain distribution measurement and application in crack prediction. *Opt. Express* **2017**, *25*, 13465–13480. [CrossRef] [PubMed]

24. Choi, Y.B.; Kim, S.W. Phase-shifting grating projection moiré topography. *Opt. Eng.* **1998**, *37*, 1005–1010. [CrossRef]

25. Kujawinska, M. Use of phase-stepping automatic fringe analysis in moiré interferometry. *Appl. Opt.* **1987**, *26*, 4712–4714. [CrossRef] [PubMed]

26. Ai, C.; Wyant, J.C. Effect of piezoelectric transducer nonlinearity on phase shift interferometry. *Appl. Opt.* **1987**, *26*, 1112–1116. [CrossRef] [PubMed]

27. Arduino UNO REV3. Available online: https://store.arduino.cc/usa/arduino-uno-rev3 (accessed on 4 September 2018).

28. ADNS 3080. Available online: https://people.ece.cornell.edu/land/courses/ece4760/FinalProjects/s2009/ncr6_wjw27/ncr6_wjw27/docs/adns_3080.pdf (accessed on 4 September 2018).

29. XBee®/XBee-PRO® RF Modules. Available online: https://www.sparkfun.com/datasheets/Wireless/Zigbee/XBee-Datasheet.pdf (accessed on 4 September 2018).

30. Lee, J.S.; Su, Y.W.; Shen, C.C. A comparative study of wireless protocols: Bluetooth, UWB, ZigBee, and Wi-Fi. In Proceedings of the 33rd Annual Conference of IEEE Industrial Electronics Society (IECON), Taipei, Taiwan, 5–8 November 2007; pp. 46–51.

sensors

MDPI

Article

A Displacement Sensor Based on a Normal Mode Helical Antenna

Songtao Xue [1,2], Zhuoran Yi [1], Liyu Xie [1,*] , Guochun Wan [3] and Tao Ding [4]

[1] Department of Disaster Mitigation for Structures, Tongji University, Shanghai 200092, China
[2] Department of Architecture, Tohoku Institute of Technology, Sendai 982-8577, Japan
[3] Department of Electronic Science and Technology, Tongji University, Shanghai 200092, China
[4] Institute of Precision Optical Engineering, School of Physics Science and Engineering, Tongji University, Shanghai 200092, China
* Correspondence: liyuxie@tongji.edu.cn; Tel.: +86-21-6598-2390

Received: 28 July 2019; Accepted: 29 August 2019; Published: 30 August 2019

check for
updates

Abstract: This paper presents a passive displacement sensor based on a normal mode helical antenna. The sensor consists of an external helical antenna and an inserting dielectric rod. First, the perturbation theory is adopted to demonstrate that both the electric intensity and magnetic intensity have a noticeable gradient change within the in-and-out entrance of the helical antenna, which will cause the sensor to experience a resonant frequency shift. This phenomenon was further verified by numerical simulation using the Ansoft high frequency structure simulator (HFSS), and results show the linear correlation between the retrieved resonant frequency and the displacement. Two sets of proposed sensors were fabricated. The experiments validated that the resonant frequency shifts are linearly proportional to the applied displacement, and the sensing range can be adjusted to accommodate the user's needs.

Keywords: displacement sensor; helical antenna; resonant frequency; perturbation theory; normal mode

1. Introduction

A civil structure is usually designed to fulfill its functionality for at least 50 years under service loadings and occasional situations, such as earthquakes, typhoons, explosions, and so on [1]. During its life span, both service loadings and occasional unexpected loadings may cause structural damage, such as cracks or fatigue and corrosion problems, which can lead to a catastrophic event after years of service. To guard against structural failure during long-term service, structural health monitoring has developed rapidly and has been applied widely over the last few decades [2–4].

Among the measurands related to a building's structural health state, its deformation in terms of strain or displacement is a significant indicator of local damage or a failed bearing capacity of individual members. Displacement sensors are either wired or wireless. The wired sensors use cables to supply energy power and transmit data to devices such as laser transducers [5] and linear variable displacement transducers (LVDTs) [6], which usually have reliable performance, good environmental resistance, and high resolution. However, the installation of wired sensors is both time- and labor-consuming due to the cable deployment. Furthermore, the wired sensing system is prone to be undermined by human activities or natural calamities.

To facilitate the deployment of sensors, various wireless sensors are developed to discard cables for data transmission using WiFi [7], the general packet radio service (GPRS), Zigbee [8,9] or other telecommunication technologies [10]. They require an analog-digital converter (ADC) for signal conversion, a microprocessor unit (MPU) for data processing, and a wireless data transceiver for data transmission [11]. However, wireless sensors still need energy to function, either with on-board batteries or energy scavenging technologies, such as solar panels or vibrometers [12,13]. These energy supply solutions will compromise the reliability of the sensors during long-term service and increase both the complexity and the cost of sensors [11,14].

Fortunately, the power lines can be eliminated by introducing passive sensing technologies, such as the surface acoustic wave (SAW)-based [15], inductive coupled [16], and radio-frequency identification (RFID) enabled sensors [17–19], as well as chipless passive wireless antenna sensors [20,21]. Because these passive sensors do not have a radio on board, they consume little power harvested from interrogation waves or need no power at all. Among all wireless sensors, patch-antenna-based wireless sensors, a dielectric substrate sandwiched between a radiation patch and a ground plane, have been widely adopted in structural health monitoring [14,22,23] due to their simple configuration, multimodality, low cost and other advantages. The antenna radiation parameters, such as resonant frequency, are sensitive to temperature [24], strain [11,25,26], displacement [14,27], and cracking [17,26,28], etc. The relationship between the antenna radiation parameters and the physical measurements can be utilized as the sensing mechanism.

Many passive antenna sensors for deformation monitoring have been developed. Lopato et al. [29] designed a strain sensor based on a circular patch antenna. Both the value and direction of the strain could be measured by analyzing two different current distributions between the first and second resonant frequency. Xue et al. [30] presented a novel crack-measuring sensor based on a rectangular patch antenna fed by a pair of microstrip lines, which formed a parallel plate capacitor as a crack-sensing unit. Patch antennas with low-profile, planar substrates can be easily deployed on the surface of relevant structures, but these antennas can also pose difficulties due to the flexible substrates when they are embedded in structures.

The helical antenna provides an alternative approach for sensing displacement passively, and it has potential merit as an embedded sensor in structures. With the shape of a spiral, the helical antenna is widely used in signal transmissions [31–33] and circuit building [33,34] for sensing physical quantities [34–36].

Simons et al. [33] proposed a miniaturized inductor/antenna system for the non-contact powering of an oscillator circuit. The inductor coil is equivalent to a helical antenna and is used to transmit signals by losing power through RF radiation from the inductor. Akira et al. [34] introduced an L-C circuit formed by a capacitance and inductance (helical antenna) for retrieving wirelessly the damage index of structures. However, besides selecting inductance as the sensing part and as the information transfer part, an extra capacitance or inductance is needed to form an L-C circuit [31,32,34,37]. These sensors are large in size and trouble-prone due to the complex circuit design. To simplify the L-C circuit structure, the planar inductor-capacitor (LC) circuit has been investigated [38,39]. However, the planar inductor-capacitor circuits will have the same difficulties as patch antennas when they are embedded in structures.

In recent years, miniaturized embedded helical antennas have been proposed and applied in human body health monitoring [35–37]. Huang et al. [35] proposed a helical-antenna-based liquid level sensor having an internal cavity filled with liquid, whose resonant frequency is sensitive to the variation of the inside material. Murphy et al. [36] used a pseudo-normal-mode helical antenna as part of a deeply implanted wireless sensor to transmit signals. After parameter optimization, results proved that this helical antenna is an excellent candidate for being implanted. However, because these helical antenna sensors in body health monitoring often favor qualitative measurement over quantitative measurement [35], they can not meet the accuracy requirement of deformation sensing in structure health monitoring.

This paper proposes a displacement sensor based on a normal mode helical antenna. This sensor consists of an external helical antenna and an internal silicon rod. When the rod is passing through the electromagnetic field inside the antenna near the in-and-out entrance, the resonant frequencies of the sensor system will vary noticeably. Because of the compact design and encapsulated structure, it can be embedded inside concrete materials. Since the major part of the sensor is a silicon rod and a helical antenna, the cost of this antenna sensor will be extremely low.

This paper is organized as follows. Section 2 introduces the concept of the displacement sensor based on a helical antenna and illustrates the sensing mechanism using the perturbation theory. This section describes one prototype of the displacement sensor based on a normal mode helical antenna. Section 3 introduces simplified principles to create a preliminary helical antenna design. In Section 4, the appropriate dimension parameters are determined for the helical antenna and interior silicon rod. Section 5 describes the fabrication of sensors and the instrumentation setup of experiments. Conclusions are then drawn and future research potential is discussed.

2. Displacement Sensor Using a Helical Antenna

A finite length helical antenna fed by wide band electromagnetic waves, as depicted in Figure 1, has a flat gradient of the electric intensity and magnetic intensity fields in its central part, while maintaining a steep gradient of intensities near both the in-and-out entrances of the spiral [40]. The electromagnetic field can be altered by the property variation of the objects inside the antenna [35,41], by object dislocation and material changing, while the altered field leads to the change of resonant frequency. Based on this principle, the authors propose a system consisting of a helical antenna and an inserted dielectric rod as illustrated in Figure 1, where both the helical antenna and dielectric rod are sharing the same central axial line. The electrical parameters of the system, such as the resonant frequencies, electric echo loss and current direction, vary when the dielectric rod is moving along the axis. The shift of its resonant frequencies can be selected as the distinguished feature representing the location of the rod. Then, a helical antenna displacement sensor is proposed.

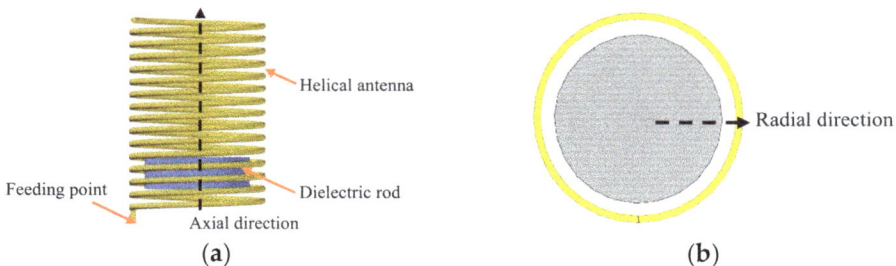

Figure 1. Concept of a displacement sensor using a helical antenna (**a**) Front view; (**b**) Top view.

2.1. Computational Electromagnetics Model of Proposed Displacement Sensor

The proposed helical antenna for displacement sensing works in normal mode. Under this working mode, the electromagnetic field radiated by the antenna is maximum in the radial direction and minimum along the axial direction. In order to simulate the electromagnetic property of the helical antenna, the method of moments (MOM) for computational electromagnetics can be exploited. The resulting electromagnetic field of the helical antenna can be determined by the finite element method, and so can the resonant frequencies of the helical antenna. For the sake of simplicity, the helical antenna can be modeled approximately by several small loops and short dipoles connected in a series, and the electromagnetic fields can be obtained by a superposition of the individual electromagnetic fields of each elemental loop and dipole [42]. However, several difficulties will occur while attempting to solve the Maxwell equations, which involve definition of boundary conditions and complex calculations. This makes the computational electromagnetics model inappropriate for the antenna simulation.

2.2. Approximation Using Perturbation Theory

Due to the similar current distribution pattern and mode of resonance, the helical antenna can be regarded as a cavity resonator for the proposed displacement sensor. In practical applications, a slight dislocation of the inside material can influence the electromagnetic field of the cavity resonator, which will consequently cause the resonant frequencies to shift. This disturbance of the cavity resonator is due to a slight change that can be approximated by the perturbation theory [41].

Near the in-and-out entrance of the antenna, the electric field intensity and the magnetic field intensity inside the helical antenna have a noticeable gradient change in the axial direction. We refer this area as the steep gradient region. Inside the middle area of the helical antenna, the intensities of the electromagnetic field have trivial differences along the axial direction, and this is the flat gradient region of the antenna. Finally, the electromagnetic field inside the helical antenna is divided into two steep gradient regions and one flat gradient region, which is shown in Figure 2.

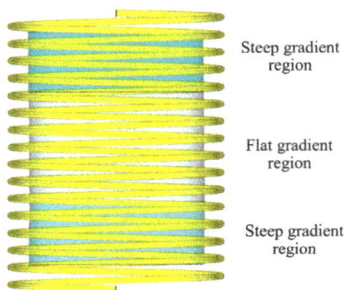

Figure 2. Steep and flat gradient regions of a helical antenna.

The dielectric rod is moving along the axial direction, and there are two circumstances moving in the steep and flat gradient regions. Using the perturbation theory, the electromagnetic field of the helical antenna with an inserting rod from the initial position to its destination can be simulated in two steps, as illustrated in Figure 3. In the first step, the intersection volume of the inserting rod between the initial state and final state remains, while a subtracted volume is removed from the top end of the dielectric rod. In the second step, the subtracted volume is added back to the bottom end of the dielectric rod; then, the integrated volume represents the moving rod in its final position. Perturbation theory is applied to each step to simulate the electromagnetic field of the helical antenna before finally retrieving the resonant frequencies shift.

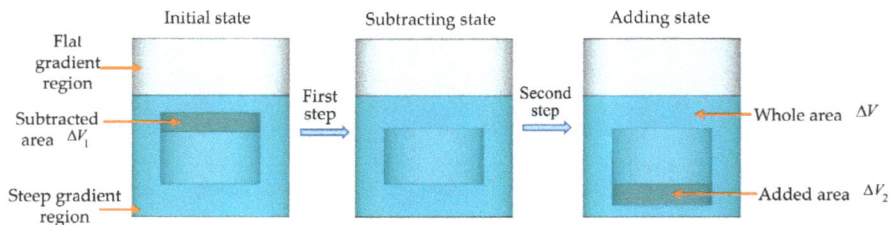

Figure 3. Perturbation theory flow chart of the dielectric rod in its initial state (**left**), subtracting state (**middle**), and adding state (**right**) in the steep gradient region.

2.2.1. Moving Rod in the Steep Gradient Region

In this circumstance, the dielectric rod remains entirely in the steep gradient region during the movement, which is shown in Figure 3.

In the first step, the Maxwell's curl equations can be written for the initial state and subtracting state of the individual helical antenna as expressed here and in [41]:

$$\nabla \times \overline{E}_0 = -jf_0\mu\overline{H}_0, \tag{1}$$

$$\nabla \times \overline{H}_0 = jf_0\varepsilon\overline{E}_0, \tag{2}$$

$$\nabla \times \overline{E} = -jf_1(\mu + \Delta\mu)\overline{H}, \tag{3}$$

$$\nabla \times \overline{H} = jf_1(\varepsilon + \Delta\varepsilon)\overline{E}, \tag{4}$$

where $\overline{E}_0$ and $\overline{H}_0$ are the electric field intensity and the magnetic field intensity of the initial state, respectively. $\overline{E}$ and $\overline{H}$ are the fields of the subtracting state. f_0 represents the resonant frequencies of the initial state and f_1 represents the resonant frequencies of the subtracting state. j is the complex vector. μ and ε are the magnetic permeability and dielectric constant of the material, respectively. $\Delta\mu$ and $\Delta\varepsilon$ are the change of permeability and dielectric constant of the material in the subtracting area, respectively.

Multiplying the conjugate of Equation (1) by $\overline{H}$, and multiplying Equation (4) by $\overline{E}_0$; then, after subtracting these two equations, the equation can be written as

$$\nabla \times (\overline{E}_0 \times \overline{H}) = jf_0\mu\overline{H} \cdot \overline{H}_0 - jf_1(\varepsilon + \Delta\varepsilon)\overline{E}_0 \cdot \overline{E}. \tag{5}$$

Similarly, Equation (6) can be obtained by using Equations (2) and (3)

$$\nabla \times (\overline{E} \times \overline{H}_0) = -jf_1(\mu + \Delta\mu)\overline{H}_0 \cdot \overline{H} + jf_0\varepsilon\overline{E}_0 \cdot \overline{E}. \tag{6}$$

Adding Equations (5) and (6), we obtain the equation by using the divergence theorem after integrating both sides over the volume V_0

$$j\int_{V_0}\{[f_0\varepsilon - \omega(\varepsilon + \Delta\varepsilon)]E_0 \cdot \overline{E} + [f_0\mu - \omega(\mu + \Delta\mu)]\overline{H}_0 \cdot \overline{H}\}dv = 0, \tag{7}$$

where V_0 refers to the whole volume of the cavity.

Rewriting Equation (7) and the resonant frequencies of the system will satisfy

$$\frac{f_1 - f_0}{f_0} = \frac{-\int_{V_0}(\Delta\varepsilon|\overline{E}_0|^2 + \Delta\mu|\overline{H}_0|^2)dv}{\int_{V_0}(\varepsilon|\overline{E}_0|^2 + \mu|\overline{H}_0|^2)dv}. \tag{8}$$

As the change of permeability and dielectric constant only exist in the subtracted area, the integral Equation (8) can be rewritten as

$$\frac{f_1 - f_0}{f_0} = \frac{-\int_{\Delta V_1}(\Delta\varepsilon|\overline{E}_0|^2 + \Delta\mu|\overline{H}_0|^2)dv}{\int_{V_0}(\varepsilon|\overline{E}_0|^2 + \mu|\overline{H}_0|^2)dv}, \tag{9}$$

where ΔV_1 refers to the subtracted area.

Analogously, in the second step, the resonant frequencies of the proposed sensor will satisfy

$$\frac{f_2 - f_1}{f_0} = \frac{\int_{\Delta V_2}(\Delta\varepsilon|\overline{E}_0|^2 + \Delta\mu|\overline{H}_0|^2)dv}{\int_{V_0}(\varepsilon|\overline{E}_0|^2 + \mu|\overline{H}_0|^2)dv}, \tag{10}$$

where ΔV_2 refers to the added area, and f_2 is the resonant frequencies of the adding state.

Based on Equations (9) and (10), we can describe the shift of the resonant frequencies of the helical antenna by the following equation

$$\frac{f_2 - f_0}{f_0} = \frac{\int_{\Delta V_2} (\Delta\varepsilon|\overline{E}_0|^2 + \Delta\mu|\overline{H}_0|^2)dv - \int_{\Delta V_1} (\Delta\varepsilon|\overline{E}_0|^2 + \Delta\mu|\overline{H}_0|^2)dv}{\int_{\Delta V_0} (\varepsilon|\overline{E}_0|^2 + \mu|\overline{H}_0|^2)dv} \tag{11}$$

Then the shift ratio of the resonant frequencies will satisfy

$$\frac{f_2 - f_0}{f_0} = \frac{\Delta h(c_1 - c_2)}{c}, \tag{12}$$

where c_1, c_2, and c can be calculated as

$$c_1 = \int_{\Delta V_2} (\Delta\varepsilon|\overline{E}_0|^2 + \Delta\mu|\overline{H}_0|^2)dv, \tag{13}$$

$$c_2 = \int_{\Delta V_1} (\Delta\varepsilon|\overline{E}_0|^2 + \Delta\mu|\overline{H}_0|^2)dv, \tag{14}$$

$$c = \int_{V_0} (\varepsilon|\overline{E}_0|^2 + \mu|\overline{H}_0|^2)dv. \tag{15}$$

Because the electromagnetic field inside the helical antenna is varied along the axial direction in the steep gradient region of the helical antenna, c_1 is different from c_2, which means there is a shift in resonant frequencies after the dielectric rod moves. This resonant frequency change, due to the dielectric rod moving in the steep gradient region, can be related to the displacement along the axial direction. Moreover, the rod moving in the steep gradient region can be treated as the sensing unit for the displacement. This phenomenon will be elaborated and verified by the numerical simulation in Section 3.

2.2.2. Moving Rod in the Flat Gradient Region

In this circumstance, the dielectric rod remains in the flat gradient region during the movement, which is shown in Figure 4.

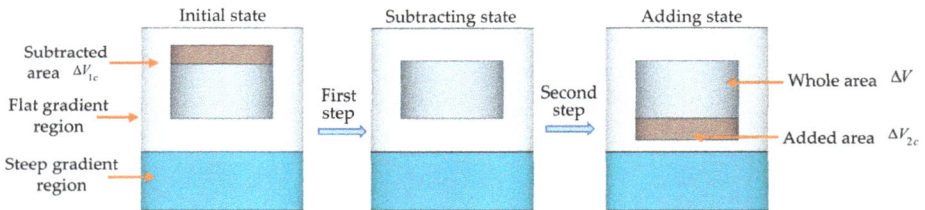

Figure 4. Perturbation theory flow chart of the dielectric rod in the initial state (left), subtracting state (middle) and adding state (right) in the flat gradient region.

Analogously, the shift ratio of the resonant frequencies will satisfy Equation (12), which can be written as

$$\frac{f_{2c} - f_0}{f_0} = \frac{\Delta h(d_1 - d_2)}{d}, \tag{16}$$

where f_{2c} is the resonant frequencies of the adding state in the flat gradient region, and d_1, d_2, and d can be calculated as

$$d_1 = \int_{\Delta V_{2c}} (\Delta\varepsilon|\overline{E}_0|^2 + \Delta\mu|\overline{H}_0|^2)dv, \tag{17}$$

$$d_2 = \int_{\Delta V_{1c}} (\Delta\varepsilon|\overline{E}_0|^2 + \Delta\mu|\overline{H}_0|^2)dv, \tag{18}$$

$$d = \int_{V_0} (\varepsilon |\overline{E}_0|^2 + \mu |\overline{H}_0|^2) dv, \tag{19}$$

where ΔV_{1c} refers to the subtracted area and ΔV_{2c} refers to the added area.

Since the electric intensity and magnetic intensity keep steady for both subtracted and added area within the flat gradient region, d_1 and d_2 in Equations (17) and (18) are almost equivalent. That is, the shift ratio of the resonant frequencies is equal to zero while the dielectric rod is moving along the axial direction in the flat gradient region. Therefore, the flat gradient region is not appropriate for displacement sensing according to perturbation theory.

3. Design of the Displacement Sensor

The perturbation theory discussed in Section 2 shows that the movement of the dielectric rod would change the resonant frequencies of the helical antenna in its steep gradient region. However, how the resonant frequencies will be influenced by the movement of the dielectric rod can not be determined analytically by the perturbation theory. Furthermore, determining the dimension parameters of the sensing system will add to the computational burden of the integration in the electromagnetic field, when involving the recursive design process.

In this paper, a helical antenna will be working around 2.4 GHz for the available testing environment in the laboratory. Initially the dimension parameters of the helical antenna will be prescribed within a narrow field according to its simplified relationship with resonant frequencies. Then the computational electromagnetics method is used to finalize the dimension of the proposed sensing system.

In the design process, the chosen internal dielectric rod material is silicon to amplify the impact of the dielectric rod movement, and copper is selected as the material of the helical antenna for its excellent electrical conductivity. The basic parameters of a helical antenna are the diameter d, number of turns n, and spacing between adjacent turns h [42]. The dimension parameters of a silicon rod are length m, diameter d_s and position of the inserting silicon rod.

3.1. Design of a Normal Mode Helical Antenna

For a normal mode helical antenna, the relationship between the total length of the wire and the wave length in a state of resonance can be described as [42]:

$$l = \frac{(2k-1)\lambda_p}{4}, \tag{20}$$

where l is the total length of the wire of the helical antenna, k is the order of resonant frequencies of the helical antenna, and λ_p is the antenna's wave length.

For the normal mode helical antenna, the resonant frequencies and the wavelengths of the antenna can be described as

$$f = \frac{c}{\lambda_p}, \tag{21}$$

where f is the resonant frequencies of the helical antenna.

Therefore, the total length of helical antenna wire can be approximately calculated by:

$$l = \pi n d, \tag{22}$$

where d and n are the diameter and the number of turns of the helical antenna, respectively.

Based on Equations (20) and (22), we can setup the relationship between resonant frequency and parameters of the helical antenna as

$$f = \frac{c(2k-1)}{4\pi n d}, \tag{23}$$

where the diameter d and number of turns n are increasing with the order of resonant frequencies when the resonant frequencies remain constant. In this paper, the working frequency of a helical antenna is set at around 2.4 GHz. To ensure the sensing range for a helical antenna, a higher order of resonant frequency is chosen for simulation and fabrication. Based on Equation (23), the setting field for dimension parameters of the helical antenna, as explained in Figure 5, can be initially determined, and these parameters listed in Table 1. These parameters will be finalized later based on the numerical simulation.

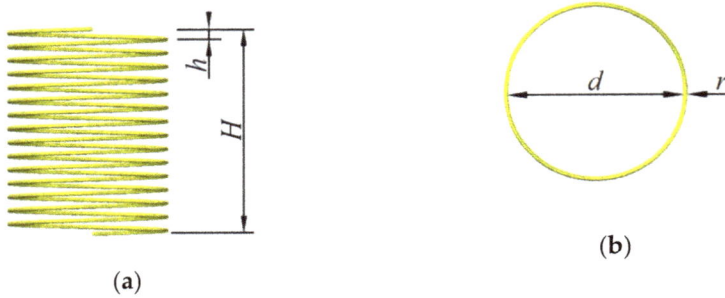

(a)

(b)

Figure 5. Parameters of the helical antenna (**a**) Front view; (**b**) Top view.

Table 1. Setting domain for parameters of the helical antenna.

Parameters	H (mm)	h (mm)	d (mm)	r (mm)	k	n	Material
Dimensions	35–57	2–3	22.5	0.5-1	17	15	Copper wire

3.2. Design of the Silicon Rod

The basic parameters of the silicon rod are shown in Figure 6, where m and d_s are the height and diameter of the silicon rod, and g is the distance of the gap between the silicon rod and the helical antenna in the radial direction.

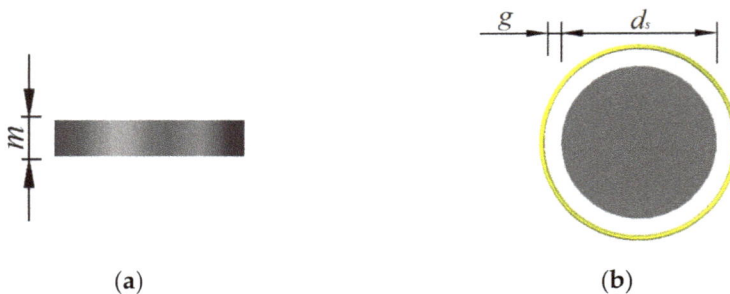

(a)

(b)

Figure 6. Parameters of the silicon rod (**a**) Front view; (**b**) Top view.

The gap g is decided by the diameter of the silicon rod, which will change the volume of the added area and subtracted area. To quantify the impact of the gap, two silicon rods with different diameters, slightly less than the inner diameter of the helical antenna, are designed for comparative study. The corresponding parameters are listed in Table 2.

Table 2. Parameters of silicon rod (Unit: mm).

Parameters	d_{s1}	g_1	m_1	d_{s2}	g_2	m_2
Dimensions	18	2.25	4	16	3.25	4

4. Modeling and Simulation

The radiation properties of the helical antenna sensor are simulated using the Ansoft high frequency structure simulator (HFSS), see Supplementary Materials. The model in the HFSS consists of a helical antenna and a coaxial dielectric rod, which is shown in Figure 7. The material of the dielectric rod is silicon, while copper is chosen as the material of the helical antenna. The sensing system is arranged inside an air cylinder with a radius of about a quarter wavelength to ensure computational accuracy of the far field radiation. The helical antenna is fed by a lumped port connected with the ground plane at the end of the helical antenna. The ground plane is set as a perfect E to ensure that the electric field is perfectly perpendicular to the surface.

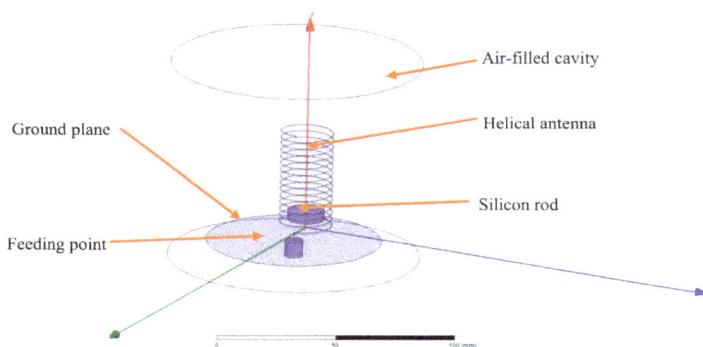

Figure 7. The schematic diagram of the helical antenna sensor.

After calculating the 17th order resonant frequency within the setting range by HFSS, four sets of proposed sensor systems with better performances are proposed. The serial number and dimension parameters are shown in Table 3.

Table 3. Serial number and dimension of the test group (Unit: mm).

Serial Number	H	h	d	r	n	d_s	g	m
TH1	35.5	2	22.5	0.5	15	18	2.25	4
TH2	35.5	2	22.5	0.5	15	16	3.25	4
TH3	57	3	22.5	1	15	18	2.25	4
TH4	57	3	22.5	1	15	16	3.25	4

Performance Simulation

The simulation proceeds in a wide range of location variations by HFSS for the TH1 group, with the location of the silicon rod moving from 5 mm above the ground plane to 27 mm. The return loss curves around the 17th order resonant frequency are acquired for each step where we insert the rod along the axis, as shown in Figure 8.

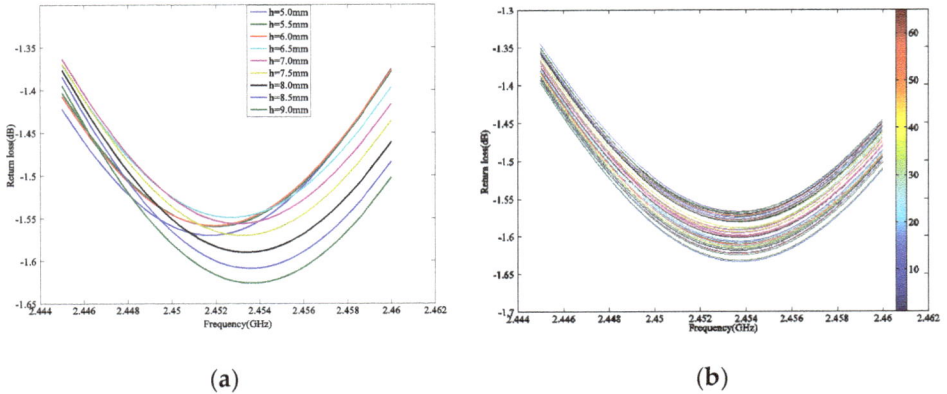

Figure 8. The return loss curves of TH1 group (**a**) The return loss curves from 5 mm to 9 mm; (**b**) The return loss curves from 9.5 mm to 25 mm.

The 17th order resonant frequency of the displacement sensor is extracted from the return loss curve at each moving step. The scatter diagram of resonant frequencies and movement of the silicon rod (which represents the displacement of the structure) is plotted in Figure 9.

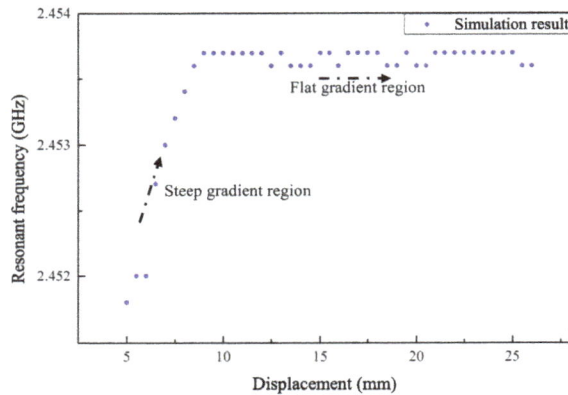

Figure 9. Relationship between resonant frequency and displacement based on the TH1 group in a wide range of location variations.

For the distance changing from 5 mm to 9.5 mm along the axial direction, the resonant frequency varies approximately linearly with the movement of the silicon rod, while the resonant frequency remains constant with the location change between 9.5 mm to 27 mm. This relationship, as shown in Figure 9, can be explained by the perturbation theory discussed in Section 2. The region between 5 mm over the ground plane to 9.5 mm inside the helical antenna can be regarded as the steep gradient region, and the region between 9.5 mm over the ground plane to 27 mm can be regarded as the flat gradient region. In this case, the height of the steep gradient region is approximately 10 percent of the total height of the helical antenna, and this area can be used for displacement sensing.

Then the simulation for each group is carried out. In this stage, the silicon rod starts to move upward from the height of 5 mm until the resonant frequency does not change and the total distance is called the measuring range. The resonant frequency of the displacement sensor is extracted from the return loss curve at each movement level. The scatter diagram of the resonant frequency and movement of the silicon rod for each group are plotted in Figure 10.

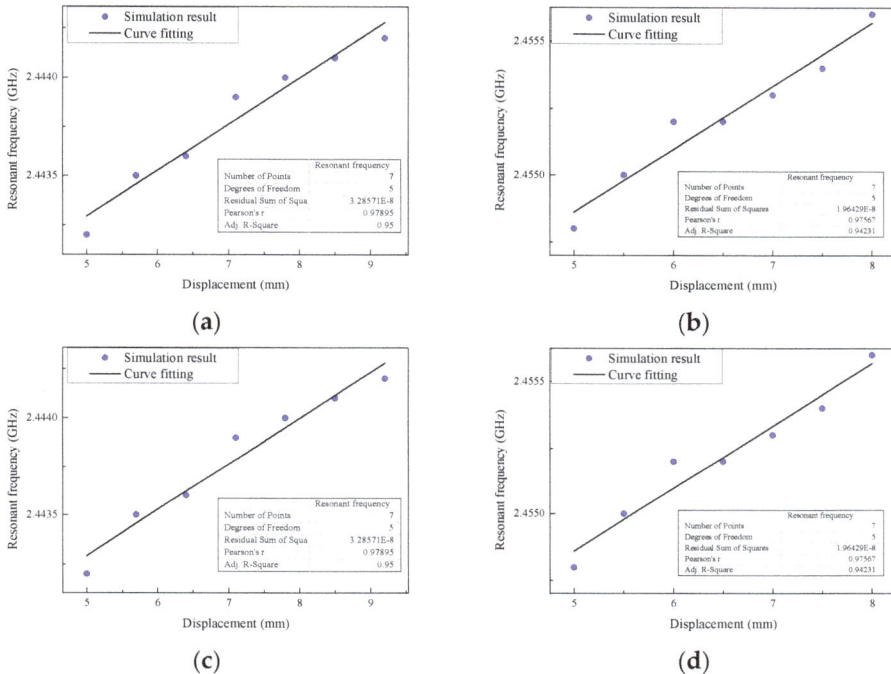

Figure 10. Relationship between resonant frequency and displacement for each group (**a**) Group TH1; (**b**) Group TH2; (**c**) Group TH3; (**d**) Group TH4.

Based on the simulation results, the sensitivity, measuring range and correlation coefficient of the fitted line in four test groups are shown in Table 4.

Table 4. Simulation results of the test groups.

Serial Number	Sensitivity (MHz/mm)	Measuring Range (mm)	Correlation Coefficient of the Fitted Line
TH1	0.330	3.0	0.9500
TH2	0.267	3.0	0.9423
TH3	0.467	4.5	0.9665
TH4	0.500	4.0	0.9501

The linearity of the displacement sensor in group TH1 and group TH3 are better than group TH2 and group TH4, respectively. Moreover, there would be a better working performance with a smaller gap between the silicon rod and helical antenna. Compared with the displacement sensor in the TH3 and TH4 groups, the measuring range in the TH1 and TH2 groups decreased. The measuring range is related to the length of the steep gradient region inside the helical antenna. As the steep gradient region usually expands with the total length of the helical antenna, the measuring range would increase. In other words, we can enlarge the sensing range by increasing the height H of the helical antenna.

However, because of the differences between numerical simulation and practical conditions in radiation, surrounding environment, boundary conditions, etc., actual experiments still need to be carried out to demonstrate the performance of the displacement sensors.

5. Experiment

For the fabricated sensors, copper was selected as the material of the helical antenna and silicon with high purity was used for making the silicon rod. The two kinds of helical antennas with different silicon rods are shown in Figure 11. The parameters of the helical antenna are the same as in the simulation stage.

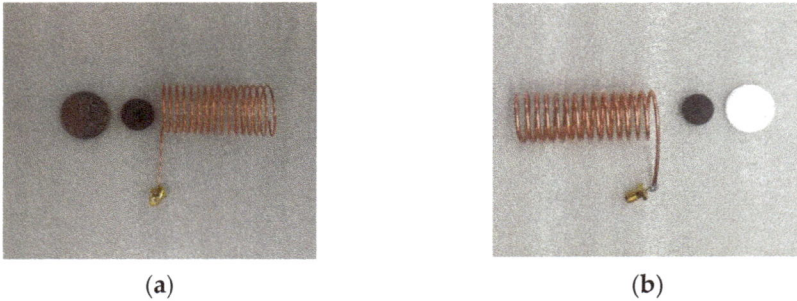

(a)   (b)

Figure 11. The manufactured displacement sensors (**a**) Helical antenna in group TH1; (**b**) Helical antenna in group TH3.

5.1. Instrumentation Setup

To verify the simulation results, four groups of tests were designed according to the parameters in Table 3. To ensure the coaxial movement of the silicon rod, the testing sensing system was established as shown in Figure 12. The helical antenna was surrounded by foam materials to keep it vertical. The silicon rod is suspended in the middle of the helical antenna by a cotton string. The other end of the cotton string is through the top beam; then, it connects with the micrometer, which controls the movement of the inserting rod. The helical antenna connects to the vector network analyzer (VNA-Rohde & Schwarz – Munich, Germany) via a tin-soldered Sub-Miniature-A (SMA) connector.

Figure 12. The experimental setup.

The experiments were carried out for each testing group. First, the silicon rod is set 5 mm beneath the feeding point in all simulation groups. The micrometer moves horizontally with 1 mm incremental steps. As the angle between the inclined cotton string and the suspended string is roughly 45 degrees, the incremental step of the silicon rod is about 0.7 mm ignoring the angle difference in its new position. Finally, the return loss curve of the sensing system was measured five times using the vector network analyzer (VNA) after the silicon rod was stabilized for at least 3 seconds.

5.2. Results and Discussion

To reduce the experimental error, the return loss curves of the displacement sensor were tested five times and averaged at each incremental step. The cubic polynomial curve was utilized to fit the return loss curve around the area of desired resonant frequencies, and the resonant frequency at the local minimum was extracted for each curve. The comparison of measured points and the fitted curve around the desired resonant frequencies is shown in Figure 13.

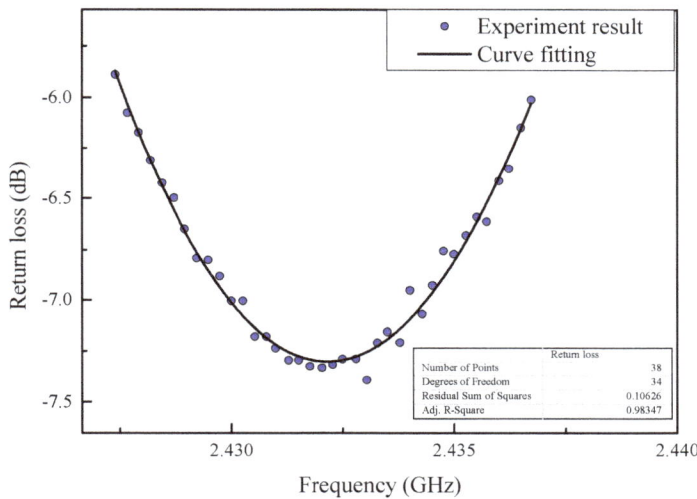

Figure 13. The comparison of measured points and fitted curve.

Based on the experimental results, the sensitivity, measuring range and correlation coefficient of the fitted line, in four test groups, are shown in Table 5. The scatter diagram of the resonant frequency and movement of the silicon rod for each group in experiment are plotted in Figure 14.

Table 5. Simulation results of the test groups.

Serial Number	Experiment			Simulation		
	Sensitivity (MHz/mm)	Measuring Range (mm)	Correlation Coefficient	Sensitivity (MHz/mm)	Measuring Range (mm)	Correlation Coefficient
TH1	0.712	3.5	0.9764	0.333	3	0.9500
TH2	0.400	4.9	0.9293	0.267	3	0.9423
TH3	1.600	3.5	0.9166	0.467	4.5	0.9665
TH4	0.650	7.7	0.9166	0.500	4	0.9501

(a)

(b)

(c)

(d)

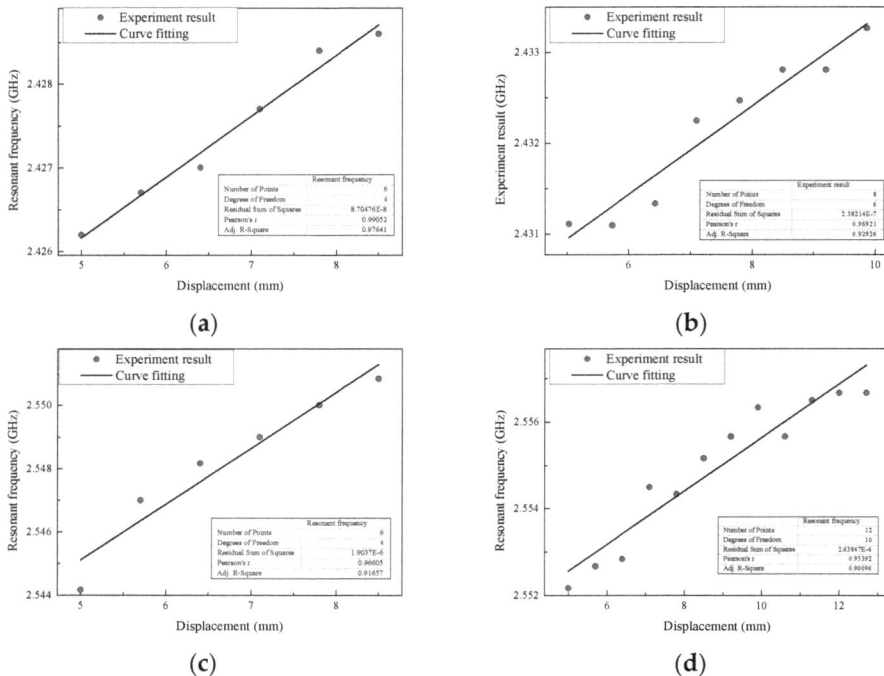

Figure 14. Resonant frequency with respect to displacement of the sensor in each group (**a**) Group TH1; (**b**) Group TH2; (**c**) Group TH3; (**d**) Group TH4.

As the correlation coefficient of the fitted line in group TH1 and group TH3 is larger than in group TH2 and group TH4, the linearity of the displacement sensor in group TH1 and group TH3 is better than group TH2 and group TH4, respectively. That is, the error caused by the gap between the silicon rod and the helical antenna will increase as the width of the gap increases, which is consistent with the simulation results.

From the experiment, it is noted that the correlation coefficient of the fitted lines of both displacement sensors is worse than that of the numerical simulation results. This difference is probably due to the following reasons.

(1) The welding of the feeding point and environmental interference may bring in some errors that are not considered in simulation.

(2) The experiments were not controlled precisely, as the cotton string has its elasticity; furthermore, the overhead hanging point may have had side movement. The angle between the inclined cotton string and the suspended string is roughly measured, as shown in Figure 15.

(3) The silicon rod may have oscillated slightly during the measurement.

For practical use, wireless interrogation is suggested to retrieve the information for the proposed displacement sensor. A RFID chip is integrated with the helical antenna, and the resonant frequency of the sensing system can be identified by finding the active frequency with the minimum interrogation energy. Data processing methods are also needed to offset the effects caused by the interference signal and environment.

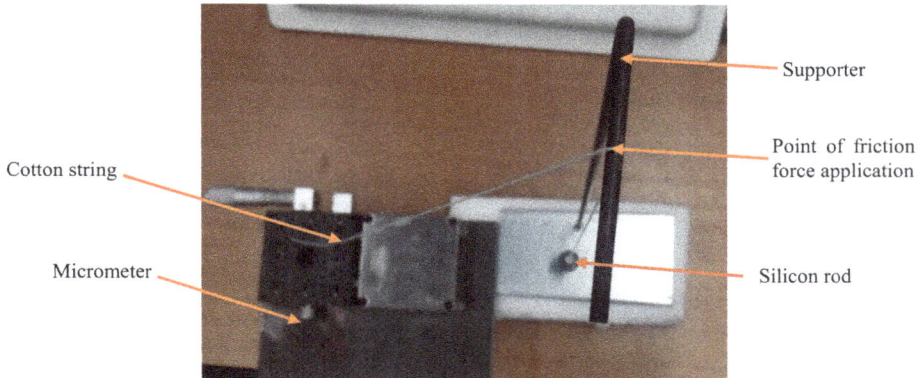

Figure 15. Connection between cotton string and supporter.

6. Conclusions

In this paper, a novel displacement sensing system based on a normal mode helical antenna was proposed. Using the theory of electromagnetic fields and the perturbation theory, the authors revealed a relationship between the resonant frequency of the displacement sensors and the movement of a silicon rod. Four combination sets of two types of silicon rods and two types of helical antennas were modeled and fabricated for numerical simulation and experiments. The steep gradient region inside the helical antenna proved to be useful for sensing displacement numerically and experimentally. The one-end region is about 10% of the total height of the helical antenna. The proposed helical sensor has a good performance for displacement sensing in a wired-testing environment. Results showed a sensitivity of 0.616 MHz/mm on average within a maximum effective measuring range of 7 mm, although the experimental sensitivity coefficient is susceptible to fabrication errors and environmental interference.

Supplementary Materials: The following are available online at http://www.mdpi.com/1424-8220/19/17/3767/s1.

Author Contributions: L.X. and S.X. conceived the idea of helical antenna sensors. Z.Y. finished the numerical simulation and performed the experiments. Z.Y. wrote the draft. L.X. edited the manuscript. G.W. and T.D. developed the methodology and analyzed the data.

Funding: This research was funded by the Key Program of Intergovernmental International Scientific and Technological Innovation Cooperation in China (grant number 2016YFE0127600), the Key Laboratory of Performance Evolution and Control for Engineering Structures (Tongji University), the Ministry of Education of the People's Republic of China (grant number 2018KF-4), and the Fundamental Research Funds for the Central Universities.

Conflicts of Interest: The authors declare no conflict of interest.

Abbreviations

Symbol	Parameter Definition
$\overline{E}_0$	electric field intensity of the initial state
$\overline{H}_0$	magnetic field intensity of the initial state
$\overline{E}$	electric field intensity of the subtracting state
$\overline{H}$	magnetic field intensity of the subtracting state
f_0	resonant frequencies of the initial state
f_1	resonant frequencies of subtracting state
μ	magnetic permeability

Symbol	Parameter Definition
ε	dielectric constant
$\Delta\mu$	change of magnetic permeability
$\Delta\varepsilon$	change of dielectric constant
j	complex vector
V_0	volume of the cavity
ΔV_1	subtracted area in the steep gradient region
ΔV_2	added area in the steep gradient region
f_2	resonant frequencies of the adding state in the steep gradient region
f_{2c}	resonant frequencies of the adding state in the flat gradient region
ΔV_{1c}	subtracted area in the flat gradient region
ΔV_{2c}	added area in the flat gradient region
d	diameter of a helical antenna
n	number of turns of a helical antenna
h	spacing between adjacent turns of a helical antenna
m	length of a silicon rod
d_s	diameter of a silicon rod
l	total length of the wire of helical antenna
k	order of resonant frequencies of the helical antenna
λ_p	wave length of helical antenna
f	resonant frequencies of the helical antenna in design occasion

References

1. Kongkitkul, W.; Hirakawa, D.; Sugimoto, T.; Kawahata, S.; Yoshida, T.; Ito, S.; Tatsuoka, F.; Li, G.; Chen, Y.; Tang, X. Post-Construction Time History of Tensile Load in Geogrid Arranged in A Full-Scale High Wall. In Proceedings of the 4th Asian Regional Conference on Geosynthetics, Shanghai, China, 17–20 June 2008; Springer: Berlin/Heidelberg, Germany, 2008; pp. 64–69.

2. Ayyildiz, C.; Erdem, H.E.; Dirikgil, T.; Dugenci, O.; Kocak, T.; Altun, F.; Gungor, V.C. Structure health monitoring using wireless sensor networks on structural elements. *Ad Hoc Netw.* **2019**, *82*, 68–76. [CrossRef]

3. Memmolo, V.; Monaco, E.; Boffa, N.; Maio, L.; Ricci, F. Guided wave propagation and scattering for structural health monitoring of stiffened composites. *Compos. Struct.* **2018**, *184*, 568–580. [CrossRef]

4. Kudela, P.; Radzienski, M.; Ostachowicz, W.; Yang, Z. Structural Health Monitoring system based on a concept of Lamb wave focusing by the piezoelectric array. *Mech. Syst. Signal Process.* **2018**, *108*, 21–32. [CrossRef]

5. Chen, L.; Zhang, D.; Zhou, Y.; Liu, C.; Che, S. Design of a high-precision and non-contact dynamic angular displacement measurement with dual-Laser Doppler Vibrometers. *Sci. Rep.* **2018**, *8*, 9094. [CrossRef] [PubMed]

6. Mandal, H.; Bera, S.K.; Saha, S.; Sadhu, P.K.; Bera, S.C. Study of a Modified LVDT Type Displacement Transducer With Unlimited Range. *IEEE Sens. J.* **2018**, *18*, 9501–9514. [CrossRef]

7. Alshamaa, D.; Mourad-Chehade, F.; Honeiné, P. Localization of Sensors in Indoor Wireless Networks: An Observation Model Using WiFi RSS. In Proceedings of the 9th IFIP International Conference on New Technologies, Mobility and Security (NTMS), Paris, France, 26–28 February 2018; pp. 1–5.

8. Jawad, H.M.; Nordin, R.; Gharghan, S.K.; Jawad, A.M.; Ismail, M. Energy-Efficient Wireless Sensor Networks for Precision Agriculture: A Review. *Sensors* **2017**, *17*, 1781. [CrossRef] [PubMed]

9. Gutiérrez, J.; Villa-Medina, J.F.; Nieto-Garibay, A.; Porta-Gándara, M.Á. Automated irrigation system using a wireless sensor network and GPRS module. *IEEE Trans. Instrum. Meas.* **2013**, *63*, 166–176. [CrossRef]

10. Kobo, H.I.; Abu-Mahfouz, A.M.; Hancke, G.P. Fragmentation-Based Distributed Control System for Software-Defined Wireless Sensor Networks. *IEEE Trans. Ind. Inform.* **2019**, *15*, 901–910. [CrossRef]

11. Cho, C.; Yi, X.; Wang, Y.; Tentzeris, M.M.; Leon, R.T. Compressive Strain Measurement Using RFID Patch Antenna Sensors. In Proceedings of the SPIE Smart Structures and Materials + Nondestructive Evaluation and Health Monitoring, San Diego, CA, USA, 10 April 2014; p. 90610X.

12. Lee, W.K.; Schubert, M.J.W.; Ooi, B.Y.; Ho, S.J.Q. Multi-Source Energy Harvesting and Storage for Floating Wireless Sensor Network Nodes With Long Range Communication Capability. *IEEE Trans. Ind. Appl.* **2018**, *54*, 2606–2615. [CrossRef]

13. He, J.; Wen, T.; Qian, S.; Zhang, Z.; Tian, Z.; Zhu, J.; Mu, J.; Hou, X.; Geng, W.; Cho, J.; et al. Triboelectric-piezoelectric-electromagnetic hybrid nanogenerator for high-efficient vibration energy harvesting and self-powered wireless monitoring system. *Nano Energy* **2018**, *43*, 326–339. [CrossRef]

14. Bhattacharyya, R.; Floerkemeier, C.; Sarma, S. Towards Tag Antenna Based Sensing—An RFID Displacement Sensor. In Proceedings of the IEEE International Conference on RFID, Orlando, FL, USA, 27–28 April 2009; pp. 95–102.

15. Huang, Y.S.; Chen, Y.Y.; Wu, T.T. A passive wireless hydrogen surface acoustic wave sensor based on Pt-coated ZnO nanorods. *Nanotechnology* **2010**, *21*, 095503. [CrossRef] [PubMed]

16. Butler, J.C.; Vigliotti, A.J.; Verdi, F.W.; Walsh, S.M. Wireless, passive, resonant-circuit, inductively coupled, inductive strain sensor. *Sens. Actuators A Phys.* **2002**, *102*, 61–66. [CrossRef]

17. Caizzone, S.; Di Giampaolo, E. Wireless Passive RFID Crack Width Sensor for Structural Health Monitoring. *IEEE Sens. J.* **2015**, *15*, 6767–6774. [CrossRef]

18. Jayawardana, D.; Kharkovsky, S.; Liyanapathirana, R.; Zhu, X. Measurement system with accelerometer integrated RFID tag for infrastructure health monitoring. *IEEE Trans. Instrum. Meas.* **2015**, *65*, 1163–1171. [CrossRef]

19. Manoj, A.J.Y.D. Wireless Sensor System for Infrastructure Health Monitoring. Ph.D. Thesis, Western Sydney University, Sydney, Australia, 2017.

20. Borgese, M.; Dicandia, F.A.; Costa, F.; Genovesi, S.; Manara, G. An Inkjet Printed Chipless RFID Sensor for Wireless Humidity Monitoring. *IEEE Sens. J.* **2017**, *17*, 4699–4707. [CrossRef]

21. Mohammad, I.; Huang, H. Monitoring fatigue crack growth and opening using antenna sensors. *Smart Mater. Struct.* **2010**, *19*, 055023. [CrossRef]

22. Lee, J.J.; Fukuda, Y.; Shinozuka, M.; Cho, S.; Yun, C.B. Development and application of a vision-based displacement measurement system for structural health monitoring of civil structures. *Smart Struct. Syst.* **2007**, *3*, 373–384. [CrossRef]

23. Park, J.; Sim, S.; Jung, H.; Spencer, B.F., Jr. Development of a Wireless Displacement Measurement System Using Acceleration Responses. *Sensors* **2013**, *13*, 8377–8392. [CrossRef] [PubMed]

24. Girbau, D.; Ramos, A.; Lazaro, A.; Rima, S.; Villarino, R. Passive Wireless Temperature Sensor Based on Time-Coded UWB Chipless RFID Tags. *IEEE Trans. Microw. Theory Tech.* **2012**, *60*, 3623–3632. [CrossRef]

25. Liu, C.; Teng, J.; Wu, N. A Wireless Strain Sensor Network for Structural Health Monitoring. *Shock. Vib.* **2015**, *2015*, 740471. [CrossRef]

26. Yi, X.; Cho, C.; Cooper, J.; Wang, Y.; Tentzeris, M.M.; Leon, R.T. Passive wireless antenna sensor for strain and crack sensing—electromagnetic modeling, simulation, and testing. *Smart Mater. Struct.* **2013**, *22*, 085009. [CrossRef]

27. Mandel, C.; Kubina, B.; Schuessler, M.; Jakoby, R. Passive Chipless Wireless Sensor for Two-Dimensional Displacement Measurement. In Proceedings of the European Microwave Conference, Manchester, UK, 10–13 October 2011; pp. 79–82.

28. Mohammad, I.; Huang, H. An Antenna Sensor for Crack Detection and Monitoring. *Adv. Struct. Eng.* **2011**, *14*, 47–53. [CrossRef]

29. Lopato, P.; Herbko, M. A Circular Microstrip Antenna Sensor for Direction Sensitive Strain Evaluation. *Sensors* **2018**, *18*, 310. [CrossRef] [PubMed]

30. Xue, S.; Xu, K.; Xie, L.; Wan, G. Crack sensor based on patch antenna fed by capacitive microstrip lines. *Smart Mater. Struct.* **2019**, *28*, 085012. [CrossRef]

31. Cook, J.D.; Marsh, B.J.; Qasimi, M.A.; Dixon, D.; Kumar, S. Wireless and Batteryless Sensor. U.S. Patent 12/239,363, 2 July 2009.

32. Selker, E.J. Methods and Apparatus for Wireless RFID Cardholder Signature and Data Entry. U.S. Patent 7,100,835, 9 September 2006.

33. Simons, R.N.; Miranda, F.A. Radio Frequency Telemetry System for Sensors and Actuators. U.S. Patent 6,667,725, 23 December 2003.

34. Mita, A.; Takahira, S. Health Monitoring of Smart Structures Using Damage Index Sensors. In Proceedings of the SPIE's 9th Annual International Symposium on Smart Structures and Materials, San Diego, CA, USA, 28 June 2002; pp. 92–100.

35. Huang, H.; Zhao, P.; Chen, P.Y.; Ren, Y.; Liu, X.; Ferrari, M.; Hu, Y.; Akinwande, D. RFID Tag Helix Antenna Sensors for Wireless Drug Dosage Monitoring. *IEEE J. Transl. Eng. Health Med.* **2014**, *2*, 1–8. [CrossRef] [PubMed]

36. Murphy, O.H.; McLeod, C.N.; Navaratnarajah, M.; Yacoub, M.; Toumazou, C. A pseudo-normal-mode helical antenna for use with deeply implanted wireless sensors. *IEEE Trans. Antennas Propag.* **2011**, *60*, 1135–1139. [CrossRef]

37. Zogbi, S.W.; Canady, L.D.; Helffrich, J.A.; Cerwin, S.A.; Honeyager, K.S.; De Los Santos, A.; Catterson, C.B. Passive and Wireless Displacement Measuring Device. U.S Patent 6,656,135, 2 December 2003.

38. Ong, J.B.; You, Z.; Mills-Beale, J.; Tan, E.L.; Pereles, B.D.; Ong, K.G. A Wireless, Passive Embedded Sensor for Real-Time Monitoring of Water Content in Civil Engineering Materials. *IEEE Sens. J.* **2008**, *8*, 2053–2058. [CrossRef]

39. Ong, K.; Grimes, C.; Robbins, C.; Singh, R. Design and application of a wireless, passive, resonant-circuit environmental monitoring sensor. *Sens. Actuators A Phys.* **2001**, *93*, 33–43. [CrossRef]

40. Labinac, V.; Erceg, N.; Kotnik-Karuza, D. Magnetic field of a cylindrical coil. *Am. J. Phys.* **2006**, *74*, 621–627. [CrossRef]

41. Pozar, D.M. *Microwave Engineering*; John Wiley & Sons: Hoboken, NJ, USA, 2009.

42. Balanis, C.A. *Antenna Theory: Analysis and Design*; John Wiley & Sons: Hoboken, NJ, USA, 2016.

sensors

MDPI

Article

Sudden Event Monitoring of Civil Infrastructure Using Demand-Based Wireless Smart Sensors

Yuguang Fu [1,*], Tu Hoang [1], Kirill Mechitov [2], Jong R. Kim [3], Dichuan Zhang [3] and Billie F. Spencer, Jr. [1]

[1] Department of Civil and Environmental Engineering, University of Illinois at Urbana-Champaign, Urbana, IL 61801, USA; tuhoang2@illinois.edu (T.H.); bfs@illinois.edu (B.F.S.J.)

[2] Department of Computer Science, University of Illinois at Urbana-Champaign, Urbana, IL 61801, USA; mechitov@illinois.edu

[3] Department of Civil and Environmental Engineering, Nazarbayev University, Astana 010000, Kazakhstan; jong.kim@nu.edu.kz (J.R.K.); dichuan.zhang@nu.edu.kz (D.Z.)

* Correspondence: yfu15@illinois.edu

Received: 18 November 2018; Accepted: 14 December 2018; Published: 18 December 2018

check for
updates

Abstract: Wireless smart sensors (WSS) have been proposed as an effective means to reduce the high cost of wired structural health monitoring systems. However, many damage scenarios for civil infrastructure involve sudden events, such as strong earthquakes, which can result in damage or even failure in a matter of seconds. Wireless monitoring systems typically employ duty cycling to reduce power consumption; hence, they will miss such events if they are in power-saving sleep mode when the events occur. This paper develops a *demand-based WSS* to meet the requirements of sudden event monitoring with minimal power budget and low response latency, without sacrificing high-fidelity measurements or risking a loss of critical information. In the proposed WSS, a programmable event-based switch is implemented utilizing a low-power trigger accelerometer; the switch is integrated in a high-fidelity sensor platform. Particularly, the approach can rapidly turn on the WSS upon the occurrence of a sudden event and seamlessly transition from low-power acceleration measurement to high-fidelity data acquisition. The capabilities of the proposed WSS are validated through laboratory and field experiments. The results show that the proposed approach is able to capture the occurrence of sudden events and provide high-fidelity data for structural condition assessment in an efficient manner.

Keywords: sudden event monitoring; wireless smart sensors; demand-based nodes; event-triggered sensing; data fusion

1. Introduction

Many civil infrastructure damage scenarios involve sudden events, such as natural disasters (e.g., earthquakes) and human-induced hazards (e.g., collisions, explosions, acts of terrorism). The occurrence of these events is generally unpredictable, and the consequences can be catastrophic. A typical example of catastrophic sudden event is found in the accidental collision between barges and a piling of railroad bridge in Mobile, Alabama, in 1993 [1]. As a result, the railroad bridge was damaged and gave way 20 min later when an Amtrak train crossed, killing 47 people. If this collision had been detected immediately and timely structural assessment of the bridge made, then the deaths of these individuals may have been prevented.

To mitigate the consequences of sudden events, the development of monitoring systems is of great importance. Traditional monitoring systems use wired sensors [2–5]. These monitoring systems not only enable sudden event detection but can also facilitate rapid condition assessment of civil infrastructure. Such wired monitoring systems require line power to operate and can be expensive for many large-scale structures, due primarily to high installation costs [6], often ranging from $5 K to $20 K per channel (e.g., the Bill Emerson Memorial Bridge monitoring system cost a total of $1.3 M for 86 sensors [4,7]).

High-fidelity wireless sensors offer tremendous opportunities to reduce costs and realize the promise of pervasive sensing for structural condition assessment. However, sudden event detection using wireless sensors remains elusive. For example, the monitoring system installed on the Golden Gate Bridge was unable to detect the three earthquakes that occurred during the three-month monitoring deployment [8]. Two main challenges to detect sudden events are apparent:

(i)   Limited power. Most wireless sensors are duty-cycled to preserve limited battery power; as a result, wireless sensors will miss the occurrence of sudden events when they are in power-saving sleep mode. Because the duty cycle is typically below 5% [9], this scenario is quite likely to occur.

(ii) Response latency. Response of wireless smart sensors (WSS) from sleep mode to data acquisition may take over a second, resulting in the loss of critical information in short-duration events (e.g., earthquakes and collisions). Moreover, even if awake, sensors may be busy with other tasks (e.g., data transmission); therefore, they will be unable to respond immediately to the occurrence of sudden events, and hence miss the short-duration events.

Addressing these challenges is critical to realizing a WSS for sudden event detection.

One intuitive strategy is to provide sustainable power for WSS to enable continuous monitoring of structures subjected to sudden events, emulating traditional wired monitoring systems. For example, in 2011, Potenza et al. [10] installed a wireless structural health monitoring (SHM) system consisting of 17 WSS on a historical church, which was damaged during the 2009 L'Aquila earthquake. The nodes were powered by the existing electrical lines, which guaranteed the continuity of operation and successfully detected several earthquakes over a 3-year monitoring period. Their strategy of using electrical lines to power WSS does not retain the inherent advantages of cable removal, and thus is not practical for other sudden event monitoring applications using WSS. Energy harvesting and wireless power transfer technologies also do not provide an efficient solution. Although technologies such as solar and wind energy harvesting have been developed and validated to power WSS for periodic monitoring [11–13], the challenge is that energy harvesting from the ambient environment is intermittent and time-varying, which is not reliable to support continuous monitoring of structures. Radio frequency (RF) energy transfer and harvesting is another wireless power technique in which WSS convert the received RF signals into electricity. The energy can be transferred reliably over a distance from a dedicated energy source to each node, or dynamically exchanged between different nodes [14]. However, the energy harvesting rate is on the order of micro-watts with low efficiency [15] and is insufficient for high-power high-fidelity monitoring of sudden events.

On the other hand, power consumption can be reduced by employing various energy-saving mechanisms, which help to mitigate, but do not fully address the challenge of limited power for WSS. For example, Jalsan et al. [16] proposed layout optimization strategies for wireless sensor networks to prolong the network lifetime by optimizing communication schemes without compromising information quality. Other examples of energy-saving mechanisms include data reduction, radio optimization, and energy-efficient routing. More detailed discussion can be found in Reference [17]. Most of these strategies are designed to reduce power consumption for wireless transmission, which does not help energy conservation for continuous sensing, because most of the power draw comes from the always-on sensor.

Recent developments in event-triggered sensing present both opportunities and challenges to realize sudden event monitoring using WSS. In event-triggered sensing, WSS only initiates measurement in response to signaling of events, which helps to save both energy and memory resources, and thus prolong the lifetime of WSS. Research has been conducted to implement low-power components (sensors and radios) that enable continuous operation and triggering mechanisms inside each sensor node. Lu et al. [18] designed the TelosW platform, which is an upgrade of the TelosB platform [19], by adding ultra-low power wake-on sensors and wake-on radios. The wake-on sensor is able to wake up the microcontroller (MCU) on occurrence of events with a predetermined threshold. Additionally, the wake-on radio can wake up the MCU when a triggering radio message is received. Similarly, Sutton et al. [20] presented a heterogeneous system architecture which included a low-power event detector circuit and low-power wake-up receivers. Although these two technologies achieve low power consumption, they do not satisfy the high-fidelity requirement of sudden event monitoring for civil infrastructure. For example, the TelosW's analog to digital converter has only 12-bit resolution. Event-triggered sensing is also developed and implemented to facilitate railway bridge monitoring, because strain cycles and vibrations induced by trains are the most important data for bridge condition assessment (e.g., fatigue), but the arrival time of trains is generally unpredictable. Bischoff et al. [21] deployed a wireless monitoring system which provided strain measurement and fatigue assessment of the Keraesjokk Railway Bridge. Each node was triggered independently by a low-power microelectromechanical systems (MEMS) accelerometer which operated continuously and detected an approaching train. Bias due to the transient start-up nature of the strain gage was removed by a post-processing technique. Liu et al. [22] developed an on-demand sensing system, named ECOVIBE, to monitor train-induced bridge vibrations. In each wireless node, a passive event detection circuit was designed to monitor bridge vibration with no power consumption, and another adaptive logical control circuit powered off the node once the designated tasks were finished. While effective for some applications, all the aforementioned approaches will lose critical data between the occurrence of the event and the time that data begins to be collected. Conversely, response times of wireless sensors from a cold boot to data acquisition are typically well over a second, making this problem particularly acute for short-duration sudden events (e.g., impacts can last only fractions of a second).

Moving the triggering mechanism to outside the sensor nodes provides a solution to address the challenge of data loss. In general, a separate trigger node or system is used to monitor the events continuously and notify of events to sensor nodes which are in power-saving mode most of the time. The trigger node/system is required to send notifications with a certain amount of time before the arrival of events at the structure, compensating the response latency of other sensor nodes. For example, an event-driven wireless strain monitoring system was implemented on a riveted steel railway bridge near Wila, Switzerland [23]. Two trigger nodes, referred to as sentinel nodes, were placed at 50 and 85 m away from the bridge, detecting approaching trains and sending alarm messages using a reliable flooding protocol to wake up sensor nodes on the bridge, before the train arrived. In a 47-day deployment, the system successfully detected 99.7% of train-crossing events. Likewise, in order to detect earthquakes and initiate seismic structural monitoring, Hung et al. [24] developed an intelligent wireless sensor network embedded with an earthquake early warning (EEW) system which was able to detect P-waves before earthquakes arrived. In addition, each sensor node was implemented with a wake-on radio which supported ultralow-power periodic listening of wake-up commands, while the main sensor node was in deep sleep mode. Once the P-wave was detected, the gateway node, integrated with the EEW system, sent wake-up commands to sensor nodes approximately 2 s ahead of earthquakes. Subsequently, sensor nodes started measurement with a latency time of only 229 ms. Despite successful detection of train-crossing and seismic events, the aforementioned methods do not provide a universal solution to address the challenge of data loss for many other sudden events, e.g., bridge impact by over-height trucks and ships which can hardly be detected ahead of impacts.

In addition, some progress has been made in addressing the challenge of response latency to sudden events, when WSS are awake but not in sensing mode. Cheng & Pakzad [8] proposed a pulse-based media access control protocol. When an earthquake occurs, a trigger message with high priority is propagated from an observation site across the WSS network to preempt current tasks; sensors will be forced to conduct measurement to capture the structural response under the earthquake. Dorvash et al. [25] developed the Sandwich node to reduce the response latency for unexpected events.

A smart trigger node continuously measures the structural response; it will broadcast a proper message across a network of Sandwich nodes in the case of occurrence of events. Sandwich nodes keep listening to the trigger message; they will preempt current tasks once the trigger message is received. Response delay of Sandwich nodes is around 8 ms to the occurrence of events. Although response delay is reduced in these two strategies, the wireless sensor's radio must always be on to listen for messages from a trigger node. Unless employing an ultralow-power wake-up radio, these strategies will result in a significant power draw.

This paper proposes a new approach for monitoring civil infrastructure subjected to sudden events, aimed at detecting sudden events of any duration and capturing complete transient response of any length. A demand-based wireless smart sensor (WSS) is developed that can capture data during the sudden event that is suitable for rapid condition assessment of civil infrastructure. As opposed to periodic monitoring, the *demand-based WSS* only wakes up and initiates sensing in response to specific conditions, such as sudden events. The results of laboratory experiments and a field experiment show that our proposed approach can capture the occurrence of sudden events and provide high-fidelity data for structural condition assessment in a timely and power-efficient manner.

2. Demand-Based WSS

As discussed in the previous section, the primary issues that must be overcome to use wireless sensors to monitor civil infrastructure subjected to sudden events are: (i) the sensor must operate on battery power, (ii) high-fidelity data appropriate for SHM application must be obtained, (iii) data surrounding the occurrence of sudden events must not be lost, and (iv) the WSS node must have sufficient computational power to translate the data collected into actionable information. This section describes a *demand-based WSS* system that can address these issues.

2.1. Ultralow-Power Trigger Accelerometer for Continuous Monitoring

To ensure that the occurrence of sudden events is not missed, the monitoring system must be continuously in an on state. A wireless node that is always on would quickly deplete its battery. Therefore, the solution proposed herein is to use an ultralow-power trigger accelerometer that can continuously monitor the vibration of structures; the data from the accelerometer is stored in a First-In-First-Out (FIFO) buffer. When an event occurs, the data in the FIFO buffer will be frozen, and the sensor triggers an interrupt signal to wake up the main sensor platform and start sensing. Such a trigger accelerometer should have low power consumption to enable continuous monitoring for several years, good sensing characteristics, including a high sampling rate and adequate resolution, and a large FIFO buffer to ensure data is not lost after the triggering event.

Trigger accelerometers in the market today were compared and the candidates that satisfied the basic needs of sudden event monitoring are listed in Table 1. The power consumption reported in the table correspond to the ultralow-noise mode of each sensor. More specifically, the ADXL362, developed by Analog Devices, consumes much less power than the other trigger accelerometers. The ADXL372, an updated high-g version of ADXL362, has a larger sampling rate and measurement range, but with a sensing resolution of only 100 mg. The LIS3DSH from STMicromechanics has high resolution of 0.06 mg, but it has a high-power draw and an inadequate FIFO buffer. Finally, the MPU6050 developed by InvenSense features a large FIFO buffer and high resolution, but it consumes substantial power.

In sum, based on application needs, the ADXL362 has been selected for this study (Figure 1); it integrates a three-axis microelectromechanical systems (MEMS) accelerometer with a temperature sensor, an analog-to-digital converter, and a Serial Peripheral Interface (SPI) digital interface. The ADXL362 consumes only 13 uA in ultralow-noise mode at 3.3 V, which theoretically could work continuously for over two years on a single coin-cell battery. A sampling rate up to 400 Hz and a resolution of 1 mg is supported, satisfying many SHM applications. The large FIFO buffer allows the sensor to save up to 512 samples, which corresponds to 1.7 s for all three axes sampled at 100 Hz.

Moreover, it has built-in logic for acceleration threshold detection; a detected event can be used as a trigger to wake up the primary sensor node.

Table 1. Comparison of trigger accelerometers in the market [26–29].

	ADXL362	ADXL372	LIS3DSH	MPU6050
Manufactures	Analog devices	Analog devices	STMicroelectronics	InvenSense
Supply voltage (V)	1.6–3.5	1.6–3.5	1.7–3.6	2.4–3.5
Power consumption (uA)	13	33	225	500
Sampling rate (Hz)	12.5~400	400~6400	3.125~1600	4~1000
Measurement range (g)	±2, ±4, ±8	±200	±2, ±4, ±8, ±16	±2, ±4, ±8, ±16
Resolution (mg)	1 mg	100 mg	0.06 mg	0.06 mg
Spectral noise (µg/√Hz)	175–350	5300	150	400
Buffer size (samples)	512	512	32	512

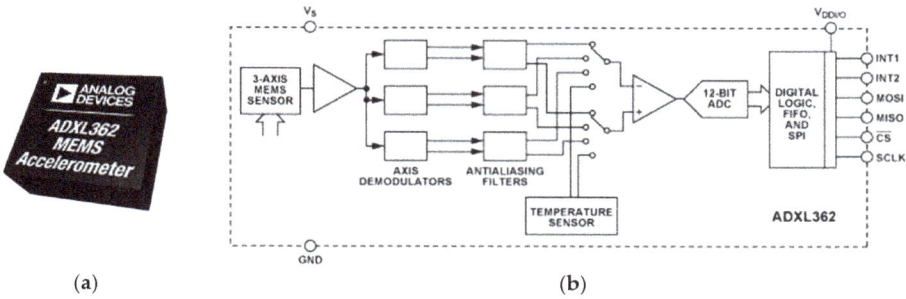

Figure 1. ADXL362: (a) sensor chip; (b) Functional block diagram [26].

2.2. High-Fidelity Sensor Platform for Sudden Event Monitoring

To provide high-quality measurement data and enable rapid condition assessment of structures subjected to sudden events, the sensor platform should have following features: (i) sensors and a data acquisition system that can obtain high-quality data at a high sampling rate for the event; (ii) powerful microcontroller to acquire and analyze sensor data in near real time. Other important features include: reliable communication, open-source software, and efficient data and power management.

A summary of the most advanced wireless sensor platforms available in the market currently is given in Table 2. The Xnode, developed by Embedor Technology, has a 24-bit Analog-to-Digital Converter (ADC), which is the best in its class. The microprocessor unit (MCU) of Waspmote (Libelium, Zaragoza, Spain) is not able to support rapid processing of large amount of data. The MCU information for the AX-3D (BeanAir, Berlin, Germany) and the G-Link-200 (LORD Sensing, Williston, VT, USA) is unavailable, and the operating systems of these platforms are proprietary. Note that several high-performance wireless sensor platforms (e.g., Imote2) are no longer commercially available, and hence not compared herein.

Table 2. Comparison of most advanced wireless sensor platforms in the market [30–34].

	Martlet	AX-3D	Waspmote	G-Link-200	Xnode
ADC resolution (bits)	12	16	16	20	24
Sampling rate (Hz)	up to 3 M	3.5 k	0.5–1 k	0–4 k	0–1.6 k
MCU Frequency (max. Hz)	80 M	NA	32 k	NA	204 M
LOS range (m)	500	650	1.6 k	2 k	1 k
Operating system	State-machine	Proprietary	Proprietary	Proprietary	FreeRTOS
Data memory	32 GB	1 million data points	16 GB	8 million data points	4 GB
Internal power source	dry cell battery	lithium-ion battery	lithium-ion battery	dry cell battery	lithium-ion battery
External power source	NA	primary cell/8–28 V DC	7 V DC	NA	5 V DC

Because of its high sensing resolution, high sampling rate, powerful microprocessor, and open-source software, the Xnode Smart Sensor [34] has been selected as the host wireless sensor platform in this study. The standard Xnode consists of three modular printed circuit boards (PCB): (i) the processor board, (ii) the radio/power board, and (iii) the sensor board (Figure 2). In particular, it employs an 8-channel, 24-bit ADC (Texas Instruments ADS131E8), allowing a maximum sampling rate up to 16 kHz and an NXP LPC4357 microcontroller operating at frequencies up to 204 MHz, which can be used to execute data-intensive on-board computation. Moreover, it implements open-source middleware services [35], which facilitates custom application development. In addition, it possesses two SPI controllers, making it possible to communicate with the selected trigger accelerometer, the ADXL362.

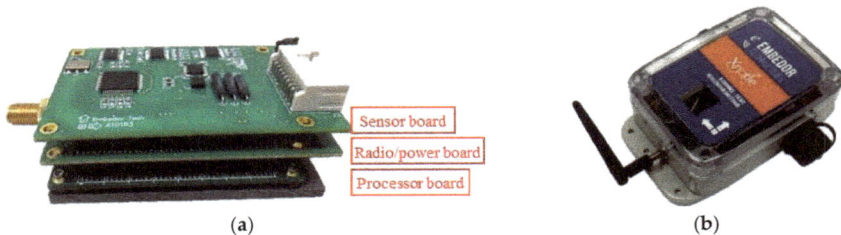

Sensor board
Radio/power board
Processor board

(a)   (b)

Figure 2. Xnode: (**a**) stacked modular boards; (**b**) weather-proof enclosure [34].

2.3. Integration of Wake-up Sensor and High-Fidelity Sensor Platform

To capture the entire event without loss of critical information, the ADXL362 accelerometer and the Xnode must be carefully integrated to build a *demand-based WSS*. This integration is discussed in the remainder of this section, in terms of hardware, software, and digital signal processing.

2.3.1. Hardware Consideration

To address the challenge of physical integration of the ADXL362 accelerometer into the Xnode, a programmable event-based switch was designed and implemented on the radio/power board of the Xnode in the *demand-based WSS*.

When a sudden event occurs, and the vibration exceeds a user-defined threshold, an interrupt pin in the ADXL362 generates a triggering signal. This signal is connected to a MOSFET to flip its state, turning on the Xnode and initiating high-fidelity sensing. When the event ends (lack of acceleration above a threshold), the other interrupt pin in the ADXL362 generates a signal to notify the Xnode to stop high-fidelity sensing. After data acquisition is completed, the triggering signal is cleared, and the MOSFET turns off the Xnode. The communication of control messages between the ADXL362 and the Xnode is carried out via SPI bus through a four-wire connection. In addition to enable event-triggered sensing, the proposed switch should be designed to retain traditional functionality for periodic monitoring. Specifically, sensor nodes are operated on low duty cycles, and the base station can access the network of nodes at random to initiate operations or measurements in the network. To achieve this goal, a real time clock, DS3231, is employed in the proposed switch. When a user-defined period passes or at a specific time of day, the DS3231 sends a triggering signal which flips the MOSFET switch and turn on the Xnode. Then the node remains awake for a short period of time to listen for messages from the base station. Once a command is obtained, the node carries out the required task (e.g., sensing, battery check). The communication of control messages between the DS3231 and the Xnode is conducted through the I2C bus.

Figure 3a illustrates the design concept of the proposed switch. Five major components are implemented, including a trigger sensor ADXL362 (U1), a real time clock DS3231 (U2), an AND gate (U3), a latch (U4), and a MOSFET (U5). Interrupt pins from the ADXL362 and the DS3231 are connected to the MOSFET through the AND gate. This circuit enables the MOSFET to be triggered by either the real time clock or the trigger sensor. In addition, a latch component is added between the AND gate and the MOSFET, to keep the power supply stable. Figure 3b shows the realized PCB for the proposed switch. The five major components, as well as companion resistors and capacitors, are all soldered on the edge of top side. In some use cases such as downloading code to the board and debugging, the sensor platform should always be on and therefore the designed switch needs to be bypassed. To achieve this goal, a 2-pin jumper is added. When the two pins on the jumper are not connected, the proposed switch works as designed, otherwise, it is bypassed.

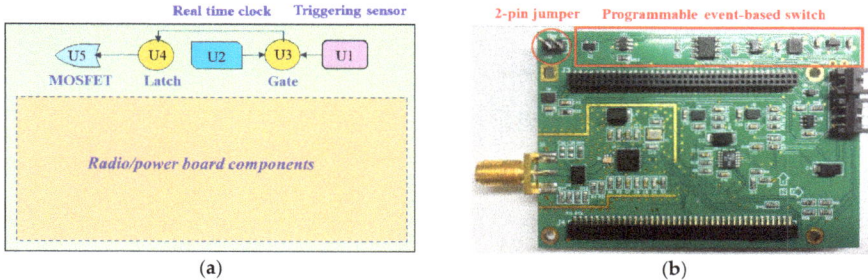

Figure 3. PCB design for the event-based switch: (**a**) design concept (**b**) realized PCB.

2.3.2. Software Consideration

In addition to hardware development of the prototype, an effective application framework is required to control the behavior of *demand-based WSS* to realize event-triggered sensing.

Figure 4 shows a flowchart of the application framework for the *demand-based WSS*. More specifically, when users turn on the physical switch of a main sensor platform, the Xnode first initializes itself and sends commands which contain configuration parameters (e.g., threshold, timers, and data buffer size) to the event-based switch discussed in the previous section. Once the commands have been received, the switch completes configuration of the device settings. Subsequently, the ADXL362 starts measurement in ultralow-noise mode, and the Xnode is turned off. If a sudden event occurs and the acceleration obtained in the ADXL362 exceeds the user-defined threshold, an interrupt pin, INT1 on the ADXL362 sends a trigger signal to turn on the Xnode. Concurrently, the ADXL362 saves 512 data samples into its FIFO buffer surrounding the onset of the event and waits for the Xnode to retrieve the data. The Xnode starts high-fidelity data acquisition using its built-in high-power high-accuracy MEMS accelerometer. When the event stops and the acceleration obtained in the ADXL362 is lower than a user-defined threshold for a certain period of time, the other interrupt pin, INT2, in the ADXL362 is triggered. Subsequently, the Xnode stops high-fidelity sensing. After sensing is completed, the Xnode reads data from the FIFO buffer of the ADXL362 and fuses it with the Xnode data. In addition, when the Xnode is busy with other tasks (e.g., data transmission of a previous event), but another sudden event occurs, the INT2 pin can be configured to interrupt undergoing tasks and force the Xnode to start high-fidelity sensing immediately. In addition, timing analysis results for each stage of a demand-based WSS are presented in the left of Figure 4.

For sudden events that are rare, e.g., earthquakes, the thresholds can be determined based on priori information about the sudden events that are monitored. The a priori information can be estimated by numerical analysis, the data in the history, or the measurement data in a preliminary test. For some events that occur frequently, e.g., railway bridge impacts from over-height vehicles, the thresholds can be determined adaptively, starting from a relatively low value during a "training phase" and then adjusted until the detection errors are minimized. A more comprehensive research regarding the triggering mechanism is the subject of future work to be addressed in the near future.

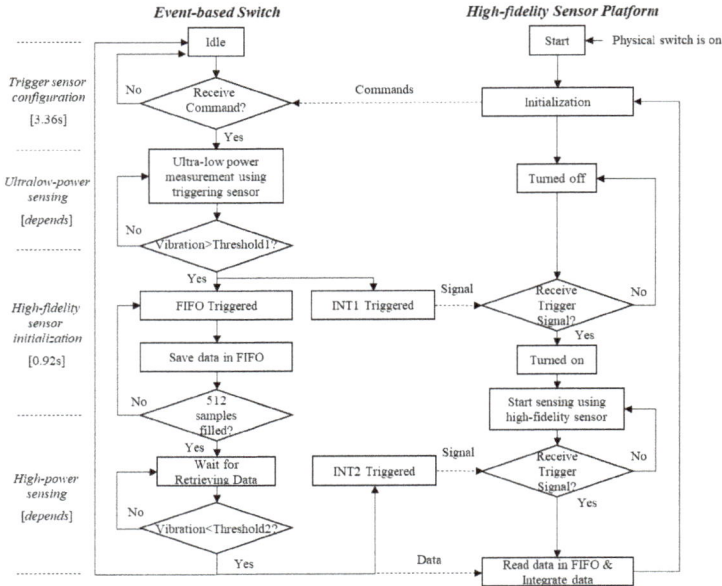

Figure 4. Flowchart of demand-based wireless smart sensors (WSS) for event-triggered sensing.

2.3.3. Data Fusion of Trigger Sensor and Xnode Acceleration Record

The objective of the *demand-based WSS* is to obtain the data from before the trigger event occurs until the structural accelerations stop. Specifically, the ADXL362 can record structural measurements surrounding the onset of a sudden event, whilst the Xnode starts sampling the data approximately 0.9 s after being triggered. Therefore, the ADXL362 data and the Xnode data must be synchronized and fused to produce a complete representation of the acceleration record. The following paragraphs describe the challenges encountered in this process, along with the associated resolutions.

To fuse the two overlapping data streams, two main challenges should be addressed, including (i) differences in the sampling rate between the ADXL362 and the Xnode, and (ii) synchronization error between the ADXL362 data and the Xnode data. More precisely, the first challenge results from the differences between the clock rates of the ADXL362 and the Xnode. The internal clock rate in the ADXL362 has a standard deviation of approximately 3%. One approach might be to calibrate the ADXL362 incorporated in each Xnode; however, this approach is not practical, as the clock rate will change with temperature, invalidating the initial calibration. The second challenge is due to the variance in start-up time of the sensing task on the Xnode. As a result, a random offset will exist between the two data streams.

To tackle the challenges identified in the previous paragraph, the beginning of the Xnode data, which is overlapped with the ADXL362 data, was utilized to calibrate the entire ADXL362 data stream. Figure 5 shows a flowchart of this approach. More specifically, the ADXL362, acc_{adxl0}, with a nominal sampling rate of 100 Hz, were first up-sampled to f_s (1000 Hz). The last 400 data points of the acc_{adxl0} were chosen as acc_{adxl1}, which was assumed to be approximately overlapped with the beginning of the Xnode data. In the meantime, the Xnode data, acc_{xnode0}, was sent through an 8-pole elliptic low-pass filter with a cutoff frequency of 50 Hz, to have the same bandwidth with acc_{adxl0}. The first 400 data points of acc_{xnode0} were considered as acc_{xnode1}. Based on the datasheet of ADXL362 [26], the clock frequency deviation from the ideal value was within the range of -10% and 10%. Therefore, to find the actual sampling frequency of acc_{adxl1}, exhaustive search was applied from 900 Hz to 1100 Hz. For Step i, the estimated sampling frequency (f_e) of the ADXL362 data was set as,

$$f_e = 1000 - df[i] \tag{1}$$

where, $df[i] = i - 100$, $i \in [0, 200]$. acc_{adxl1} was resampled from f_e to f_s using resampling-based approach [36]. Then, to estimate the synchronization error between acc_{adxl1} and acc_{xnode1}, the cross-correlation between the two data segments was calculated. The optimal offset, $SE[i]$, was obtained, for which the cross-correlation reaches its maximum value. Afterwards, acc_{adxl1} was shifted by $SE[i]$, and then the data fusion error (Err) was calculated as,

$$Err[i] = \|acc_{adxl1} - acc_{xnode1}\|_2 \tag{2}$$

where, $\| \ \|_2$ represents Euclidean norm. After completing these steps, the best estimations of sampling frequency f_a and synchronization error SE_a were obtained in the step that achieves minimal Err. Subsequently, f_a and SE_a were applied to calibrate the original data set, acc_{adxl0}. Finally, acc_{adxl0} and acc_{xnode0} were combined and down-sampled to 100 Hz for ensuing analysis.

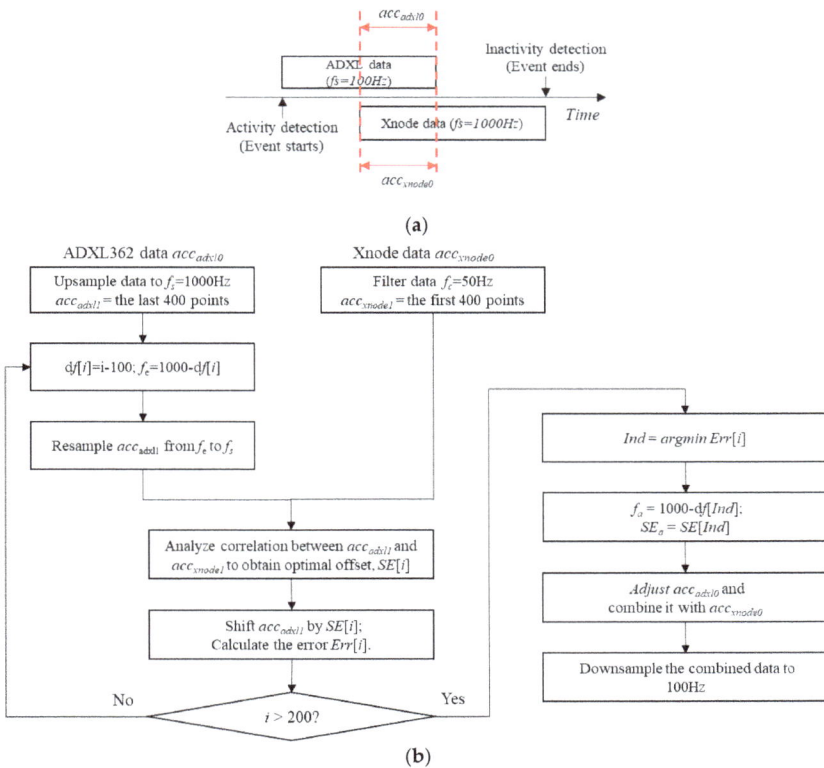

Figure 5. Post-sensing data fusion: (**a**) illustration of two data sources, (**b**) flowchart of data fusion strategy.

3. Validation of the Demand-Based WSS Performance

To validate the performance of the *demand-based WSS*, laboratory tests were carried out for data fusion and earthquake monitoring. The detailed test setup and results are presented in this section.

The performance of the *demand-based WSS* is discussed, in terms of power consumption, sensing characteristics, and data quality for sudden event monitoring.

3.1. Validation of Data Fusion

A lab test was conducted to illustrate the challenge of data fusion between the ADXL362 data and the Xnode data. Specifically, a *demand-based WSS* was located at 10th floor of an 18-story building model shown in Figure 6. The ADXL362 was configured to capture samples at 100 Hz, starting at 0.2 s before the triggered event and continuing until 1.5 s after the event. The event-triggering threshold was set to 150 mg, at which time, the Xnode was turned on and 1000 Hz high-fidelity measurement was started. To reduce false positives, two consecutive data points exceeding the threshold were required to cause triggering. In addition, a wired piezoelectric accelerometer, model PCB353B33, was installed on the same floor and sampled at a frequency of 128 Hz. The acceleration from these sensors served as reference data. A sudden event was simulated by manually shaking the building model in horizontal direction.

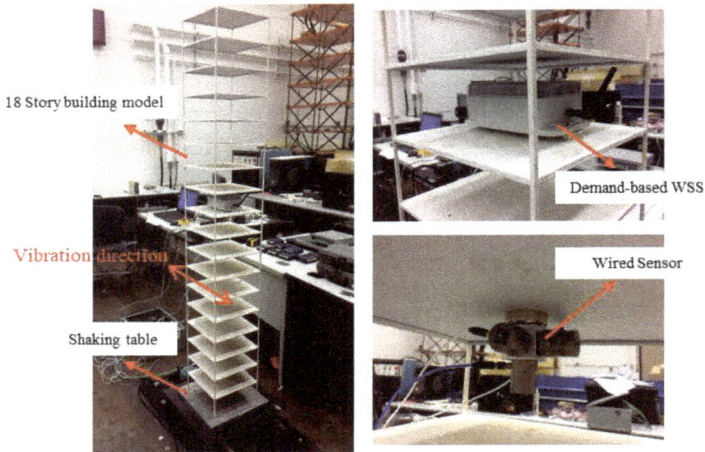

Figure 6. Experiment setup for a *demand-based WSS*.

To make a direct comparison in the time domain, the Xnode data and wired sensor data were sent through an 8-pole elliptic low-pass filter with a cutoff frequency of 50 Hz, as displayed in Figure 7. The direction of acceleration measurement was the same with the vibration direction specified in Figure 6. The vibration exceeded the threshold at 0.7 s, and the event-based switch turned on the Xnode. The Xnode required 0.92 s for initialization. Fortunately, as shown in Figure 7b, acceleration data stored in the FIFO buffer of the ADXL362 was recorded during this period. Specifically, the ADXL362 data can be divided into three parts: (i) Part 1 is the pre-triggered data which is around 0.28 s in length; (ii) Part 2 is the data that cover the time where the Xnode is initializing; (iii) Part 3 is where the ADXL362 data overlaps with the Xnode data. The length of the data in Part 3 is approximately 0.6 s. As shown in Figure 7b, the data obtained from the ADXL362 does not match well with the reference data from the wired sensors, because the sampling frequency of the ADXL362 was slightly smaller than 100 Hz, which illustrates the first challenge mentioned in the Section 2.3.3. In addition, the time offset between the Xnode data (blue line in Figure 7a) and the ADXL362 data (red line in Figure 7b) must be accurately estimated to fuse these two data streams, which illustrates the second challenge of data fusion.

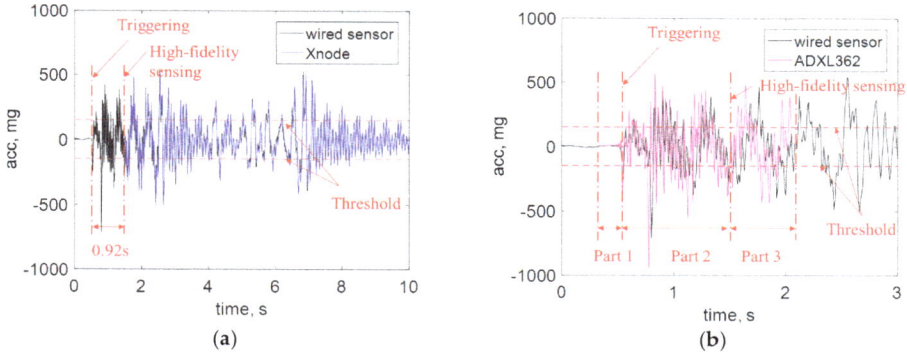

Figure 7. Time history data comparison: (**a**) Xnode measurement, (**b**) ADXL362 data buffer.

The data fusion strategy discussed in the previous section is applied to the test data. Figure 8 shows a comparison of time history data between sensor data from wired sensors and the fused data from the *demand-based WSS*. The excellent agreement demonstrates the ability of the proposed strategy to seamlessly capture the structural response subjected to a sudden event.

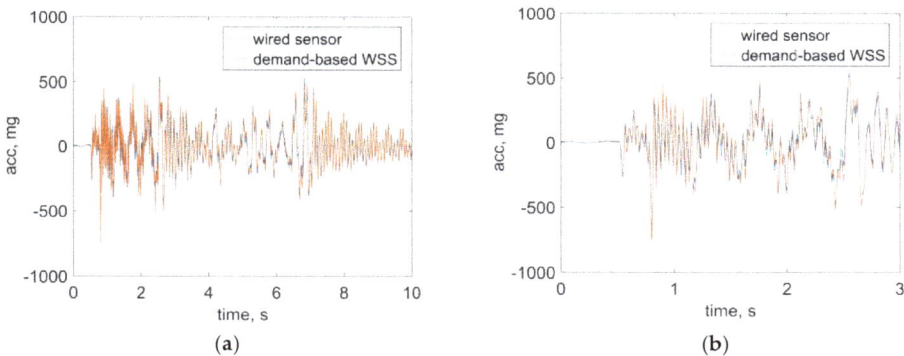

Figure 8. Results of data fusion: (**a**) time history data, (**b**) zoomed view (ADXL362 data).

3.2. Earthquake Monitoring

As a typical sudden event, an earthquake is transient and unpredictable, and the consequences can be catastrophic. Continuous efforts are required to develop cost-effective earthquake monitoring systems to mitigate the effect of earthquakes. *Demand-based WSS* have a significant potential to enable earthquake detection and rapid condition assessment of civil infrastructure, which was validated through a lab test in this section.

The test setup was the same with that in Section 3.1, as shown in Figure 6. The structure model was mounted on a uniaxial shaking table. This shaking table can simulate earthquakes in one horizontal direction, driving a 15 kg mass at 2.5 g with a maximum stroke of ±7.5 cm. The El Centro earthquake excitation was generated by the shaking table to represent a sudden event. The detection threshold for the event-based switch in *demand-based WSS* was configured as follows: the onset of event was detected when the acceleration was above 80 mg over 0.02 s, and the end of event was detected when the acceleration was below 40 mg over 5 s. Other configuration parameters are the same with the test in Section 3.1, such as sampling frequencies and filter parameters.

A segment of 90 s recorded time history is shown in Figure 9a. As can be seen in the zoomed view of the time history data (Figure 9b,c), the ground motion started at 10.7 s, but the vibration in the beginning was very small. From 11.79 s to 11.80 s, two consecutive acceleration samples obtained by the trigger accelerometer exceeded 80 mg. As a result, the event-based switch turned on the *demand-based WSS* immediately and the WSS started high-fidelity measurement. The acceleration became smaller than 40 mg after 54.50 s. Approximately 5 s later, the event-based switch stopped the high-fidelity sensing. Furthermore, Figure 9d shows the power spectral density (PSD) in the frequency domain. The excellent agreement between the results of wired sensors and the *demand-based WSS* in the both time and frequency domain demonstrates the ability of the proposed WSS to detect the earthquake and capture the accurate structural response during earthquakes in a timely and efficient manner.

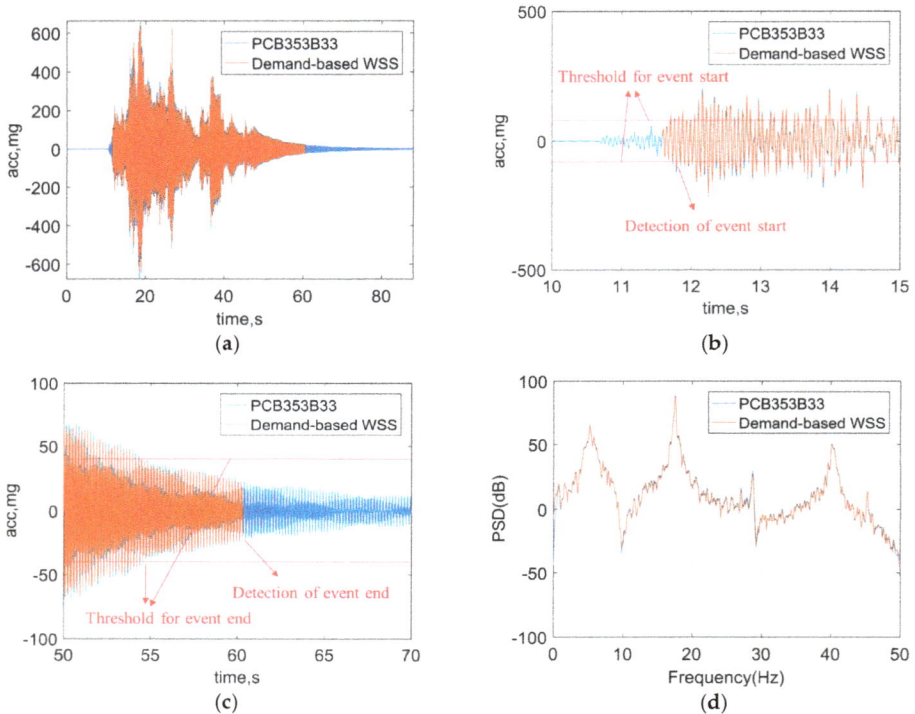

Figure 9. Test results of earthquake monitoring: (**a**) time history data, (**b**) zoomed view of time history data for event start, (**c**) zoomed view of time history data for event end, (**d**) power spectral density (PSD) data.

3.3. Evaluation and Discussion

In the lab tests described in previous sections, the attractive performance of the *demand-based WSS* was successfully validated to detect sudden events and provide high-quality sensing data for SHM analysis.

(1) Power consumption tests showed that the proposed WSS has a current draw of only 365 μA when no sudden event occurred, but the power consumption of the original Xnode sensor platform is approximately 170 mA during sensing. Considering that sudden events are rare and short-duration, most of the time the demand-based WSS deployed on a structure is in low-power measurement mode. Therefore, if using a 3.7 V DC, 10,000 mAh, rechargeable lithium polymer battery, employing the proposed WSS can extend the lifetime of always-on monitoring from three days to over three years using a single lithium battery. This feature helps to successfully

detect the occurrence of sudden events with minimal power budget in long-term monitoring. In addition, the current draw in each operation associated with duration for a demand-based WSS was determined experimentally and shown in Figure 10, in which the majority component of power consumption is sensing.

(2) The data obtained from the *demand-based WSS* is high-quality, matching well with the data from wired piezoelectric accelerometers. In particular, high-fidelity sensing enables 24-bit sensing resolution and over 1 kHz sampling rates. This feature helps to conduct structural condition assessments accurately under sudden events.

(3) The test results show that, when an event occurs, a seamless transition from the low-power sensing to high-fidelity measurement is carried out, without losing any data about the event.

These three features demonstrate that the proposed WSS satisfies the demands of sudden event monitoring.

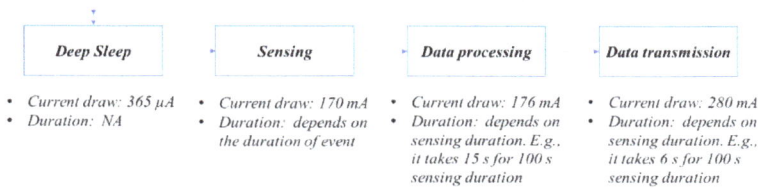

Deep Sleep	**Sensing**	**Data processing**	**Data transmission**
• Current draw: 365 µA • Duration: NA	• Current draw: 170 mA • Duration: depends on the duration of event	• Current draw: 176 mA • Duration: depends on sensing duration. E.g., it takes 15 s for 100 s sensing duration	• Current draw: 280 mA • Duration: depends on sensing duration. E.g., it takes 6 s for 100 s sensing duration

Figure 10. Flowchart of event-triggered sensing regarding current draw and duration for each operation.

4. Field Application

To further validate system performance, a field test was conducted on a steel railroad bridge north of Champaign, Illinois. Having vibration data while in-service trains traverse the bridge is useful to assess the bridge condition [37]. Train events have similar features to sudden natural events, e.g., unpredictability due to uncertain train schedules, but occur more frequently and therefore provide a convenient test platform. A *demand-based WSS* was deployed on the bottom side of a bridge girder. Simultaneously, wired sensors, model PCB353B33, were selected as reference sensors and deployed close to the WSS (see Figure 11). A detection threshold was configured to be the same as the test in Section 3.2. To avoid signal saturation, the measurement range of the trigger accelerometer was set to the maximum value of 8 g. At 10:52:06 a.m. on 7 May 2019, an Amtrak passenger train passed by the bridge.

Figure 11. Field application of the *demand-based WSS*.

Figure 12a–d shows the raw acceleration data of the bridge in vertical direction. The train came to the bridge at 121 s and left at 128 s. The event had a short duration of 7 s, and it was successfully detected by the *demand-based WSS*. Figure 12e shows the PSD data. Some slight discrepancies between the data from two sensors are possibly due to the different locations of the sensors. In sum, good agreement can be observed between the wired sensor and the *demand-based WSS* both in the time and frequency domain, demonstrating that the new WSS can capture the sudden event and obtain high-fidelity measurement in real applications.

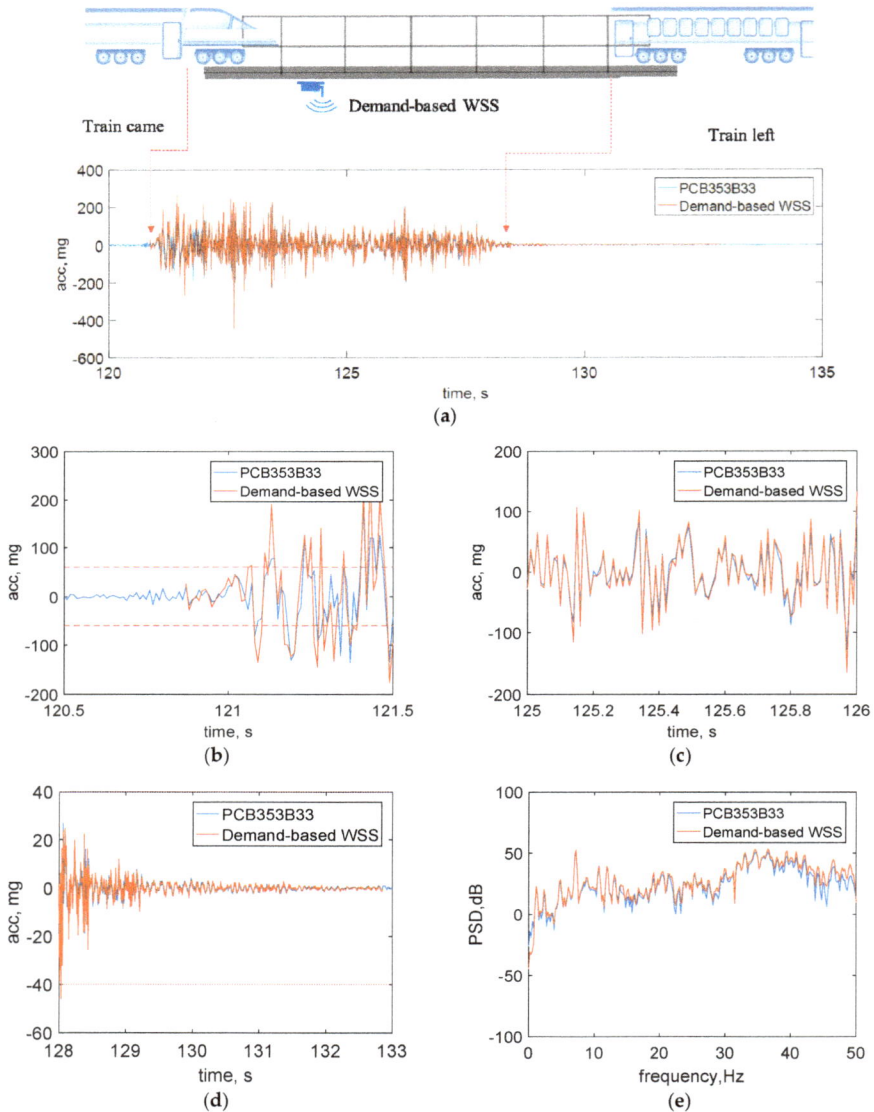

Figure 12. Test results on a railroad bridge: (**a**) time history data, (**b**) zoomed view of event starts (ADXL362 data), (**c**) zoomed view of event data (Xnode data), (**d**) zoomed view of event ends (Xnode data), (**e**) PSD data.

5. Conclusions

This paper presented the design of *demand-based WSS* to meet the application requirements of sudden event monitoring. The proposed WSS mainly consist of a unique programmable event-based switch and a powerful high-fidelity WSS platform. In particular, the event-based switch is built on a trigger accelerometer which allows the new WSS to measure the structural response in ultralow-power in long term, so as not to miss sudden events. In addition, the software of event-triggered sensing and data fusion is implemented. The performance of the proposed WSS is evaluated through the lab tests of earthquake monitoring and a field application on a railroad bridge. The test results show that the proposed WSS can continuously monitor structural response with minimal power budget, and hence detect the occurrence of the sudden event with the smallest delay. Besides detecting sudden events, the proposed WSS have the excellent features of high sampling rates and sensing resolution, which finally helps to provide high-quality data in sudden events for rapid condition assessment of civil infrastructure. Moreover, the proposed WSS are powerful and versatile not only for sudden events (e.g., earthquakes), but also for autonomous monitoring of other general events (e.g., bridge/highway overloads).

For large-scale structures, one demand-based WSS is not sufficient, and a network of nodes are needed for a meaningful characterization of the structural response. When subjected to a sudden event, each node may be triggered independently to initiate measurement at slightly different times due to varying response levels in the structure. Future work will address the challenges encountered for a network of demand-based WSS under sudden events. For example, one critical issue is to synchronize data from different sensor nodes without introducing delay of event-triggered sensing.

Author Contributions: Conceptualization, Y.F.; methodology, Y.F. and T.H.; software, Y.F. and T.H.; validation, Y.F., T.H., K.M. and B.F.S.J.; formal analysis, Y.F.; investigation, Y.F. and T.H.; resources, K.M. and B.F.S.J.; data curation, Y.F.; writing—original draft preparation, Y.F.; writing—review and editing, T.H., K.M., J.R.K., D.Z., B.F.S.J.; visualization, Y.F.; supervision, K.M. and B.F.S.J.; project administration, B.F.S.J. and J.R.K.; funding acquisition, J.R.K. and D.Z.

Funding: This research was funded in part by the Nazarbayev University Research Fund under Grant #SOE2017003, ZJU-UIUC Institute Research under Grant #ZJU083650, Federal Railroad Administration under Grant #DTFR53-17-C00007, as well as China Scholarship Council scholarship.

Acknowledgments: The authors gratefully acknowledge the support of this research by the Federal Railroad Administration, Canadian National Railway, and Sandro Scola, as well as the China Scholarship Council.

Conflicts of Interest: The authors declare no conflict of interest.

References

1. Garner, A.C.; Huff, W.A. The wreck of Amtrak's Sunset Limited: News coverage of a mass transport disaster. *Disasters* **1997**, *21*, 4–19. [CrossRef] [PubMed]
2. Yamazaki, F.; Katayama, T.; Yoshikawa, Y. On-line damage assessment of city gas networks based on dense earthquake monitoring. In Proceedings of the 5th U.S. National Conference on Earthquake Engineering, Chicago, IL, USA, 10–14 July 1994; Volume 4, pp. 829–837.
3. Yamazaki, F.; Motomura, H.; Hamada, T. Damage assessment of expressway networks in Japan based on seismic monitoring. In Proceedings of the 12th World Conference on Earthquake Engineering, Auckland, New Zealand, 30 January–4 February 2000.
4. Celebi, M. Real-time seismic monitoring of the New Cape Girardeau Bridge and preliminary analyses of recorded data: An overview. *Earthq. Spectra* **2006**, *22*, 609–630. [CrossRef]
5. Okada, K.; Nakamura, Y.; Saruta, M. Application of earthquake early warning system to seismic-isolated buildings. *J. Disaster Res.* **2009**, *4*, 242–250. [CrossRef]
6. Rice, J.A.; Mechitov, K.; Sim, S.H.; Nagayama, T.; Jang, S.; Kim, R.; Spencer, B.F.; Agha, G.; Fujino, Y. Flexible smart sensor framework for autonomous structural health monitoring. *Smart Struct. Syst.* **2010**, *6*, 423–438. [CrossRef]
7. Caicedo, J.M.; Clayton, E.; Dyke, S.J.; Abe, M.; Tokyo, J. Structural health monitoring for large structures using ambient vibrations. In Proceedings of the ICANCEER Conference, Hong Kong, China, 15–20 August 2002.

8.   Cheng, L.; Pakzad, S.N. Agility of wireless sensor networks for earthquake monitoring of bridges. In Proceedings of the 2009 IEEE Sixth International Conference on Networked Sensing Systems (INSS), Pittsburgh, PA, USA, 17–19 June 2009.

9.   Guo, S.; Kim, S.M.; Zhu, T.; Gu, Y.; He, T. Correlated flooding in low-duty-cycle wireless sensor networks. In Proceedings of the 2011 19th IEEE International Conference on Network Protocols, Vancouver, BC, Canada, 17–20 October 2011; pp. 383–392.

10. Potenza, F.; Federici, F.; Lepidi, M.; Gattulli, V.; Graziosi, F.; Colarieti, A. Long-term structural monitoring of the damaged Basilica S. Maria di Collemaggio through a low-cost wireless sensor network. *J. Civ. Struct. Health Monit.* **2015**, *5*, 655–676. [CrossRef]

11. Miller, T.I.; Spencer, B.F., Jr.; Li, J.; Jo, H. *Solar Energy Harvesting and Software Enhancements for Autonomous Wireless Smart Sensor Networks*; NSEL Report No. 022; University of Illinois at Urbana-Champaign: Urbana, IL, USA, 2010.

12. Jang, S.; Jo, H.; Cho, S.; Mechitov, K.A.; Rice, J.A.; Sim, S.H.; Jung, H.-J.; Yun, C.-B.; Spencer, B.F., Jr.; Agha, G. Structural health monitoring of a cable-stayed bridge using smart sensor technology: Deployment and evaluation. *Smart Struct. Syst.* **2010**, *6*, 439–459. [CrossRef]

13. Park, J.W.; Jung, H.J.; Jo, H.; Spencer, B.F., Jr. Feasibility study of micro wind turbines for powering wireless sensors in a cable-stayed bridge. *Energies* **2012**, *5*, 3450–3464. [CrossRef]

14. Lu, X.; Wang, P.; Niyato, D.; Han, Z. Resource allocation in wireless networks with RF energy harvesting and transfer. *IEEE Netw.* **2015**, *29*, 68–75. [CrossRef]

15. Lu, X.; Wang, P.; Niyato, D.; Kim, D.I.; Han, Z. Wireless networks with RF energy harvesting: A contemporary survey. *IEEE Commun. Surv. Tutor.* **2015**, *17*, 757–789. [CrossRef]

16. Jalsan, K.E.; Soman, R.N.; Flouri, K.; Kyriakides, M.A.; Feltrin, G.; Onoufriou, T. Layout optimization of wireless sensor networks for structural health monitoring. *Smart Struct. Syst.* **2014**, *14*, 39–54. [CrossRef]

17. Rault, T.; Bouabdallah, A.; Challal, Y. Energy efficiency in wireless sensor networks: A top-down survey. *Comput. Netw.* **2014**, *67*, 104–122. [CrossRef]

18. Lu, G.; De, D.; Xu, M.; Song, W.Z.; Cao, J. TelosW: Enabling ultra-low power wake-on sensor network. In Proceedings of the 2010 IEEE Seventh International Conference on Networked Sensing Systems (INSS), Kassel, Germany, 15–18 June 2010.

19. Polastre, J.; Szewczyk, R.; Culler, D. Telos: Enabling ultra-low power wireless research. In Proceedings of the 4th International Symposium on Information Processing in Sensor Networks, Boise, ID, USA, 15 April 2005; p. 48.

20. Sutton, F.; Da Forno, R.; Gschwend, D.; Lim, R.; Gsell, T.; Beutel, J.; Thiele, L. A Heterogeneous System Architecture for Event-triggered Wireless Sensing. In Proceedings of the 2016 15th ACM/IEEE International Conference on Information Processing in Sensor Networks (IPSN), Vienna, Austria, 11–14 April 2016.

21. Bischoff, R.; Meyer, J.; Enochsson, O.; Feltrin, G.; Elfgren, L. Event-based strain monitoring on a railway bridge with a wireless sensor network. In Proceedings of the 4th International Conference on Structural Health Monitoring of Intelligent Infrastructure, Zurich, Switzerland, 22–24 July 2009; Volume 2224, p. 7482.

22. Liu, Y.; Voigt, T.; Wirström, N.; Höglund, J. ECOVIBE: On-Demand Sensing for Railway Bridge Structural Health Monitoring. *IEEE Internet Things J.* **2018**. [CrossRef]

23. Popovic, N.; Feltrin, G.; Jalsan, K.E.; Wojtera, M. Event-driven strain cycle monitoring of railway bridges using a wireless sensor network with sentinel nodes. *Struct. Control Health Monit.* **2017**, *24*, e1934. [CrossRef]

24. Hung, S.L.; Ding, J.T.; Lu, Y.C. Developing an energy-efficient and low-delay wake-up wireless sensor network-based structural health monitoring system using on-site earthquake early warning system and wake-on radio. *J. Civ. Struct. Health Monit.* **2018**, 1–13. [CrossRef]

25. Dorvasha, S.; Lib, X.; Pakzada, S.; Chengb, L. Network architecture design of an agile sensing system with sandwich wireless sensor nodes. *Proc. SPIE* **2012**, *8345*, 83450H.

26. Analog Devices. ADXL362 Datasheet. 2016. Available online: http://www.analog.com/media/en/technical-documentation/data-sheets/ADXL362.pdf (accessed on 18 November 2018).

27. Analog Devices. ADXL372 Datasheet. 2017. Available online: http://www.analog.com/media/en/technical-documentation/data-sheets/adxl372.pdf (accessed on 18 November 2018).

28. STMicroelectronics. LIS3DSH Datasheet. 2017. Available online: https://www.st.com/resource/en/datasheet/lis3dsh.pdf (accessed on 18 November 2018).

29. InvenSense. MPU-6050 Datasheet. 2013. Available online: https://www.invensense.com/products/motion-tracking/6-axis/mpu-6050/ (accessed on 18 November 2018).

30. Kane, M.; Zhu, D.; Hirose, M.; Dong, X.; Winter, B.; Häckell, M.; Lynch, J.P.; Wang, Y.; Swartz, A. Development of an extensible dual-core wireless sensing node for cyber-physical systems. In Proceedings of the Sensors and Smart Structures Technologies for Civil, Mechanical, and Aerospace Systems 2014, San Diego, CA, USA, 9–13 March 2014; Volume 9061, p. 90611U.

31. BeanAir. AX-3D Datasheet. 2016. Available online: http://www.beanair.com/wa_files/Datasheet-Wireless-Accelerometer-BeanDevice-AX-3D.pdf (accessed on 18 November 2018).

32. Libelium. Waspmote v15 Datasheet. 2016. Available online: http://www.libelium.com/downloads/documentation/waspmote_datasheet.pdf (accessed on 18 November 2018).

33. LORD Sensing. G-Link-200 Datasheet. 2017. Available online: http://www.microstrain.com/sites/default/files/g-link-200_datasheet_8400-0102_2_0.pdf (accessed on 18 November 2018).

34. Spencer, B.F., Jr.; Park, J.W.; Mechitov, K.A.; Jo, H.; Agha, G. Next Generation Wireless Smart Sensors toward Sustainable Civil Infrastructure. In Proceedings of the Sustainable Civil Engineering Structures and Construction Materials (SCESCM 2016), Bali, Indonesia, 5–7 September 2016.

35. Fu, Y.G.; Mechitov, K.A.; Hoskere, V.; Spencer, B.F., Jr. Development of RTOS-based wireless SHM system: Benefits in applications. In Proceedings of the International Conference on Smart Infrastructure and Construction, Cambridge, UK, 27–29 June 2016.

36. Nagayama, T.; Spencer, B.F., Jr. *Structural Health Monitoring Using Smart Sensors*; Newmark Structural Engineering Laboratory, University of Illinois at Urbana-Champaign: Urbana, IL, USA, 2007.

37. Moreu, F.; Kim, R.E.; Spencer, B.F., Jr. Railroad bridge monitoring using wireless smart sensors. *Struct. Control Health Monit.* **2017**, *24*, e1863. [CrossRef]

sensors

MDPI

Article

Characterization of a Patch Antenna Sensor's Resonant Frequency Response in Identifying the Notch-Shaped Cracks on Metal Structure

Liang Ke, Zhiping Liu * and Hanjin Yu

School of Logistics Engineering, Wuhan University of Technology, 1178 Heping Ave., Wuhan 430063, China; keliang@whut.edu.cn (L.K.); yuhanjinwh@whut.edu.cn (H.Y.)
* Correspondence: lzp@whut.edu.cn; Tel.: +86-189-8616-5189

Received: 29 November 2018; Accepted: 24 December 2018; Published: 30 December 2018

check for
updates

Abstract: Patch antenna sensor is a novel sensor that has great potential in structural health monitoring. The two resonant frequencies of a patch antenna sensor are affected by the crack on its ground plane, which enables it to sense the crack information. This paper presents a detailed characterization of the relationship between the resonant frequencies of a patch antenna sensor and notch-shaped cracks of different parameters, including the length, the orientation, and the center location. After discussing the principle of crack detection using a patch antenna sensor, a parametric study was performed to understand the response of the sensor's resonant frequencies to various crack configurations. The results show that the crack parameters affect the resonant frequencies in a way that can be represented by the crack's cutting effect on the sensor's current flow. Therefore, we introduced a coefficient φ to comprehensively describe this interaction between the crack and the current distribution of the antenna radiation modes. Based on the definition of coefficient φ, an algorithm was proposed for predicting the resonant frequency shifts caused by a random notch-shaped crack and was verified by the experimental measurements. The presented study aims to provide the foundation for the future use of the patch antenna sensor in tracking the propagation of cracks of arbitrary orientation and location in metal structures.

Keywords: patch antenna; sensor; structural health monitoring; crack identification; resonant frequency

1. Introduction

Metal structures are widely used in the fields such as mechanical equipment, civil infrastructure, aerospace facilities, and offshore platforms. In the long-term load bearing process, a variety of damage can be seen on metal structures. When the damage accumulates to a certain level, the structure's load-carrying and anti-fatigue capacity will be impaired, which may lead to extremely serious consequences. To ensure the safe operation and prolong the lifespan of metal structures, a common method is to use sensors to monitor the structure's health status. Since fatigue-related cracking is the major form of structural damage [1], a number of techniques have been developed for the purpose of crack identification. Vibration analysis [2,3] can be used for crack detection because the presence of crack would change the structure's dynamic properties such as natural frequency and mode shape, which can give clues about the crack location and magnitude. However, the vibration characteristics are usually not sensitive to small-size local cracks. Strain-based analysis is another method to detect crack, and this is based on the fact that the emergence or extension of cracks would dramatically disturb the strain distribution in its vicinity. A very common tool to achieve this is an optical fiber sensor. In most cases, the structure's strain distribution is obtained first by the optical fiber sensor deployed on structure surface or embedded internally, and then compared with the pre-set non-destructive strain field to extract the crack information [4–6]. The optical fiber sensor is especially suited for large-scale crack detecting and monitoring, but its high cost and fragility have limited the application. Eddy current technique [7,8] is also developed to identify the surface or subsurface crack of the metal structures. Eddy-current inspection by Hall sensor can easily recognize the crack existence but can hardly give quantitative information. In addition, the lift-off effect is a big challenge for eddy-current detection of complicated surfaces. Since ultrasonic waves propagating in a structure would be reflected, refracted, or diffracted by defects such as cracks, the ultrasound-based inspection has become a useful technique for crack detection [9–11]. Ultrasonic testing is outstanding for detecting internal cracks due to its strong penetrating ability, but the drawback is the need for excitation devices and coupling agents. Acoustic emission (AE) [12,13] is another sound-based non-destructive testing method. AE refers to the phenomenon that a crack or other defect can trigger a sudden release of the stored elastic energy and thus generate a transient elastic wave. The AE signal can be collected by AE sensors (e.g., PZT patches) deployed on the structure surface, and can be analyzed to extract the crack information. Nevertheless, AE testing can hardly determine the crack shape and size, so a retesting is often needed. Infrared thermal imaging (ITM) [14,15] is an effective way of crack detection as well. Fundamentally, the thermal diffusion process would be interfered by cracks in the structure, which leads to temperature discontinuity on the observed surface. Therefore, the crack can be recognized by analyzing the recorded thermal images. The advantage of ITM is that it enables fast and full-field testing in the camera range, but the excitation devices and the infrared cameras might increase the system's cost and complexity.

Patch antenna sensor is a novel structural health monitoring (SHM) sensor that appears in recent years. It can sense the crack in metal structures and has great application prospect due to its advantages of simple configuration, light weight, easy fabrication, low cost, etc. The idea of using patch antenna sensor to identify cracks was first proposed by Deshmukh et al. [16] in 2009. They demonstrated that the sensor's resonant frequency dropped linearly when a crack on the ground plane propagated perpendicular to the current path. The observed sensitivity was 29.6 MHz/mm, and a sub-millimeter detection resolution could be achieved. Mohammad and Huang [17] tested the antenna sensor's resonant frequency response to crack using a double cantilever beam and they acquired three crack sensitivities: 2.5 MHz/mm before the crack reached the patch area, 48.7 MHz/mm when the crack was underneath the patch, and 4.7 MHz/mm after the crack tip passed the patch edge. The same team also found that the patch antenna sensor can detect not only crack growth but also crack opening [18]. Later, Xu and Huang presented a four-element antenna sensor array to detect crack growth at multiple locations [19]. A wireless interrogation method was developed by implementing a light-activated RF switch, and the maximum interrogation distance of different measurement configurations could be estimated by a power budget model [20]. Yi et al. conducted an emulated crack test and a fatigue crack test to characterize the RFID-based patch antenna sensor's performance with the presence of the crack. The experimental results show that the sensor is capable of measuring sub-millimeter crack and tracking crack propagation and that remote interrogation distance can be as far as 24 inches [21,22]. Cook et al. [23] investigated the effect of non-linear shaped (i.e., rectangular and pie-shaped) cracks on the patch antenna sensor's resonant frequency. It is worth noting that a pie-shaped crack can decrease the resonant frequency of the radiation mode perpendicular to it but cause the other resonant frequency to increase. Cho C et al. proposed a frequency doubling scheme, which consists of a transmitting antenna, a diode-integrated matching network and a receiving antenna, for wireless interrogation of the patch antenna sensor, and measured the relationship between the antenna sensor's resonant frequency and crack width [24]. Zhang J et al. utilized a circular patch antenna sensor with an open rectangular window for crack monitoring and presented that the antenna sensor could be useful for detecting crack depth [25].

These published studies are mainly focused on quantifying the patch antenna sensor's behavior under the crack that coincides with the centerlines of the ground plane. No attention was paid to cracks that are more complex. Although Mohammad et al. [26] found it possible to detect crack orientation by introducing the ratio of the sensor's two normalized resonant frequency shifts as an indicator, the quantitative relationship between the sensor's resonant frequency and the crack information was not clarified. In this paper, we presented a more comprehensive and detailed characterization of the patch antenna sensor's performance in sensing notch-shaped cracks. First, the crack detection mechanism was discussed, and then a patch antenna sensor was designed, simulated, fabricated and tested to understand the resonant frequency responses to various crack configurations. Besides, an algorithm was proposed for predicting both resonant frequency shifts caused by a random notch-shaped crack, which was verified by the experimental measurements. Owing to the fact that it is difficult to perform fatigue experiments to generate cracks of different characteristics for this study, we adopted a Computer Numerical Control (CNC) machine to cut slots on the sensor's ground plane to imitate cracks in realistic metal structures. This methodology is accessible and reasonable for the lab study and can shed light upon the real scenario as well. Because wireless interrogation of patch antenna has been successfully achieved by other researchers and is not the main focus of our work, we utilized cable connection for stability and convenience in the measurement.

2. Principle of Operation

2.1. Crack Detection Mechanism of Patch Antenna Sensor

As illustrated in Figure 1a, a rectangular patch antenna consists of a metal radiation patch, a dielectric substrate, and a conductive ground plane. These three parts form an electromagnetic (EM) resonator that works at two fundamental radiation modes: the TM_{010} mode with current flowing along the patch length direction and the TM_{001} mode with current flowing along the patch's width direction. When fed with a multi-frequency EM signal, the patch antenna radiates the frequency component that matches its resonant frequency while reflects the other back. This radiation characteristic can be represented by a S_{11} curve shown in Figure 1b, which describes the relationship between the patch antenna's return loss and the incident wave frequency. Accordingly, the patch antenna's two resonant frequencies, denoted as f_{010} and f_{001}, can be extracted from its S_{11} curve at the point where the return loss is a local minimum. From the transmission line theory [27], f_{010} and f_{001} are mainly determined by the geometric size of the radiation patch according to

$$f_{mnp} = \frac{c}{2\pi\sqrt{\varepsilon_{re}}}\sqrt{\left(\frac{m\pi}{h}\right)^2 + \left(\frac{n\pi}{L}\right)^2 + \left(\frac{p\pi}{W}\right)^2}, \quad (m = 0; n = 0, 1; p = 0, 1) \qquad (1)$$

where c is the velocity of light in vacuum, ε_{re} is the effective dielectric constant of the substrate, h is the thickness of the substrate, and L and W are the geometric length and width of the patch, respectively.

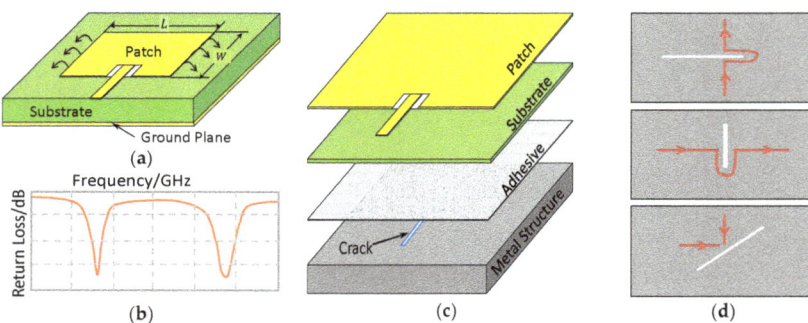

Figure 1. Crack detection mechanism of the patch antenna sensor. (**a**) configuration of a patch antenna; (**b**) S_{11} curve representing the radiation characteristics of a path antenna; (**c**) patch antenna used on metal structure as a sensor; (**d**) impact of cracks on the sensor's current pattern.

Considering the patch antenna's configuration and the fact that any good conductor can be a ground plane, we can create a SHM sensor by bonding a substrate and a radiation patch to the surface of metal structure using adhesive, as shown in Figure 1c. In this case, the metal structure serves as the ground plane, and the formed patch antenna sensor can perceive certain physical quantity changes of the metal structure. For example, a surface crack in the metal structure (see Figure 1c) would cause a local conductivity loss and therefore disturb the current path on the ground plane, leading to a resonant frequency shift of the patch antenna sensor. As demonstrated in Figure 1d, a crack parallel to the patch length would significantly increase the current path of TM_{010} mode and thus change f_{010}, and a crack parallel to the patch width would do the same to TM_{001} mode and f_{001}. A crack with a certain inclination would affect both radiation modes and resonant frequencies. Characterizing the patch antenna sensor's resonant frequency response to crack would give us means to inversely identify the crack information from the measured resonant frequencies.

2.2. Analytical Expression of Antenna Sensor's Current Distribution

To understand the antenna sensor's behavior with a cracked ground plane, the current density on its ground plane and the patch was derived first. For the coordinate system shown in Figure 2, the EM field inside the patch antenna could be expressed as Equation (2) according to the 'cavity model' and the vector potential method [27,28]:

$$E_z = -j\frac{(k^2 - k_z{}^2)}{\omega\mu\epsilon}A_{mnp}\cos(k_x x)\cos(k_y y)\cos(k_z z)$$

$$E_x = -j\frac{k_x k_z}{\omega\mu\epsilon}A_{mnp}\sin(k_x x)\cos(k_y y)\sin(k_z z)$$

$$E_y = -j\frac{k_y k_z}{\omega\mu\epsilon}A_{mnp}\cos(k_x x)\sin(k_y y)\sin(k_z z) \tag{2}$$

$$H_z = 0$$

$$H_x = \frac{k_y}{\mu}A_{mnp}\cos(k_x x)\sin(k_y y)\cos(k_z z)$$

$$H_y = -\frac{k_x}{\mu}A_{mnp}\sin(k_x x)\cos(k_y y)\cos(k_z z)$$

where ω is the angular frequency of time-harmonic field, while μ and ϵ are the permeability and the dielectric constant of the substrate, respectively. A_{mnp} is the amplitude of the introduced vector potential for TM_{mnp} mode. k_x, k_y, k_z are the wavenumbers along the x, y, z directions, and are calculated as:

$$k_z = \frac{m\pi}{h}, m = 0, 1, 2, \ldots$$

$$k_y = \frac{n\pi}{L}, n = 0, 1, 2, \ldots \tag{3}$$

$$k_x = \frac{p\pi}{W}, p = 0, 1, 2, \ldots$$

For the TM_{010} mode, $k_x = k_z = 0$ and $k_y = \pi/L$, so its EM field components should be written as (where E_0 and H_0 are the maximum values of sinusoidal E and H)

$$E_z = E_0 \cos\left(\frac{\pi}{L}y\right)$$

$$H_x = H_0 \sin\left(\frac{\pi}{L}y\right) \tag{4}$$

$$E_x = E_y = H_z = H_y = 0$$

For the TM_{001} mode, $k_y = k_z = 0$ and $k_x = \pi/W$, and its EM field components should be written as:

$$E_z = E_0 \cos\left(\frac{\pi}{W}x\right)$$

$$H_y = H_0 \sin\left(\frac{\pi}{W}x\right) \tag{5}$$

$$E_x = E_y = H_z = H_x = 0$$

After the EM field is determined, the current density on the ground plane (or the patch) can be calculated based on [29]:

$$J = \vec{n} \times H \tag{6}$$

where $\vec{n}$ is the unit normal vector of the ground plane (or the patch). Since H for TM_{010} mode is along the x direction, the corresponding current density is along the y direction and should be expressed as:

$$J_y = J_{010} \sin\left(\frac{\pi}{L}y\right) \tag{7}$$

Similarly, the current density for TM_{001} mode is along the x direction and the analytical expression is

$$J_x = J_{001} \sin\left(\frac{\pi}{W}x\right) \tag{8}$$

The current flows for TM_{010} mode and TM_{001} mode are depicted in Figure 2a,b where the length of the arrows indicates the magnitude of the current density. It is obvious that the maximum current appears in the middle of the ground plane and decreases gradually to zero at the edges.

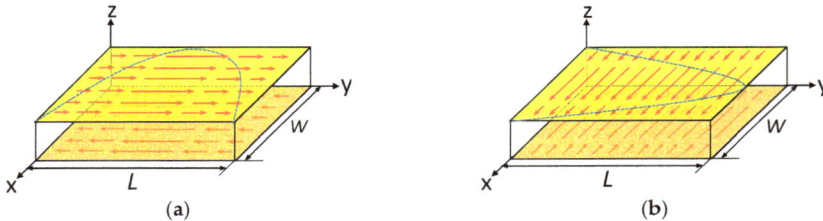

Figure 2. Current distribution of different antenna radiation modes. (**a**) current density of the TM_{010} mode; (**b**) current density of the TM_{001} mode.

3. Sensor Design and Simulation

3.1. Sensor Design Parameters

As shown in Figure 3, a rectangular patch antenna sensor was designed using the procedure described in [16]. The FR4 plane with a thickness of 0.5 mm and a dielectric constant of 4.4 was chosen as the substrate. The initial resonant frequencies were selected to be $f_{010} = 1.8$ GHz and $f_{001} = 2.5$ GHz because the frequency range of our vector network analyzer (VNA) is 300 kHz–3 GHz. For the selected substrate, this resulted in a 39.62 mm long and 27.89 mm wide radiation patch, which were rounded up to be 40 mm and 28 mm for the convenience of manufacturing. In addition, the patch is fed at a proper position with a transmission line, through which the incident signal can be applied to excite the radiation modes. The detailed parameters for each part of the designed antenna sensor can be found in Table 1.

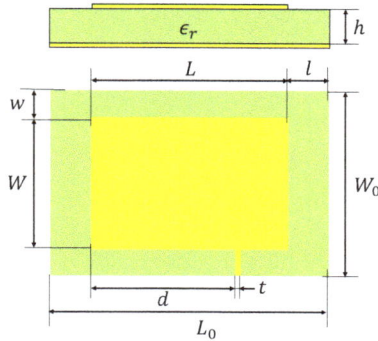

Figure 3. Antenna sensor design parameters.

Table 1. Detailed parameters of the designed antenna sensor.

Symbol	Physical Quantity	Selected Value
f_{010}, f_{001}	Designed initial resonant frequencies	1.80 GHz, 2.54 GHz
ϵ_r	Substrate dielectric constant	4.4
h	Substrate thickness	0.5 mm
L	Radiation patch length	40 mm
W	Radiation patch width	28 mm
t	Transmission line width	1 mm
d	Transmission line position	29.5 mm
L_0	Length of substrate/ground plane	64 mm
W_0	Width of substrate/ground plane	44 mm
l	Horizontal distance from patch edge to substrate edge	12 mm
w	Vertical distance from patch edge to substrate edge	8 mm

3.2. Simulation Model

The designed sensor was modeled in commercial EM simulation software HFSS^{TM} (Ansoft, Pittsburgh, PA, USA) to characterize its crack sensing ability. As shown in Figure 4a, the entire antenna sensor is confined in an air box whose surfaces are set as radiation boundaries. The patch and the ground plane are treated as perfect electrical conductors. The antenna sensor is excited at the end of the transmission line with a 50 Ω lumped port. On the ground plane, a 0.6 mm wide slot was introduced to imitate the crack. Since the crack could appear randomly, three parameters were predefined to quantitatively describe it. In the coordinate system shown in Figure 4b, a crack is represented by a line-shaped slot. The midpoint P and the length s of the slot denote the crack position and crack length, respectively, and the crack orientation is defined as the angle θ between the slot and the x-axis. As such, an arbitrary crack on the ground plane can be presented by three parameters—the position P, length s, and orientation θ.

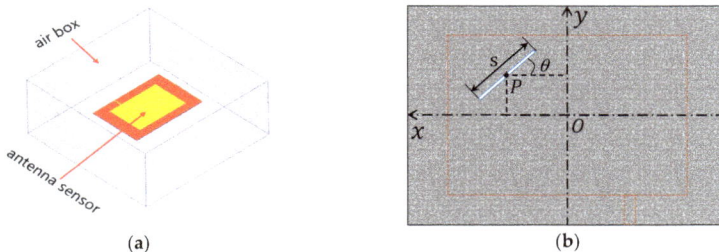

(a)

(b)

Figure 4. *Cont.*

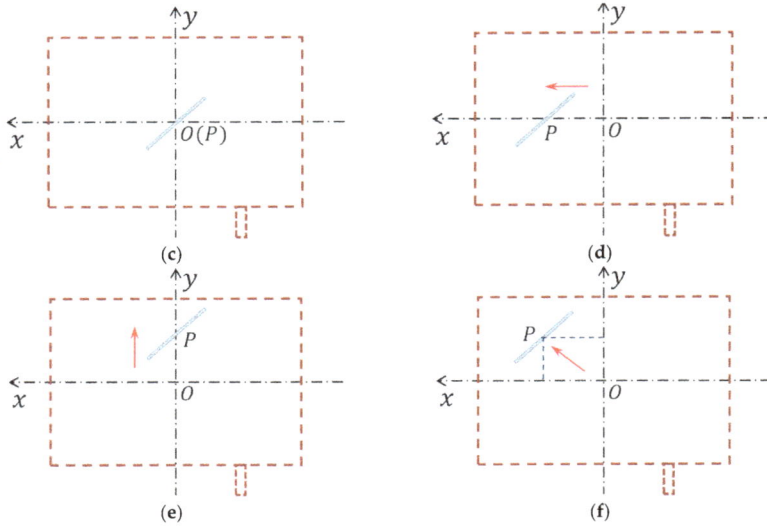

Figure 4. Simulation model. (**a**) entire model in HFSS^TM; (**b**) quantitative description of the crack; (**c**) cracks located at the center of the ground plane; (**d**) crack position moving leftwards; (**e**) crack position moving upwards; (**f**) crack position moving leftwards and upwards simultaneously.

In order to study the resonant frequency response to crack, the antenna sensor was modeled with the following procedure. Firstly, the crack position P was fixed at the center of the ground plane to investigate the variation of resonant frequency with the crack length and orientation change (see Figure 4c)). After that, the same investigation was conducted by moving the crack position P leftwards along the x-axis (see Figure 4d) and upwards along the y-axis (see Figure 4e), respectively, with a step of 3 mm. Finally, the crack position P was moved simultaneously in both directions (see Figure 4f), and the corresponding resonant frequency shifts were analyzed. In the entire modeling process, the crack length increased with a step of 1 mm and the crack profile was limited inside the area under the patch. Considering the symmetry of the antenna geometry and the current density, the crack orientation θ was assigned to vary from 0° to 90° with an increment of 15°.

3.3. Simulated Results

The simulated response of the sensor's resonant frequency to the cracks at the coordinate origin is shown in Figure 5a,b. When the crack orientation θ is 0°, f_{010} almost remains constant. In other directions, f_{010} drops with the increase of the crack length, and the closer θ approaches 90° the greater the drop is. Conversely, f_{001} is unchanged when the crack orientation is 90° but declines fastest with the crack growth in 0° direction. These observations could be explained by the interaction between the crack and the current flow on the ground plane. As shown in Figure 6a,b, the cracks in 0° direction or 90° direction only perturbs the current of one of the two modes while has no influence on the other, therefore causing only one of the resonant frequencies to decrease. When the crack of the same length gradually varies from 0° direction to 90° direction, its effective cutting length on TM_{010} mode current gets bigger (see Figure 6a) whereas that on TM_{001} mode drops down (see Figure 6b). This accounts for the different rate in resonant frequency reduction resulted from the crack orientation change.

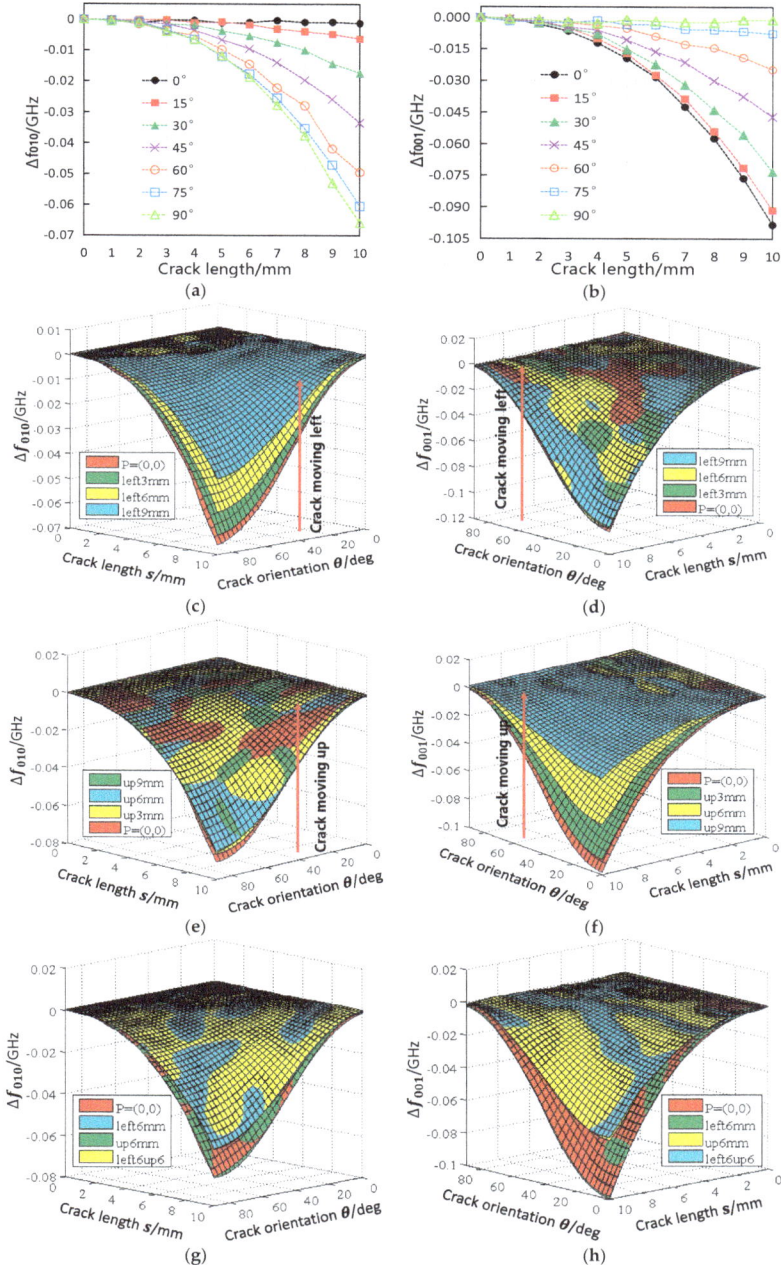

Figure 5. Simulated results. (**a**) f_{010} and (**b**) f_{001} response to cracks at the center of the ground plane; (**c**) f_{010} and (**d**) f_{001} response with the crack position moving leftwards; (**e**) f_{010} and (**f**) f_{001} response with the crack position moving upwards; (**g**) f_{010} and (**h**) f_{001} response with the crack position moving leftwards and upwards simultaneously.

For the crack configuration in Figure 4d, the simulated results are plotted in Figure 5c,d. The surface 'P = (0,0)' represents the sensor's resonant frequency response to the crack whose position is at the coordinate origin, and the other surfaces are the corresponding results of the cracks moving leftwards by 3, 6, and 9 mm. In the process of the crack moving leftwards, the trend of the f_{010} change keeps constant but the amplitude of the change decreases in sequence; the f_{001} response remains approximately the same as when the crack is at the coordinate origin. The reason for this is shown in Figure 6c,d. When moving to the left, the crack enters the low-density area of TM_{010} mode current from the high-density area. As a result, the cutting effect of the crack on TM_{010} mode current is gradually weakened and the amount of f_{010} shift declines accordingly. In comparison, the magnitude of f_{001} shift does not change because the crack is still in the high-density area of TM_{001} mode current when moving along the x-axis.

When the crack moves upwards as in Figure 4e, the antenna sensor's resonant frequency responses are shown in Figure 5e,f. In this case, the response of f_{010} to the crack stays unchanged all the way while the shift of f_{001} gets weaker. Similar to the scenario where the crack moves leftwards, this is caused by the current density change of the region being cut by the crack (see Figure 6e,f).

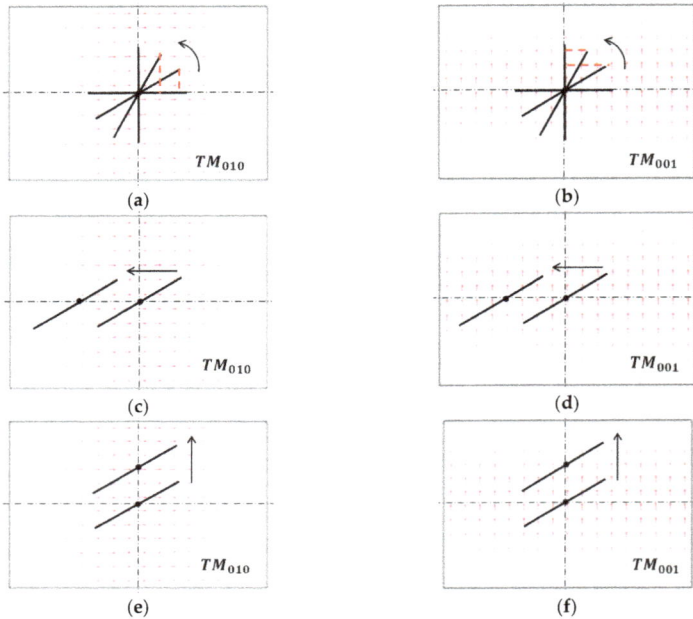

Figure 6. Interaction between the crack and the sensor's current flow. (**a,b**) cutting effect of the cracks at ground plane center on both current flows; (**c,d**) cutting effect of the crack moving leftwards; (**e,f**) cutting effect of the crack moving upwards.

To understand the sensor's resonant frequency behavior with the crack position changes shown in Figure 4f, the results of four selected crack configurations, i.e., crack at the coordinate origin, crack moving to the left by 6 mm, crack moving up by 6 mm, and crack moving leftwards and upwards simultaneously by 6 mm, are compared in Figure 5g,h. It can be seen that the f_{010} response of 'left6up6' is almost the same as that of 'left6mm' and the f_{001} response of 'left6up6' and 'up6mm' roughly overlap. This indicates that crack movements along the x-axis and y-axis independently affect the resonant frequency f_{010} and f_{001}, respectively.

4. Algorithm for Predicting the Crack-Caused Resonant Frequency Shifts

4.1. Definition of the Current Cutting Effect Coefficient φ

According to the above simulation and analysis, the influence of the crack on the antenna sensor's resonant frequencies is governed by the crack's cutting effect on the current flow of both modes. Generally, this cutting effect leads to the resonant frequency decrease, and the decrease becomes more significant when the cutting effect is more intense. The intensity of this cutting effect is related to the crack position, length, and orientation, so we introduced a coefficient φ to comprehensively describe it. The definition of φ is

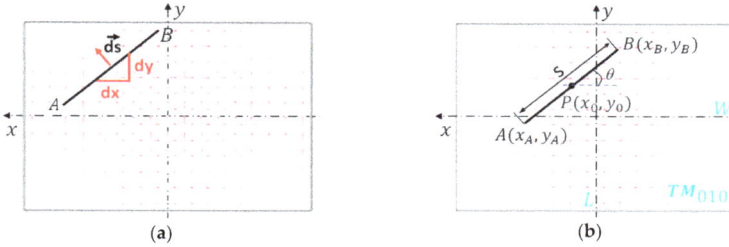

Figure 7. Definition of crack's cutting effect intensity φ on the antenna current. (a) definition; (b) calculation of φ_{010}.

As shown in Figure 7a, line AB represents an arbitrary crack, and $\overrightarrow{ds}$ is one element on it. The element's cutting effect intensity $d\varphi$ can be expressed by vector point multiplication as:

$$d\varphi = \overrightarrow{ds} \cdot \overrightarrow{j_s} \tag{9}$$

where $\overrightarrow{j_s}$ is the current density vector at the element point. Therefore, the crack's total cutting effect intensity φ is the integration of $d\varphi$ along the entire crack path, i.e.,

$$\varphi = \int_A^B \overrightarrow{j_s} \cdot \overrightarrow{ds} \tag{10}$$

The calculation of formula (10) is illustrated in Figure 7b. Denote the crack position, length and orientation as $P(x_0, y_0)$, s and θ, respectively, and the coordinates of A and B can be written as

$$A(x_A, y_A) = A(x_0 - \frac{s\cos\theta}{2}, y_0 - \frac{s\sin\theta}{2}) \tag{11}$$

$$B(x_B, y_B) = B(x_0 + \frac{s\cos\theta}{2}, y_0 + \frac{s\sin\theta}{2}) \tag{12}$$

From Equations (7) and (8), the current distribution of two modes in the coordinate system of Figure 7b should be expressed as

$$\overrightarrow{J}_{010}(x, y) = J_{010}\sin\left[\frac{\pi}{L}(x + \frac{L}{2})\right]\overrightarrow{e}_x \tag{13}$$

$$\overrightarrow{J}_{001}(x, y) = J_{001}\sin\left[\frac{\pi}{W}(y + \frac{W}{2})\right]\overrightarrow{e}_y \tag{14}$$

where J_{010} and J_{001} are the maximum values of the current densities of TM_{010} and TM_{001} mode. $\vec{e}_x$ and $\vec{e}_y$ are the unit vectors along the x-axis and y-axis. The cutting effect intensity of crack AB on TM_{010} mode current is then

$$\varphi_{010} = \int_A^B \vec{J}_{010}(x,y) \cdot \vec{ds} = \int_{y_A}^{y_B} J_{10} \sin\left[\frac{\pi}{L}\left(x + \frac{L}{2}\right)\right] dy \tag{15}$$

For $0° \leq \theta < 90°$, the integration path AB can be written as

$$y = \tan\theta x + y_0 - x_0 \tan\theta \tag{16}$$

By changing the integral variable to dx, Equation (15) can be calculated as

$$\varphi_{010} = \int_{x_A}^{x_B} J_{010} \sin\left[\frac{\pi}{L}\left(x + \frac{L}{2}\right)\right] \tan\theta dx = \frac{J_{10}L\tan\theta}{\pi}\left(\sin\frac{\pi x_B}{L} - \sin\frac{\pi x_A}{L}\right) \tag{17}$$

For $\theta = 90°$, the crack is perpendicular to the current of TM_{010} mode. Therefore, its cutting effect intensity should be $\varphi_{010} = sJ_{010}$. Overall, the cutting effect coefficient of crack AB on TM_{010} mode current is defined as

$$\varphi_{010} = sJ_{010} \ (\theta = 90°) \tag{18}$$

$$\varphi_{010} = \frac{J_{010}L\tan\theta}{\pi}\left(\sin\frac{\pi x_B}{L} - \sin\frac{\pi x_A}{L}\right), \ (0° \leq \theta < 90°)$$

Following a similar process, the cutting effect coefficient of crack AB on TM_{001} mode current can be derived as

$$\varphi_{001} = sJ_{001} \ (\theta = 0°) \tag{19}$$

$$\varphi_{001} = \frac{J_{001}W}{\pi\tan\theta}\left(\sin\frac{\pi y_B}{W} - \sin\frac{\pi y_A}{W}\right), \ (0° < \theta \leq 90°)$$

4.2. Relationship between Resonant Frequency Shift and Coefficient φ

After coefficient φ was defined, we investigated the relationship between the sensor's resonant frequency shift and φ using the simulation model shown in Figure 8a,b. The cracks were located at the coordinate origin, growing from 0 mm to 20 mm in six orientations. The obtained Δf–φ relationship are presented in Figure 8c,d, in which the vertical axis is the Δf value from HFSS and the horizontal axis is the φ value calculated by Mathematica. It can be seen that Δf_{001}–φ_{001} relationship keeps the same regardless of the crack configuration. By contrast, Δf_{010}–φ_{010} relationship of each orientation remains unified when φ_{010} is small but begins to deviate when φ gets to a certain value. Such a difference between Δf_{010}–φ_{010} and Δf_{001}–φ_{001} is probably related to the antenna sensor's feeding way and feeding position.

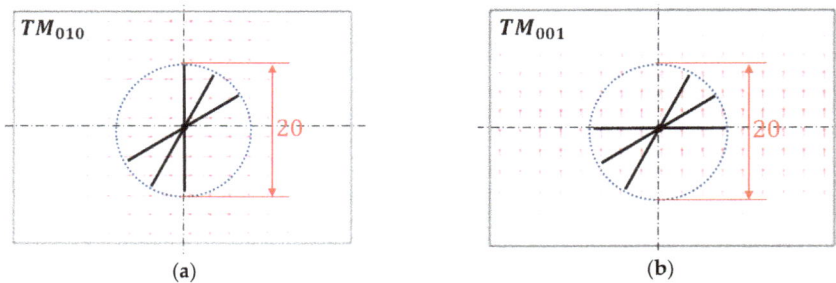

(a)

(b)

Figure 8. *Cont.*

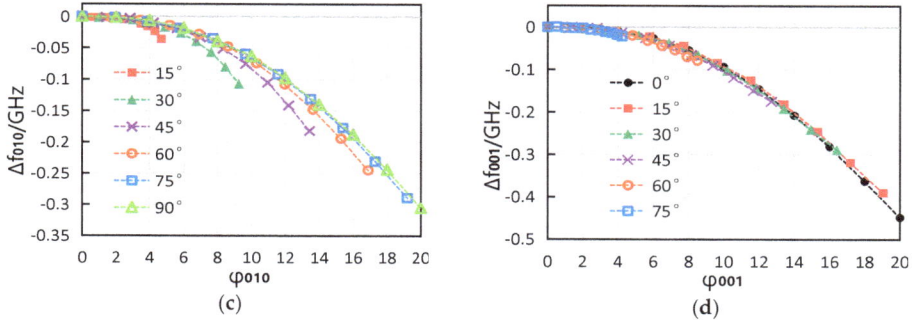

Figure 8. Investigation on the relationship between the sensor's resonant frequency shift and coefficient φ. (**a,b**) simulation model; (**c**) $\Delta f_{010}-\varphi_{010}$ relationship; and (**d**) $\Delta f_{001}-\varphi_{001}$ relationship.

4.3. Process of the Algorithm

Based on the above analysis, we present an algorithm for predicting resonant frequency shifts caused by an arbitrary crack. The fundamental idea is to first calculate the φ of the crack and then to get the resonant frequency shifts according to the $\Delta f - \varphi$ relationship. In this paper, we select the $\Delta f_{010}-\varphi_{010}$ relationship of the crack at $90°$ orientation and the $\Delta f_{001}-\varphi_{001}$ relationship of the crack at $0°$ orientation as standard $\Delta f - \varphi$ relationships of two radiation modes. The process of the algorithm is shown in Figure 9, which will be validated by experimental data in the following content.

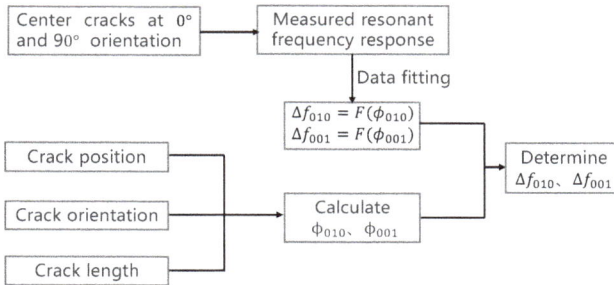

Figure 9. Algorithm for predicting the crack-caused resonant frequency shifts.

5. Experimental Validation

5.1. Sensor Fabrication and Experiment Setup

The antenna sensor was fabricated by the chemical etching process shown in Figure 10. First, a 0.6 mm thick FR4 copper clad laminate was cut into 64mm × 44mm pieces. The designed patch shape was then printed on a PCB pattern transfer paper and transferred to one side of the laminate, followed by dipping it into a ferric chloride solution to etch the unwanted copper. After the etching is done, acetone was used to wash off the ink covering the patch and the SMA connector was soldered.

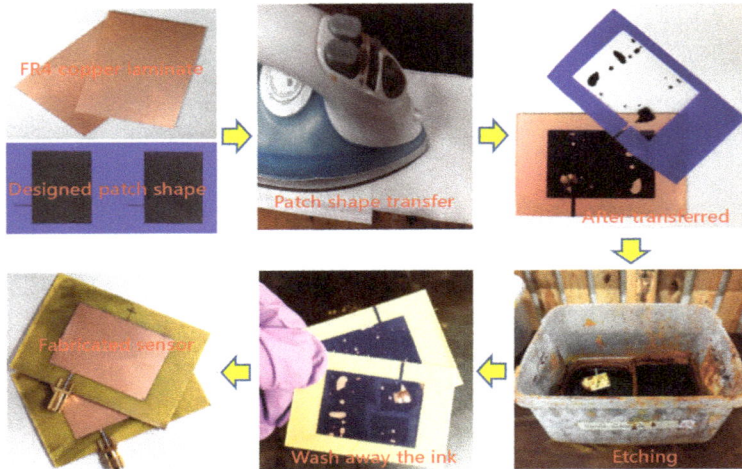

Figure 10. Fabricating process of the antenna sensor.

An experimental setup (see Figure 11) was established to test the antenna sensor's crack sensing behavior. The sensor specimen was clamped on the work plane of a CNC machine (Click N Carve 84015, Rockler, Medina, MN, USA) and connected to a vector network analyzer (VNA) (Hewlett Packard 85047A, Keysight, Santa Rosa, CA, USA) through the SMA connector and the coaxial cable. The CNC machine was used to cut the cracks of different lengths and orientations at different positions of the sensor's ground plane. For each tested crack configuration, the step of the crack length increase was set to be 2 mm. This was achieved by controlling the blade to cut into the ground plane at the midpoint of the scheduled crack profile (i.e., crack position P) and then move 1 mm further to the left and the right, respectively. After the cutting was done, the sensor specimen was removed from the CNC machine and the actual crack length was measured using a ruler before the resonant frequencies were acquired by the VNA. The VNA was set to sweep over a span of 40 MHz with 1601 frequency points in every measurement, which results in a 25 kHz frequency resolution.

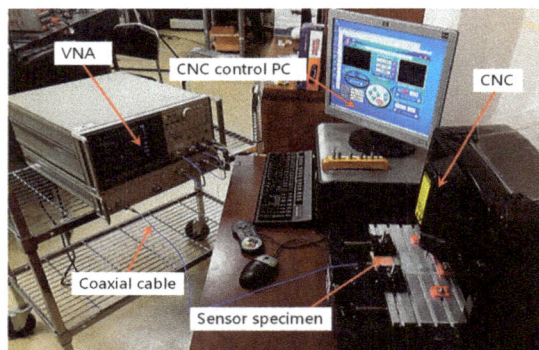

Figure 11. Experiment setup.

5.2. Experiment Results

The measured data of sensors with cracks located at the center of the ground plane is plotted in Figure 12. Cracks in seven directions were tested, and the length for all the cracks was increasing from 0 to 20 mm with an interval of 2 mm. As predicted by the simulation, both resonant frequencies decreased with the crack growth, and the equal-length cracks in different orientations led to different amount of frequency shifts. It could be noticed that some data points were missing because the S_{11} curve degraded (which only showed one or no resonant frequency peak) at certain crack lengths and directions. This was probably due to the impedance mismatch of the antenna sensor in such crack circumstances.

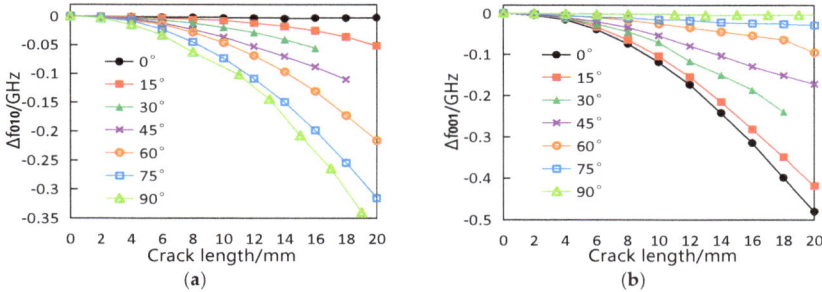

Figure 12. Measured resonant frequency response to cracks located at the center of the ground plane. (a) f_{010} results and (b) f_{001} results.

To efficiently validate the effect of crack movement on the antenna sensor's resonant frequencies, the crack position was moved away from the coordinate origin to $P = (5,5)$ and $P = (10,10)$ respectively, and only two typical directions, i.e., 30° and 60°, were tested. For the convenience of comparing the results of different cracks, the crack length was increased from 0 mm to 14 mm with an increment of 2 mm. The observed resonant frequency responses are shown in Figure 13. Some data points were missing because the crack tips were out of the patch area. It is obvious that, for either f_{010} or f_{001}, the rate of the resonant frequency decrease becomes smaller as the crack position moves further away from the coordinate origin. Qualitatively, this behavior agrees with the effect of crack position movements on the resonant frequency response revealed by simulation. However, such phenomenon is not prominent when the crack is short (e.g., less than 4 mm in length). This might be contributed by the fact that the resonant frequency shifts are so small at the beginning of the crack growth that the measurement errors (either in measuring the crack length or the resonant frequencies) could have significant influence.

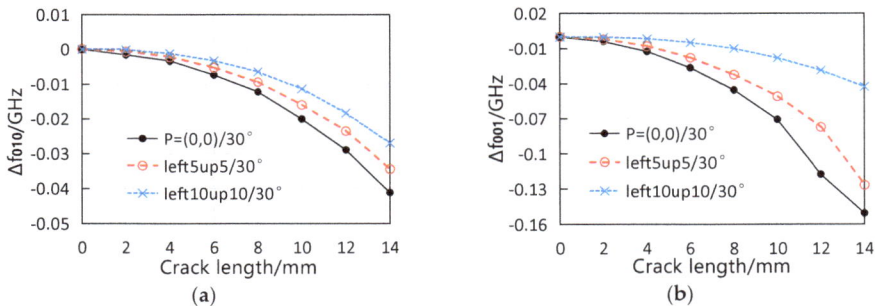

Figure 13. *Cont.*

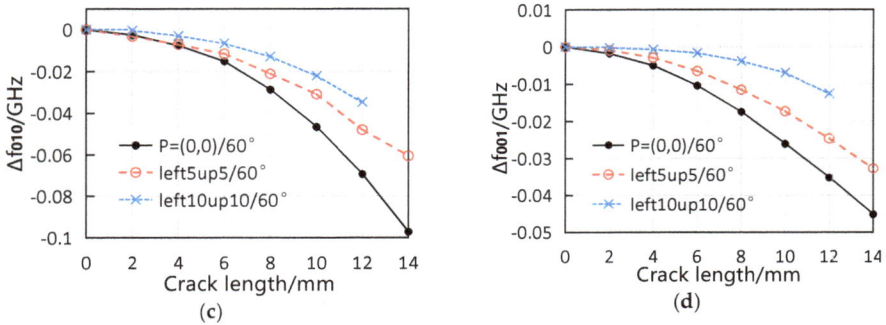

Figure 13. Measured resonant frequency response with the crack moving to $P = (5,5)$ and $P = (10,10)$. (a) f_{010} and (b) f_{001} response at 30° direction; (c) f_{010} and (d) f_{001} response at 60° direction.

5.3. Case Study—Validation of the Proposed Algorithm

In order to evaluate the effectiveness of the proposed algorithm, the fundamental idea is to measure the resonant frequency shifts of the antenna sensor with an arbitrary crack, and then compare the results with the corresponding values predicted by the algorithm. Since the crack could not be really arbitrary in the test, we selected two crack configurations to conduct the case study. One crack was located at $P = (7,3)$, growing in the direction of $\theta = 40°$; the other was located at $P = (4,2)$, growing in the direction of $\theta = 55°$. The crack length was assigned to increase 2 mm per step to collect more data points.

Prior to the measurement, the $\Delta f_{010} = F(\varphi_{010})$ was obtained by cubic polynomial fitting the measured data of the crack that was located at the coordinate origin and propagates in 90° direction (i.e., the '90°' curve in Figure 12a) Similarly, the $\Delta f_{001} = F(\varphi_{001})$ was the cubic polynomial fitting of the measured data of the crack located at the coordinate origin and propagating in 0° direction (i.e., the '0°' curve in Figure 12b. The acquired functions are shown as the Equations (20) and (21), and the corresponding R^2 values are 0.9999 and 0.9998, respectively. To calculate the predicted resonant frequency shifts, the coefficient φ_{010} and φ_{001} of a certain crack were calculated first according to the Equations (18) and (19), and then Δf_{010} and Δf_{001} could be gained from the Equations (20) and (21).

$$\Delta f_{010} = F(\varphi_{010}) = 4.98 \times 10^{-6}\varphi_{010}{}^3 - 0.001\varphi_{010}{}^2 + 0.00015\varphi_{010} - 0.0008 \tag{20}$$

$$\Delta f_{001} = F(\varphi_{001}) = 1.44 \times 10^{-5}\varphi_{001}{}^3 - 0.0016\varphi_{001}{}^2 + 0.0033\varphi_{001} - 0.0021 \tag{21}$$

The comparison between the measured resonant frequency shifts and their predicted counterparts are shown in Figure 14. At most data points that are observed, the prediction is in good agreement with the measurement although a slight difference can be seen. The discrepancy might come from the errors in the process of measuring the crack length and extracting the resonant frequency from the S_{11} curve. Another source of the discrepancy might be the ideal assumption in the proposed algorithm. When calculating the predicted resonant frequency, the coefficient φ is based on the current density function (13) and (14), which should be slightly different from the real current density of the tested antenna sensor because of the antenna's fringing effect in practical scenario. It is also observed that, for f_{010} in both case studies, the discrepancy between the predicted resonant frequency shift and the measured one becomes considerable when the crack is longer than 20 mm. This can be explained by the inconsistency of Δf_{010}–φ_{010} relationships (as shown in Figure 8c): since we take the Δf_{010}–$\varphi_{010}\varphi_{010}$ relationship of the crack at 90° direction as the standard in the algorithm, the predicted Δf_{010} for the crack at other directions would not be accurate if the crack's φ_{010} is relatively large. Moreover, it can be anticipated that the algorithm-predicted Δf_{010} would be more precise for the crack whose orientation is closer to 90°. Generally, the proposed algorithm works well in predicting the resonant frequency shifts but may lose the accuracy in predicting Δf_{010} of long cracks. This performance is likely to be improved by modifying the $\Delta f_{010} - \varphi_{010} \varphi_{010}$ in the future study.

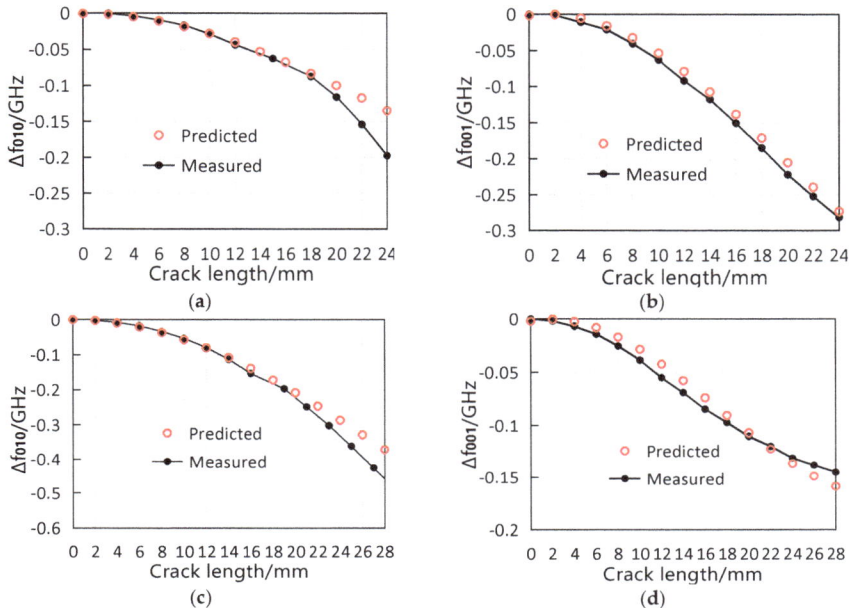

Figure 14. Evaluation of the proposed algorithm. (**a**,**b**) comparison between the measured and the predicted resonant frequency response for case $P = (7,3)$ and $\theta = 40°$; (**c**,**d**) comparison between the measured and the predicted resonant frequency response for case $P = (4,2)$ and $\theta = 55°$.

6. Conclusions

This study characterized the response of a patch antenna sensor's resonant frequency to notch-shaped cracks. Both the simulation and experiment demonstrate that the crack position, length and orientation exert influence on the resonant frequencies in a way that could be represented by the interaction between the crack and the current distribution of the sensor's ground plane. As a result, we presented an algorithm for predicting the resonant frequency shifts caused by a random notch-shaped crack. The experimental tests show that this algorithm works well in most cases but might be inaccurate in predicting Δf_{010} when the crack is of a considerable length. Based on this study, the behavior of patch antenna sensor's resonant frequencies in identifying notch-shaped cracks could be substantially understood and quantitatively described, which contributes to the research and development of patch antenna sensors for SHM purposes. The subsequent work in this field should be focused on developing the inversion algorithm for crack identification and monitoring.

Author Contributions: Conceptualization, Z.L.; Methodology, L.K. and H.Y.; Data curation, L.K.; Formal analysis, L.K.; Investigation, L.K. and H.Y.; Writing—original draft preparation, L.K.; Writing—review and editing, L.K.

Funding: This study was funded by the Fundamental Research Funds for the Central Universities of China, grant number 2017-JL-006.

Acknowledgments: We really thank Haiying Huang from University of Texas at Arlington for providing some of the experimental equipment and for her kind help in performing this study.

Conflicts of Interest: The authors declare no conflict of interest.

References

1. Chee-Hoe, F.; Marian, W.; William, F.D. Novel dynamic fatigue-testing device: Design and measurements. *Meas. Sci. Technol.* **2006**, *17*, 2218. [CrossRef]
2. Yan, Y.J.; Cheng, L.; Wu, Z.Y.; Yam, L.H. Development in vibration-based structural damage detection technique. *Mech. Syst. Signal Proc.* **2007**, *21*, 2198–2211. [CrossRef]
3. Jassim, Z.A.; Ali, N.N.; Mustapha, F.; Abdul Jalil, N.A. A review on the vibration analysis for a damage occurrence of a cantilever beam. *Eng. Fail. Anal.* **2013**, *31*, 442–461. [CrossRef]
4. Liu, X.J.; Qing-Xu, Y.U. Experimental study of distributed optical fiber sensor for crack detection. *J. Optoelectron. Laser* **2005**, *16*, 779–782.
5. Ravet, F.; Briffod, F.; Glisic, B.; Nikles, M.; Inaudi, D. Submillimeter crack detection with brillouin-based fiber-optic sensors. *IEEE Sens. J.* **2009**, *9*, 1391–1396. [CrossRef]
6. Glisic, B.; Inaudi, D. Development of method for in-service crack detection based on distributed fiber optic sensors. *Struct. Health Monit.* **2012**, *11*, 161–171. [CrossRef]
7. Tian, G.Y.; Sophian, A.; Taylor, D.; Rudlin, J. Multiple sensors on pulsed eddy-current detection for 3-D subsurface crack assessment. *IEEE Sens. J.* **2005**, *5*, 90–96. [CrossRef]
8. Hughes, R.R.; Dixon, S. Eddy-current crack detection at frequencies approaching electrical resonance. *AIP Conf. Proc.* **2014**, *1581*, 1366–1373. [CrossRef]
9. Zhang, W.; Wu, W.; Sun, X.; Xiao, L.; Qu, W. Damage Detection of closed crack in a metallic plate using nonlinear ultrasonic time reversal method. *J. Sens.* **2013**, *2013*, 10. [CrossRef]
10. Sohn, H.; Lim, H.J.; DeSimio, M.P.; Brown, K.; Derriso, M. Nonlinear ultrasonic wave modulation for online fatigue crack detection. *J. Sound Vib.* **2014**, *333*, 1473–1484. [CrossRef]
11. Glushkov, Y.V.; Glushkova, N.V.; Yekhlakov, A.V. A mathematical model of the ultrasonic detection of three-dimensional cracks. *J. App. Math. Mech.* **2002**, *66*, 141–149. [CrossRef]
12. Chia, C.C.; Lee, J.-R.; Bang, H.-J. Structural health monitoring for a wind turbine system: A review of damage detection method. *Meas. Sci. Technol.* **2008**, *19*, 122001.
13. Nair, A.; Cai, C.S. Acoustic emission monitoring of bridges: Review and case studies. *Eng. Struct.* **2010**, *32*, 1704–1714. [CrossRef]
14. Li, X.; Liu, Z.; Jiang, X.; Lodewijks, G. Method for detecting damage in carbon-fibre reinforced plastic-steel structures based on eddy current pulsed thermography. *Nondestruct. Test. Eval.* **2018**, *33*, 1–19. [CrossRef]
15. Liu, Z.; Fang, G.; Jiang, L.; Jiang, X.; Lodewijks, G. Design and experimental study of a novel excitation coil based on pulsed eddy current thermography. *Insight Nondestruct. Test. Cond. Monit.* **2017**, *59*, 491–499.
16. Deshmukh, S.; Mohammad, I.; Tentzeris, M.; Wu, T.; Huang, H. Crack detection and monitoring using passive wireless sensor. In Proceedings of the ASME 2009 Conference on Smart Materials, Adaptive Structures and Intelligent Systems, Oxnard, CA, USA, 21–23 September 2009.
17. Mohammad, I.; Huang, H. An antenna sensor for crack detection and monitoring. *Adv. Struct. Eng.* **2011**, *14*, 47–53. [CrossRef]
18. Mohammad, I.; Huang, H. Monitoring fatigue crack growth and opening using antenna sensors. *Smart Mater. Struct.* **2010**, *19*, 055023. [CrossRef]
19. Xu, X.; Huang, H. Multiplexing passive wireless antenna sensors for multi-site crack detection and monitoring. *Smart Mater. Struct.* **2012**, *21*, 015004. [CrossRef]
20. Deshmukh, S.; Huang, H. Wireless interrogation of passive antenna sensors. *Meas. Sci. Technol.* **2010**, *21*, 035201. [CrossRef]
21. Yi, X.; Cho, C.; Fang, C.; Cooper, J.; Lakafosis, V.; Vyas, R.; Wang, Y.; Leon, R.T.; Tentzeris, M.M. Wireless strain and crack sensing using a folded patch antenna. Proceedings of 6th European Conference on Antennas and Propagation (EUCAP 2012), Prague, Czech Republic, 26–30 March 2012; pp. 1678–1681.
22. Yi, X.; Cho, C.; Cooper, J.; Wang, Y.; Tentzeris, M.M.; Leon, R. T Passive wireless antenna sensor for strain and crack sensing—Electromagnetic modeling, simulation, and testing. *Smart Mater. Struct.* **2013**, *22*, 085009. [CrossRef]
23. Cook, B.S.; Shamim, A.; Tentzeris, M.M. Passive low-cost inkjet-printed smart skin sensor for structural health monitoring. *IET Microw. Antennas Propag.* **2013**, *6*, 1536–1541. [CrossRef]
24. Cho, C.; Yi, X.; Li, D.; Wang, Y.; Tentzeris, M.M. Passive wireless frequency doubling antenna sensor for strain and crack sensing. *IEEE Sens. J.* **2016**, *16*, 5725–5733. [CrossRef]

25. Zhang, J.; Huang, B.; Zhang, G.; Yun Tian, G. Wireless passive ultra high frequency RFID antenna sensor for surface crack monitoring and quantitative analysis. *Sensors* **2018**, *18*, 2130. [CrossRef] [PubMed]
26. Mohammad, I.; Gowda, V.; Zhai, H.; Huang, H. Detecting crack orientation using antenna sensors. *Meas. Sci. Technol.* **2011**, *7981*, 765–768. [CrossRef]
27. Balanis, C. *Antenna Theory: Analysis and Design*; John Wiley and Sons: New York, NY, USA, 2005; Volume 1.
28. Ghosh, D.K.; Ghosh, S.; Chattopadhyay, S.; Nandi, S.; Chakraborty, D.; Anand, R.; Raj, R.; Ghosh, A. Physical and quantitative analysis of compact rectangular microstrip antenna with shorted non-radiating edges for reduced cross-polarized radiation using modifi ed cavity model. *IEEE Antennas Propag. Mag.* **2014**, *56*, 61–72. [CrossRef]
29. Barkeshli, K.; Volakis, J.L. Electromagnetic scattering from thin strips. I. Analytical solutions for wide and narrow strips. *IEEE Trans. Educ.* **2004**, *47*, 100–106. [CrossRef]

![sensors logo] *sensors*

MDPI

Article

Recent Advances in Piezoelectric Wafer Active Sensors for Structural Health Monitoring Applications

Hanfei Mei [1,*] , Mohammad Faisal Haider [1] , Roshan Joseph [1], Asaad Migot [1,2] and Victor Giurgiutiu [1]

1   Department of Mechanical Engineering, University of South Carolina, 300 Main Street, Columbia, SC 29208, USA; haiderm@email.sc.edu (M.F.H.); rjoseph@email.sc.edu (R.J.); amigot@email.sc.edu (A.M.); victorg@mailbox.sc.edu (V.G.)
2   Department of Mechanical Engineering, College of Engineering, Thi-Qar University, Nasiriyah 64001, Iraq
*   Correspondence: hmei@email.sc.edu; Tel.: +1-803-724-8029

Received: 28 November 2018; Accepted: 16 January 2019; Published: 18 January 2019

check for updates

Abstract: In this paper, some recent piezoelectric wafer active sensors (PWAS) progress achieved in our laboratory for active materials and smart structures (LAMSS) at the University of South Carolina: http://www.me.sc.edu/research/lamss/ group is presented. First, the characterization of the PWAS materials shows that no significant change in the microstructure after exposure to high temperature and nuclear radiation, and the PWAS transducer can be used in harsh environments for structural health monitoring (SHM) applications. Next, PWAS active sensing of various damage types in aluminum and composite structures are explored. PWAS transducers can successfully detect the simulated crack and corrosion damage in aluminum plates through the wavefield analysis, and the simulated delamination damage in composite plates through the damage imaging method. Finally, the novel use of PWAS transducers as acoustic emission (AE) sensors for in situ AE detection during fatigue crack growth is presented. The time of arrival of AE signals at multiple PWAS transducers confirms that the AE signals are originating from the crack, and that the amplitude decay due to geometric spreading is observed.

Keywords: structural health monitoring; piezoelectric wafer active sensors; active sensing; passive sensing; damage detection; acoustic emission

---

1. Introduction

Structural health monitoring (SHM) is an emerging interdisciplinary research field, which aims at detecting damage and providing a diagnosis of structural health [1–5]. Among the SHM technologies, Lamb wave, a type of ultrasonic guided waves propagating between two parallel surfaces without much energy loss, is suitable for the large-area inspection of complicated structures [6,7]. Piezoelectric wafer active sensors (PWAS) were developed by our LAMSS group as convenient enablers for generating and receiving Lamb waves in structures for SHM applications [8].

Depending on the type of application, PWAS can be utilized for (i) active sensing of far-field damage using pulse-echo, pitch-catch, and phased-array methods, (ii) active sensing of near-field damage using the electromechanical impedance method, and (iii) passive sensing of acoustic emissions at the tip of advancing cracks and low-velocity impacts [8]. The main advantage of PWAS transducers over conventional ultrasonic probes is their low cost and light weight. They can be permanently bonded on the host structures in large quantities, and achieve real-time monitoring of the structural health status. These PWAS transducers can also be used in a harsh environment (e.g., high temperature, nuclear radiation, or a vacuum environment). A proper sensor characterization would be required before its installation on the host structure in a harsh environment.

In recent years, many researchers have explored the capability of PWAS for SHM applications, such as the characterization of PWAS [9–13], impact localization [14–19], acoustic emission (AE) detection [20,21], and damage detection in isotropic and composite plates [22–27]. These studies facilitate the understanding of PWAS-based SHM applications.

In order to use the PWAS transducer as an SHM transducer in harsh environments, PWAS material properties should be investigated, and PWAS transducers should be defect-free after high temperature and radiation exposure. Baptista et al. [9] conducted an experimental study of the effect of temperature on the electrical impedance of the piezoelectric sensors, and they found that the temperature effects were strongly frequency-dependent. Similarly, Haider et al. [10] investigated the irreversible change in PWAS electromechanical (E/M) impedance and admittance signature under high-temperature exposure. It was concluded that the changes in anti-resonance and resonance frequencies have a linear relationship with temperature. In addition, the sensitivity characterization of high temperature piezoelectric transducers was performed by using resonance analysis [11]. Sinclair and Chertov [12] presented a comprehensive literature study of the radiation endurance of a piezoelectric transducer.

General PWAS sensing technology can be cast into two methodological categories: passive sensing and active sensing. PWAS passive sensing methods only record events passively, which happens during the period of interest. By analyzing the recorded signal, diagnosis can be made on the health status of the structure. Examples of PWAS passive sensing methods can be found in the impact localization [15,18] and AE detection [20,21]. Park et al. [15] proposed a new technique for predicting the impact location on an anisotropic plate by analyzing the geometric shape of the wavefront, and it did not require prior knowledge of the material properties. Qiu et al. [18] conducted an impact localization on an aircraft composite wing box using piezoelectric sensor networks, and achieved promising results. Apart from impact localization using PWAS transducers, preliminary work has also been performed to capture fatigue AE hits [20,21].

In contrast to PWAS passive sensing, PWAS active sensing methods interrogate the structures with defined excitations and record the corresponding response. One common active sensing method in SHM applications is the pitch-catch experiment, where one PWAS transducer acts as the transmitter and sends out Lamb waves, and another PWAS transducer acts as the receiver and picks up the sensing signal. The subtraction between the pristine and damaged responses may indicate the presence and severity of the damage. To achieve more complicated diagnostic approaches for SHM applications, several sensors may work together in a systematically designed manner, forming a sensor network. Advanced damage imaging techniques have been developed by using the phased array [28,29] and sparse array [30–49]. Substantial research has focused on the research of damage imaging algorithms for damage localization and characterization, using a sparse PWAS array, including tomography [30–32], the delay-and-sum imaging method [33–35], the time-reversal imaging approach [36–38], correlation-based imaging algorithm [39,40], probabilistic and statistical imaging methods [41–43], and the minimum variance imaging method [44–46]. Recently, Kudela et al. [48] proposed a Lamb wave-focusing method to detect and visualize the crack in an aluminum plate, and the damage imaging resolution was improved compared to the original delay-and-sum algorithm. Xu et al. [49] developed a weighted sparse reconstruction-based anomaly imaging method for damage detection on composite plates, and anomaly imaging with fewer artifacts was achieved.

Moreover, various methods have been used for wavefield analysis. In this active sensing method, one PWAS transducer was used to excite Lamb waves propagating in the structures, and the wavefield was measured by a scanning laser Doppler vibrometer (SLDV). The damage can be unveiled through wavefield analysis, such as the wavefield amplitude profile [49], frequency-wavenumber filtering [50–53], standing wave filtering [54–56], zero-lag cross-correlation imaging [57,58], local wavenumber analysis [59–61], and wavenumber adaptive image filtering [62–64]. Staszewski et al. [49] used the amplitude profiles of the wavefield to detect the delamination damage in composites.

Sohn et al. [54] proposed a standing wave filter to isolate the standing waves from the propagating waves for delamination detection in composites. He and Yuan [57] employed the zero-lag cross-correlation (ZLCC) imaging condition for damage imaging in a composite plate, using a single piezoelectric wafer for excitation. In recent years, various wavenumber imaging methods were applied for impact-induced delamination detection. Girolamo et al. [58] applied the ZLCC imaging condition for visualizing impact damage in a honeycomb composite panel. Rogge and Leckey [59] presented a local wavenumber domain analysis to process the wavefield, and they demonstrated that it could be used to quantify the delamination depth and size. Tian et al. [60,61] improved the damage visualization algorithm by using filtering reconstruction imaging and spatial wavenumber imaging. Kudela et al. [64] studied the relation between impact energy and BVID detectability, using wavenumber adaptive image filtering, and they found that damage caused by the impact of 10 J or higher could be successfully detected.

In this paper, some recent PWAS progress achieved in our LAMSS group is reported, including studies of (a) PWAS endurance in harsh temperature and radiation environments; (b) PWAS active sensing of various damage types in aluminum and composite structures; (c) PWAS passive sensing of acoustic emission (AE) signals from fatigue crack growth in aluminum coupons. First, the characterization of PWAS materials is conducted. The endurance of the PWAS after exposure to high temperature and nuclear radiation was also assessed for harsh environmental applications. Next, applications of PWAS, using active sensing methods for detecting the simulated crack and corrosion damage in aluminum plates, and the simulated delamination damage in composite plates are conducted. Finally, the novel use of PWAS transducers as AE sensors for in situ AE detection during fatigue crack growth is presented.

2. Characterization of PWAS

In this section, PWAS transducers made with piezoelectric material lead zirconate titanate ($PbZrO_3TiO_3$ or PZT) were investigated. PZT exhibits large electromechanical coupling coefficients and piezoelectric constants [65]. For PZT-PWAS, APC 850 type transducers [66] were used in this research. The PZT-PWAS transducer is circular in shape, and the diameter is 7 mm. The wafer has a PZT thin film with Ag electrodes on both sides. The thickness of the PZT-PWAS transducer is 0.2 mm. Energy-dispersive spectroscopy (EDS) was done on the PWAS transducer to obtain chemical compositions of the PWAS. Figure 1 shows the EDS spectrum of PZT-PWAS. The figure confirms that PZT-PWAS contains lead (Pb), zirconium (Zr), titanium (Ti), oxygen (O), and silver (Ag) electrode.

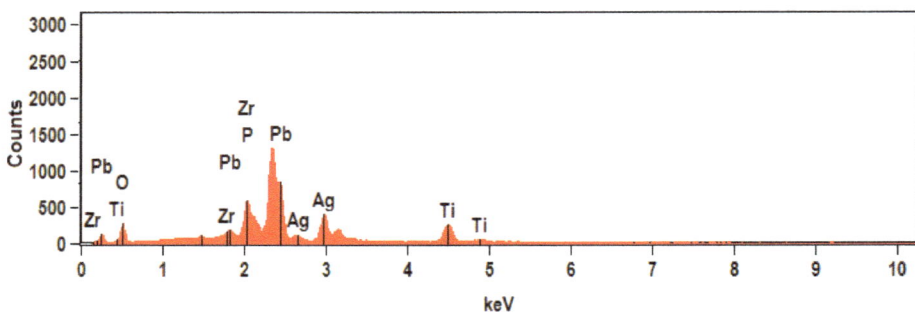

Figure 1. Energy-dispersive spectroscopy (EDS) spectrum of the lead zirconate titanate (PZT) piezoelectric wafer active sensors (PWAS).

The significance of SHM has been emphasized in many fields, such as dry cask storage canister (nuclear-spent fuel storage), pressure vessel and pipe (PVP), turbine blade and so on; where attention is being drawn to the successful implementation of SHM techniques, due to temperature variation and radiation. The PZT material used in a PWAS transducer is a ferroelectric material. For most ferroelectric materials, the existence of a domain structure or domain wall makes a significant influence on the material properties. In a PZT solid-solution system, the material properties may be changed, due to the change in the domain size and the domain wall motion [67–69]. In addition, the PWAS transducers are susceptible to damage after exposure to harsh environments. A scanning electron microscopy (SEM) was also done to visualize the microstructure of the PWAS. Figure 2 shows the cross-section of PZT-PWAS: (a) at room temperature; (b) after exposure to 250 °C temperature; and (c) after exposure to 225 kGy radiation. There are no significant changes in the microstructure or PZT grain after exposure to high temperature and radiation. Hence, PWAS can be integrated into the SHM system for successful damage detection.

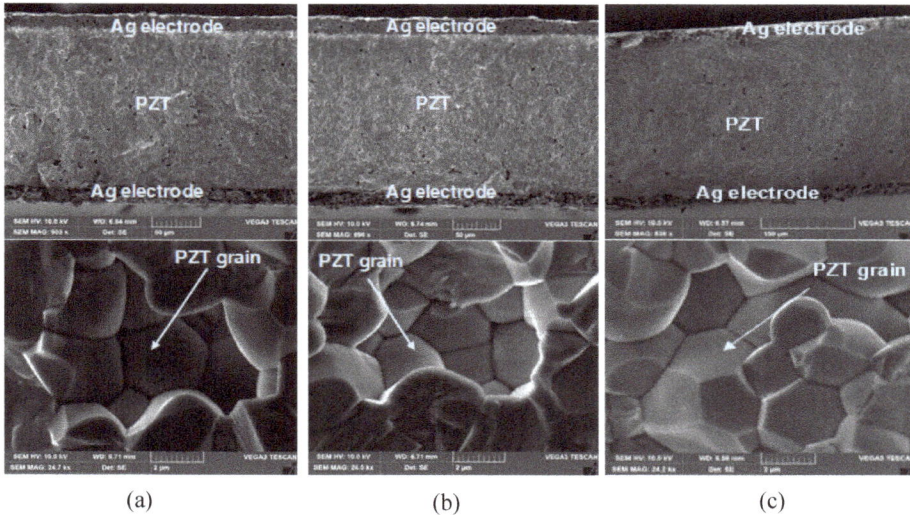

(a)   (b)   (c)

Figure 2. Scanning electron microscopy (SEM) image of the cross-section of PZT-PWAS: (**a**) at room temperature; (**b**) after exposure to 250 °C temperature; (**c**) after exposure to 225 kGy nuclear radiation.

3. Applications of PWAS for Detecting Damage in Isotropic Plates

In this section, damage detections on aluminum plates using a circular PWAS and a long PWAS were performed.

3.1. Corrosion Damage Detection in an Aluminum Plate Using a Circular PWAS

In this experiment, a 2.032 mm 2024-T3 aluminum plate was examined. The material properties are given in Table 1. A simulated corrosion damage was made on the 2024-T3 aluminum plate to study the damage detection by using the PWAS active sensing method, as shown in Figure 3.

Table 1. Material properties of 2024-T3 aluminum.

Young's Modulus (*E*)	Poisson's Ratio (*v*)	Density (*ρ*)
73.1 GPa	0.33	2780 kg/m^3

To excite Lamb waves in the aluminum plate, a circular PWAS (APC 850, 7 mm in diameter and 0.2 mm thick) was bonded on the top surface as the excitation source. Figure 4 shows the schematic of the experimental setup. The function generator was used to generate a three-count Hanning window modulated tone burst with the center frequency of 200 kHz, which was amplified to 50 Vpp by the power amplifier and applied to the circular PWAS.

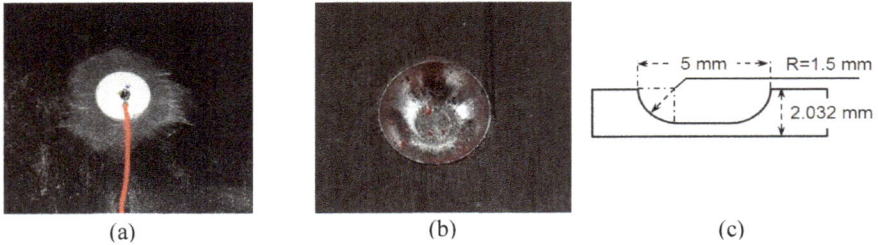

Figure 3. Experimental setup: (**a**) a circular PWAS; (**b**) corrosion damage; (**c**) schematic of the corrosion damage.

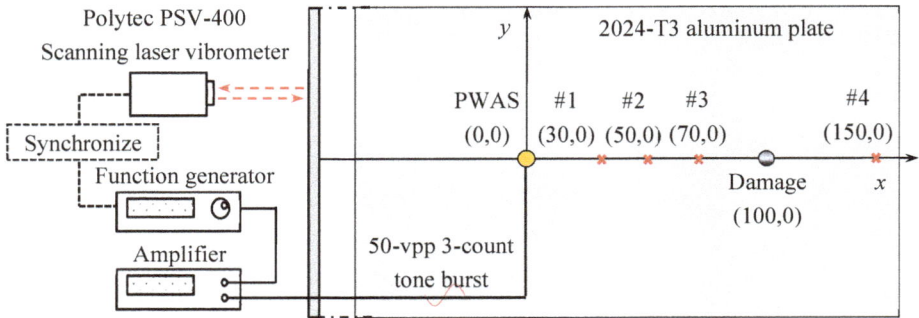

Figure 4. Schematic of an experimental setup using PWAS and scanning laser Doppler vibrometer (SLDV) for damage detection.

Under electrical excitation, the PWAS generates Lamb waves in the aluminum plate. Lamb waves propagate with an out-spreading pattern, interact with damage, undergo scattered and mode conversion, and are finally picked up by a Polytec PSV-400-M2 scanning laser Doppler vibrometer (SLDV). The quantity measured by the SLDV is the out-of-plane velocity of the bottom surface. Reflective tape was used to improve the signal quality. In the experiment, a line scan and an area scan on the specimen surface were carried out to detect the corrosion damage. The locations of the circular PWAS, corrosion damage, and special recording points are illustrated in Figure 4.

3.1.1. Damage Detection by using a SLDV Line Scan

In the experiment, the time-space Lamb wave data was obtained from the SLDV line scan, as shown in Figure 5a. To obtain the experimental frequency–wavenumber dispersion curves, the measured time–space wavefield $u(t,x)$ was transformed into the frequency–wavenumber domain by applying a two-dimensional (2D) fast Fourier transform (FFT). Figure 5b,c show the time–space wavefield and the corresponding frequency–wavenumber dispersion curves, the wave transmission and reflection due to the corrosion damage can be clearly observed. Hence, the damage can be easily detected by using the wavefield analysis.

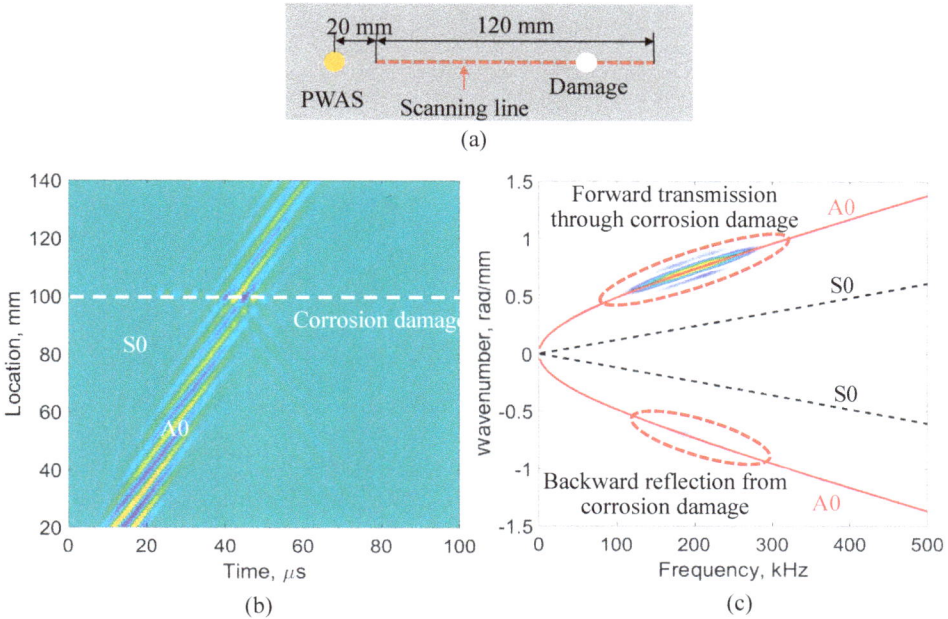

Figure 5. SLDV line scan for a 2.032-mm 2024-T3 aluminum plate: (**a**) schematic of the SLDV line scan; (**b**) time–space wavefield at 200 kHz; (**c**) frequency–wavenumber dispersion curves.

3.1.2. Damage Detection by Using an SLDV Area Scan

In this section, an SLDV area scan was conducted to measure the wave interaction with the corrosion damage under PWAS excitation. Figure 6 shows a transient spatial wavefield in the plate. At 30 µs, the fast propagating S0 mode with a long wavelength, and the slowly propagating A0 mode with a short wavelength could be identified. The mode-converted A0 waves could be noticed, propagating with a short wavelength from S0 interaction with the damage. At 50 µs, after A0 waves interacted with damage, the scattered A0 waves could be observed, as well as the shadow behind the damage. Therefore, the damage could be visualized from the measured wavefield.

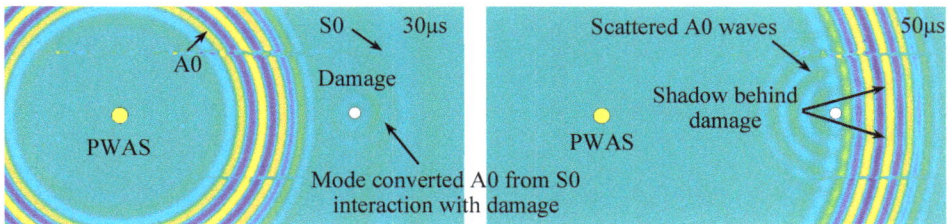

Figure 6. Spatial wavefield in the aluminum plate with simulated corrosion damage, showing S0 mode and A0 mode waves interacting with the corrosion damage.

Figure 7 shows the waveforms at various sensing locations for the 200 kHz excitation. The signals at location #1, #2, and #3 show that the scattered A0 wave amplitude increases when the sensing location moves closer to the damage. The signal at location #4 shows the mode-converted A0 wave packet from the S0 interaction with the damage.

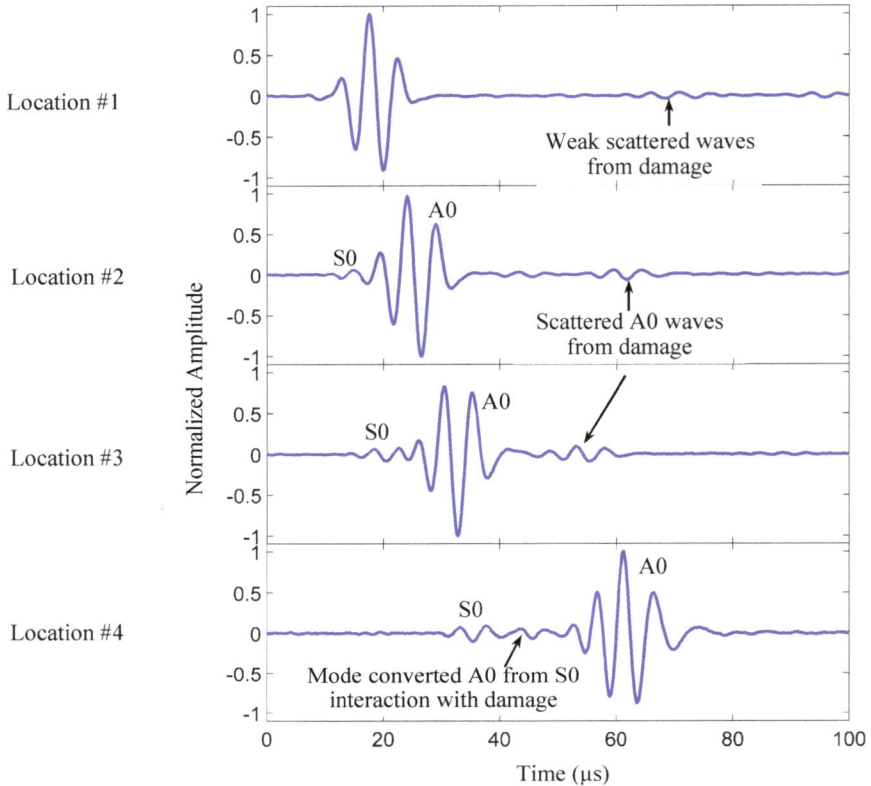

Figure 7. 200 kHz signals of the damaged plate at locations #1 through #4, shown in Figure 4.

3.2. Crack Detection in a Stiffened Aluminum Plate Using Long PWAS

This section describes an experimental procedure to analyze the scattered Lamb wave to detect a horizontal crack at the root of the stiffener in an aluminum plate.

3.2.1. Experimental Procedure

Two aluminum plates with a pristine stiffener and cracked stiffener were manufactured for the experimental study. Electrical discharge machining (EDM) method was used to create a crack along the entire length of the stiffener. The height and width of the stiffener is 8.47 mm. The thickness of the plate is 4.23 mm. For the crack stiffener, the crack width is half of the stiffener width, and the crack is present along the entire length of the stiffener (Figure 8). The experimental setup and the plates with pristine stiffener and cracked stiffener are shown in Figure 8. Two 60 mm × 5 mm × 0.2 mm PWAS transducers were bonded in a straight line on the top and bottom surfaces of the plate to create a line source. On both plates, the PWAS transducers were bonded 200 mm away from the stiffener. Two PWAS were excited simultaneously in opposite phase to generate A0 Lamb wave, selectively. The excitation signal is a three-count tone burst at 150 kHz generated by a Tektronix AFG3052C dual channel function generator. A power amplifier was used to amplify the excitation signal, which strengthens the reflected and transmitted signal from the discontinuity or damage. To create non-reflecting boundary condition, absorbing clay was applied all around the plate boundaries.

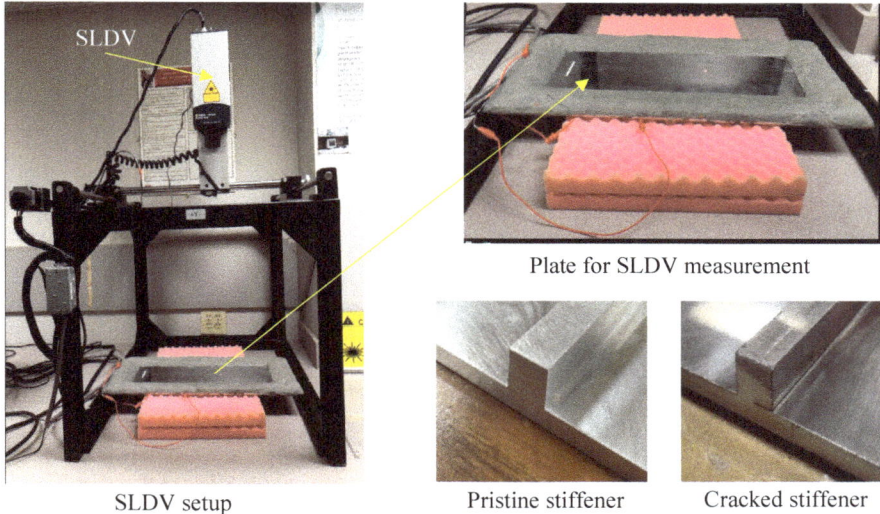

Figure 8. Experimental setup for the SLDV to measure the out-of-plane velocity of the scattered wavefields.

3.2.2. Experimental Results

The wavefields were measured using SLDV. Figure 9a is the schematic of the SLDV measurement. Figure 9b–e show the reflected and transmitted wavefields for the pristine stiffener and cracked stiffener, respectively. It can be found that the long PWAS transducers can successfully generate the straight crested A0 mode Lamb wave. It also shows a minimal reflection from the plate edges, due to the use of absorbing clay. In addition, these figures show that the scattered wavefields are also straight crested waves after interacting with the discontinuity, as expected.

Figure 10a shows the schematic diagram of the sensing locations. Sensing location 1 is 170 mm before the stiffener and the sensing location 2 is 200 mm after the stiffener. Figure 10b–e show the experimental scattered waveforms for the pristine and cracked stiffener, respectively. The corresponding fast Fourier transform (FFT) of the incident and scattered waves for the pristine and cracked stiffener are given in Figure 11.

The reflected and transmitted Lamb waves show a similar pattern in the time–domain signals for the pristine and cracked stiffener. However, the frequency response of the scattered signals shows a clear change in the amplitude of the scattered wavefields, due to the presence of the crack. Also, the shifting of the frequency spectrum due to the presence of the crack is an important phenomenon to note. The transmitted A0 Lamb wave has clear anti-resonance at 150 kHz for the cracked stiffener. Such information may be useful for crack detection in complex geometry.

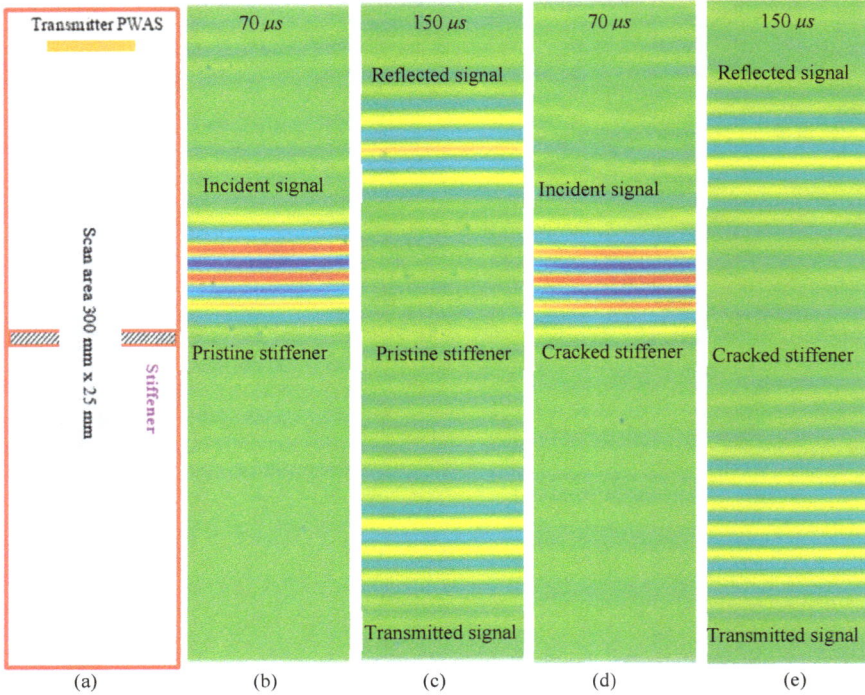

Figure 9. Experimentally measured scattered wavefields using SLDV: (**a**) scan schematic; (**b**) incident wave and (**c**) reflected and transmitted waves from pristine stiffener; (**d**) incident wave, and (**e**) reflected and transmitted waves from the cracked stiffener.

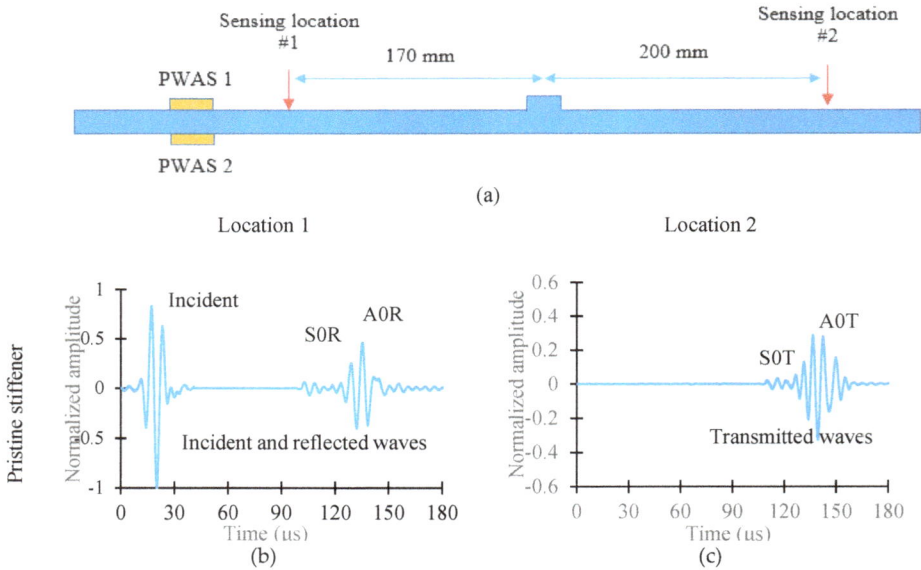

Figure 10. *Cont.*

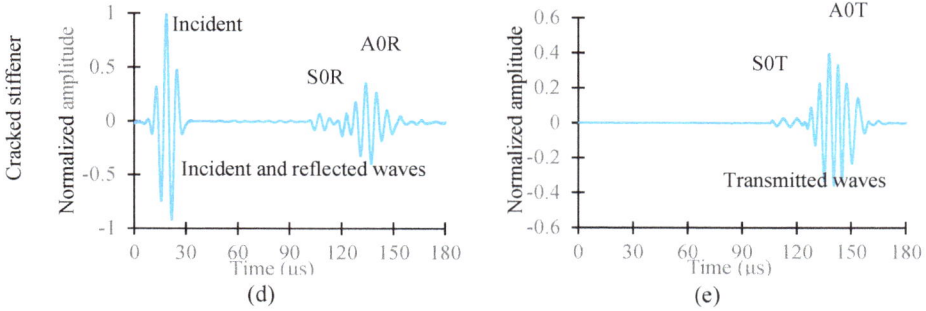

Figure 10. (**a**) Schematic diagram of the sensing locations; (**b**) incident and reflected Lamb waves from pristine stiffener; (**c**) transmitted Lamb waves from the pristine stiffener; (**d**) incident and reflected Lamb waves from cracked stiffener; (**e**) transmitted Lamb waves from cracked stiffener.

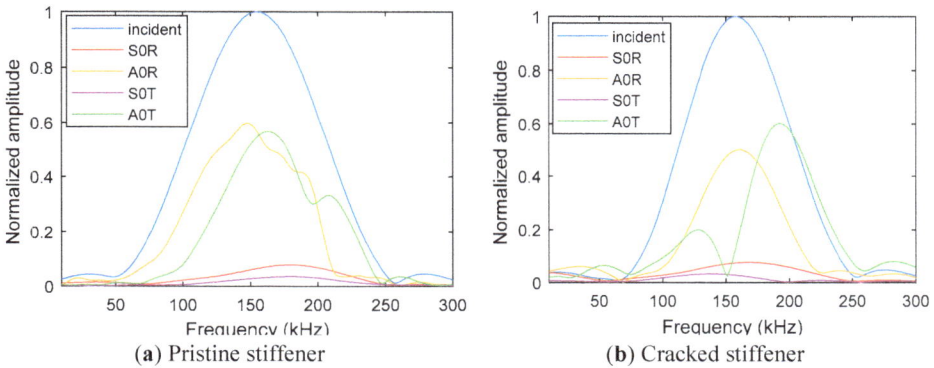

Figure 11. FFT of the incident and scattered Lamb waves: (**a**) pristine stiffener; (**b**) cracked stiffener.

4. Applications of PWAS for Detecting Damage in Anisotropic Plates

In this section, a 1.6 mm thick in-house cross-ply carbon fiber-reinforced polymer (CFRP) composite plate with a stacking sequence of $[0/90]_{2s}$ was examined. The delamination is generated by inserting a 50 mm diameter Teflon film between the first ply and second ply during the ply lay-up process. Figure 12 shows the configuration of the composite plate. The engineering elastic properties of the unidirectional prepreg are given in Table 2.

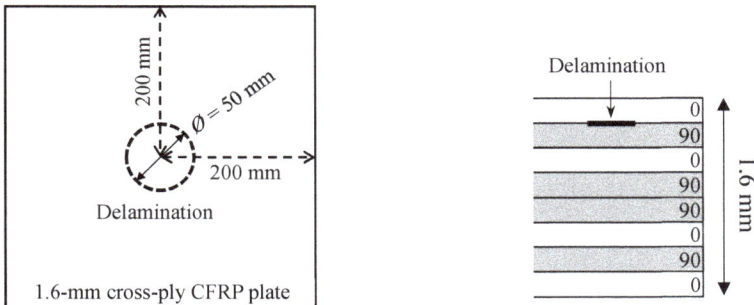

Figure 12. Schematic of a 1.6 mm cross-ply carbon fiber-reinforced polymer (CFRP) composite plate with circular delamination by inserting Teflon.

Table 2. Engineering constants of the unidirectional prepreg.

E_{11}	E_{22}	E_{33}	v_{12}	v_{13}	v_{23}	G_{12}	G_{13}	G_{23}	ρ
140.8 GPa	11.3 GPa	11.3 GPa	0.31	0.31	0.5	5.7 GPa	5.7 GPa	3.4 GPa	1640 kg/m^3

A network of PWAS transducers and the damage imaging method were used to detect and quantify the delamination in the cross-ply composite plate. In recent work [70,71], an improved imaging method was developed to obtain accurate results of localization and sizing damages in metallic plates and composite laminates. Here, some more recent results in this direction are presented. The gist of our methodology [70,71] is to perform a point-by-point detection and localization of damage as a first step before using the imaging methods. Four sensors are distributed to make a cross sign on the area of interest. From the scattered waves of pulse-echo experiments, the difference in the time-of-flight (TOF) values of scattered waves of sensors are determined. If these difference values are close to zero, the damage is at the center of the area of interest. If there is difference in TOF of scattered waves, the damage location is close to the sensor that has less TOF. Based on the location of damage and sensors, the directions of the incident and scattered waves are determined. Then, the direction-dependent group velocities of incident and scattered waves for all the individual sensing paths on the composite plate are determined using the semi-analytical finite element (SAFE) approach [72].

4.1. Experimental Setup

The experimental setup of using SHM techniques to detect the delamination in the composite plate is shown in Figure 13. Eight PWAS transducers were bonded onto the plate surface to form a sensor network. The diameter of each PWAS transducer is 7 mm, and the thickness is 0.2 mm. The clay was applied to the plate edges to absorb boundary reflections. An Agilent 33120A function generator was used to generate the excitation signal. The response signals were recorded by an oscilloscope.

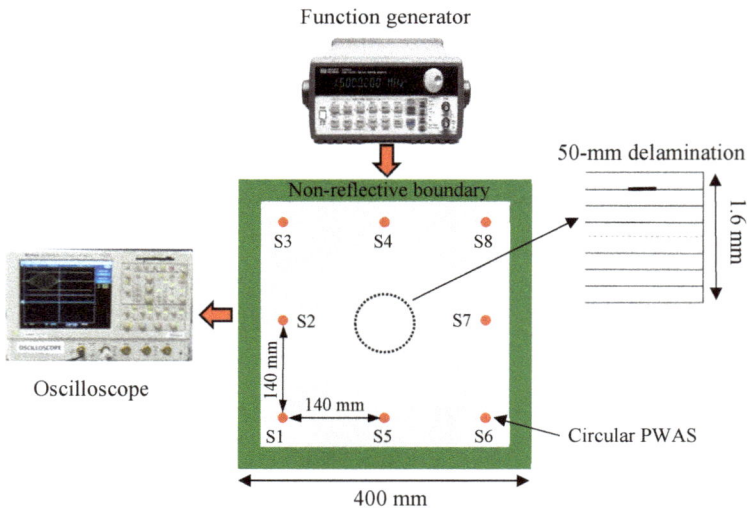

Figure 13. Experimental setup on the 1.6 mm cross-ply composite plate.

The excitation signal was a three-count tone burst signal at a center frequency of 330 kHz. This frequency was chosen based on the tuning curve. Experimental tuning curves of the cross-ply CFRP composite plate in the 0° and 45° directions are shown in Figure 14. It can be observed that only a single mode (S0 mode) is dominant around the 330 kHz frequency in the 0° direction. However, SH0 mode was also observed as strong as S0 mode in the 45° direction. The excitation of the SH0 wave in the off-axial direction is due to the anisotropic behavior of the composite plate already reported by Giurgiutiu [73].

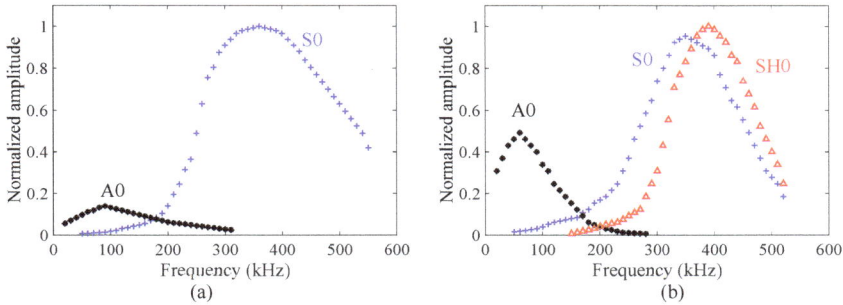

Figure 14. Experimental tuning curves of the cross-ply CFRP composite plate in the: (**a**) 0° directions; (**b**) 45° direction.

4.2. Experimental Results

The signals were collected by using pulse-echo and pitch-catch modes. First, the pulse-echo experiments were conducted for the PWAS S2, S4, S5, and S7 to detect and localize the damage using the proposed method in Section 4. Figure 15a shows the pulse-echo signals of the sensing paths S2-S2, S4-S4, S5-S5, and S7-S7. It can be noted that all of the signals have strong scattered S0 waves with the same TOF values. Hence, the damage is located at the center, as shown in Figure 15b.

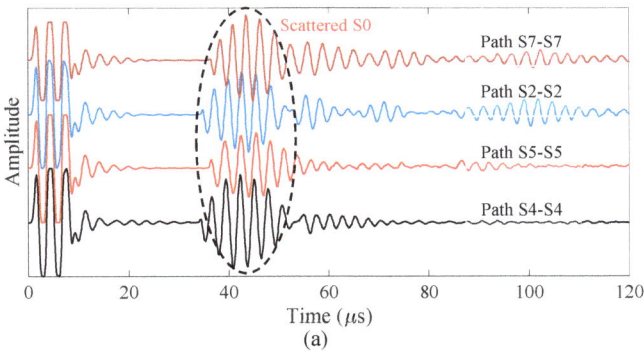

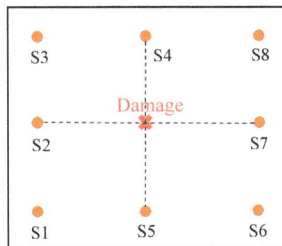

Figure 15. The new methodology for the localization of damage point: (**a**) pulse-echo signals of the sensing paths S2-S2, S4-S4, S5-S5, and S7-S7; (**b**) determining the location of damage.

In the pitch-catch mode, each PWAS transducer acted as a transmitter, whereas the rest of them in the network acted as receivers. For demonstration, only one set of signals is shown in Figure 16. In this set, PWAS S2 is the transmitter and PWAS S4 is the receiver. It can be observed that both S0 and SH0 are dominant in this sensing path (45° direction), which agrees with the experimental tuning curves in the 45° direction (Figure 14). The incident waves were determined as the S0 and SH0 modes, based on the group-velocity dispersion curve. Figure 17 shows the group-velocity directivity plot at 330 kHz. The curves indicate that Lamb waves propagate with various velocities in different directions. It can be observed that S0 mode has the highest group velocity, while A0 mode has the lowest group velocity. SH0 mode possesses a group velocity between A0 and S0, and it shows self-crossing behavior, as reported in Glushkov [74]. The scattered SH0 wave, due to the delamination, can be determined by calculating the TOF of incident path (PWAS S2 to damage) and damage path (damage to PWAS S4) using the corresponding group velocities and distances.

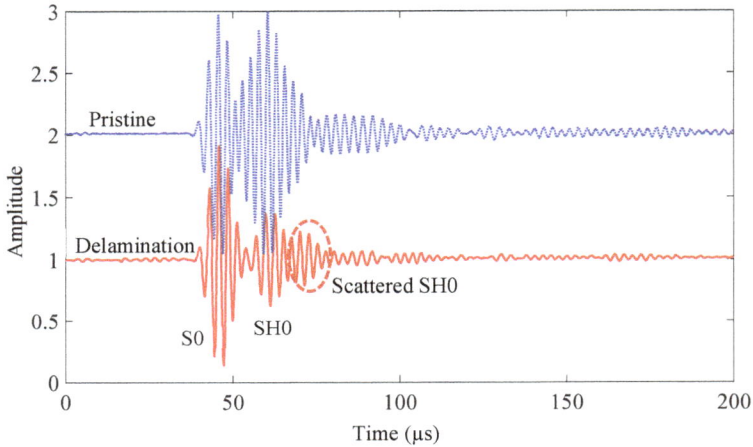

Figure 16. Comparison between the pristine signal and the delamination signal (Scattered S0 is overlapped with incident waves).

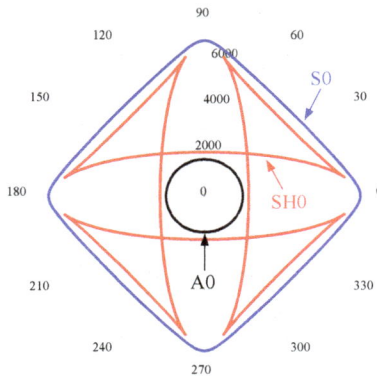

Figure 17. Group velocity directivity curves at 330 kHz in a 1.6 mm cross-ply CFRP plate.

Based on the time of flight (TOF) of scattered SH0 waves, the imaging method can be used to detect and quantify the delamination in the plate. The basic idea of the imaging method is to divide the interested area into pixels, and to find the field values of these pixels for each sensing path. The TOF of every pixel can be determined by using Equation (1):

$$t_{ij} = \frac{\sqrt{(x_T - x_i)^2 + (y_T - y_j)^2}}{v_{g1}} + \frac{\sqrt{(x_i - x_R)^2 + (y_j - y_R)^2}}{v_{g2}} \tag{1}$$

where t_{ij} is the TOF of every pixel, and x_T, y_T, x_R, y_R, x_i, y_j are the coordinates of the transmitter PWAS, receiver PWAS, and pixel, respectively. v_{g1} and v_{g2} are the group velocities of the incident path (transmitter PWAS to damage) and damage path (damage to receiver PWAS), respectively. These group velocities are determined from Figure 17, based on the direction of the incident and scattered waves. When the pixels lie on the damage orbit of a particular sensing path, which means $t_{ij} = t_d$ (t_d is the TOF of scattered wave). In this case, the field value of these pixels is maximum.

Figure 18 gives the damage orbit for a certain sensing path. To determine the size of the delamination area, the sensing paths of multiple transmitters around the area of interest were used to get multiple intersection points that represent the damage edges. A new methodology was implemented to visualize the delamination without setting a threshold, using a combination of summation and multiplication algorithms. In this methodology, the summation algorithm [71] is used to extract the individual image of all the sensing paths for each PWAS transmitter. These images have strong intersection points, which represent the delamination edges, and the rest are undesirable orbits. The multiplication algorithm is used to fuse all of these individual images, to obtain the final image for the delamination, which does not require setting of a threshold. Figure 19 shows the final imaging result of the delamination. It can be found that the results of the imaging method match well with the real delamination. The edges of the 50 mm delamination can be observed.

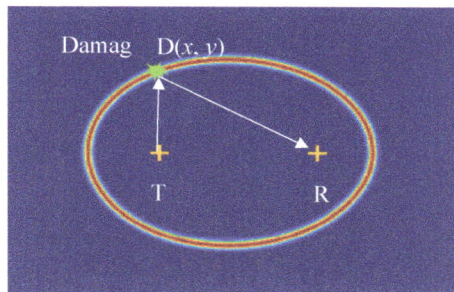

Figure 18. Damage orbit with the transmitter and receiver PWAS as foci.

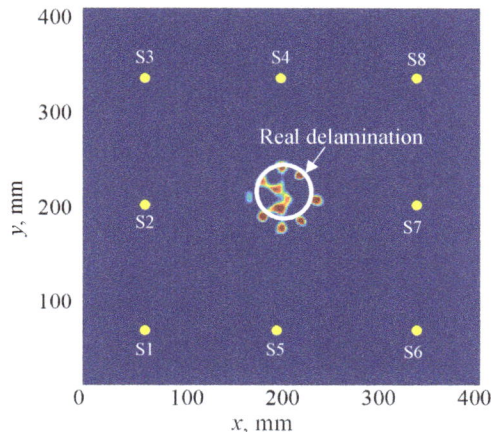

Figure 19. The imaging result of the 50 mm delamination in the cross-ply composite plate.

5. PWAS as an AE Sensor for In Situ AE Detection during a Fatigue Crack Event

Application of PWAS as an acoustic emission (AE) sensor is presented in this section. AE events in metallic or composite materials can be generated, due to various phenomena such as friction, plastic deformation, crack growth event, etc. Of all the causes, fatigue crack growth is a problem of great importance in metals.

To understand the sensing capability of PWAS to detect AE signals due to a crack growth event, PWAS transducers were used as an in situ sensor on a high-strength aluminum 2024-T3 specimen while undergoing fatigue crack growth. The aluminum specimen was 103 mm in width, 305 mm in, and 1 mm thick. A 1 mm hole was drilled at the geometric center of the specimen. Fatigue loading was applied on the specimen to generate a pre-crack of 14 mm tip-to-tip length. For the pre-crack generation, fatigue loading from 13.85 kN to 1.38 kN at a frequency of 10 Hz was applied. After initiating the crack approximately up to 14 mm tip to tip, two PWAS transducers at 6 mm and 25 mm from the hole, as well as two S9225 sensors at 6 mm and 25 mm from the hole were installed on the aluminum plate, as shown in Figure 20. S9225 is a commonly used and commercially available AE sensor. The S9225 sensor was used to make a comparison between PWAS and S9225 in this experiment. The test specimen installed with PWAS and S9225 transducers was mounted on the MTS machine. The fatigue loading was continued to vary between 13.85 kN and 1.38 kN with a loading rate of 2 Hz, and simultaneous AE measurements were performed. The experimental setup for capturing the AE signal from the fatigue crack growth is presented in Figure 20. AE signals during the crack growth were captured by using PWAS and S9225 sensors. The sensors are connected to the acoustic preamplifier. The acoustic preamplifier is a bandpass filter, which can filter out signals between 30 kHz to 700 kHz. The preamplifier is then connected to a four-channel Mistras AE system for processing the signals.

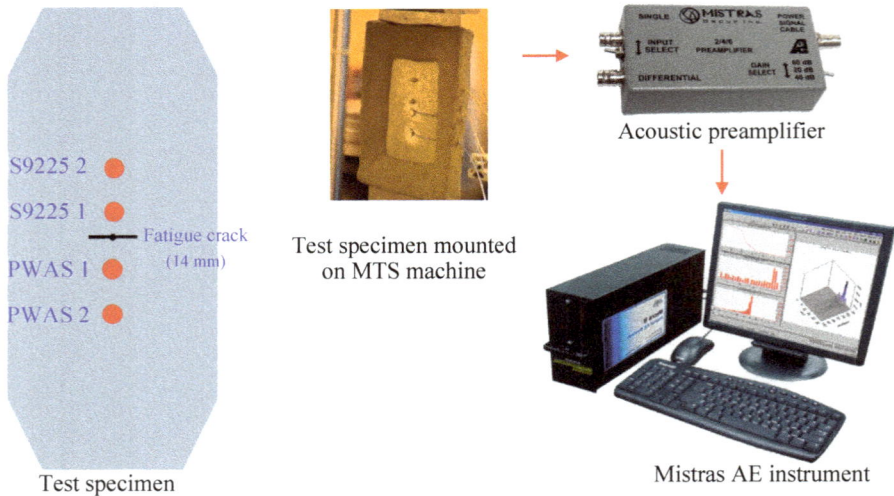

Figure 20. Experimental setup for capturing AE signals during the fatigue crack event.

The crack growth on the specimen due to 2000 fatigue loading cycles from 14 mm to 16 mm is presented in Figure 21. The initial crack is presented in the green box, and the final crack in the blue box. The locations of the initial crack tips were marked using the blue lines, and the locations of the final crack tips were marked using red lines. The crack grew by approximately 2 mm. AE signals were captured simultaneously during the fatigue crack growth.

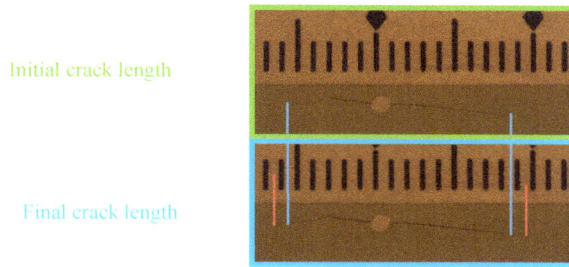

Figure 21. The initial tip-to-tip crack length was approximately 14 mm. After 2000 fatigue cycles, an advancement in the crack length of 2 mm was observed.

A particular AE event captured by PWAS 1 and S9225 1 sensors is presented in Figure 22. The relative time of arrival of the signals at the PWAS 1 and S9225 1 sensors installed equidistant from the crack was obtained from the Mistras AE system. The time of arrival of signals at PWAS 1 and S9225 1 was found to be the same, which confirmed that the signals corresponded to an AE event happening that was equidistant from the sensors, which is the crack location. If the AE event was happening at a different location other than the crack, the time of arrival at the PWAS 1 and S9225 1 sensors would be different. In this way, it was confirmed that the AE signal captured was originating from the crack.

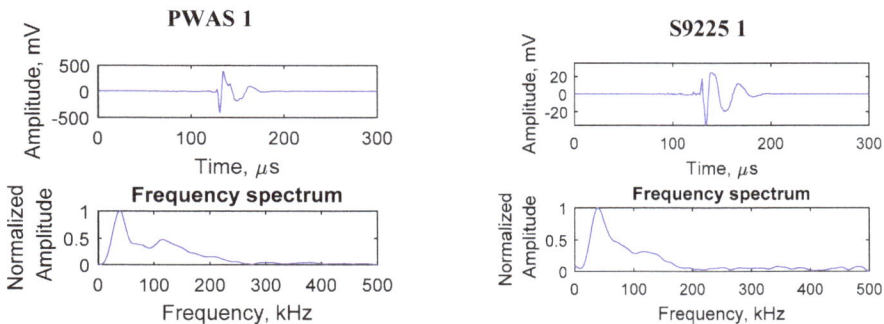

Figure 22. Signals captured by the PWAS 1 and S9225 1 sensors, due to a particular AE event at the crack.

Several AE signals originating from the crack were captured during this crack growth event, as shown in Figure 23. From the time of arrival of the AE signals at PWAS 1 and PWAS 2, signals corresponding to the same AE event were identified. The sequence of arrival, as well as the amplitude of the AE signals, are presented in Table 3. The sequence of arrival confirms that the AE signals were originating from the crack. Geometric spreading caused the amplitude of AE signal to decay while propagating. This was also observed from the corresponding amplitude values of the signals at PWAS 1 and PWAS 2. The AE signals were found to have a similar signature by comparing the waveform, as well as the frequency spectrum of the signals. It can be observed that the frequency spectrum maintains its signature when reaching PWAS 2.

From the experimental investigation presented in this section, it can be concluded that PWAS can be used as an in situ sensor for detecting AE signals during the fatigue crack growth. Fatigue loading was applied on the specimen, to grow the crack, and simultaneous measurement of AE signals was performed. The time of arrival of AE signals at multiple sensors confirms that the AE signals are originating from the crack. The amplitude decay of AE signals due to geometric spreading was also observed through AE measurements at PWAS transducers located at 6 mm and 25 mm from the crack.

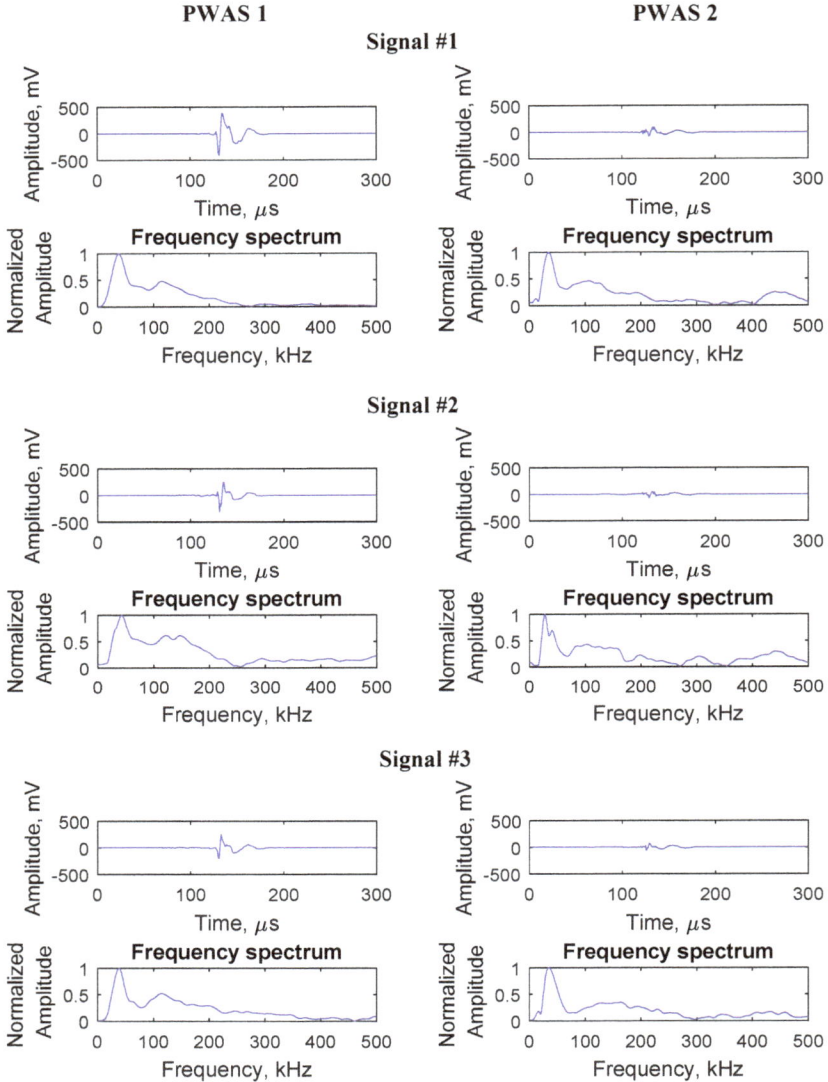

Figure 23. Typical AE signals captured during fatigue crack growth. Three sets of captured signals are presented. For each set, the signal at PWAS 1 and the corresponding signal at PWAS 2 are presented. The AE signals seem to maintain their signature at PWAS 1 and 2.

Table 3. Sequence of arrival and amplitude of the AE signals at PWAS 1 and PWAS 2.

Signal #	Sequence of Arrival		Amplitude of Signal (dB)	
	PWAS 1	PWAS 2	PWAS 1	PWAS 2
1	1	2	62	51
2	1	2	59	49
3	1	2	58	48

6. Conclusions

In this paper, PWAS endurance in harsh temperatures and radiation environments, and PWAS active and passive sensing methods for SHM applications were demonstrated. First, the characterization of PWAS was conducted. No significant change in the microstructure or PZT grain after exposure to high temperature and radiation was found through the SEM technique. Proper characterization of PWAS allows a SHM system to infer the integrity of the transducers and separate transducer-flawed signals from structural defects in a harsh environment. For PWAS active sensing methods, PWAS was successfully used to detect the simulated crack and corrosion damage in aluminum plates, and simulated delamination damage in the composite plates. The frequency response of the scattered signals showed a clear change in the amplitude, due to the presence of the crack. The improved imaging method does not require a threshold to obtain the final image for the damage visualization. For the PWAS passive sensing methods, PWAS was successfully used as AE sensors for in situ AE detection during fatigue crack growth. The time of arrival of AE signals at multiple sensors confirms that the AE signals were originating from the crack. The amplitude decay of the AE signals due to geometric spreading was validated.

Author Contributions: In this article, the formal analysis was done by H.M. and M.F.H. The detailed methodology was provided by H.M., M.F.H., A.M., R.J., and V.G. H.M. prepared the original draft; review and editing were done by H.M., M.F.H., A.M., R.J., and V.G.

Funding: This work was supported by the National Aeronautics and Space Administration (NASA) grant number NNL15AA16C, Air Force Office of Scientific Research (AFOSR) grant number FA9550-16-1-0401, US Department of Energy (DOE) grant number DE-NE 0000726, Office of Nuclear Energy (ONE) grant number DE-NE 0008400, and Office of Naval Research (ONR) grant number N00014-17-1-2829.

Acknowledgments: The authors would like to acknowledge the help of the Electron Microscopy Center, USC for SEM.

References

1. Farrar, C.R.; Worden, K. An introduction to structural health monitoring. *Philos. Trans. R. Soc. A* **2006**, *365*, 303–315. [CrossRef] [PubMed]
2. Su, Z.; Ye, L.; Lu, Y. Guided Lamb waves for identification of damage in composite structures: A review. *J. Sound Vib.* **2006**, *295*, 753–780. [CrossRef]
3. Ciang, C.C.; Lee, J.R.; Bang, H.J. Structural health monitoring for a wind turbine system: A review of damage detection methods. *Meas. Sci. Technol.* **2008**, *19*, 122001. [CrossRef]
4. Fan, W.; Qiao, P. Vibration-based damage identification methods: A review and comparative study. *Struct. Health Monit.* **2011**, *10*, 83–111. [CrossRef]
5. Mitra, M.; Gopalakrishnan, S. Guided wave based structural health monitoring: A review. *Smart Mater. Struct.* **2016**, *25*, 053001. [CrossRef]
6. Achenbach, J. *Wave Propagation in Elastic Solids*; North-Holland Publishing Company: London, UK, 1973.
7. Rose, J.L. *Ultrasonic Waves in Solid Media*; Cambridge University Press: New York, NY, USA, 1999.
8. Giurgiutiu, V. *Structural Health Monitoring with Piezoelectric Wafer Active Sensors*, 2nd ed.; Elsevier Academic Press: Amsterdam, The Netherlands, 2007.
9. Baptista, F.G.; Budoya, D.E.; de Almeida, V.A.; Ulson, J.A.C. An experimental study on the effect of temperature on piezoelectric sensors for impedance-based structural health monitoring. *Sensors* **2014**, *14*, 1208–1227. [CrossRef]
10. Haider, M.F.; Giurgiutiu, V.; Lin, B.; Yu, L. Irreversibility effects in piezoelectric wafer active sensors after exposure to high temperature. *Smart Mater. Struct.* **2017**, *26*, 095019. [CrossRef]
11. Bilgunde, P.N.; Bond, L.J. Resonance analysis of a high temperature piezoelectric disc for sensitivity characterization. *Ultrasonics* **2018**, *87*, 103–111. [CrossRef]
12. Sinclair, A.N.; Chertov, A.M. Radiation endurance of piezoelectric ultrasonic transducers—A review. *Ultrasonics* **2015**, *57*, 1–10. [CrossRef]

13. Giurgiutiu, V.; Postolache, C.; Tudose, M. Radiation, temperature, and vacuum effects on piezoelectric wafer active sensors. *Smart Mater. Struct.* **2016**, *25*, 035024. [CrossRef]

14. Si, L.; Baier, H. Real-time impact visualization inspection of aerospace composite structures with distributed sensors. *Sensors* **2015**, *15*, 16536–16556. [CrossRef] [PubMed]

15. Park, W.H.; Packo, P.; Kundu, T. Acoustic source localization in an anisotropic plate without knowing its material properties—A new approach. *Ultrasonics* **2017**, *79*, 9–17. [CrossRef] [PubMed]

16. De Simone, M.E.; Ciampa, F.; Boccardi, S.; Meo, M. Impact source localisation in aerospace composite structures. *Smart Mater. Struct.* **2017**, *26*, 125026. [CrossRef]

17. Ren, Y.; Qiu, L.; Yuan, S.; Su, Z. A diagnostic imaging approach for online characterization of multi-impact in aircraft composite structures based on a scanning spatial-wavenumber filter of guided wave. *Mech. Syst. Signal Process.* **2017**, *90*, 44–63. [CrossRef]

18. Qiu, L.; Deng, X.; Yuan, S.; Huang, Y.; Ren, Y. Impact monitoring for aircraft smart composite skins based on a lightweight sensor network and characteristic digital sequences. *Sensors* **2018**, *18*, 2218. [CrossRef]

19. Sen, N.; Kundu, T. A new wave front shape-based approach for acoustic source localization in an anisotropic plate without knowing its material properties. *Ultrasonics* **2018**, *87*, 20–32. [CrossRef] [PubMed]

20. Bhuiyan, M.Y.; Giurgiutiu, V. The signatures of acoustic emission waveforms from fatigue crack advancing in thin metallic plates. *Smart Mater. Struct.* **2017**, *27*, 015019. [CrossRef]

21. Bhuiyan, M.Y.; Bao, J.; Poddar, B.; Giurgiutiu, V. Toward identifying crack-length-related resonances in acoustic emission waveforms for structural health monitoring applications. *Struct. Health Monit.* **2018**, *17*, 577–585. [CrossRef]

22. Ebrahimkhanlou, A.; Dubuc, B.; Salamone, S. Damage localization in metallic plate structures using edge-reflected lamb waves. *Smart Mater. Struct.* **2016**, *25*, 085035. [CrossRef]

23. Qiu, L.; Yuan, S.; Bao, Q.; Mei, H.; Ren, Y. Crack propagation monitoring in a full-scale aircraft fatigue test based on guided wave-Gaussian mixture model. *Smart Mater. Struct.* **2016**, *25*, 055048. [CrossRef]

24. Dziendzikowski, M.; Kurnyta, A.; Dragan, K.; Klysz, S.; Leski, A. In situ Barely Visible Impact Damage detection and localization for composite structures using surface mounted and embedded PZT transducers: A comparative study. *Mech. Syst. Signal Process.* **2016**, *78*, 91–106. [CrossRef]

25. Shen, Y.; Cesnik, C.E. Hybrid local FEM/global LISA modeling of damped guided wave propagation in complex composite structures. *Smart Mater. Struct.* **2016**, *25*, 095021. [CrossRef]

26. Hong, X.; Liu, Y.; Liufu, Y.; Lin, P. Debonding Detection in Hidden Frame Supported Glass Curtain Walls Using the Nonlinear Ultrasonic Modulation Method with Piezoceramic Transducers. *Sensors* **2018**, *18*, 2094. [CrossRef] [PubMed]

27. Dziendzikowski, M.; Niedbala, P.; Kurnyta, A.; Kowalczyk, K.; Dragan, K. Structural Health Monitoring of a Composite Panel Based on PZT Sensors and a Transfer Impedance Framework. *Sensors* **2018**, *18*, 1521. [CrossRef]

28. Yu, L.; Giurgiutiu, V. In situ 2-D piezoelectric wafer active sensors arrays for guided wave damage detection. *Ultrasonics* **2008**, *48*, 117–134. [CrossRef] [PubMed]

29. Ostachowicz, W.; Kudela, P.; Malinowski, P.; Wandowski, T. Damage localisation in plate-like structures based on PZT sensors. *Mech. Syst. Signal Process.* **2009**, *23*, 1805–1829. [CrossRef]

30. Zhao, X.; Royer, R.L.; Owens, S.E.; Rose, J.L. Ultrasonic Lamb wave tomography in structural health monitoring. *Smart Mater. Struct.* **2011**, *20*, 105002. [CrossRef]

31. Wang, D.; Zhang, W.; Wang, X.; Sun, B. Lamb-wave-based tomographic imaging techniques for hole-edge corrosion monitoring in plate structures. *Materials* **2016**, *9*, 916. [CrossRef]

32. Rao, J.; Ratassepp, M.; Lisevych, D.; Hamzah Caffoor, M.; Fan, Z. On-Line Corrosion Monitoring of Plate Structures Based on Guided Wave Tomography Using Piezoelectric Sensors. *Sensors* **2017**, *17*, 2882. [CrossRef]

33. Cai, J.; Shi, L.; Yuan, S.; Shao, Z. High spatial resolution imaging for structural health monitoring based on virtual time reversal. *Smart Mater. Struct.* **2011**, *20*, 055018. [CrossRef]

34. Lu, G.; Li, Y.; Wang, T.; Xiao, H.; Huo, L.; Song, G. A multi-delay-and-sum imaging algorithm for damage detection using piezoceramic transducers. *J. Intell. Mater. Syst. Struct.* **2017**, *28*, 1150–1159. [CrossRef]

35. Cai, J.; Wang, X.; Zhou, Z. A signal domain transform method for spatial resolution improvement of Lamb wave signals with synthetically measured relative wavenumber curves. *Struct. Health Monit.* **2018**. [CrossRef]

36. Wang, C.H.; Rose, J.T.; Chang, F.K. A synthetic time-reversal imaging method for structural health monitoring. *Smart Mater. Struct.* **2004**, *13*, 415. [CrossRef]

37. Agrahari, J.K.; Kapuria, S. A refined Lamb wave time-reversal method with enhanced sensitivity for damage detection in isotropic plates. *J. Intell. Mater. Syst. Struct.* **2016**, *27*, 1283–1305. [CrossRef]

38. Zeng, L.; Lin, J.; Huang, L. A modified Lamb wave time-reversal method for health monitoring of composite structures. *Sensors* **2017**, *17*, 955. [CrossRef] [PubMed]

39. Quaegebeur, N.; Ostiguy, P.C.; Masson, P. Correlation-based imaging technique for fatigue monitoring of riveted lap-joint structure. *Smart Mater. Struct.* **2014**, *23*, 055007. [CrossRef]

40. He, J.; Yuan, F.G. Lamb-wave-based two-dimensional areal scan damage imaging using reverse-time migration with a normalized zero-lag cross-correlation imaging condition. *Struct. Health Monit.* **2017**, *16*, 444–457. [CrossRef]

41. Su, Z.; Cheng, L.; Wang, X.; Yu, L.; Zhou, C. Predicting delamination of composite laminates using an imaging approach. *Smart Mater. Struct.* **2009**, *18*, 074002. [CrossRef]

42. Zhou, C.; Su, Z.; Cheng, L. Quantitative evaluation of orientation-specific damage using elastic waves and probability-based diagnostic imaging. *Mech. Syst. Signal Process.* **2011**, *25*, 2135–2156. [CrossRef]

43. Wu, Z.; Liu, K.; Wang, Y.; Zheng, Y. Validation and evaluation of damage identification using probability-based diagnostic imaging on a stiffened composite panel. *J. Intell. Mater. Syst. Struct.* **2015**, *26*, 2181–2195. [CrossRef]

44. Hall, J.S.; Michaels, J.E. Minimum variance ultrasonic imaging applied to an in situ sparse guided wave array. *IEEE Trans. Ultrason. Ferroelectr. Frequency Control* **2010**, *57*, 2311–2323. [CrossRef] [PubMed]

45. Hall, J.S.; Fromme, P.; Michaels, J.E. Guided wave damage characterization via minimum variance imaging with a distributed array of ultrasonic sensors. *J. Nondestruct. Eval.* **2014**, *33*, 299–308. [CrossRef]

46. Hua, J.; Lin, J.; Zeng, L.; Luo, Z. Minimum variance imaging based on correlation analysis of Lamb wave signals. *Ultrasonics* **2016**, *70*, 107–122. [CrossRef] [PubMed]

47. Kudela, P.; Radzienski, M.; Ostachowicz, W.; Yang, Z. Structural health monitoring system based on a concept of Lamb wave focusing by the piezoelectric array. *Mech. Syst. Signal Process.* **2018**, *108*, 21–32. [CrossRef]

48. Xu, C.B.; Yang, Z.B.; Zhai, Z.; Qiao, B.J.; Tian, S.H.; Chen, X.F. A weighted sparse reconstruction-based ultrasonic guided wave anomaly imaging method for composite laminates. *Compos. Struct.* **2018**, *209*, 233–241. [CrossRef]

49. Staszewski, W.J.; Mahzan, S.; Traynor, R. Health monitoring of aerospace composite structures–Active and passive approach. *Compos. Sci. Technol.* **2009**, *69*, 1678–1685. [CrossRef]

50. Ruzzene, M. Frequency–wavenumber domain filtering for improved damage visualization. *Smart Mater. Struct.* **2007**, *16*, 2116. [CrossRef]

51. Michaels, T.E.; Michaels, J.E.; Ruzzene, M. Frequency–wavenumber domain analysis of guided wavefields. *Ultrasonics* **2011**, *51*, 452–466. [CrossRef]

52. Tian, Z.; Yu, L. Lamb wave frequency–wavenumber analysis and decomposition. *J. Intell. Mater. Syst. Struct.* **2014**, *25*, 1107–1123. [CrossRef]

53. Liu, Z.; Chen, H.; Sun, K.; He, C.; Wu, B. Full non-contact laser-based Lamb waves phased array inspection of aluminum plate. *J. Vis.* **2018**, *21*, 751–761. [CrossRef]

54. Sohn, H.; Dutta, D.; Yang, J.Y.; Park, H.J.; DeSimio, M.; Olson, S.; Swenson, E. Delamination detection in composites through guided wave field image processing. *Compos. Sci. Technol.* **2011**, *71*, 1250–1256. [CrossRef]

55. An, Y.K.; Park, B.; Sohn, H. Complete noncontact laser ultrasonic imaging for automated crack visualization in a plate. *Smart Mater. Struct.* **2013**, *22*, 025022. [CrossRef]

56. Park, B.; An, Y.K.; Sohn, H. Visualization of hidden delamination and debonding in composites through noncontact laser ultrasonic scanning. *Compos. Sci. Technol.* **2014**, *100*, 10–18. [CrossRef]

57. He, J.; Yuan, F.G. A quantitative damage imaging technique based on enhanced CCRTM for composite plates using 2D scan. *Smart Mater. Struct.* **2016**, *25*, 105022. [CrossRef]

58. Girolamo, D.; Chang, H.Y.; Yuan, F.G. Impact damage visualization in a honeycomb composite panel through laser inspection using zero-lag cross-correlation imaging condition. *Ultrasonics* **2018**, *87*, 152–165. [CrossRef]

59. Rogge, M.D.; Leckey, C.A. Characterization of impact damage in composite laminates using guided wavefield imaging and local wavenumber domain analysis. *Ultrasonics* **2013**, *53*, 1217–1226. [CrossRef] [PubMed]

60. Tian, Z.; Yu, L.; Leckey, C.; Seebo, J. Guided wave imaging for detection and evaluation of impact-induced delamination in composites. *Smart Mater. Struct.* **2015**, *24*, 105019. [CrossRef]

61. Tian, Z.; Yu, L.; Leckey, C. Delamination detection and quantification on laminated composite structures with Lamb waves and wavenumber analysis. *J. Intell. Mater. Syst. Struct.* **2015**, *26*, 1723–1738. [CrossRef]

62. Kudela, P.; Radzieński, M.; Ostachowicz, W. Identification of cracks in thin-walled structures by means of wavenumber filtering. *Mech. Syst. Signal Process.* **2015**, *50*, 456–466. [CrossRef]

63. De Marchi, L.; Marzani, A.; Moll, J.; Kudela, P.; Radzieński, M.; Ostachowicz, W. A pulse coding and decoding strategy to perform Lamb wave inspections using simultaneously multiple actuators. *Mech. Syst. Signal Process.* **2017**, *91*, 111–121. [CrossRef]

64. Kudela, P.; Radzienski, M.; Ostachowicz, W. Impact induced damage assessment by means of Lamb wave image processing. *Mech. Syst. Signal Process.* **2018**, *102*, 23–36. [CrossRef]

65. Newnham, R.E.; Xu, Q.C.; Kumar, S.; Cross, L.E. Smart ceramics. *Ferroelectrics* **1990**, *102*, 259–266. [CrossRef]

66. APC 850 Materials Data Sheet. Available online: http://www.americanpiezo.com (accessed on 10 October 2018).

67. Herbiet, R.; Robels, U.; Dederichs, H.; Arlt, G. Domain wall and volume contributions to material properties of PZT ceramics. *Ferroelectrics* **1989**, *98*, 107–121. [CrossRef]

68. Xu, F.; Trolier-McKinstry, S.; Ren, W.; Xu, B.; Xie, Z.L.; Hemker, K.J. Domain wall motion and its contribution to the dielectric and piezoelectric properties of lead zirconate titanate films. *J. Appl. Phys.* **2001**, *89*, 1336–1348. [CrossRef]

69. Zhang, Q.M.; Wang, H.; Kim, N.; Cross, L.E. Direct evaluation of domain-wall and intrinsic contributions to the dielectric and piezoelectric response and their temperature dependence on lead zirconate-titanate ceramics. *J. Appl. Phys.* **1994**, *75*, 454–459. [CrossRef]

70. Faisal Haider, M.; Migot, A.; Bhuiyan, M.; Giurgiutiu, V. Experimental Investigation of Impact Localization in Composite Plate Using Newly Developed Imaging Method. *Inventions* **2018**, *3*, 59. [CrossRef]

71. Migot, A.; Bhuiyan, Y.; Giurgiutiu, V. Numerical and experimental investigation of damage severity estimation using Lamb wave–based imaging methods. *J. Intell. Mater. Syst. Struct.* **2018**. [CrossRef]

72. Mei, H.; Giurgiutiu, V. Guided wave excitation and propagation in damped composite plates. *Struct. Health Monit. Int. J.* **2018**. [CrossRef]

73. Giurgiutiu, V. Predictive simulation of guide wave structural health monitoring in metallic and composite structures. In Proceedings of the 9th European Workshop on Structural Health Monitoring EWSHM 2018, Manchester, UK, 10–13 July 2018.

74. Glushkov, E.; Glushkova, N.; Eremin, A.; Lammering, R. Group velocity of cylindrical guided waves in anisotropic laminate composites. *J. Acoust. Soc. Am.* **2014**, *135*, 148–154. [CrossRef]

![sensors logo] **sensors**

MDPI

Article

Performance Comparison of Three Fibre-Based Reflective Optical Sensors for Aero Engine Monitorization

Rubén Fernández-Bello [1,*], Josu Amorebieta [1], Josu Beloki [2], Gotzon Aldabaldetreku [1], Iker García [3], Joseba Zubia [1] and Gaizka Durana [1]

[1] Communications Engineering Department, University of the Basque Country UPV/EHU, Ingeniero Torres Quevedo Plaza 1, E-48013 Bilbao, Spain; josu.amorebieta@ehu.eus (J.A.); gotzon.aldabaldetreku@ehu.eus (G.A.); joseba.zubia@ehu.eus (J.Z.); gaizka.durana@ehu.eus (G.D.)
[2] Fundación Centro de Tecnologías Aeronáuticas (CTA), Bizkaia Technological Park, E-48170 Zamudio, Spain; josu.beloki@ctabef.com
[3] AOTECH, Advanced Optical Technologies S.L., E-48002 Bilbao, Spain; igarcia@aotech.es
* Correspondence: ruben.fernandez@ehu.eus; Tel.: +34-94-601-7305

Received: 12 April 2019; Accepted: 11 May 2019; Published: 15 May 2019

✔ check for updates

Abstract: Among the different available optical technologies, fibre bundle-based reflective optical sensors represent an interesting alternative for parameter monitorization in aero engines. Tip clearance is one of the parameters of great concern for engine designers and engineers. In the framework of this optical technology, three fibre-based reflective optical sensors have been compared. Two of them are custom designed and based on the same geometrical fibre arrangement, whereas the third one is commercially available and relies on a different geometrical arrangement of the fibres. Their performance has been compared in clearance measurements carried out during an experimental program followed at a transonic wind tunnel for aero turbines. The custom-designed solution that operates in the most sensitive part of its response curve proved to be by far the most reliable tool for clearance measurements. Its high resolution opens up the possibility to detect small blade features such as cracks, reflectivity changes, etc. that otherwise could not be tracked. These results show that the detection of unexpected features on blade tips may have an important effect on how the clearance is calculated, ultimately giving rise to corrective actions.

Keywords: fibre bundle; reflective optical sensor; tip clearance; turbine; aero engine

1. Introduction

Nowadays, the presence of turbines has notoriously increased in critical sectors such as military, civil transport, power plants, etc., which has derived from the interest of developing increasingly more efficient designs in order to improve both their efficiency, and durability.

The efficiency of turbines and their lifespan has been a field of study for years [1,2]. Those studies have resulted in a progressive improvement and refinement of turbines performance and design. The aforementioned improvement is related to different elements and/or parameters such as the tip clearance (TC), which is defined as the air gap between the most prominent part of any of the rotor blades and the inner part of the casing, also known as abradable [3]. Lowering TC allows extracting as much energy as possible from the incoming fluid. As the TC decreases, the sealing of the turbine stage improves, avoiding undesired air-leaks that reduce efficiency and performance. Overall, the benefits from TC reduction are lower fuel consumption [4,5], the decrease of contaminant emissions [6,7], and the increase of available payload [5], which are very interesting topics in the aeronautical industry, especially for civil aviation, which make the engines economically more profitable.

However, keeping the TC as low as possible may cause the blades to scratch the abradable. Even if the abradable is a material designed to withstand this situation, excessive friction may lead to a premature wear of certain engine parts and a reduction of the lifecycle of the engine. That effect would increase the running and maintenance costs [4]. In order to avoid this fact, active TC control systems have been developed [8].

Tip clearance measurement systems are required to constantly monitor the TC parameter to provide designers and engineers with reliable and precise information that could avoid potential issues or could be helpful for new turbine designs. Currently, several TC measurement systems are commercially available, which are based on different sensors: capacitive [9–11], eddy current [12–16], pneumatic [17], strain gauges [18], electromechanical [19], microwave [20–24], and optical sensors [25–27].

Each sensor type has different characteristics. On the one hand, both capacitive and eddy-current sensors have poor frequency response. On the other hand, the capacitive sensors are known for being low cost and robust, whereas the eddy-current sensors require targets with magnetic materials. Pneumatic sensors are insensitive to contamination, but they require considerable hardware. Strain gauges, along with their required wires, are usually installed in the blades, interacting with the air flow passing around and changing the mass balance of the turbine [28]. Electromechanical sensors have high resolution, but they only provide the measure of the closest blade of the rotor, and do not provide blade identification. Microwave sensors are not sensitive to contamination, but they need complex circuitry. Finally, optical sensors have fast response and small dimensions, and thus they are able to provide more detailed information, but they can be affected by dirt [29].

According to the conditions and limitations of measuring the TC in an aero turbine, where small sensor volume and high bandwidth are desirable, the use of an optical sensor is very appropriate. Among the available optical solutions for distance measurements, the most common techniques are based on Doppler effect, interferometry, triangulation, intensity modulation, and time of flight.

The Doppler effect-based technique offers high positioning and temporal resolution over metallic and non-metallic blades [30]. Optical coherence tomography uses the interference back-reflected light from the blade tips and a frequency-shifted reference with variable time delay, making use of a low-coherence light source [31]. The triangulation technique emits a light beam towards the target with certain angle, and according to the target position, the reflected ray would hit the receiving sensor in a different position. However, this technique could experience problems with abrupt shapes [32,33]. Intensity modulation-based techniques rely on the dispersion of the laser beam energy. A laser beam is emitted, and as it travels, the light power per area decreases. So, the longer the beam travels the less power per area is sensed by a given photodetector. However, any increment in the power of the light source would be incorrectly interpreted as a target approach, even if the distance did not change. To avoid this, a source light with a stable power is needed. There is also a more complex variant, based on the Gaussian dispersion of the power of the traveling beam, where various fibre groups at different radial distances are needed to measure how the beam intensity is radially distributed [15,29,34–38]. This technology has more tolerance to the target angle compared to the intensity-based technique, but the sensor is more expensive and complex to build. The time of flight method relies on the time the light beam needs to travel from the sensor to the target and back. This technique is robust and reliable, but its accuracy is strongly dependent on the slope of the leading and trailing edge of the blade [17]. Therefore, after commenting about different techniques, the Gaussian dispersion-based technique is the best candidate for its light source power variation immunity and target angle variation tolerance.

The paper presents the performance comparison of three different optical sensors based on the Gaussian dispersion technique. First, the configuration of each of them is explained. Then, the experimental setup is described, as well as the test program carried out in a turbine rig at "Fundación Centro de Tecnologías Aeronáuticas" (CTA) facilities. Afterwards, the most relevant experimental data obtained from the three optical sensors are compared and discussed. Finally, the most relevant conclusions are summarized.

2. Materials and Methods

2.1. Description of the Setup

The optical sensors (OS) under comparison all have optical fibre-based sensing heads. One is commercially available and the other two are based on custom-designed optical fibre bundles. The cross-section of each of them is shown in Figure 1. For the sake of clarity, from now on, from left to right they will be referred to as OS 1, OS 2, and OS 3. On the one hand, OS 1 (Philtec model RC171) consists of several transmitting and receiving fibres arranged in adjacent semi-circular pattern where the fibres are distributed randomly. Its specifications are available in Reference [39]. On the other hand, the two custom-designed sensors share the same geometrical structure, which is explained in References [3,36]. They are both based on two independents receiving fibre rings that are located at different radii surrounding the central transmitting fibre. They were designed based on the mathematical approach shown in References [37,38] but including the design constraints (maximum allowed diameter for the cross-section of the fibre bundle, expected tip clearance working range, etc.) imposed by the turbine specifications in order to have the most sensitive performance in their respective linear regions (Figure 4). With these considerations in mind, the manufactured custom designs that fulfilled the previously mentioned boundaries are shown in Figure 1. For both cases, OS 2 and OS 3, the transmitting fibre is a single-mode fibre with a numerical aperture (NA) of 0.12, and the inner ring of fibres (of radius $R_{i2} = R_{i3} = 200$ μm) is formed by five multimode fibres with an NA of 0.22. The OS 2 outer ring (of radius $R_{o2} = 930$ μm) is formed by 17 multimode fibres with a core diameter of 300 μm and an NA of 0.22, whereas the OS 3 (with a radius $R_{o3} = 1800$ μm) is formed by 30 multimode fibres with core diameters of 300 μm with an NA of 0.22. The main difference between both configurations lies in the number of the receiving fibres and ring radius on the outer ring: OS 3 has a bigger outer ring radius with a larger number of fibres on it than in the case of OS 2.

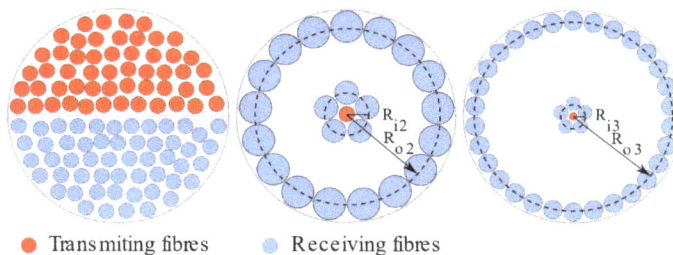

Figure 1. Cross-sections of fibre bundles from the three optical sensors under comparison: OS 1 (left), OS 2 (middle), and OS 3 (right). Images not to scale. R_i and R_o stand for inner and outer radius for OS 2 and OS 3 sensors.

On the one hand, OS 1 integrates its own light source and a light-to-voltage converter by means of proprietary optoelectronics. On the other hand, OS 2 and OS 3 share a common continuous-wave (CW) laser light source (@660 nm wavelength) via a 50:50 optical fibre coupler. The light launches onto the passing blade by the central transmitting fibre, after being reflected on it, is gathered by each of the receiving fibre rings, the inner ring and outer ring, and converted into two voltage levels (V_1 and V_2) at photodetectors PD 1 and PD 2, respectively (please refer to Figure 2). The ratio of V_2 to V_1 is considered to minimize undesirable effects related to intensity fluctuations of the light source and reflectivity variations of the target surface [3]. Finally, all the voltage outputs are acquired simultaneously by a shared acquisition board.

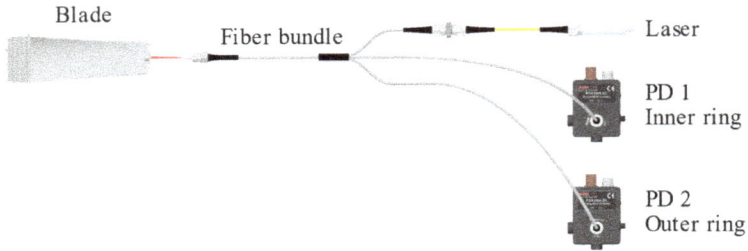

Figure 2. Schematic representation of the light-to-voltage conversion for OS 2 and OS 3. PD stands for photodetector. PD 1 and PD 2 are connected to inner and outer rings, respectively.

2.2. Assembly, Calibration, and Tests Program

The sensors were installed in an aircraft turbine rig at the facilities of CTA in order to calibrate them specifically for the rig under test before the TC measurements were done. For that purpose, they were fixed into three of the four radially oriented inlets (see Figure 3a) that gave access to the platform located at the middle of the blade tip profile. The blade tip detail and the light path along it are shown in Figure 3b. The radial positioning was carried out with a precision positioner (Figure 3c).

The calibration curves of the optical sensors are shown in Figure 4. The shaded area represents the region of interest at which each sensor worked during the rig tests. The distance shown in Figure 4 represents the distance from the blade platform (Figure 3b) to the tip of the sensor installed in the engine casing. Notice that OS 1 and OS 3 are operating in their positive and more sensitive front slope region, whereas OS 2 does it in its less sensitive back slope region.

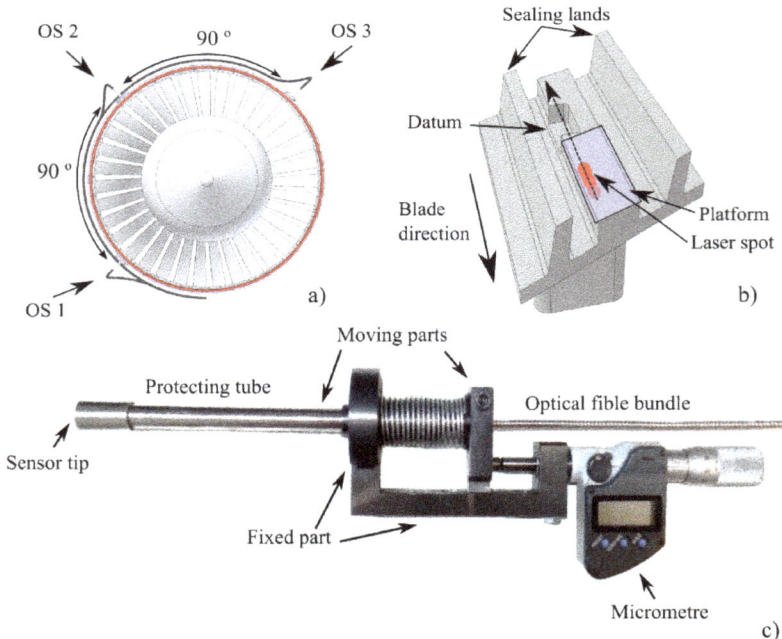

Figure 3. (**a**) Location of OS 1, OS 2, and OS 3 in the turbine casing. (**b**) Detail of a typical blade tip. (**c**) Side view of the tool used to locate precisely each sensor head with respect to the blade tips.

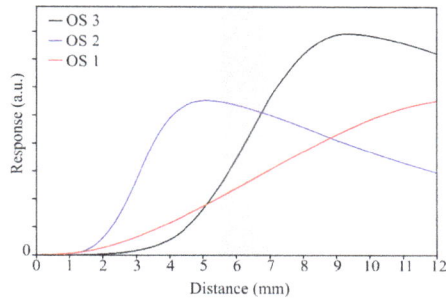

Figure 4. Calibration curve of each sensor. The shaded region shows the working region of interest.

The tests were carried out in a wind tunnel commonly used for turbine testing [3]. Briefly, the test rig consisted of two stages: the first stage was formed by a 3.7 MW air compressor with a temperature control section, and the second stage was formed by a vacuum pump of 5 MW. In the pipe which connected both stages, an air flow rate up to 18 kg/s could be generated, a pressure up to 450 kPa, and a temperature up to 573 °C. The turbine was connected to a hydraulic brake to simulate changing loads.

The tests lasted from eight to ten continuous hours per day and they were repeated for two weeks in consecutive days. Every test point was recorded, since each one implied a change in the working conditions (pressure, rpm, etc.). All the sensors data were acquired by one acquisition board. Additionally, a once per revolution (OPR) signal for synchronisation and blade identification purposes was acquired. From the gathered data, the most representative results along with the corresponding discussion were analysed and the results along with a discussion are presented and explained in the following section.

3. Results and Discussion

From the raw signals of the three sensors, different levels of blade information and detail may be obtained (see Figure 5). On the one hand, OS 1 shows a smoothed signal that, as we will see, is responsible for the slight delay observed in signal response to specific blade profile features. This fact makes it harder to detect and synchronise the different events found during a complete shaft turn. On the other hand, OS 2 and OS 3 are able to detect specific features, such as the datum and inter-blade spacing, more clearly and without delays, which results in an easier posterior data processing.

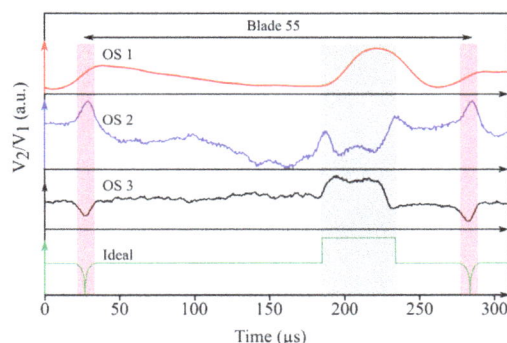

Figure 5. Comparison of sensors signals for a typical blade. The lower curve represents the ideal signal response. The datum and inter-blade spacing are highlighted with grey and red-shaded areas, respectively.

The high resolution and sensitivity offered by OS 3 makes it possible to determine the passage of each blade precisely. The steep rising and falling edges in the waveform enable us to precisely determine the boundaries of the datum, and ultimately, their arrival times. More specifically, the derivative of the raw waveform yields a sequence of peaks associated with the leading and trailing boundaries of the datum and the inter-blade spacing that, after setting a convenient threshold value to the derivative, allows us to determine the passing times of the datum boundaries precisely. Therefore, within a complete turbine turn, it is possible to define as many instantaneous speeds as number of blades (94 in the turbine rig under test). For the discussion that follows, we define the instantaneous speed as the average speed (in rpm) over one turbine turn. Hence, the OPR signal directly gives it, whereas in the case of OS 1, OS 2, and OS 3, all blade-to-blade instantaneous speeds within one full turn must be averaged. If we consider the passing time of the trailing or leading edge of the datum as the reference value for defining the blade passing time, we can determine the instantaneous speed of the turbine. Figure 6a shows the turbine instantaneous speed determined according to the signals of each three optical sensors, using both the leading and trailing edges of the datum. It should be pointed out the inaccuracy of sensors OS 1 and OS 2 when trying to approach the reference rpm value given by the OPR signal. In fact, OS 1 and OS 2 using the datum leading edge to extract the blade-to-blade speed produced an output mean speed reading with an error of 0.4% and 2.9%, respectively, whereas using the datum trailing edge produced a speed error of 2.4% and 16%. In addition to that, the obtained result varies quite substantially depending on whether the leading or trailing edge has been considered for the determination of the arrival time. This is particularly remarkable in the case of OS 1, for which the leading edge converged to the expected value given by the OPR signal, whereas the trailing edge falls far apart from it. Regarding OS 3, a closer look at its performance revealed a very accurate behaviour irrespective of the chosen datum edge (see Figure 6b). Observe that all curves followed the same trend over eight revolutions of the rotor, but with the distinctive feature of the curve obtained from the OPR signal that has an edgy behaviour in contrast to the other two curves. The smoothed behaviour of the latter was a consequence of the averaging process of the 94 instantaneous speeds involved in a complete revolution, but gives a more realistic view of the actual speed behaviour of the turbine. This kept the mean error for the OS 3 below 0.009% and the maximum error around 0.07%.

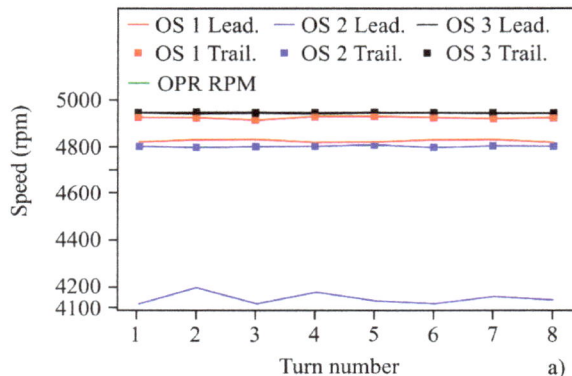

Figure 6. *Cont.*

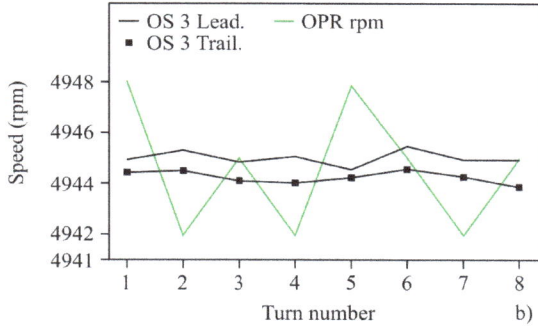

Figure 6. (a) Turbine instantaneous speed over several shaft turns measured by different optical sensors. The reference speed was defined by the OPR signal. (b) Comparison of the turbine instantaneous speed values given by OS 3 and the OPR signal.

In order to appreciate further the accuracy provided by OS 3 against OS 1 and OS 2, a statistical analysis based on data from 52 complete turbine revolutions was carried out. More specifically, for each complete turbine turn, 94 blade-to-blade instantaneous speeds were determined and compared with the reference speed value set by the OPR signal. For the sake of graphical clarity, Figure 7 only shows the frequency distribution of the speed deviation corresponding to OS 3. The trailing edge of the datum was considered to define the passing time of each blade. Although not shown in Figure 7, data corresponding to OS 1 and OS 2 spread out over a wide range of values, whereas data from OS 3 were very concentrated around the nominal value. Additionally, outlier values also existed in the cases of OS 1 and OS 2.

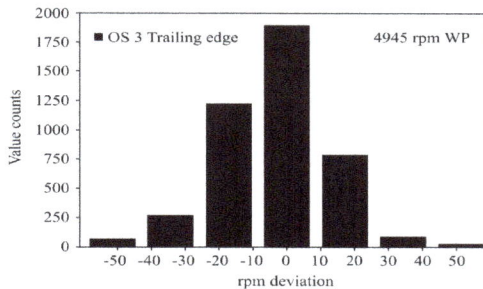

Figure 7. Histogram representing the deviation of the calculated instantaneous speed with respect to the nominal speed (4945 rpm) for the OS 3 sensor using the datum trailing edge. Data obtained from 52 consecutive shaft turns.

A short summary of the statistical analysis is presented in Table 1. It highlights the lower performance of OS 1 and OS 2 compared to OS 3. The capability of the OS 3 to extract the blade-to-blade rotor speed, even along the duration of each turn, is outstanding. Similar results were obtained using the leading edge of the datum.

Table 1. Statistical analysis of OS 1, OS 2, and OS 3 working at 4945 rpm using the trailing edge of the datum feature.

Sensor	Mean (rpm)	Standard Deviation of the Mean (rpm)
OS 1	−21.37	±3.06
OS 2	−141.74	±7.13
OS 3	0.09	±0.40

These observations are a direct consequence of the waveform quality of the acquired signals, among which OS 3 excels over the rest.

The high resolution and sensitivity of OS 3 also provides the ability to identify unexpected features in the blades through the corresponding waveform analysis. Two representative examples of this ability are shown in Figure 8.

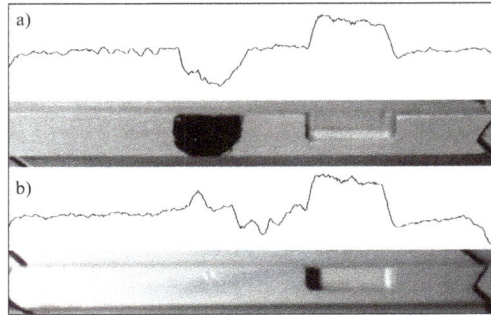

Figure 8. Waveform signals corresponding to OS 3 in two different cases: (**a**) a blade with a black-painted area on it. (**b**) A blade with variable reflectivity along it.

The first example (see Figure 8a) shows the waveform signal corresponding to a blade that has a black marker dot on it. The waveform shows a clear drop in light intensity as a consequence of the lower reflectivity of that blackened area. Something similar occurs in the second example (see Figure 8b) where the surface reflectivity varies along the long flat platform and the waveform responds accordingly. It is worthy of note that the waveform response to the passage of standard blades (blades without any unexpected features) is quite constant in the long and short platform areas of the blades. The comparison between the waveform response to a standard blade and to a partially blackened blade is shown in Figure 9. At this point, it is important to bring to the reader's attention that the signal comparative has not been extended to OS 1 and OS 2 as they struggle to distinguish between a standard blade and a blade with a certain feature on its tip. Coming back to Figure 9, it should also be pointed out that the signal response to the passage of a typical blade, which is quite constant over the long and short platforms of the blade, has a low-amplitude, high-frequency fluctuation superposed on it. Although at first sight it may seem a random noise contribution to the actual waveform signal, it actually represents the response to a surface with a certain roughness level on it.

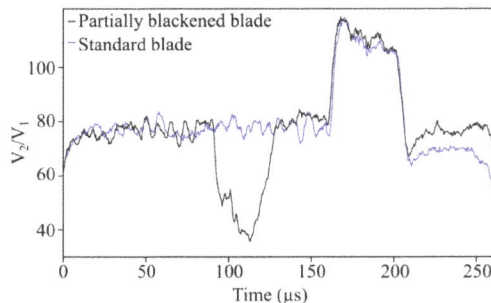

Figure 9. Comparison between a standard blade (blue curve) and a blade with an unexpected feature on it (black curve). In this case, the feature corresponds to a small blackened area on the long platform of the blade.

As evidence of the previous statement, Figure 10 shows the response of OS 3 to the passage of a typical blade over 1100 turns of the turbine. The almost perfect overlapping of the successive signals suggests that the low-amplitude, high-frequency fluctuations were mainly related to surface features and irregularities of the blade tip, which again confirm the low noise and high stability of the sensor. As a quantitative measure of it, Table 2 shows the average and standard deviation values of the signal at different points distributed in an equidistant way along the length of a typical blade (points A, B, C, D, and E in Figure 10). Observe that the standard deviation did not exceed—in any case—5% of the corresponding average value.

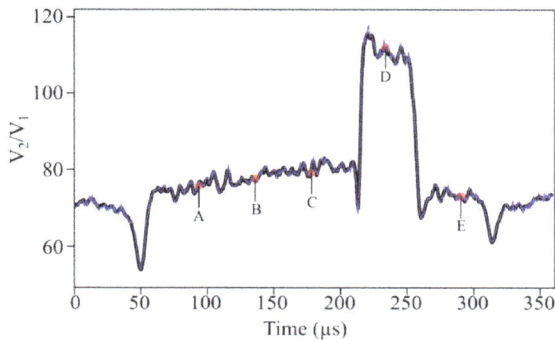

Figure 10. Signal response of OS 3 to the passage of a typical blade over 1100 revolutions. The black curve represents the average response curve, whereas the blue curve corresponds to a single revolution.

Table 2. Signal stability over 1100 turbine turns at five points distributed equally along the length of a typical blade.

Point	Mean Value and Standard Deviation (V_2/V_1)	Relative Standard Deviation (%)
A	74.21 ± 0.53	3.5
B	77.91 ± 0.54	2.8
C	79.28 ± 0.64	3.3
D	114.62 ± 0.73	1.3
E	73.05 ± 0.64	4.6

In the tests carried out at the CTA facilities, the TC was the parameter of interest to be monitored. The TC is defined as the air gap existing between the abradable coating and the most prominent part of each blade. Its determination requires a specific processing of the signal. For the analysis that follows, we will concentrate on OS 3 as its performance stands out over the other two optical sensors. The aforementioned signal treatment (for a standard blade) consists in averaging the data from a predefined section of it. More specifically, the average is carried out over the set of data corresponding to the long platform of each blade where, as commented previously, the signal level remains quite constant. However, in the case of the few non-standard blades, i.e., blades with unexpected features, such as shown in Figure 8, the TC value gets miscalculated due to the effect of those features on the signal shape and level. In those cases, some corrective actions may be taken to compensate for the harmful effects on the TC value. In order to make it clearer, let us consider blade number 54. The numbering refers to the number assigned to each blade according to its arrival position starting to count from the OPR signal. Therefore, blade 54 is the 54th blade to be detected by OS 3. In the case of that blade, if we do not consider any corrective action, its TC value results to be the minimum blade TC among all blades, i.e., the turbine TC. However, if we consider a corrective action based on excluding those data points coming from the problematic region of the blade, so that we end up with a smaller dataset for the averaging process, the new TC value increases, and the blade ceases to be the critical one (see Table 3).

Table 3. Tip clearance values at different rpms for a non-standard blade (blade 54), without and with the correction factor.

Rpm	TC without Correction (mm)	TC with Correction (mm)
2406	0.2	0.6
3627	0.3	0.5
4945	−0.2	0.4

As a proof of the stability and low noise brought by OS 3, Figure 11 shows the TC map measured with OS 3 corresponding to a certain engine working point (WP) at 3627 rpm. The variability of the TC values over time (20,000 engine revolutions), shown as vertical error bars, give rise to a worst case of 20 µm within the same WP, and the average value over the 94 blade is below 7 µm.

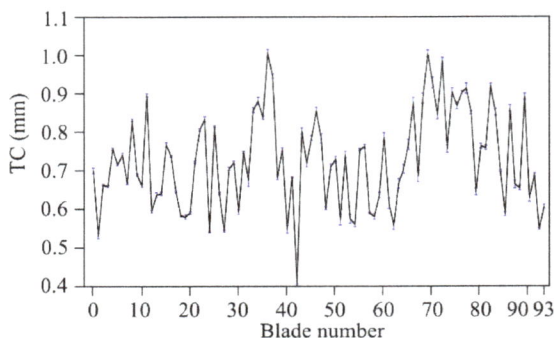

Figure 11. Tip clearance values of individual blades at 3627 rpm with error bars that account for the standard deviation over 20,000 revolutions.

The TC mapping of all blades corresponding to three WPs of the turbine in ascending order of rpm is shown in Figure 12. On those non-standard blades, for the correct determination of the corresponding TC, the aforementioned corrective action was applied. Results indicated that the higher the rotational speed was, the lower the TC. This was due to the centrifugal forces that make the blades stretch, and therefore, approach the casing.

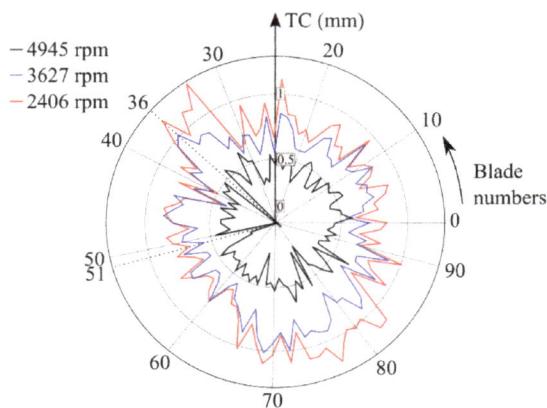

Figure 12. TC values of each of the 94 blades obtained with OS 3 at three different WPs.

4. Conclusions

We have reported on a performance comparison between three intensity-modulated optical fibre-based sensors (one commercial with semi-circular fibre distribution, OS 1, and two custom-designed with concentrical ring distribution, OS 2 and OS 3) for aero engine applications. More specifically, the sensors have been tested in the wind tunnel available at the CTA facilities for TC measuring during turbine testing.

The OS 2 and OS 3 custom-designed sensors both had a central transmitting single-mode fibre and an inner ring formed by five multimode receiving fibres (core diameters of 200 μm and NAs of 0.22). On the other hand, their outer rings were built using larger multimode fibres (core diameters of 300 μm and NAs of 0.22), in each case with different amounts of multimode fibres in order to meet different working distance specifications. Each sensor had different calibration curves with two linear regions (front slope and back slope). The sensors worked in different regions depending on the distance to the target. The custom-designed sensor OS 3 was designed with the help of a custom simulation program developed within the research group. That design was developed in such a way that the sensor would be working in the front slope region of its response curve for the expected target distances on this test. During the test, OS 3 offered the best performance due to the higher stability, lower noise level, higher resolution and sensitivity, and great repeatability, which allowed for more accurate measurements. The TC values over 20,000 engine revolutions were registered showing a mean variability below 7 μm for each individual blade turn after turn. The signal provided by the OS 3 can be straightforwardly related to the physical shape of the blade, detecting the different features as the interlock and the datum. That feature makes it possible an easier and more reliable development of different software-based solutions, such as the blade-by-blade speed detection system that has been proven here, whose results provided a smoother rotor speed behaviour. That makes the OS 3 sensor stand out as the most interesting sensor of the test.

Both custom-designed sensors, OS 3 and OS 2 (even though the signal in the latter is noisier), can detect small surface imperfections as scratches or dirt that are not recognized by the OS 1 sensor. This capability enables detecting blades that cause anomalous TC output value readings. In such situations, the software can be adapted to overcome that problem. Thus, the probed ability to detect surface irregularities in certain blades makes it also useful for health monitoring analysis.

Our optical bundle is a great sensor for high speed TC measurement systems. It can also be customized for different applications where a small distance has to be measured. Therefore, we believe that this sensor may be appealing in high-precision applications, where non-contact measurement, small dimensions, and electromagnetic immunity are required, such as aeronautical engines, gas and oil facilities, etc.

Author Contributions: The original draft of the paper was written by R.F.-B. and J.A. and reviewed by G.A., I.G. and G.D.; the software was developed by R.F.-B., J.A. and G.D.; data was curated by R.F.-B. and J.A., and validated by J.B. and I.G.; data visualization was fulfilled by R.F.-B. and J.A.; the project was supervised by G.A. and G.D.; the methodology, project administration and resources were provided by J.B., J.Z. and G.D.; the investigation was carried out by R.F.-B., J.A. and G.D.; the funding was acquired J.Z. and G.D.

Funding: This work has been funded in part by the Fondo Europeo de Desarrollo Regional (FEDER), in part by the Ministerio de Economía y Competitividad under project RTI2018-094669-B-C31, and in part by the Gobierno Vasco/Eusko Jaurlaritza IT933-16 and ELKARTEK KK-2018/00078, μ4Indust and SMARTRESNAK). The work of Josu Amorebieta is supported in part by a PhD fellowship from the Universidad del País Vasco/Euskal Herriko Unibertsitatea (UPV/EHU), Vicerrectorado de Euskera y Formación Continua.

Conflicts of Interest: The authors declare no conflict of interest.

References

1. Holeski, D.E.; Futral, S.M., Jr. *Effect of Rotor Tip Clearance on the Performance of a 5-Inch Single-Stage Axial-Flow Turbine*; NASA Scientific and Technical Publications: Springfield, VA, USA, March 1969.
2. Venter, S.J.; Kröger, D.G. The effect of tip clearance on the performance of an axial flow fan. *Energy Convers. Manag.* **1992**, *33*, 89–97. [CrossRef]

3. Durana, G.; Amorebieta, J.; Fernández, R.; Beloki, J.; Arrospide, E.; García, I.; Zubia, J. Design, fabrication and testing of a high-sensitive fibre sensor for tip clearance measurements. *Sensors* **2018**, *18*, 2610. [CrossRef] [PubMed]

4. Wiseman, M.W.; Guo, T.-H. An investigation of life extending control techniques for gas turbine engines. In Proceedings of the 2001 American Control Conference (Cat. No.01CH37148), Arlington, VA, USA, 25–27 June 2001; Volume 5, pp. 3706–3707. [CrossRef]

5. Lattime, S.B.; Steinetz, B.M. High-Pressure-Turbine Clearance Control Systems: Current Practices and Future Directions. *J. Propuls. Power* **2004**, *20*, 302–311. [CrossRef]

6. Miller, K.; Key, N.; Fulayter, R. Tip Clearance Effects on the Final Stage of an HPC. In Proceedings of the 45th AIAA/ASME/SAE/ASEE Joint Propulsion Conference & Exhibit, Denver, CO, USA, 2–5 August 2009. [CrossRef]

7. Neuhaus, L.; Neise, W. Active Control to Improve the Aerodynamic Performance and Reduce the Tip Clearance Noise of Axial Turbomachines. In Proceedings of the 11th AIAA/CEAS Aeroacoustics Conference, Monterey, CA, USA, 23–25 May 2005. [CrossRef]

8. Steinetz, B.M.; Taylor, S.; Jay, O.; DeCastro, J.A. *Seal Investigations of an Active Clearance Control System Concept*; NASA/TM—2006-214114; NASA Glenn Research Center: Cleveland, OH, USA, 2006.

9. Sheard, A.G. Blade by Blade Tip Clearance Measurement. *Int. J. Rotating Mach.* **2011**, *2011*, 516128. [CrossRef]

10. Haase, W.C.; Haase, Z.S. High-Speed, capacitance-based tip clearance sensing. In Proceedings of the IEEE Aerospace Conference, Big Sky, MT, USA, 2–9 March 2013; pp. 1–8. [CrossRef]

11. Ye, D.C.; Duan, F.J.; Guo, H.T.; Li, Y.; Wang, K. Turbine blade tip clearance measurement using a skewed dual-beam fiber optic sensor. *Opt. Eng.* **2012**, *51*, 081514. [CrossRef]

12. Sheard, A.; O'Donnell, S.; Stringfellow, J. High Temperature Proximity Measurement in Aero and Industrial Turbomachinery. *J. Eng. Gas Turbines Power* **1999**, *121*, 167–173. [CrossRef]

13. Roeseler, C.; von Flotow, A.; Tappert, P. Monitoring blade passage in turbomachinery through the engine case (no holes). In Proceedings of the IEEE Aerospace Conference, Big Sky, MT, USA, 9–16 March 2002; Volume 6, pp. 3125–3129. [CrossRef]

14. Du, L.; Zhu, X.; Zhe, J. A high sensitivity inductive sensor for blade tip clearance measurement. *Smart Mater. Struct.* **2014**, *23*. [CrossRef]

15. Cao, S.Z.; Duan, F.J.; Zhang, Y.G. Measurement of rotating blade tip clearance with fiber-optic probe. *J. Phys. Conf. Ser.* **2006**, *48*, 873–877. [CrossRef]

16. Vakhtin, A.B.; Chen, S.J.; Massick, S.M. Optical probe for monitoring blade tip clearance. In Proceedings of the 47th AIAA Aerospace Sciences Meeting Including the New Horizons Forum and Aerospace Exposition, Orlando, FL, USA, 5–8 January 2009. [CrossRef]

17. Dhadwal, H.S.; Kurkov, A.P. Dual-Laser Probe Measurement of Blade-Tip Clearance. *ASME J. Turbomach.* **1999**, *121*, 481–485. [CrossRef]

18. Sinha, J.K. *Vibration Engineering and Technology of Machinery*; Springer International Publishing: Cham, Switzerland, 2015.

19. Davidson, D.P.; DeRose, R.D.; Wennerstrom, A.J. The Measurement of Turbomachinery Stator-to-Drum Running Clearances. In Proceedings of the ASME 1983 International Gas Turbine Conference and Exhibit, Phoenix, AZ, USA, 27–31 March 1983; ASME: New York, NY, USA, 1983; Volume 1, p. V001T01A054. [CrossRef]

20. Schicht, A.; Schwarzer, S.; Schmidt, L.P. Tip clearance measurement technique for stationary gas turbines using an autofocusing millimeter-wave synthetic aperture radar. *IEEE Trans. Instrum. Meas.* **2012**, *61*, 1778–1785. [CrossRef]

21. Violetti, M.; Skrivervik, A.K.; Xu, Q.; Hafner, M. New microwave sensing system for blade tip clearance measurement in gas turbines. In Proceedings of the 2012 IEEE Sensors, Taipei, Taiwan, 28–31 October 2012; pp. 1–4. [CrossRef]

22. Szczepanik, R.; Przysowa, R.; Spychała, J.; Rokicki, E.; Kaźmierczak, K.; Majewski, P. *Application of Blade-Tip Sensors to Blade-Vibration Monitoring in Gas Turbines*; INTECH Open Access Publisher: Rijeka, Croatia, 2012.

23. Chivers, J.W.H. Microwave Interferometer. U.S. Patent 4,359,683, 16 November 1982.

24. Woolcock, S.C.; Brown, E.G. Checking the Location of Moving Parts in a Machine. U.S. Patent 4,346,383, 24 August 1982.

25. Kempe, A.; Schlamp, S.; Rósben, T.; Haffner, K. Spatial and Temporal High-Resolution Optical Tip-Clearance Probe for Harsh Environments. In Proceedings of the 13th International Symposium on Applications of Laser Techniques to Fluid Mechanics, Lisbon, Portugal, 26–29 June 2006.

26. García, I.; Zubia, J.; Durana, G.; Aldabaldetreku, G.; Illarramendi, M.A.; Villatoro, J. Optical Fiber Sensors for Aircraft Structural Health Monitoring. *Sensors* **2015**, *15*, 15494–15519. [CrossRef]

27. López-Higuera, J.M. (Ed.) *Handbook of Optical Fibre Sensing Technology*; Wiley: New York, NY, USA, 2002; ISBN 978-0-471-82053-6.

28. García, I.; Beloki, J.; Zubia, J.; Durana, G.; Aldabaldetreku, G. Turbine-blade tip clearance and tip timing measurements using an optical fiber bundle sensor. In Proceedings of the SPIE 87883H Optical Measurement Systems for Industrial Inspection VIII, Munich, Germany, 13 May 2013. [CrossRef]

29. García, I.; Beloki, J.; Zubia, J.; Aldabaldetreku, G.; Illarramendi, M.A.; Jiménez, F. An Optical Fiber Bundle Sensor for Tip Clearance and Tip Timing Measurements in a Turbine Rig. *Sensors* **2013**, *13*, 7385–7398. [CrossRef] [PubMed]

30. Pfister, T.; Büttner, L.; Czarske, J.; Krain, H.; Schodl, R. Turbo machine tip clearance and vibration measurements using a fiber optic laser Doppler position sensor. *Meas. Sci. Technol.* **2006**, *17*, 1693–1705. [CrossRef]

31. Kempe, A.; Schlamp, S.; Rösgen, T.; Haffner, K. Low-coherence interferometric tip-clearance probe. *Opt. Lett.* **2003**, *28*, 1323–1325. [CrossRef] [PubMed]

32. Barranger, J.P.; Ford, M.J. Laser-Optical Blade Tip Clearance Measurement System. *J. Eng. Power* **1981**, *103*, 457–460. [CrossRef]

33. Matsuda, Y.; Tagashira, T. Optical Blade-Tip Clearance Sensor for Non-Metal Gas Turbine Blade. *J. Gas Turbine Soc. Jpn.* **2001**, *29*, 479–484.

34. García, I.; Zubia, J.; Berganza, A.; Beloki, J.; Arrue, J.; Illarramendi, M.A.; Mateo, J.; Vázquez, C. Different Configurations of a Reflective Intensity-Modulated Optical Sensor to Avoid Modal Noise in Tip-Clearance Measurements. *J. Lightwave Technol.* **2015**, *33*, 2663–2669. [CrossRef]

35. Ma, Y.-Z.; Zhang, Y.-K.; Li, G.-P.; Liu, H.-G. *Tip Clearance Optical Measurement for Rotating Blades*; MSIE: Harbin, China, 2011; pp. 1206–1208.

36. García, I.; Przysowa, R.; Amorebieta, J.; Zubia, J. Tip-Clearance Measurement in the First Stage of the Compressor of an Aircraft Engine. *Sensors* **2016**, *16*, 1897. [CrossRef] [PubMed]

37. Xie, S.; Zhang, X.; Wu, B.; Xiong, Y. Output characteristics of two-circle coaxial optical fiber bundle with regard to three-dimensional tip clearance. *Opt. Express* **2018**, *26*, 25244–25256. [CrossRef] [PubMed]

38. Xie, S.; Zhang, X.; Xiong, Y.; Liu, H. Demodulation technique for 3-D tip clearance measurements based on output signals from optical fiber probe with three two-circle coaxial optical fiber bundles. *Opt. Express* **2019**, *27*, 12600–12615. [CrossRef] [PubMed]

39. Philtec RC171. Available online: http://www.webcitation.org/78Bh7gnm9 (accessed on 7 May 2019).

sensors

MDPI

Article

Concrete Crack Detection and Monitoring Using a Capacitive Dense Sensor Array

Jin Yan [1,*], Austin Downey [2], Alessandro Cancelli [1], Simon Laflamme [1], An Chen [1], Jian Li [3] and Filippo Ubertini [4]

[1] Department of Civil, Construction and Environmental Engineering, Iowa State University, 813 Bissell Road, Ames, IA 50011, USA; acancell@iastate.edu (A.C.); laflamme@iastate.edu (S.L.); achen@iastate.edu (A.C.)
[2] Department of Mechanical Engineering, University of South Carolina, 300 Main St, Columbia, SC 29208, USA; austindowney@sc.edu
[3] Department of Civil, Environmental and Architectural Engineering, University of Kansas, Lawrence, KS 66045, USA; jianli@ku.edu
[4] Department of Civil and Environmental Engineering, University of Perugia, 06125 Perugia, Italy; filippo.ubertini@unipg.it
* Correspondence: yanjin@iastate.edu

Received: 22 March 2019; Accepted: 16 April 2019; Published: 18 April 2019

check for
updates

Abstract: Cracks in concrete structures can be indicators of important damage and may significantly affect durability. Their timely identification can be used to ensure structural safety and guide on-time maintenance operations. Structural health monitoring solutions, such as strain gauges and fiber optics systems, have been proposed for the automatic monitoring of such cracks. However, these solutions become economically difficult to deploy when the surface under investigation is very large. This paper proposes to leverage a novel sensing skin for monitoring cracks in concrete structures. This sensing skin is constituted of a flexible electronic termed soft elastomeric capacitor, which detects a change in strain through changes in measured capacitance. The SEC is a low-cost, durable, and robust sensing technology that has previously been studied for the monitoring of fatigue cracks in steel components. In this study, the sensing skin is introduced and preliminary validation results on a small-scale reinforced concrete beam are presented. The technology is verified on a full-scale post-tensioned concrete beam. Results show that the sensing skin is capable of detecting, localizing, and quantifying cracks that formed in both the reinforced and post-tensioned concrete specimens.

Keywords: crack; strain; distributed dense sensor network; structural health monitoring

---

1. Introduction

Cracks that manifest in concrete structures can be caused by a combination of poor construction practices, deleterious chemical reactions such as corrosion and alkali-aggregate reactions, construction overloads, cyclic freezing and thawing damage [1]. Cracks may represent the full extent of the damage or may point to problems of a larger scale. Their gravity depends on the type of structural system and the nature of cracking. If located at critical locations and of significant sizes, these cracks will decrease the capacity of the component and affect the durability and safety of the structure. A survey of cracks generally aids practitioners in evaluating and managing maintenance actions for a given structural system by providing information on the affected area, severity of the cracks, and their possible effect on structural integrity.

Various evaluation techniques can be leveraged during an inspection to determine the location and extent of cracking, and to evaluate the general condition of the concrete. These methods include visual inspections and nondestructive evaluation techniques such as impact-echo [2], ultrasonic [3], acoustic emission [4], and ground penetrating radar [5] methods. The advantage of leveraging nondestructive evaluation techniques during inspections is in their quantitative nature that can validate the subjective judgment of an inspector, but they yet require highly trained agents and expensive equipment. It must also be noted that inspections are typically conducted at fixed intervals, and it follows that they do not guarantee that critical damage will be detected timely.

A solution is the implementation of automated monitoring solutions, also known as structural health monitoring (SHM). Conventional SHM approaches to crack monitoring include resistive strain gauges [6], vibrating wires [7], and linear variable differential transformers [8]. While each technology has demonstrated success in certain conditions, they are limited by relatively small gauge lengths that impede their practicality for the monitoring of large-scale surfaces. Recently, electro-mechanical impedance (EMI) techniques using piezoelectric transducers (PZT) have been studied for real-time crack monitoring and early-damage detection [9]. These sensors exhibit a high sensitivity to crack growth. However, the disadvantage of PZTs is that they have low interface compatibility and poor durability when used for monitoring concrete structures [10]. Fiber optic sensors (FOS) that can be multiplexed for a long distance and immune to electromagnetic interference have gained popularity since the 1990s [11] to map cracks over large areas for both surface [12] and embedded [13] applications. Nevertheless, FOS technologies are still expensive to deploy, can be brittle, are challenging to bond onto surfaces, and embedment is limited to new retrofits and constructions.

Novel surface strain sensing technologies, or sensing skins, with excellent durability, sensitivity, and cost-effectiveness for geometrically large systems, have gained popularity in the research community as an organic step beyond FOS. These include strain sensing sheets based on large area electronics and integrated circuits [14–16], electrical impedance tomography (EIT) [17,18], and multifunctional materials [19,20]. The authors have previously proposed sensing skin technology based on soft elastomeric capacitors (SECs) that act as large-area strain gauges. The SECs offer unique advantages for crack detection and monitoring over traditional sensing technologies due to their low-cost [21], high durability to environmental conditions [22], and mechanical robustness [23]. Previous investigations have experimentally evaluated the feasibility of using SECs for fatigue crack localization and quantification in steel bridges [24].

This study aims at extending previous research efforts on the SECs to the monitoring of cracks in concrete. In particular, the performance of SECs at localizing and assessing flexural crack development in concrete infrastructures through strain measurements is evaluated on a small-scale and a full-scale concrete beam using a network of strip-shaped SECs. The remainder of the paper is organized as follows. Section 2 introduces the sensing principle of SECs and presents a verification of the sensing principle on a small reinforced concrete beam. Section 3 presents and discusses experimental results conducted on a full-scale post-tensioned concrete beam to validate the performance of the technology. Section 4 concludes the paper.

2. Soft Elastomeric Capacitor Technology

This section provides a background on the SEC technology, including its fabrication process and electromechanical model, and presents validation results on a small-scale concrete beam.

2.1. Sensor Fabrication

The SEC is a low-cost, robust, and highly scalable thin-film strain sensor that consists of a flexible parallel plate capacitor. A given change in a monitored surface's geometry (i.e., strain) is transduced into a measurable change in the SEC's capacitance. An SEC is presented in Figure 1 with its key components annotated. The SEC is constituted from a styrene–ethylene/butylene–styrene (SEBS) block copolymer arranged in three layers. The inner layer (dielectric) is filled with titania to increase both its durability and permittivity, while the outer layers (conductors) are filled with carbon black to provide conductivity. The carbon black-filled outer layers also provide enhanced UV light protection, therefore enhancing the sensor's environmental durability [22]. The fabrication process of the SEC is covered in more detail in [23]. An electromechanical model that relates a change in the monitored structure geometry (i.e., strain) to a change in the sensor's capacitance (C) can be derived from the parallel plate capacitor equation:

$$C = e_0 e_r \frac{A}{h} \tag{1}$$

where $e_0 = 8.854$ pF/m is the vacuum permittivity, e_r is the polymer's relative permittivity, $A = d \cdot l$ is the sensor area of width d and length l (as annotated in Figure 1a), and h is the thickness of the dielectric.

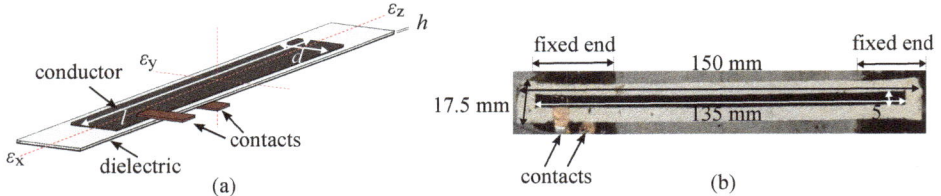

Figure 1. (**a**) Schematic representation of an SEC; (**b**) picture of an SEC with active sensing area measuring 135 mm × 5 mm.

Equation (1) can be specialized for the sensor configuration of interest to this paper, where the sensor is glued at each end and free-standing in the middle, as shown in Figure 1b, undergoing uniaxial strain ($\varepsilon = \varepsilon_x$):

$$\varepsilon = \lambda \frac{\Delta C}{C_0} = \frac{\Delta l}{l_0} \tag{2}$$

where l_0 is the unstrained length of the SEC, C_0 the initial unstrained capacitance, ΔC the incremental change in capacitance and λ the gauge factor. In the sensor configuration of interest, the gauge factor is a function of the sensor geometry. The sensor's general overall dimensions used in this study are 150 mm × 17.5 mm, with the active sensing area measuring 135 mm × 5 mm (Figure 1b). The next subsection characterizes such SEC's response to determine the gauge factor λ.

2.2. Sensor Response Characterization

The electromechanical response of an end-bonded SEC (as shown in Figure 1b) was investigated by applying an axial 0.12 Hz cyclic excitation on a free-standing specimen using a servo-hydraulic testing machine, as shown in Figure 2a. During the test, the SEC's capacitance was recorded at 24 Hz using a custom-built data acquisition device (DAQ), and the displacement response was recorded at 600 Hz from the dynamic testing machine. Figure 2 presents the results of the electromechanical test, comparing the measured strain (black line in Figure 2b) and the corresponding change in capacitance measured by the SEC. Figure 2c reports the change in capacitance as a function of the change in strain. As shown in Figure 2c, the strain and measured change in capacitance have a linear relationship that when fitted with linear least squares regression can be used to obtain the gauge factor $\lambda = 0.78$ over the tested range 0–0.7% strain. Figure 2c shows the capacitance error bound ($\pm 0.00075\ \Delta C/C_0$), equivalent to a resolution of 9.6 $\mu\varepsilon$.

Note that the capacitance is expected vary linearly with temperature and humidity [25]. However, the sensor's response with respect to strain (i.e., gauge factor) will remain constant. A thorough study of environmental effects in terms of sensor weatherability and long-term signal stability could be found in Reference [22].

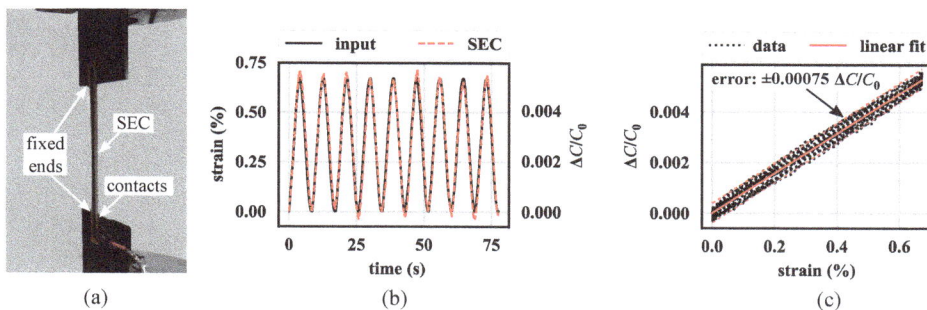

Figure 2. (**a**) Experimental gauge factor characterization test setup; (**b**) capacitance time history response subject to cyclic strain input; and (**c**) sensitivity and linearity of the sensor.

2.3. Small-Scale Prototyping Test

To investigate the feasibility of the proposed approach, an experimental campaign was conducted on two small-scale reinforced concrete beams and results published in a conference proceeding [26]. This subsection presents typical results to validate the sensor's capabilities. The testing specimens were subjected to a three-point bending test to study the detectability of bending cracks using an SEC array. The dimension of the small-scale reinforced concrete beam was 61 cm × 15 cm × 15 cm. The specimen is shown in Figure 3a and was equipped with an array of four sensors identified as SEC A, B, C and D. SECs B and C were both placed at midspan but at different heights to study additional crack assessment capability. The remaining two SECs were placed symmetrically around the midspan.

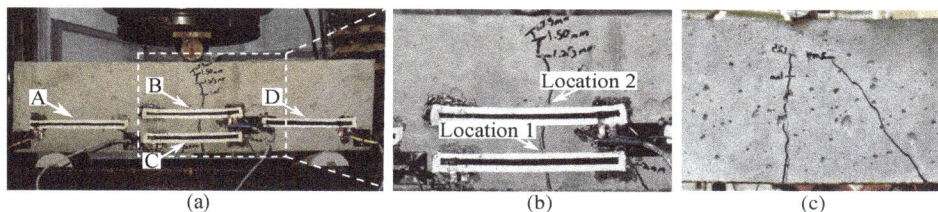

Figure 3. (**a**) Crack patterns on the small-scale reinforced concrete beam; (**b**) an enlarged view of the cracks on the front side of the specimen; and (**c**) an enlarged view of the cracks on the back side of the specimen showing a shear crack.

Figure 3b is a picture of the final crack pattern, and Figure 4 plots a time history of data. Note that the magnitude of strain in Figure 4 is very high, because it represents the strain experienced by the sensor spanning the crack opening. The gray dashed lines in Figure 4 show when the machine paused and resumed to produce incremental loads. The testing machine did not maintain a stable load while paused. This was confirmed through the analysis of the load-displacement curve of the specimen, along with the normalized crack width amplitude obtained by averaging the crack width at the top and bottom edges of the SEC (Figure 3a). For low levels of displacement, a single crack formed initially at the bottom-center of the beam. This single crack formation was confirmed through the slight drop in the load-displacement curve (the blue solid line between the third and fourth gray dashed line from the left in Figure 4a), along with the opening of the crack at mid-span (termed Location 1, illustrated with green dots, and Location 2 on top of Location 1 illustrated with orange triangles in Figure 4a). The flexural crack propagated to the compression zone, with a decrease of stiffness, up to a 2 mm displacement. There is a loss of capacity around 1.9 mm (second blue solid line in Figure 4a) which was produced by a shear crack opening on the back side of the concrete specimen (Figure 3c). From then up to 3.6 mm, the load remained constant. This behavior was almost completely captured by the SEC network installed on the specimen. As shown in Figure 4b, SECs A, B, and D have a decrease in slope that matches the one associated with the formation of the first flexural crack. At the initialization of the backside shear crack, SEC B was able to capture this behavior as a drop in its capacitance followed with an unstable capacitance growth (the second blue solid line from the left in Figure 4b), which may be associated with stress redistribution. One can also observe that SEC C, mounted at the bottom of the tension zone, should have the highest strain value/relative capacitance change, but it has the lowest change in relative capacitance at the initial stage. This could be attributed to strain transfer at the sensor interface caused by the installation procedure, which cannot be quantified directly, but does not significantly affect its capability to measure crack [27].

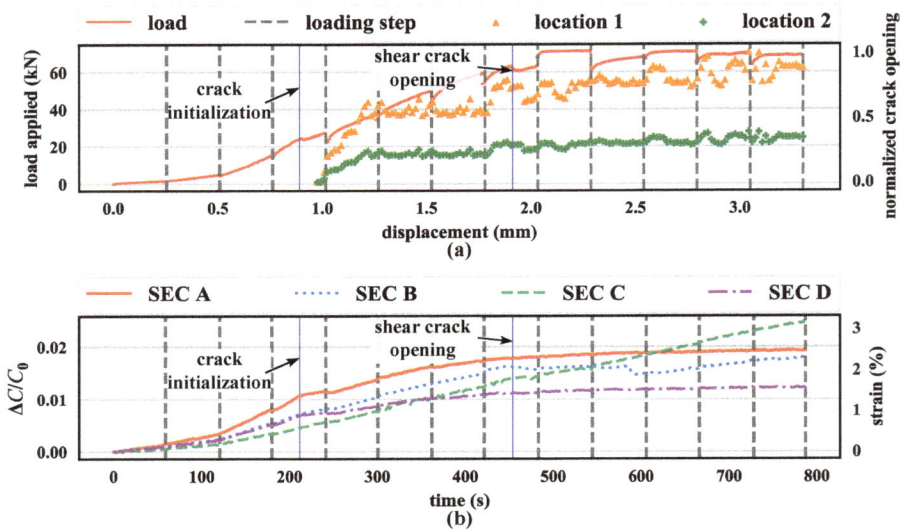

Figure 4. Time series results from the small-scale reinforced concrete beam: (**a**) load-displacement curve and normalized crack opening widths histories; and (**b**) relative capacitance and computed strain histories for all four SECs.

3. Verification on Full-Scale Post-Tensioned Concrete Beam

This section presents and discusses results on an experimental verification conducted on a full-scale post-tensioned concrete beam.

3.1. Experimental Structural System

The full-scale post-tensioned concrete beam of interest is part of a structural system consisting of two parallel full-scale beams connected by a deck. The structural system was tested within the scope of an unrelated research project. The installation of a network of SECs was allowed onto the side surface of one of the beams. During testing, the structural system was subjected to several damage scenarios, which included the removal of the post-tension in one of the girders and a large number of damage states ranging from incipient to severe damages and until failure.

Each beam has a rectangular cross-section of 254 mm × 508 mm, a length of 10 m, and a clear span of 9.6 m, as illustrated in Figure 5. The beam dimensions were designed to limit the first natural frequency of the full system below 12 Hz, which is a typical value for short-span bridges. A 391 kN post-tension force was applied to both beams before the casting of the connecting deck to avoid early cracking under self-weight. A circular plastic duct of 63.5 mm was installed in the cross section with its center located at 178 mm from the bottom face of the beam to accommodate a single post-tension bar of 25.4 mm diameter. The sections were reinforced using 6 reinforcement bars with a diameter of 25.4 mm, with their positions indicated in Figure 5 to provide high ductility of the girders. The deck connecting the two girders had a width of 3 m, thickness of 9 cm, and length equal to that of the beams (10 m). The deck was designed to ensure that cracks would only form on the girders, except around ultimate strength. The deck was reinforced using two layers of reinforcement bars of 12.7 mm diameter under both the bottom and top surfaces. The beams were positioned at 1.4 m from each other, leaving an overhang of 70 cm on each side as shown in Figure 5b. The structural system was loaded using hydraulic actuators (Figure 5b) installed over a beam transmitting the actuator force to the beams' centerline. The girders and the deck were casted in Iowa State University's Structures Laboratory using a self-compacting concrete with specified compressive strength of 41 MPa and 28 MPa, respectively. During tests, the measured compressive strength for the girders and deck were 48 MPa and 28 MPa, respectively.

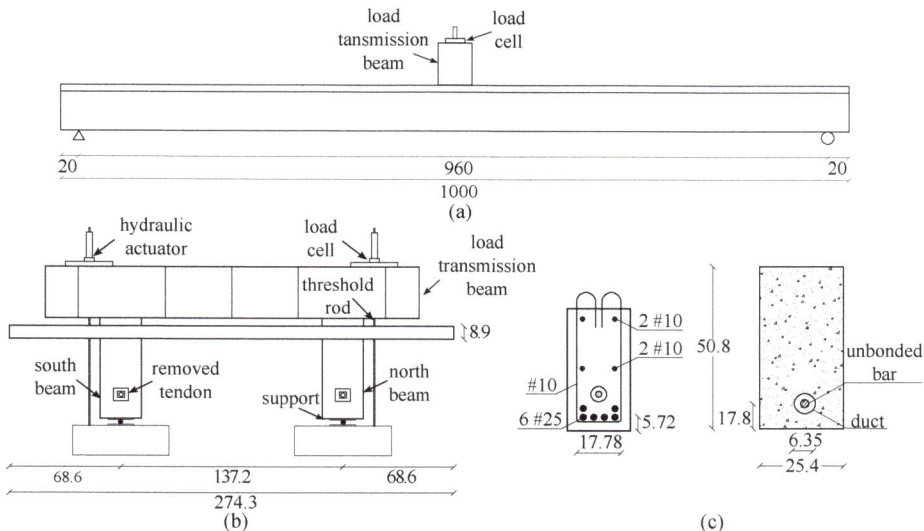

Figure 5. Detailing of the tested specimen: (**a**) elevation view; (**b**) typical cross-section; and (**c**) reinforcement distribution and cross-section design of a beam (dimensions are in cm).

The experiment was conducted in two sequential phases. In the first phase, the prestress was released at one side of the beam (Figure 6) to provide differential damage. In the second phase, a load was applied using the pair of hydraulic force actuators installed on top of load transmission beam (Figure 6), and a load cell was used to constantly monitor the applied load. Loading and unloading sequences were designed to generate damage to the beams in increasing severity. Visual observations were conducted during the unloading phases. Figure 7 is a plot of the loading history.

3.1.1. Dense Sensor Network Instrumentation Strategy

In this investigation, an SEC array was instrumented on a post-tensioned concrete beam which post-tension was released. A total of 20 SECs were placed in order to cover 2.84 m of the beam centered around the beam's midspan to study the evolution of spatial crack distribution. Based on the validation results from the small-scale beams (Section 2), the array was designed in an overlapping staggered pattern, as shown in Figure 8, to improve the probability of detection of all cracks. SECs were placed at 127 mm and 76 mm along the vertical direction from the bottom surface of the beam with a 280 mm spacing. Each SEC was pre-stretched and affixed onto the concrete substructure at two ends using a thin layer of an off-the-shelf epoxy (JB Kwik). It follows that the effective strain-sensitive portion of each sensor was narrowed to the unglued section. The numbering scheme of the SECs is shown in Figure 8. Capacitance data collected at 24 Hz using a customized DAQ (Figure 8) driven by a LabVIEW code.

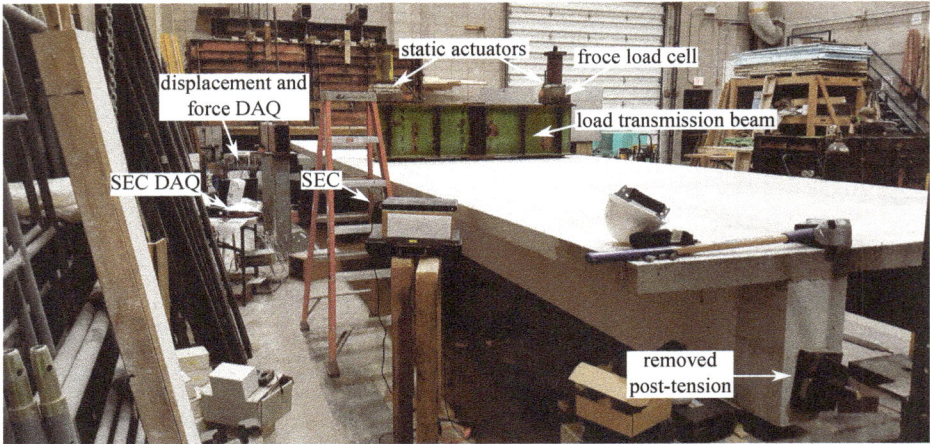

Figure 6. Picture showing the experimental setup of the large-scale post-tension beam test setup.

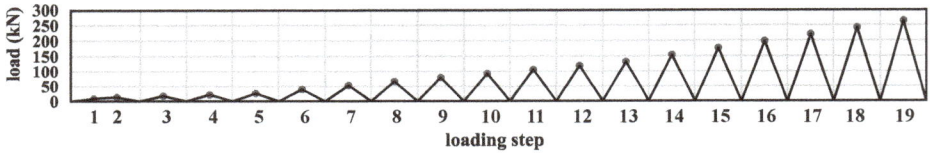

Figure 7. Loading-unloading sequence for the load test.

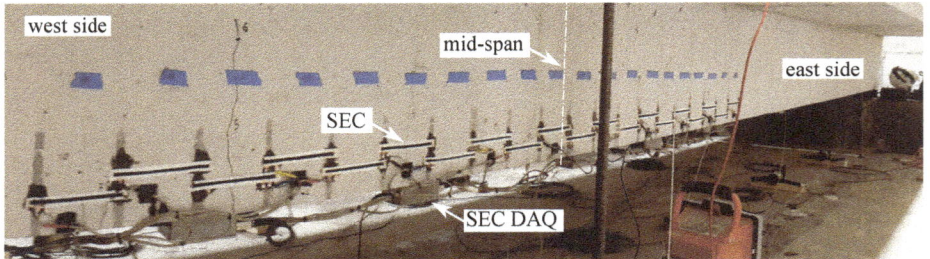

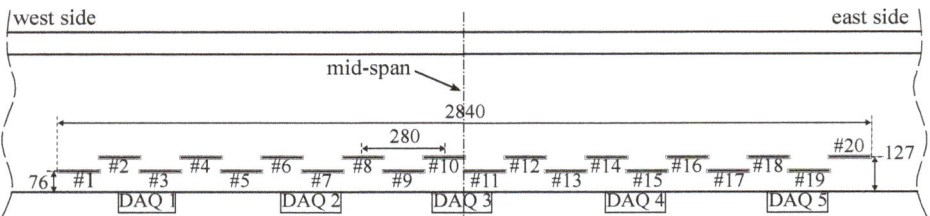

Figure 8. Sensor instrumentation around the midspan of the beam: (**a**) picture; and (**b**) schematic of sensor locations, partial beam elevation (dimensions are in mm).

3.2. Results and Discussion

Figure 9 is a picture showing the crack pattern at the end of loading step 19, with visible cracks traced using a black marker. The beam experienced uniformly distributed transverse and shear cracks at both two ends. The area under the loading point experienced shorter cracks, which is as expected, given the pressure from the flange restraining the crack growth. Cracks were observed in the loading zone under SEC #9, and #12 after loading step 2, under SECs #6, #7, #15, #16, #18, and #20 after loading step 3, and under SEC #1, #3 after loading step 4. Subsequent loading steps 5 to 17 induced a uniform formation and growth of flexural cracks along the span of the girder. Rapid formation of shear cracks crossing different SECs at both ends, from SEC #1 to #7 and from SEC #16 to #20, occurred after loading step 18 and 19.

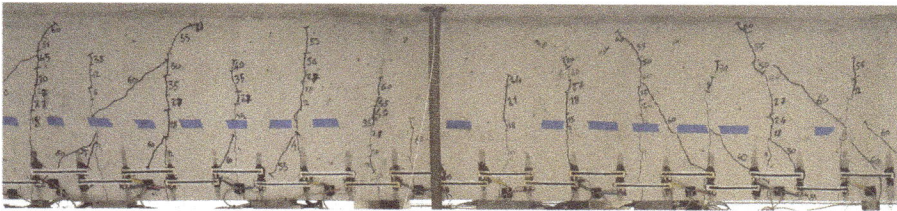

Figure 9. Picture of crack pattern before reaching ultimate strength after loading step 19.

Figure 10 plots the time evolution of the relative changes in capacitance $\Delta C/C_0$ during the loading test, where the gray dashed lines indicate when the machine paused after each incremental loading step was produced. Negative $\Delta C/C_0$ values indicate compression, while conversely a positive value indicates tension. Results are presented after the replacement of outliers with averaged values and the application of a low-pass Butterworth filter with a cutoff frequency of 2 Hz. SECs #16, #19 and #20 experienced corrupted data after a crack formed under the epoxy adhering the sensor to the beam.

SEC #8 exhibited a negative capacitance change and an inverse loading-unloading shape. This disagreement between the capacitance change and loading was caused by localized compressive strains induced by the splitting of the specimen along the flexural crack. This splitting in the specimen caused the right-hand-side portion of the crack under SEC #7 to move towards the right, thus resulting in compressive loading of the concrete under SEC #8. This behavior was confirmed when the flexural crack opened under SEC #9 at loading step 8, at which point the amplitude of the maximum cracks began to increase significantly. A similar, yet smaller in magnitude, compression behavior was observed from SEC #17 that was located between two cracks. Under loading step 5, the relative capacitance did not change significantly and the compressive effect induced from the flange at the center of the beam was captured by most SECs. The loading-unloading patterns in relative capacitance changes were observable in SEC readings after a crack formed under that particular SEC.

In order to evaluate the performance of the SECs at quantifying crack openings, the relationship between crack growth and relative capacitance change was investigated. Three features were extracted from time series data to associate with crack length: (1) maximum relative change in capacitance, taken as the peak-to-peak amplitude in signal for each load step; (2) residual relative change in capacitance, taken as the difference between the maximum reading during a given load step and the capacitance left after unloading; and (3) average relative change in capacitance, taken as the average change in capacitance over a loading step. Figure 11 plots the maximum relative change in capacitance (green dotted line), residual relative change in capacitance (orange dashed line), average relative change in capacitance (gray line) against crack growth (black dots) from each loading step (L1 to L19) and SEC. The red dashed lines indicate the visible shear cracks initiation underneath the corresponding SECs. The crack length was measured as a straight line between the extremities of a crack, and normalized by the height of the web. An agreement is observed between the three features and crack growth.

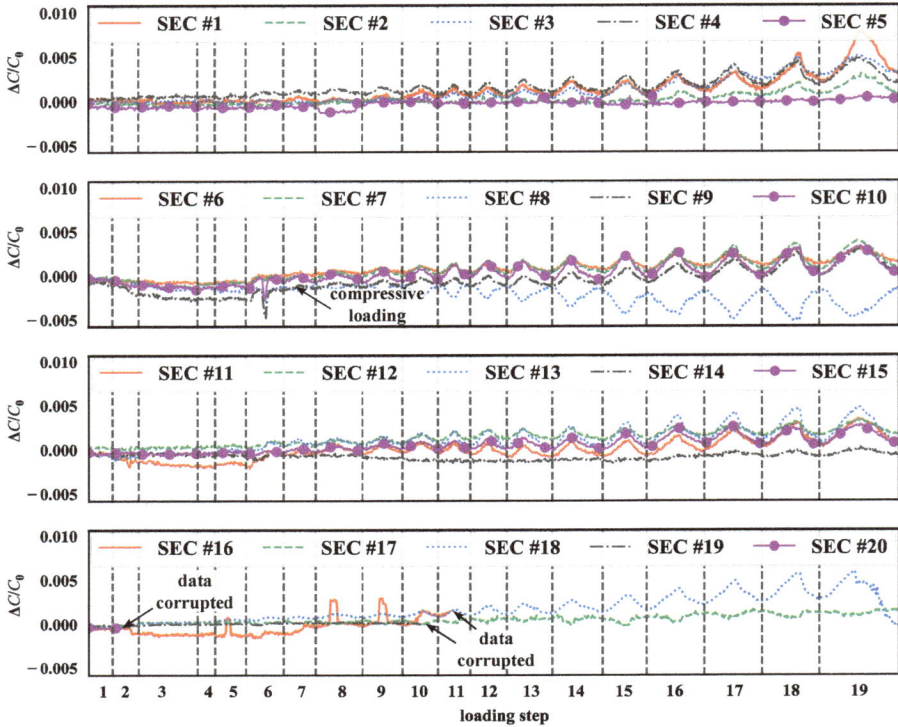

Figure 10. Time histories of strain measurements for all SECs.

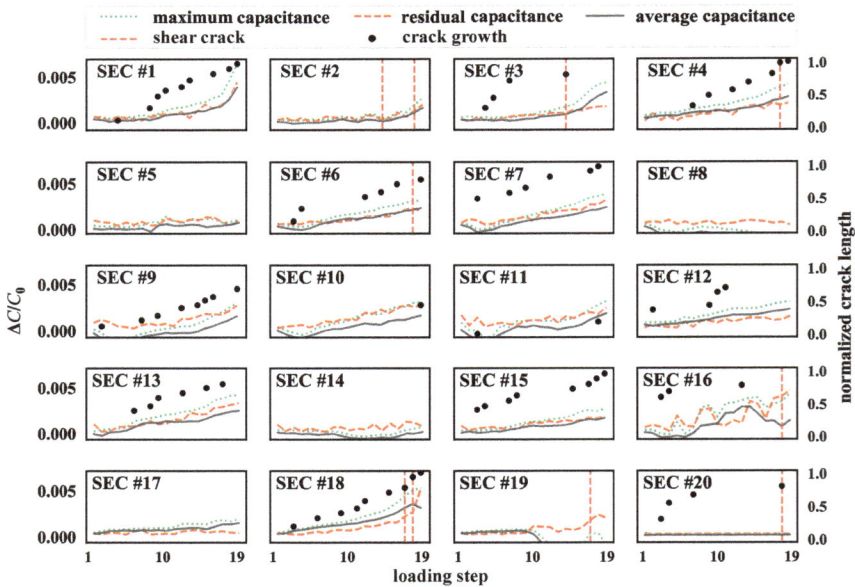

Figure 11. Comparison of maximum, residual, and average relative change in capacitance measured by the SECs and normalized crack growth length under each loading step (loading step 1 to 19).

Figure 12 illustrates a compilation of the normalized crack lengths and the feature "maximum relative change in capacitance" extracted from each loading step for all the SECs. It can be observed that the feature is generally consistent with the location of a crack and its normalized length. At small strain (low loadings), this relationship is harder to distinguish. A significant change in the feature is associated with a rapid growth of the shear crack and final fracture, shown as tall red bars at the two corners of Figure 12b (SEC #1 and #18). Both small and negative changes in the feature indicate that no crack formed.

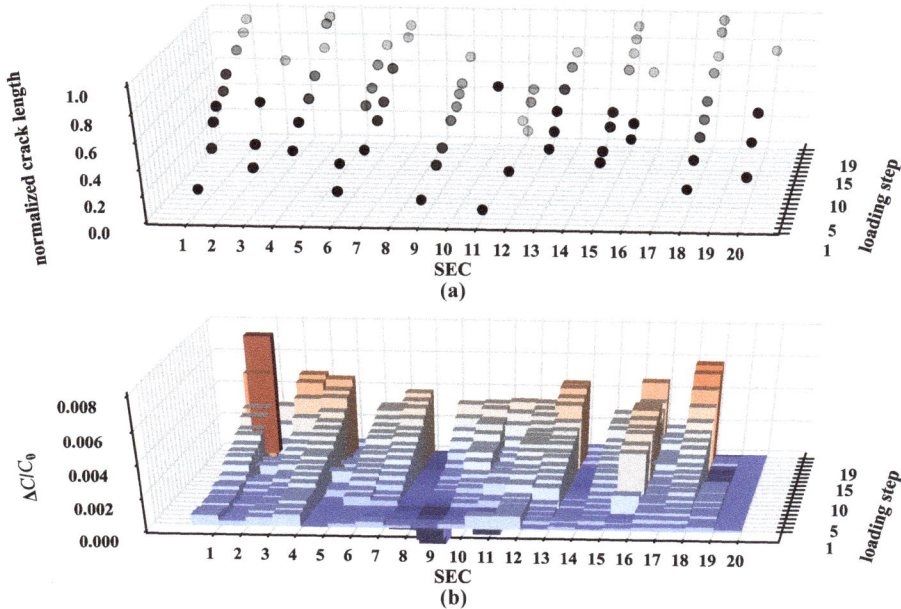

Figure 12. (a) Normalized crack length as a function of SEC location and loading step; (b) maximum relative change in capacitance change as a function of SEC location and loading step.

4. Conclusions

This paper presented the study of a novel sensing skin for the detection, localization, and quantification of cracks in concrete. The sensing skin, constituted from an array of soft elastomeric capacitors (SECs), is an inexpensive, durable, and robust sensing solution that can be leveraged to measure strain over large surfaces. The strain measurement values are collected in the form of discrete point values among the network. The spatio–temporal comparison of strains enables the detection, localization, and quantification of cracks.

The sensing skin was first introduced and validation results preliminarily conducted on small-scale reinforced concrete beams were presented. Initial characterization of a free-standing SEC led to a gauge factor that was used to map electrical signal to strain and thus, cracks. Time series measurements from the SECs and visual observations from crack growth were in agreement. The sensing capability was further studied by deploying a sensor network of 20 SECs onto the surface of a full-scale post-tensioned concrete beam. A bending test with a loading and unloading sequence was conducted until structural failure, and data from the SECs and visual observation of cracks collected. Results demonstrated that data collected from the distributed SEC network correlated with crack-induced damage. The extraction of time series features, among which the maximum relative change in capacitance at each loading step, showed good agreement with the observed normalized crack length.

Overall, the SEC network showed capable of detecting, localizing, and quantifying cracks in concrete. The application of dense networks of SECs could provide a cost-effective monitoring solution for real-time, long-term crack monitoring on civil structures. Future research will include the influence of the gauge length on the accuracy of SEC's final configuration, bonding sensitivity between the concrete and sensing materials, optimal dense sensor network design for multi-crack detection, and crack initialization characterization algorithms.

Author Contributions: J.Y. performed the data analysis and developed the data processing techniques, performed experimental activities and led the write-up; A.D. designed the experiments, established the experimental protocol, and led the experimental activities; A.C. (Alessandro Cancelli) designed the large-scale test specimen and performed experimental activities; S.L. led the investigation; A.C. (An Chen), J.L. and F.U. provided scientific guidance; all authors proof-read the manuscript.

Funding: This work is partially funded by the Air Force of Scientic Research (AFOSR) under award number FA9550-17-1-0131, the American Society for Nondestructive Testing, and by USDOT pooled fund study TPF-5(328) that includes the following participating state Departments of Transportation (DOTs): Iowa, Kansas, Minnesota, North Carolina, Pennsylvania, Texas, and Oklahoma. Their support is gratefully acknowledged.

Conflicts of Interest: The authors declare no conflict of interest.

References

1. Tang, S.; Yao, Y.; Andrade, C.; Li, Z. Recent Durability Studies on Concrete Structure. *Cem. Concr. Res.* **2015**, *78*, 143–154. [CrossRef]
2. Zhang, J.K.; Yan, W.; Cui, D.M. Concrete Condition Assessment Using Impact-Echo Method and Extreme Learning Machines. *Sensors* **2016**, *16*, 447. [CrossRef]
3. Hwang, E.; Kim, G.; Choe, G.; Yoon, M.; Gucunski, N.; Nam, J. Evaluation of Concrete Degradation Depending on Heating Conditions by Ultrasonic Pulse Velocity. *Constr. Build. Mater.* **2018**, *171*, 511–520. [CrossRef]
4. ElBatanouny, M.K.; Ziehl, P.H.; Larosche, A.; Mangual, J.; Matta, F.; Nanni, A. Acoustic Emission Monitoring for Assessment of Prestressed Concrete Beams. *Constr. Build. Mater.* **2014**, *58*, 46–53. [CrossRef]
5. Eisenmann, D.; Margetan, F.J.; Koester, L.; Clayton, D. Inspection of a Large Concrete Block Containing Embedded Defects Using Ground Penetrating Radar. *AIP Conf. Proc.* **2016**, *1706*, [CrossRef]
6. Gribniak, V.; Arnautov, A.K.; Kaklauskas, G.; Tamulenas, V.; Timinskas, E.; Sokolov, A. Investigation on Application of Basalt Materials as Reinforcement for Flexural Elements of Concrete Bridges. *Balt. J. Road Bridge Eng.* **2015**, *10*, 201–206. [CrossRef]
7. Park, H.; Shin, Y.; Choi, S.; Kim, Y. An Integrative Structural Health Monitoring System for the Local/Global Responses of a Large-Scale Irregular Building under Construction. *Sensors* **2013**, *13*, 9085–9103. [CrossRef]
8. Zhou, J.; Xu, Y.; Zhang, T. A Wireless Monitoring System for Cracks on the Surface of Reactor Containment Buildings. *Sensors* **2016**, *16*, 883. [CrossRef] [PubMed]
9. Yu, L.; Tian, Z.; Ziehl, P.; ElBatanouny, M. Crack Detection and Evaluation in Grout Structures with Passive/Active Methods. *J. Mater. Civ. Eng.* **2016**, *28*, 04015168. [CrossRef]
10. Han, B.; Yu, X.; Ou, J. *Self-Sensing Concrete in Smart Structures*; Butterworth-Heinemann: Oxford, UK, 2014.
11. Glisic, B. *Fibre Optic Methods for Structural Health Monitoring*; John Wiley and Sons: Chichester, UK, 2007.
12. Barrias, A.; Casas, J.; Villalba, S. A Review of Distributed Optical Fiber Sensors for Civil Engineering Applications. *Sensors* **2016**, *16*, 748. [CrossRef]
13. Barrias, A.; Casas, J.; Villalba, S. Embedded Distributed Optical Fiber Sensors in Reinforced Concrete Structures—A Case Study. *Sensors* **2018**, *18*, 980. [CrossRef] [PubMed]
14. Tung, S.T.; Yao, Y.; Glisic, B. Sensing Sheet: The Sensitivity of Thin-Film Full-Bridge Strain Sensors for Crack Detection and Characterization. *Meas. Sci. Technol.* **2014**, *25*, 075602. [CrossRef]
15. Zhang, B.; Wang, S.; Li, X.; Zhang, X.; Yang, G.; Qiu, M. Crack Width Monitoring of Concrete Structures Based on Smart Film. *Smart Mater. Struct.* **2014**, *23*, 045031. [CrossRef]
16. Zhou, Z.; Zhang, B.; Xia, K.; Li, X.; Yan, G.; Zhang, K. Smart Film for Crack Monitoring of Concrete Bridges. *Struct. Health Monit.* **2010**, *10*, 275–289. [CrossRef]
17. Hallaji, M.; Seppänen, A.; Pour-Ghaz, M. Electrical Impedance Tomography-Based Sensing Skin for Quantitative Imaging of Damage in Concrete. *Smart Mater. Struct.* **2014**, *23*, 085001. [CrossRef]

18. Gupta, S.; Gonzalez, J.G.; Loh, K.J. Self-Sensing Concrete Enabled by Nano-Engineered Cement-Aggregate Interfaces. *Struct. Health Monit.* **2016**, *16*, 309–323. [CrossRef]

19. Downey, A.; D'Alessandro, A.; Ubertini, F.; Laflamme, S. Automated Crack Detection in Conductive Smart-Concrete Structures Using a Resistor Mesh Model. *Meas. Sci. Technol.* **2018**, *29*, 035107. [CrossRef]

20. Schumacher, T.; Thostenson, E.T. Development of Structural Carbon Nanotube-Based Sensing Composites for Concrete Structures. *J. Intell. Mater. Syst. Struct.* **2013**, *25*, 1331–1339. [CrossRef]

21. Downey, A.; Laflamme, S.; Ubertini, F. Experimental wind tunnel study of a smart sensing skin for condition evaluation of a wind turbine blade. *Smart Mater. Struct.* **2017**. [CrossRef]

22. Downey, A.; Pisello, A.L.; Fortunati, E.; Fabiani, C.; Luzi, F.; Torre, L.; Ubertini, F.; Laflamme, S. Durability Assessment of Soft Elastomeric Capacitor Skin for SHM of Wind Turbine Blades. *Proc. SPIE* **2018**, *10599*, 105991J. [CrossRef]

23. Laflamme, S.; Kollosche, M.; Connor, J.J.; Kofod, G. Robust Flexible Capacitive Surface Sensor for Structural Health Monitoring Applications. *J. Eng. Mech.* **2013**, *139*, 879–885. [CrossRef]

24. Kong, X.; Li, J.; Collins, W.; Bennett, C.; Laflamme, S.; Jo, H. Sensing Distortion-Induced Fatigue Cracks in Steel Bridges with Capacitive Skin Sensor Arrays. *Smart Mater. Struct.* **2018**, *27*, 115008. [CrossRef]

25. Harrey, P.; Ramsey, B.; Evans, P.; Harrison, D. Capacitive-type Humidity Sensors Fabricated Using the Offset Lithographic Printing Process. *Sens. Actuators B Chem.* **2002**, *87*, 226–232. [CrossRef]

26. Yan, J.; Austin Downey, A.C.S.L.; Chen, A. Detection and Monitoring of Cracks in Reinforced Concrete Using an Elastic Sensing Skin. *Struct. Congr.* **2019**, preprint.

27. Gerber, M.; Weaver, C.; Aygun, L.; Verma, N.; Sturm, J.; Glišić, B. Strain Transfer for Optimal Performance of Sensing Sheet. *Sensors* **2018**, *18*, 1907. [CrossRef] [PubMed]

sensors

MDPI

Article

Research on Damage Detection of a 3D Steel Frame Model Using Smartphones

Botao Xie [1,2], Jinke Li [1,2] and Xuefeng Zhao [1,2,*]

[1]   School of Civil Engineering, Dalian University of Technology, Dalian 116024, China;
      botaoxie@mail.dlut.edu.cn (B.X.); jinkeli@mail.dlut.edu.cn (J.L.)
[2]   State Key Laboratory of Coastal and Offshore Engineering, Dalian University of Technology, Dalian 116024, China
[*]   Correspondence: zhaoxf@dlut.edu.cn; Tel.: +86-411-8470-6261

Received: 7 January 2019; Accepted: 9 February 2019; Published: 12 February 2019

check for updates

Abstract: Smartphones which are built into the suite of sensors, network transmission, data storage, and embedded processing capabilities provide a wide range of response measurement opportunities for structural health monitoring (SHM). The objective of this work was to evaluate and validate the use of smartphones for monitoring damage states in a three-dimensional (3D) steel frame structure subjected to shaking table earthquake excitation. The steel frame is a single-layer structure with four viscous dampers mounted at the beam-column joints to simulate different damage states at their respective locations. The structural acceleration and displacement responses of undamaged and damaged frames were obtained simultaneously by using smartphones and conventional sensors, while the collected response data were compared. Since smartphones can be used to monitor 3D acceleration in a given space and biaxial displacement in a given plane, the acceleration and displacement responses of the Y-axis of the model structure were obtained. Wavelet packet decomposition and relative wavelet entropy (RWE) were employed to analyze the acceleration data to detect damage. The results show that the acceleration responses that were monitored by the smartphones are well matched with the traditional sensors and the errors are generally within 5%. The comparison of the displacement acquired by smartphones and laser displacement sensors is basically good, and error analysis shows that smartphones with a displacement response sampling rate of 30 Hz are more suitable for monitoring structures with low natural frequencies. The damage detection using two kinds of sensors are relatively good. However, the asymmetry of the structure's spatial stiffness will lead to greater RWE value errors being obtained from the smartphones monitoring data.

Keywords: damage detection; smartphones; steel frame; shaking table tests; wavelet packet decomposition

1. Introduction

Structural health monitoring (SHM) technology has been successfully implemented to many structural applications and could be used to understand the environmental actions, loads, and behaviors of a structure subjected to various actions through solving a reverse problem, as well as to deduce structural safety and damage information using a measured structural response [1–6]. With the development of structural health monitoring, wireless sensing technology that can wirelessly collect and transmit data has received more and more attention and will become an inevitable trend [7–10]. However, the form and configuration of conventional wireless sensors mainly include a data acquisition module, a microprocessor module, a wireless communication module, and an energy module. These sensors employ the collaborative work of each module to complete the monitoring task; however, the professional standards of these numerous modules and operations also limit the generalization of their application. For this reason, the development of smartphones with a variety of different sensors, highly integrated processors, efficient network communication technology, data storage, and a large number of users provides the possibility for the comprehensive application of such smartphones to SHM.

A milestone in the development of smartphones was the first iPhone (iPhone 2G) launched by Apple in 2007 [11]. Since then, with the embedment of a variety of different sensors into one single device to enable more applications, smartphones have been applied to human health monitoring [12], vehicle maintenance services [13], motion recognition [14], seismic sensing [15,16], and optical biosensing [17]. The development of smartphones also offers opportunities for SHM [18,19]. Yu and Zhao [18] proposed the concept for SHM using smartphones in civil infrastructures. Morgenthal and Höpfner [20,21] investigated the potential use of smartphones for monitoring transient displacement. They also studied the possibilities and limitations of using smartphones for measuring oscillations. Cimellaro et al. [22] developed a system for the rapid assessment of damage using smartphones and the Web to collect information on damaged houses. Sharma and Arabinda [23] presented some civil engineering projects which can use smartphones to improve accuracy and efficiency. Akinwande et al. [24] developed a system for real-time pothole detection and traffic monitoring using smartphones and the Machine learning method. Feng et al. [25] and Ozer et al. [26] developed a crowdsourcing platform for SHM and a post-event damage assessment app. Zhao et al. [27–29] developed acceleration monitoring software, Orion-CC, and displacement monitoring software, D-viewer, which were applied for SHM of the Xinghai Bay Cross-Sea Bridge. Peng et al. [30] developed the smartphone software E-Explorer using mobile Bluetooth communication technology to realize information delivery in emergency events such as earthquakes. Min et al. [31] developed dynamic displacement monitoring software, RINO, which shows comparable accuracy with conventional laser displacement sensor under experimentally validated data. Ozer et al. [32] used collocated smartphone cameras and accelerometers to monitor a small-scale multistory laboratory model's displacement and acceleration responses and compared this with conventional sensors. Kong et al. [33] used smartphones to monitor two directions' acceleration responses of a building so as to demonstrate the potential usage as a way to monitor health states of buildings. As can be seen from the aforementioned studies, the research of SHM based on smartphones has developed rapidly and received considerable attention from various countries. Nevertheless, these research results are mainly concentrated on large structures such as bridges. Research on the use of smartphones to monitor building response under extreme events such as earthquakes remains limited and is thus the focus of this study.

This paper applied smartphones and conventional sensors to monitor a three-dimensional (3D) steel frame which was subjected to earthquake excitation on a shaking table. The structural acceleration and displacement responses were monitored and compared separately. Smartphones can also be used to monitor the three-dimensional acceleration in a given space and the biaxial displacement in a given plane; therefore, the acceleration and displacement responses of the Y-axis of the steel frame were obtained. Based on wavelet packet decomposition, the acceleration data monitored by smartphones and conventional sensors when the structure is in undamaged and different damaged states were analyzed. The relative wavelet entropy (RWE) was used to evaluate the damage of the structure.

2. Experimental Details

2.1. Monitoring Software on Smartphones

The development of two smartphone apps by previous studies, namely Orion-CC [27] and D-viewer [29], were used to monitor the structural responses. These mobile apps are built for an iOS 7.0 or higher platform and are currently available for free on the iTunes Store.

Orion-CC can access the internal accelerometer and gyroscope of an Apple iPhone, which can obtain not only acceleration and angle data, however also cable force. In this work, the smartphones which were preloaded with Orion-CC were used to monitor the acceleration response of the steel frame.

D-viewer is an app for monitoring structural dynamic displacement, using the built-in camera and the aid of a laser pointer to track and recognize a moving spot in order to determine relative displacement. When we used D-viewer, it first processed the color images captured by the camera into gray images and then processed the gray images into binarization images. A black circle printed on white paper was needed for purposes of calibration. The diameter of the black circle as the calibration reference could be set to determine the ratio between the actual size and the pixel size of the black circle in advance. During the monitoring process, D-Viewer monitors the coordinates of the laser spot centroid and displays the X and Y coordinate values of the laser spot in real time while also storing the laser spot coordinate values of each frame image in real time.

2.2. Steel Frame Details

The structure used in this test was a single-layer three-dimensional steel frame, which was connected by two planar steel frames (frame 1 and frame 2) through two rigid beams to form an integral frame. The experimental schematic illustration is shown in Figure 1. The frame was 500 mm in length, 400 mm in width, and 400 mm in height, and each beam-column joint was installed with a viscous damper to simulate the damage of the frame (highlighted by the green circle). A removable rigid beam was installed in front of the damper (highlighted by the yellow box). When the beams were installed, the rotary damper did not function, so the structure was in an elastic state. However, because the mass of the rigid beam is about 0.75 kg and the overall mass of the structure is about 30 kg, once the rigid beam was removed, the structural mass' loss was small, however the stiffness was greatly reduced, the dampers were rotated, and damping of the structure could be gradually increased. Therefore, in this study, removing the rigid beam was employed to simulate the structural damage states. The basis of the structure was two rigid plates that were bolted onto a dual-axis XY Shake Table III that can support and excite loads up to 100 kg. The connection was fixed to ensure that the frame did not undergo horizontal shear translation during the vibration process.

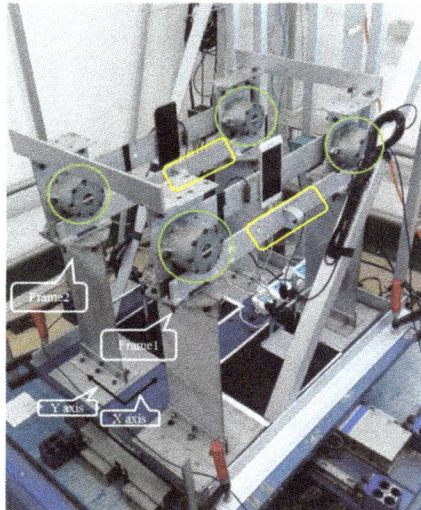

Figure 1. Schematic illustration of the test frame model.

2.3. Instrumentation Layout

The sensor device in the steel frame is shown in Figure 2. Two uniaxial piezoelectric accelerometers (PAs) were respectively mounted on the viscous dampers' articulated beams of frame 1 and frame 2, wherein the PAs were bonded to the magnetic bearings which were instrumented on the rear plane of the articulated beams. The articulated beams were firmly connected so that the PAs and smartphones would be able to measure the acceleration responses of the test structure. The sensitivity of the accelerometer was 500 mV/g with a maximum output voltage of 6 V. In the front plane of the damper articulated beams, two smartphones (SP1 and 2) which were pre-installed with Orion-CC were attached to monitor the acceleration response of the structure and the results were compared with those of the piezoelectric accelerometers.

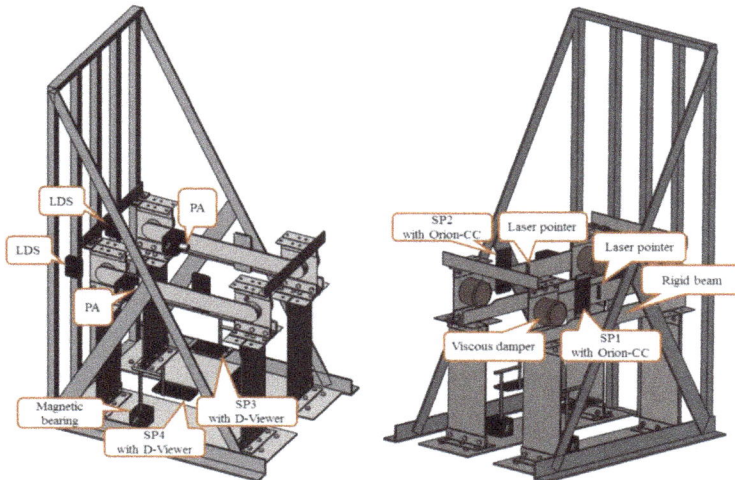

Figure 2. The various sensors instrumented in different locations in the steel frame testbed are shown.

The triangular support frame was also attached to the shaking table so that it could move with the table. Two laser displacement sensors (LDSs) were mounted on the triangular support frame's vertical columns and it was ensured that the outer planes of the two vertical columns were in the same planes as the two magnetic bearings on the rear plane of the damper articulated beams of frame 1 and frame 2. Then, the laser spot could be launched into the center of the magnetic bearing to monitor the relative displacement of the structure. The LDSs had a measuring range of 160 to 450 mm and an output voltage of 0 to 5V. Meanwhile, two smartphones (SP3 and 4), which were pre-installed with the D-Viewer software, were fixed at the bottom of frame 1 and frame 2, and the laser pointers were fixed on the damper articulated beams as well. When the frame began to vibrate, the laser spot that was projected on the pedestal moved laterally. Each smartphone's camera recorded images of the laser spot and analyzed the position of the spot in each image to obtain the relative displacement of the frame. Finally, both the PAs and LDSs were connected to a data acquisition (DAQ) system.

2.4. Test Plan and Damage Cases

Four cases were considered in this study. The first was the undamaged case—at this point in time, the tops of the two columns of frame 1 and frame 2 were mounted with beams so that the beam-column connections were considered to be rigid and the dampers did not work. The second was the damaged 1 case, wherein the rigid beam of frame 1 was removed. The viscous dampers could be rotated to affect the structural response, and the elastoplastic behavior of frame 1 was simulated. The third was the damaged 2 case, wherein the rigid beam of frame 2 was removed and it also simulated the elastoplastic behavior of frame 2. In the last case, the rigid beams of frame 1 and frame 2 were removed simultaneously. Table 1 shows the information of all the damage cases.

Table 1. Damage cases of the test used in this study.

Damage Case	State of the Structure
Undamaged	Rigid beams were mounted
Damaged 1	Rigid beam of frame 1 was removed
Damaged 2	Rigid beam of frame 2 was removed
Damaged 3	Rigid beams of frame 1 and frame 2 were removed simultaneously

The M_w 6.7 Northridge earthquake occurred on 17 January, 1994 in the San Fernando Valley, southern California and caused a large number of structural collapses and casualties [34]. Since then, many researchers have regarded this seismic wave as one of the input waves for structural seismic resistance. In this paper, the Northridge wave was employed as input, representing earthquake excitation. Since the magnitude of the input excitation was differentiated according to the peak displacement of the seismic waves, the excitation in this study was divided into two working conditions—one peak of 1 cm and another of 2 cm. The characteristics of the seismic waves are shown in Table 2, and Figure 3 shows the waveform of the seismic wave.

Table 2. Earthquake records.

Earthquake (abbr.)	Peak Displacement (cm)	Direction of Excitation
Northridge (Nr)	1	Unidirectional
	2	Unidirectional

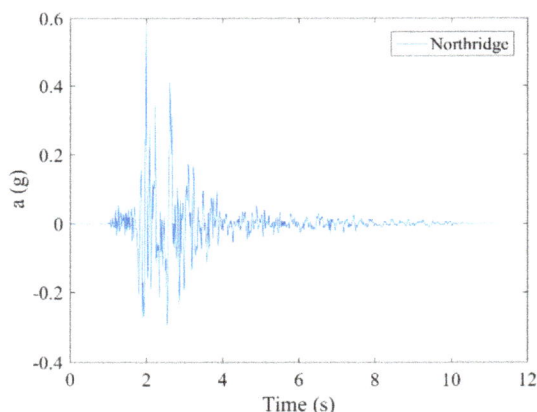

Figure 3. Schematic illustration of an earthquake wave.

3. Steel Frame Response Comparison

3.1. Acceleration Time-History Comparison

In this study, the PAs and the SPs were used to monitor the top acceleration responses of frame 1 and frame 2, and the monitoring results were compared. Due to space limitations, Figure 4 shows the results of the comparative analysis, which were used in the acceleration data of frame 1 and 2 in the different damage cases under the excitation of Nr-1 cm. As can be seen in Figure 4, the acceleration data obtained by the SPs and the PAs are well matched with each other. At the same time, by performing fast Fourier transform (FFT) on the time domain data, the power spectral density (PSD) function corresponding to the case shown in Figure 4 was obtained, as shown in Figure 5. It can be seen that the frequency domain results also have a good match. These results show that when the structure is subjected to extreme events such as earthquakes, the structure can be monitored for vibration using smartphones.

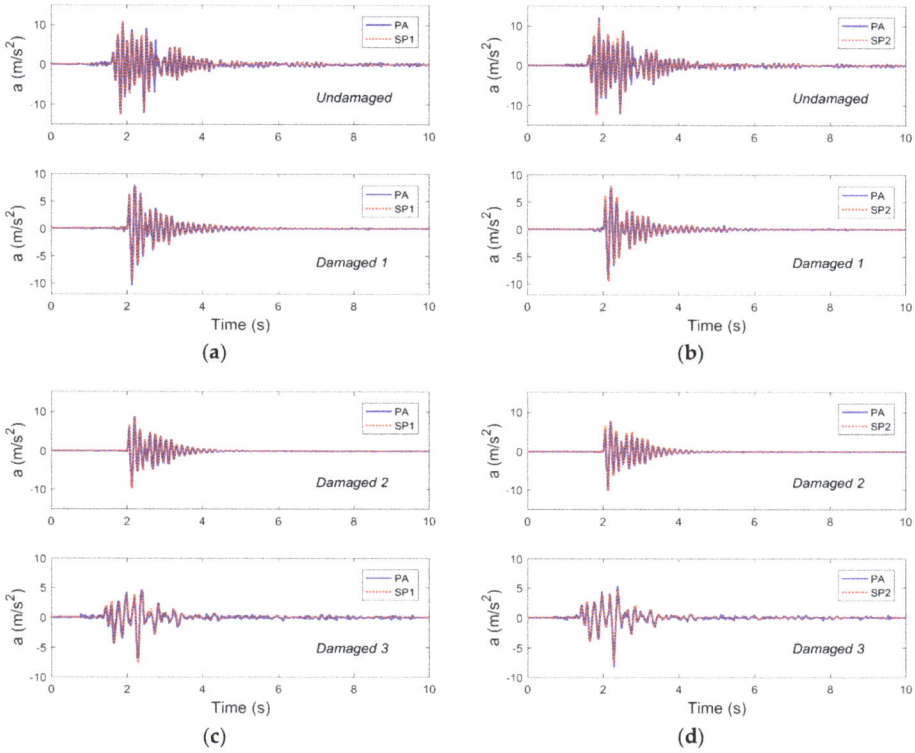

(a)   (b)

(c)   (d)

Figure 4. The acceleration time-history responses, as collected by the piezoelectric accelerometers (PAs) and smartphones (SPs) for four damage cases subjected to Nr-1 cm excitation, are compared. Representative results from (**a**) frame 1 and (**b**) frame 2 in the undamaged and damaged 1 cases, as well as (**c**) frame 1 and (**d**) frame 2 in the damaged 2 and damaged 3 cases are overlaid.

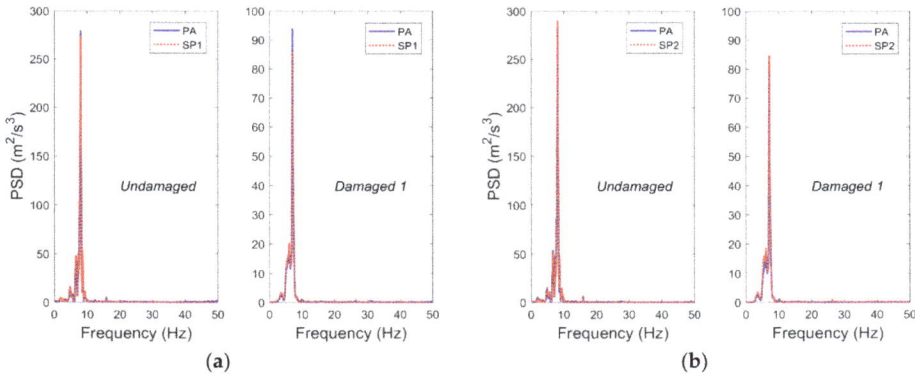

(a)   (b)

Figure 5. *Cont.*

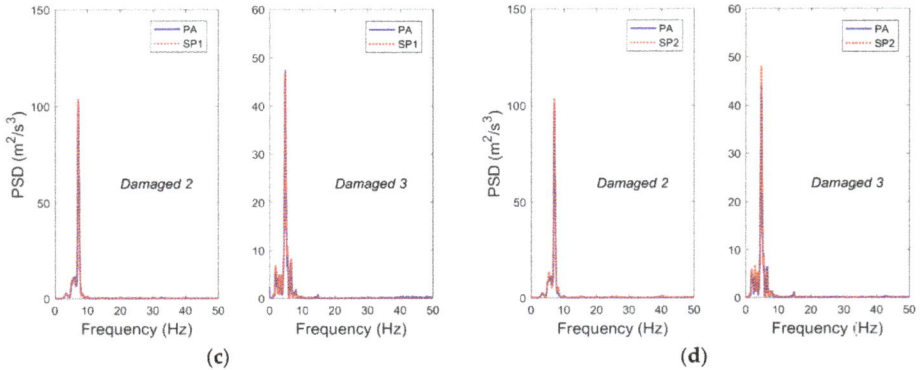

Figure 5. Using the acceleration time-history measurements from Figure 4, the power spectral density functions of (**a**) frame 1 and (**b**) frame 2 in the undamaged and damaged 1 cases, as well as (**c**) frame 1 and (**d**) frame 2 in the damaged 2 and damaged 3 cases are compared.

Table 3 shows the acceleration response peaks of the structures under different damage conditions monitored by the two sensors under different excitation. It can be seen from Table 3 that although some numerical anomalies are present, the acceleration peak errors monitored by the two sensors are generally within 5%. This further illustrates the reliability of vibration monitoring using smartphones.

Table 3. Peak values of acceleration under different excitation levels by two kinds of sensors.

Damage Case	Earthquake Excitation	Frame 1			Frame 2		
		PA (m/s^2)	SP (m/s^2)	Error	PA (m/s^2)	SP (m/s^2)	Error
Undamaged	Nr-1 cm	12.73	12.74	0.08%	12.26	12.52	2.16%
	Nr-2 cm	27.28	28.16	3.20%	27.22	25.60	−5.98%
Damaged 1	Nr-1 cm	10.46	9.75	−6.75%	9.29	9.56	2.85%
	Nr-2 cm	18.98	19.08	0.52%	17.97	18.96	−5.51%
Damaged 2	Nr-1 cm	9.64	9.86	−2.30%	10.11	10.16	0.48%
	Nr-2 cm	19.51	20.02	−2.61%	20.26	21.28	5.03%
Damaged 3	Nr-1 cm	6.87	7.63	−11.13%	8.36	7.58	−9.38%
	Nr-2 cm	14.47	15.32	5.82%	16.50	16.53	0.18%

Table 4. First modal frequency results obtained using the PAs and SPs.

Damage Case	Sensor Type	Steel Frame Model	
		Frame 1 (Hz)	Frame 2 (Hz)
Undamaged	SP	8.05	8.05
	PA	8.00	8.05
Damaged 1	SP	7.05	7.10
	PA	7.05	7.05
Damaged 2	SP	7.10	7.10
	PA	7.15	7.15
Damaged 3	SP	4.80	4.75
	PA	4.85	4.80

The peak-picking algorithm was used to process the structural acceleration response data under different damage cases, and the basic modal frequency of the structure was obtained. Table 4 shows the results of the analysis. It can be seen from Table 4 that the first modal frequency of the structures in the different damage cases recognized by the two kinds of sensors are substantially the same. Meanwhile, as the damage increases, the first modal frequency of the structure gradually decreases, which is reasonable because the removed rigid beam at the corresponding location reduces the overall stiffness of the structure.

In this study, although the direction of the seismic excitation only occurred in the plane X direction, smartphones could simultaneously monitor the structural acceleration responses in the plane Y direction. Figure 6 shows the Y-direction acceleration responses of the structure in the different damage cases.

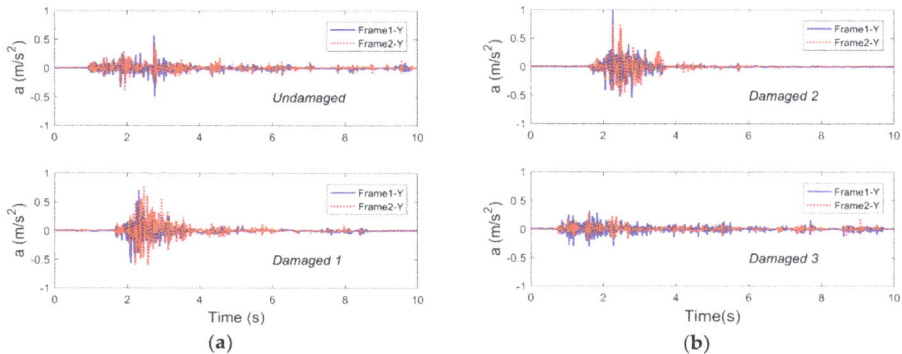

Figure 6. The Y-axis acceleration time-history responses as collected by smartphones subjected to Nr-1 cm are shown (**a**) in the undamaged and damaged 1 cases and (**b**) in the damaged 2 and 3 cases.

It can be seen that although the amplitude is small, the vibration trend and amplitude of frames 1 and 2 in the Y direction are similar, which may be because the columns of the model were formed by 4-mm-thick, 150-mm-wide, and 400-mm-tall steel plates. Such a column has a small outer plane (X-direction) stiffness and a large in-plane (Y-direction) stiffness, which limits the vibration and displacement of the structure in the Y direction. Despite all this, smartphones can monitor the acceleration response in the Y direction of the structure. Compared with the traditional triaxial acceleration sensors, smartphones not only can simultaneously monitor the acceleration responses in three directions, however they are also cheaper and more popular than the sensors.

3.2. Displacement Time-History Curves Comparison

According to Section 2.1, the relative displacement of the vertex was obtained using a video image processing method recorded by smartphones, and the reference displacement measurement was collected using the LDS mounted on the top side of the structure. Also limited by space, Figure 7 shows the displacement comparison results for the representative structures monitored by the two kinds of sensors in the four cases subjected to Nr-1 cm excitation. It can be seen from Figure 7 that the displacement time-history data monitored by the smartphones are basically the same as that of the LDSs. Furthermore, compared with the case of structural damage, the data monitored by the two kinds of sensors differ greatly in peak value in the case where the structure was not damaged. This could be due to two reasons. The first reason is the effect of the sampling rate, where the sampling rate of the LDS is 100 Hz, while the sampling rate of the smartphones equipped with the D-Viewer app is 30 Hz, and the difference in sampling rate may cause measurement errors. The second reason may be that the smartphones are subject to ambient illumination, affecting the target reference (i.e., laser spot) and installation methods when recording and processing video images, thus resulting in measurement errors.

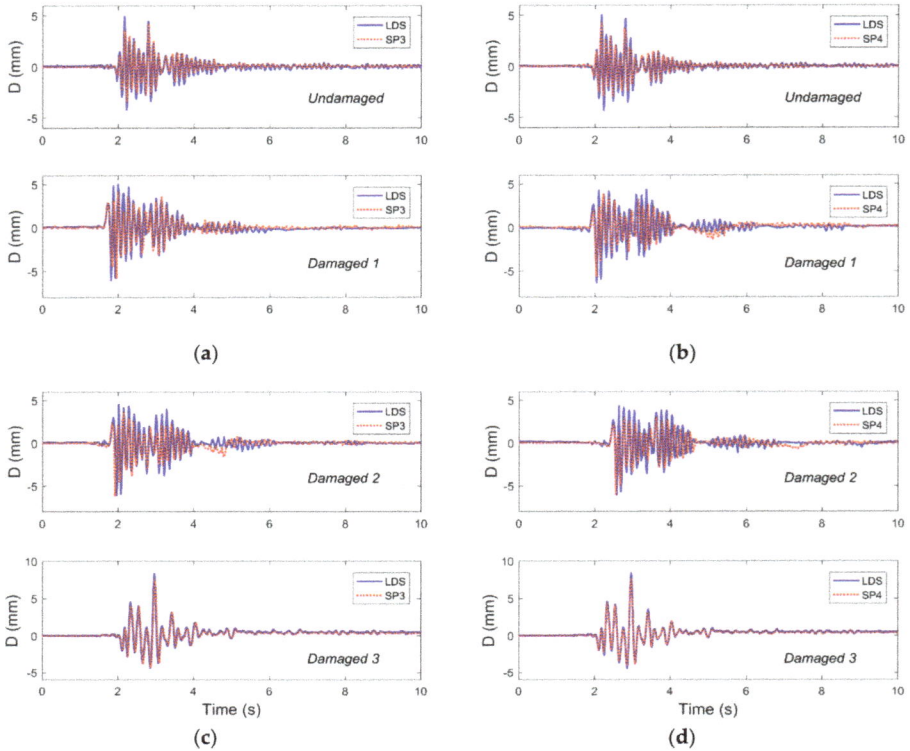

Figure 7. The displacement time-history responses as collected by the LDSs (laser displacement sensors) and SPs with Nr-1 cm excitation for four damage cases are compared and found to show good agreement. Representative results are shown from (**a**) frame 1 and (**b**) frame 2 in the undamaged case and damaged 1 case, as well as from (**c**) frame 1 and (**d**) frame 2 in the damaged 2 and 3 cases.

In order to analyze the difference in displacement data monitored by the two kinds of sensors, Table 5 shows the displacement response peaks of the structures with different damage cases subjected to different excitation levels. It can be seen from the table that the displacement data monitored by smartphones are more error-prone than the data monitored by LDSs in the undamaged case. This may be because the smartphone's sampling rate is only 30 Hz. As can be seen from Table 4 above, the first model frequency of the undamaged structure is 8.05 Hz. Considering the influence of higher order frequencies on the structural displacement response, the sampling rate of 30 Hz may not satisfy the sampling theorem and results in the data's error. When the structure is damaged, the natural frequency of the structure is gradually reduced, and the sampling rate of 30 Hz gradually satisfies the sampling demand so that the error of the monitoring result is gradually reduced. In summary, although the comparison results have some errors, they still verified the use of smartphones for displacement monitoring and indicated that smartphones could be used to monitor the interlayer displacement of structures. Compared with traditional sensors, smartphones with a displacement response sampling rate of 30 Hz are more suitable for monitoring structures with low natural frequencies.

3.3. Y-Axis Displacement Time-History Curves of the Steel Frame

The D-Viewer software could measure the displacement not only in the X direction of the structure, however also in the Y direction, so that the torsional response of the structure during the vibration process could be analyzed. Due to the small displacement of the structure in the Y direction and the influence of factors such as the environment on the smartphone, some signal-to-noise ratios of the displacement data were poor. However, some good data can be used to explain some problems. Figure 8 shows the displacement of the structure in the Y-axis in undamaged and damaged states.

Table 5. Peak values of displacement under different excitations by two kinds of sensors.

Damage Case	Earthquake Excitation	Frame 1			Frame 2		
		LDS (mm)	SP (mm)	Error	LDS (mm)	SP (mm)	Error
Undamaged	Nr-1 cm	4.931	4.194	−14.95%	5.026	4.24	−15.64%
	Nr-2 cm	11.15	9.074	−18.62%	11.7	11	−5.98%
Damaged 1	Nr-1 cm	6.066	5.763	-5.00%	6.412	5.636	−12.10%
	Nr-2 cm	13.34	10.4	-22.04%	12.48	12.23	−2.00%
Damaged 2	Nr-1 cm	6.105	6.142	0.61%	6.108	6.024	−1.38%
	Nr-2 cm	13.07	9.693	−25.84%	12.95	9.795	−24.36%
Damaged 3	Nr-1 cm	8.321	7.603	−8.63%	8.158	7.703	−5.58%
	Nr-2 cm	16.94	16.24	−4.13%	16.95	16.35	−3.54%

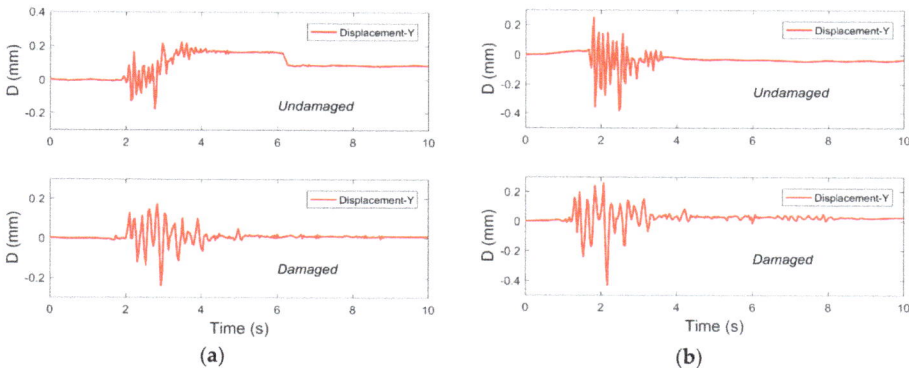

Figure 8. Displacement of the frame in the Y direction for (**a**) the Northridge-1 cm case and (**b**) the Northridge-2 cm case.

It can be seen that although the Y-axis displacement response of the single-layer structure model is small, smartphones still could monitor the Y-axis displacement. This may be due to the small difference in stiffness between the two planar frames, resulting in a torsional effect and displacement response in the Y direction. This result shows that when the structure is under the action of extreme events such as earthquakes, the application of smartphones can simultaneously monitor the displacement response in both directions of the structure plane, thus reducing the number of sensors required.

3.4. Different Frequencies of LDSs Compared with SPs

In order to verify the influence of different sampling rates on the displacement data, the data recorded by the LDSs were resampled at 100 Hz, 50 Hz, and 25 Hz. Compared with the displacement data recorded by the smartphones, the results are shown in Figure 9. It can be seen that in the undamaged case, the data collected by smartphones are closest to the data recorded by the LDS at a sampling rate of 25 Hz. At the same time, in the damaged case, the changes in the sampling rate of the LDS are not significantly different from the data recorded by the smartphones. This shows that the difference in sampling rate has a greater influence on structures that have a higher fundamental mode frequency. The data in Table 6 further illustrates this conclusion.

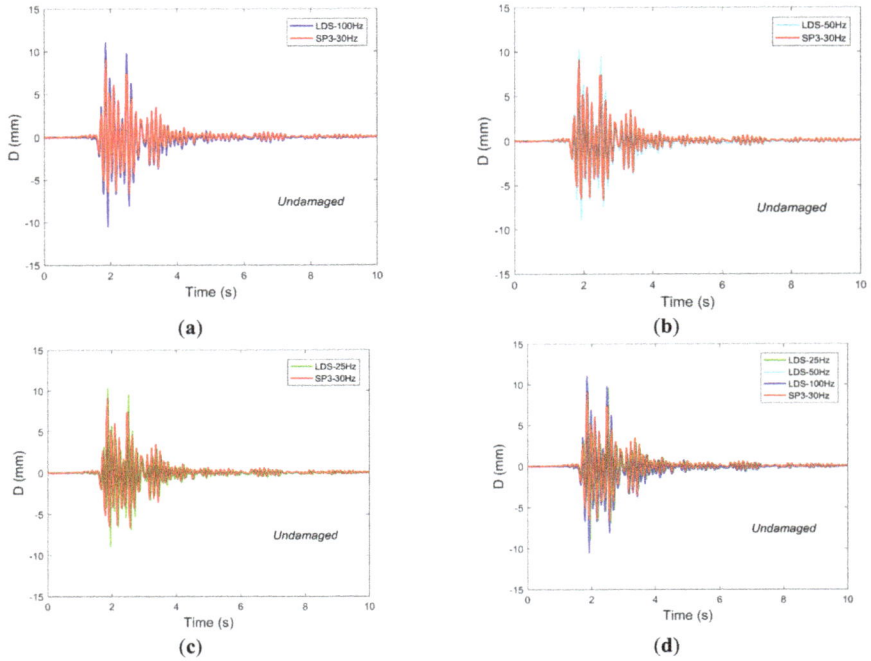

Figure 9. The displacement time-history responses as collected by SPs compared with LDSs using different sampling rates of (**a**) 100 Hz, (**b**) 50 Hz, and (**c**) 25 Hz. (**d**) A comparison of all sampling rates.

Table 6. Peak values displacement of SPs and LDSs with different frequencies.

Damage Case	Earthquake Excitation	Frame Model and Difference	SP (mm)	100 Hz (mm)	50 Hz (mm)	25 Hz (mm)
Undamaged	Nr-1 cm	Frame 1	4.194	4.931	4.549	4.489
		Error		−14.95%	−7.80%	−6.57%
		Frame 2	4.24	5.036	4.764	4.482
		Error		−15.64%	−11.00%	−5.40%
	Nr-2 cm	Frame 1	9.074	11.15	10.15	9.858
		Error		−18.62%	−18.62%	−7.95%
		Frame 2	11.00	11.70	11.70	9.81
		Error		−5.98%	-5.98	12.15%
Damaged 1	Nr-1 cm	Frame 1	5.763	6.066	6.066	5.092
		Error		−5.00%	−5.00%	13.18%
		Frame 2	5.636	6.412	6.159	6.159
		Error		−12.10%	−8.49%	-8.49%
	Nr-2 cm	Frame 1	10.4	13.34	12.82	12.17
		Error		−22.04%	−18.88%	−14.54%
		Frame 2	12.23	12.48	12.48	12.48
		Error		−2.00%	−2.00%	−2.00%

Table 6. *Cont.*

Damage Case	Earthquake Excitation	Frame Model and Difference	SP (mm)	100 Hz (mm)	50 Hz (mm)	25 Hz (mm)
Damaged 2	Nr-1 cm	Frame 1	6.142	6.105	5.993	5.993
		Error		0.61%	2.49%	2.49%
		Frame 2	6.024	6.108	6.108	5.436
		Error		−1.38%	−1.38%	10.82%
	Nr-2 cm	Frame 1	9.693	13.07	13.07	9.265
		Error		−25.84%	−25.84%	4.62%
		Frame 2	9.795	12.95	12	10.91
		Error		−24.36%	−18.38%	−10.22%
Damaged 3	Nr-1 cm	Frame 1	7.603	8.32	8.32	8.32
		Error		−8.62%	−8.62%	−8.62%
		Frame 2	7.703	8.368	8.368	8.368
		Error		−7.95%	−7.95%	−7.95%
	Nr-2 cm	Frame 1	16.24	16.94	16.85	16.85
		Error		−4.13%	−3.62%	−3.62%
		Frame 2	16.35	16.95	16.95	16.95
		Error		−3.54%	−3.54%	−3.54%

4. Damage Detection Results and Discussion

4.1. Wavelet Packet Analysis Background

Wavelet packet analysis (WPA) is a method of vibration signal processing proposed by Wu and Du in 1996 [35]. With the development of WPA, structural damage identification based on the energy of wavelet packet nodes of structural dynamic response has been widely studied [36–38].

According to Parseval's theorem, the energy in the time-domain is equal to that of the frequency-domain. When damage occurs, the energy corresponding to each frequency is redistributed and the structural response of each frequency band changes.

In order to extract structural damage information from structural response signal, here we suppose that a signal, $S(t)$, can be expressed by wavelet packet decomposition:

$$S(t) = \sum_{j=1}^{2^{k-1}} S_{kj}(t) \tag{1}$$

where $S_{kj}(t)$ is the sub-signal with an orthogonal frequency band and k indicates the layer number of the tree structure of wavelet decomposition.

The energy of these sub-signals E_j can be expressed as:

$$E_j = \left\| S_{kj}(t) \right\|_2 = \sum_k \left| S_{kj}(t) \right|^2 \tag{2}$$

In consequence, the total energy E_{tol} can be obtained by:

$$E_{tol} = \left\| S(t) \right\|_2 = \sum_{j=1}^{N} E_j \tag{3}$$

The wavelet energy ratio $\{Pj\}$ for the jth scale is considered as a normalized value:

$$P_j = \frac{Ej}{E_{tol}} \tag{4}$$

Clearly,

$$\sum_{j=1}^{N} Pj = 1 \tag{5}$$

In this study, a damage index S_{WT} based on relative wavelet entropy (RWE) was employed and is formulated as follows [36]:

$$S_{WT}(p|q) = \sum_{j<0} p_j \cdot \ln\left[\frac{p_j}{q_j}\right] \tag{6}$$

where $\{p\}$ is the set of damaged structural wavelet energy ratio vectors and $\{q\}$ is the set of undamaged structural wavelet energy ratio vectors. Note that the RWE is positive and vanished only if $\{p_j\} = \{q_j\}$.

4.2. Wavelet Packet Decomposition of Acceleration Time-Histories

De-noising of acceleration results was achieved by decomposing the data using three-order WPA with 'db3' to obtain eight frequency bands. The data were decomposed into wavelet packets to obtain the energy distribution of each frequency band. P_j was obtained using the acceleration data monitored by the SPs and that obtained by the PAs, as shown in Figure 10. It can be seen from the figure that the energy distribution of the three damaged systems changes at each frequency band compared with the undamaged system and is mainly concentrated on frequency bands 1 and 2. Compared with the damaged 3 system, the energy distribution change of each frequency band in the damaged 1 and damaged 2 systems are more similar. This could be caused by the same damage degree of the damaged 1 and 2 cases. Due to the rigid connection of frame 1 and frame 2, the response of the structure was forced to be similar; thus, the changes of energy distribution in each frequency band were similar. This is also consistent with the change in the fundamental mode frequency of the structure.

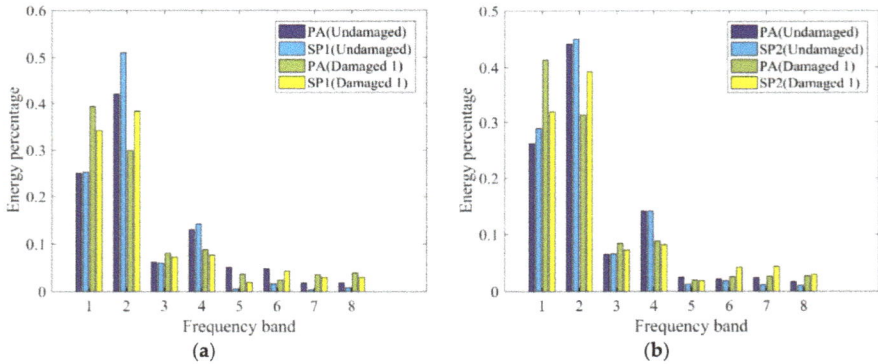

Figure 10. *Cont.*

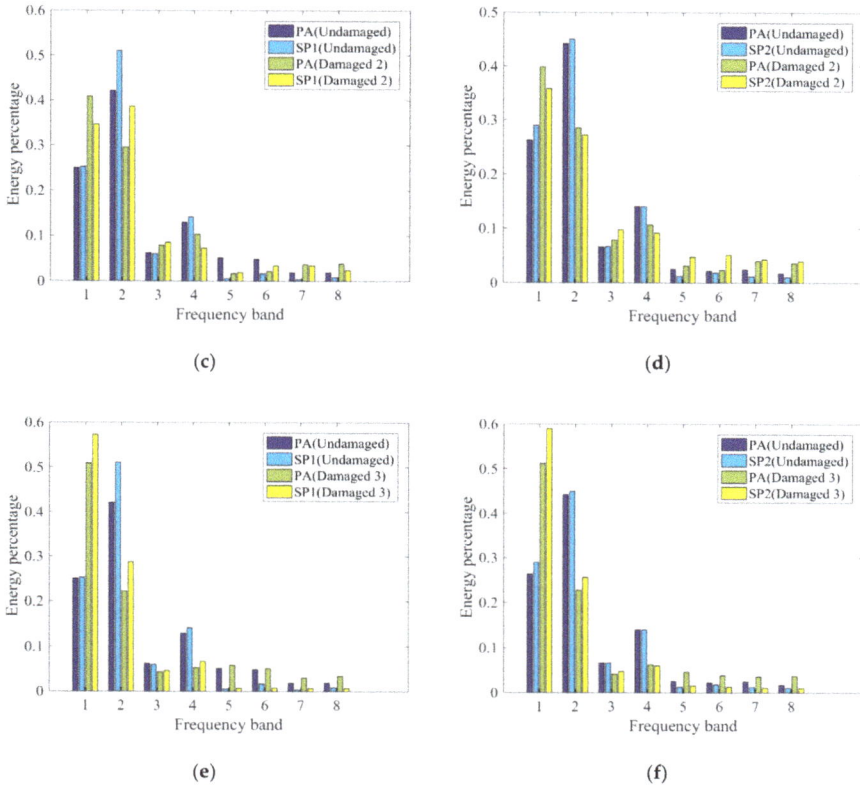

Figure 10. p_j calculated for (**a**) frame 1 and (**b**) frame 2 in the damaged 1 case, (**c**) frame 1 and (**d**) frame 2 in the damaged 2 case, and (**e**) frame 1 and (**f**) frame 2 in the damaged 3 case when structure is subjected to Nr-1 cm excitation.

Table 7 shows the relative wavelet entropy coefficients of the various acceleration data for the structure with different damage cases. As can be seen from the table, for the damaged 1 and damaged 2 cases, the structure's RWE values do not differ greatly, and the RWE of the frame from which the rigid beam was removed is larger than that of the frame with the beam. For the damaged 3 case, the structure's RWE value is significantly increased. This indicates that the damage level of the structure was increased. This is also consistent with the change of the fundamental mode frequency of the structure. Meanwhile, the contrast of wavelet analysis results using two kinds of sensors are relatively good. For the damaged 1 and damaged 2 cases, the structure's RWE values' errors are larger than the damaged 3 case. This is probably because the structures of damage 1 and damage 2 only removed one rigid beam so that the structural rigidity was asymmetrical. When the structure vibrated, the eccentricity caused the acceleration responses of the two kinds of sensors attached to the two sides of the damper articulated beam to be different in different frequency bands. However, on the contrary, the structure of the damage 3 removed two rigid beams at the same time and the spatial stiffness of the structure was symmetrical, so the energy distributions of the acceleration responses monitored by the two kinds of sensors were similar in different frequency bands and the RWE values' errors are relatively small. Despite all this, these results demonstrate the feasibility of using smartphones and corresponding damage index methods (such as RWE) for damage detection.

Table 7. The relative wavelet entropy (RWE) index for different cases.

Damage Case	Frame Model	Earthquake Excitation	Different Sensors		
			PA	SP	Error
Damaged 1	Frame 1	Nr-1 cm	0.086	0.102	18.26%
	Frame 2		0.073	0.0577	−21.39%
	Frame 1	Nr-2 cm	0.114	0.096	−15.57%
	Frame 2		0.114	0.052	−54.00%
Damaged 2	Frame 1	Nr-1 cm	0.110	0.102	-7.61%
	Frame 2		0.079	0.136	71.59%
	Frame 1	Nr-2 cm	0.121	0.149	23.10%
	Frame 2		0.129	0.127	-0.86%
Damaged 3	Frame 1	Nr-1 cm	0.202	0.224	10.80%
	Frame 2		0.213	0.199	-6.61%
	Frame 1	Nr-2 cm	0.198	0.202	1.97%
	Frame 2		0.201	0.215	6.70%

5. Conclusions

In this paper, a three-dimensional steel frame was subjected to shaking table tests simulating earthquake excitation. Three damaged cases were introduced by removing the different rigid beams in the frame to cause the viscous damper mounted at the beam-column joint to rotate. The response data of the undamaged structure and damaged structures were monitored using both conventional sensors and smartphones which were pre-installed with measurement acceleration (Orion-CC) and displacement (D-viewer) software. The data monitored by smartphones and conventional sensors were compared, and the causes of data errors were analyzed. Wavelet packet decomposition and relative wavelet entropy were employed to identify and evaluate structural damage. The conclusions are summarized as follows:

(1) The acceleration responses acquired by smartphones and piezoelectric acceleration sensors were matched quite well. In order to obtain the change of the basic modal frequency of the frame in the different damage cases, the peak-picking method was applied. The results show that the first modal frequency of the structures in the different damage cases recognized by the two kinds of sensors are substantially the same. Meanwhile, as the damage increases, the first modal frequency of the structure gradually decreases.

(2) The results of the comparison of the displacement acquired by smartphones and LDS are basically good. The influence of the sampling rate of the two kinds of sensors on the monitoring results was analyzed. The results show that compared with traditional sensors, smartphones with a displacement response sampling rate of 30 Hz are more suitable for monitoring structures with low natural frequencies.

(3) Wavelet packet analysis was used to analyze the acceleration data, and the damage index based on RWE was obtained under different damage cases. The results demonstrate that the contrast of wavelet analysis results using two kinds of sensors are relatively good. However, the asymmetry of the structure's spatial stiffness will lead to greater RWE value errors being obtained from the smartphones monitoring data. Despite all this, the structural damage could be detected using smartphones.

In summary, this study demonstrates the feasibility of using smartphones to monitor the response of a building structure subjected to extreme events such as earthquakes. It is worth noting that because the conclusions of this paper are drawn from the model test analysis, the practical application of smartphones is not included in this paper and will be the focus of the next research. At the same time, how to improve the sample rate of the displacement collected by the smart phone through the camera should be studied.

Author Contributions: Conceptualization, B.X. and Z.X.; Methodology, B.X. and J.L.; Software, X.Z.; Formal Analysis, B.X. and J.L.; Writing-Original Draft Preparation, B.X.; Writing-Review & Editing, B.X., J.L., and X.Z.

Funding: This research was funded by the National Key research and development programs (grant number 2016YFE0202400) and National Key R&D Program of China during the Thirteenth Five-Year Plan Period (grant number 2016YFC0802002).

Conflicts of Interest: The authors declare no conflict of interest.

References

1. Ou, J.; Li, H. Structural Health Monitoring in mainland China: Review and Future Trends. *Struct. Heal. Monit.* **2010**, *9*, 219–231.

2. Li, H.-N.; Li, D.-S.; Ren, L.; Yi, T.-H.; Jia, Z.-G.; Li, K.-P. Structural health monitoring of innovative civil engineering structures in Mainland China. *Struct. Monit. Maint.* **2016**, *3*, 1–32. [CrossRef]

3. Worden, K.; Cross, E.J. On switching response surface models, with applications to the structural health monitoring of bridges. *Mech. Syst. Signal Process.* **2018**, *98*, 139–156. [CrossRef]

4. Raghavan, A. Guided-Wave Structural Health Monitoring. Ph.D. Thesis, The University of Michigan, Ann Arbor, MI, USA, 2007.

5. Haider, M.F.; Giurgiutiu, V. Analysis of axis symmetric circular crested elastic wave generated during crack propagation in a plate: A Helmholtz potential technique. *Int. J. Solids Struct.* **2018**, *134*, 130–150. [CrossRef]

6. Lorenzoni, F.; Caldon, M.; da porto, F.; Modena, C.; Aoki, T. Post-earthquake controls and damage detection through structural health monitoring: applications in l'Aquila. *J. Civ. Struct. Heal. Monit.* **2018**, *8*, 217–236. [CrossRef]

7. Hill, J.; Szewczyk, R.; Woo, A.; Hollar, S.; Culler, D.; Pister, K. System architecture directions for networked sensors. *Continuum (N. Y).* **2009**, *23*, 93–104.

8. Spencer, B.F.; Ruiz-Sandoval, M.E.; Kurata, N. Smart sensing technology: Opportunities and challenges. *Struct. Control Heal. Monit.* **2004**, *11*, 349–368. [CrossRef]

9. Hoult, N.A.; Fidler, P.R.A.; Hill, P.G.; Middleton, C.R. Long-Term Wireless Structural Health Monitoring of the Ferriby Road Bridge. *J. Bridg. Eng.* **2010**, *15*, 153–159. [CrossRef]

10. Cho, S.; Giles, R.K.; Spencer, B.F., Jr. System identification of a historic swing truss bridge using a wireless sensor network employing orientation correction Soojin. *Struct. Control Heal. Monit.* **2015**, *22*, 255–272. [CrossRef]

11. Goggin, G. Adapting the mobile phone: The iPhone and its consumption. *Continuum (N. Y).* **2009**, *23*, 231–244. [CrossRef]

12. Lau, S.L.; König, I.; David, K.; Parandian, B.; Carius-Düssel, C.; Schultz, M. Supporting patient monitoring using activity recognition with a smartphone. In Proceedings of the Proceedings of the 2010 7th International Symposium on Wireless Communication Systems (ISWCS'10), York, UK, 19–22 September 2010; pp. 810–814.

13. Wan, J.; Zhang, D.; Sun, Y.; Lin, K.; Zou, C.; Cai, H. VCMIA: A novel architecture for integrating vehicular cyber-physical systems and mobile cloud computing. *Mob. Networks Appl.* **2014**, *19*, 153–160. [CrossRef]

14. Ketabdar, H.; Polzehl, T. Fall and emergency detection with mobile phones. In Proceedings of the eleventh international ACM SIGACCESS conference on Computers and accessibility (ASSETS '09), Pittsburgh, PA, USA, 25–28 October 2009; p. 241.

15. Reilly, J.; Dashti, S.; Ervasti, M.; Bray, J.D.; Glaser, S.D.; Bayen, A.M. Mobile Phones as Seismologic Sensors: Building the iShake System. *IEEE Trans. Autom. Sci. Eng.* **2013**, *10*, 242–251. [CrossRef]

16. Dashti, S.; Bray, J.D.; Reilly, J.; Glaser, S.; Bayen, A.; Mari, E. Evaluating the reliability of phones as seismic monitoring instruments. *Earthq. Spectra* **2014**, *30*, 721–742. [CrossRef]

17. Chun, H.J.; Han, Y.D.; Park, Y.M.; Kim, K.R.; Lee, S.J.; Yoon, H.C. An optical biosensing strategy based on selective light absorption and wavelength filtering from chromogenic reaction. *Materials* **2018**, *11*, 388. [CrossRef]

18. Yu, Y.; Zhao, X.; Ou, J. A new idea: Mobile structural health monitoring using Smart phones. In Proceedings of the ICICIP 2012—2012 3rd International Conference on Intelligent Control and Information Processing, Dalian, China, 15–17 July 2012; pp. 714–716.

19. Kotsakos, D.; Sakkos, P.; Kalogeraki, V.; Gunopulos, D. SmartMonitor: using smart devices to perform structural health monitoring. *Proc. VLDB Endow.* **2013**, *6*, 1282–1285. [CrossRef]

20. Morgenthal, G.; Höpfner, H. The application of smartphones to measuring transient structural displacements. *J. Civ. Struct. Heal. Monit.* **2012**, *2*, 149–161. [CrossRef]

21. Höpfner, H.; Morgenthal, G.; Schirmer, M.; Naujoks, M.; Halang, C. On measuring mechanical oscillations using smartphone sensors: possibilities and limitation. *ACM SIGMOBILE Mob. Comput. Commun. Rev.* **2013**, *17*, 29–41. [CrossRef]

22. Cimellaro, G.P.; Scura, G.; Renschler, C.S.; Reinhorn, A.M.; Kim, H.U. Rapid building damage assessment system using mobile phone technology. *Earthq. Eng. Eng. Vib.* **2014**, *13*, 519–533. [CrossRef]

23. Sharma, A. Smartphone as a Real-time and Participatory Data Collection Tool for Civil Engineers. *Int. J. Mod. Comput. Sci.* **2014**, *2*, 22–27.

24. Akinwande, V.; Bello, O.W.; Akinwande, V. Automatic and real-time Pothole detection and Traffic monitoring system using Smartphone Technology Automatic and real-time Pothole detection and Traffic monitoring system using Smartphone Technology. In Proceedings of the International Conference on Computer Science Research and Innovations (CoSRI 2015), Ibadan, Nigeria, 20–22 August 2015.

25. Feng, M.; Fukuda, Y.; Mizuta, M.; Ozer, E. Citizen sensors for SHM: Use of accelerometer data from smartphones. *Sensors* **2015**, *15*, 2980. [CrossRef]

26. Ozer, E.; Feng, M.Q.; Feng, D. Citizen sensors for SHM: Towards a crowdsourcing platform. *Sensors* **2015**, *15*, 14591. [CrossRef]

27. Zhao, X.; Han, R.; Ding, Y.; Yu, Y.; Guan, Q.; Hu, W.; Li, M.; Ou, J. Portable and convenient cable force measurement using smartphone. *J. Civ. Struct. Heal. Monit.* **2015**, *5*, 481–491. [CrossRef]

28. Zhao, X.; Yu, Y.; Hu, W.; Jiao, D.; Han, R.; Mao, X.; Li, M.; Ou, J. Cable force monitoring system of cable stayed bridges using accelerometers inside mobile smart phone. In Proceedings of the Sensors and Smart Structures Technologies for Civil, Mechanical, and Aerospace Systems 2015, San Diego, CA, USA, 9–12 March 2015; p. 94351H.

29. Zhao, X.; Liu, H.; Yu, Y.; Zhu, Q.; Hu, W.; Li, M.; Ou, J. Displacement monitoring technique using a smartphone based on the laser projection-sensing method. *Sens. Actuators A Phys.* **2016**, *246*, 35–47. [CrossRef]

30. Peng, D.; Zhao, X.; Zhao, Q.; Yu, Y. Smartphone based public participant emergency rescue information platform for earthquake zone—"E-Explorer". In Proceedings of the International Conference on Vibroengineering, Nanjing, China, 26–28 September 2015; Volume 5.

31. Min, J.; Gelo, N.J.; Jo, H. Real-time Image Processing for Non-contact Monitoring of Dynamic Displacements using Smartphone Technologies. In Proceedings of the Sensors and Smart Structures Technologies for Civil, Mechanical, and Aerospace Systems, Las Vegas, NV, USA, 21–24 March 2016.

32. Ozer, E.; Feng, D.; Feng, M.Q. Hybrid motion sensing and experimental modal analysis using collocated smartphone camera and accelerometers. *Meas. Sci. Technol.* **2017**, *28*, 105903. [CrossRef]

33. Kong, Q.; Allen, R.M.; Kohler, M.D.; Heaton, T.H.; Bunn, J. Structural Health Monitoring of Buildings Using Smartphone Sensors. *Seismol. Res. Lett.* **2018**, *89*, 594–602. [CrossRef]

34. Miller, D.K. Lessons learned from the Northridge earthquake. *Eng. Struct.* **1998**, *20*, 249–260.

35. Wu, Y.; Du, R. Feature extraction and assessment based on wavelet packet transform. *Mech. Syst. Signal Process.* **1996**, *10*, 29–53. [CrossRef]

36. Ren, W.X.; Sun, Z.S. Structural damage identification by using wavelet entropy. *Eng. Struct.* **2008**, *30*, 2840–2849. [CrossRef]

37. Yan, Y.J.; Yam, L.H. Online detection of crack damage in composite plates using embedded piezoelectric actuators/sensors and wavelet analysis. *Compos. Struct.* **2002**, *58*, 29–38. [CrossRef]

38. Zhang, L.; Wang, C.; Song, G. Health status monitoring of cuplock scaffold joint connection based on wavelet packet analysis. *Shock Vib.* **2015**, *2015*. [CrossRef]

sensors

MDPI

Article

Feature Extraction and Mapping Construction for Mobile Robot via Ultrasonic MDP and Fuzzy Model

Zhili Long, Ronghua He, Yuxiang He, Haoyao Chen and **Zuohua Li** *

Harbin Institute of Technology Shenzhen, Shenzhen 518055, China; longzhili@hit.edu.cn (Z.L.);
he.ronghua@foxmail.com (R.H.); heyuxiang1991@foxmail.com (Y.H.); hychen5@hit.edu.cn (H.C.)
* Correspondence: lizuohua@hit.edu.cn; Tel.: +86-134-2094-8695

Received: 10 September 2018; Accepted: 26 October 2018; Published: 29 October 2018

check for
updates

Abstract: This paper presents a modeling approach to feature classification and environment mapping for indoor mobile robotics via a rotary ultrasonic array and fuzzy modeling. To compensate for the distance error detected by the ultrasonic sensor, a novel feature extraction approach termed "minimum distance of point" (MDP) is proposed to determine the accurate distance and location of target objects. A fuzzy model is established to recognize and classify the features of objects such as flat surfaces, corner, and cylinder. An environmental map is constructed for automated robot navigation based on this fuzzy classification, combined with a cluster algorithm and least-squares fitting. Firstly, the platform of the rotary ultrasonic array is established by using four low-cost ultrasonic sensors and a motor. Fundamental measurements, such as the distance of objects at different rotary angles and with different object materials, are carried out. Secondly, the MDP feature extraction algorithm is proposed to extract precise object locations. Compared with the conventional range of constant distance (RCD) method, the MDP method can compensate for errors in feature location and feature matching. With the data clustering algorithm, a range of ultrasonic distances is attained and used as the input dataset. The fuzzy classification model—including rules regarding data fuzzification, reasoning, and defuzzification—is established to effectively recognize and classify the object feature types. Finally, accurate environment mapping of a service robot, based on MDP and fuzzy modeling of the measurements from the ultrasonic array, is demonstrated. Experimentally, our present approach can realize environment mapping for mobile robotics with the advantages of acceptable accuracy and low cost.

Keywords: feature extraction; mapping construction; fuzzy classification; rotary ultrasonic array

1. Introduction

Currently, intelligent control of indoor robots has become an essential goal. Robots require the ability to acquire environmental information and detect their own location [1]. Because most environments that robots face in real life cannot be known in advance, it is necessary to recognize the features of environmental objects and construct a surrounding map for the robot to navigate. Therefore, the technology of simultaneous localization and mapping (SLAM) has become a key development trend for indoor robots [2].

Environmental modeling can be divided into laser, visual, infrared, and ultrasound methods according to the sensor type. Laser modeling has advantages such as high accuracy, high speed, and broad detection range. However, lasers cannot accurately detect some transparent materials such as glass [3]. Thus, the laser method is mainly applied in specialized environment modeling. Visual sensors can collect huge amounts of environmental information, but the associated data processing is more time-consuming [4]. Infrared sensors are very sensitive to light in basic operation mode [5], which often results in errors in the signal to the robots. Ultrasonic sensors, which detect the environment by transmitting and receiving ultrasonic waveforms, have the advantages of sufficient detection range, accurate resolution, and stable controllability, as they are not influenced by light or the material composition of objects. Moreover, ultrasonic sensors have the competitive advantage of low cost compared with laser and visual sensors. Therefore, ultrasonic sensors are a potential solution for environment mapping for indoor mobile service robots.

The current modeling approaches to environment mapping by ultrasonic sensing include grid mapping [6,7], topological mapping [8], and geometric feature mapping [9]. Compared with other mapping approaches, feature mapping can visualize more features for secondary data processing and the mapping results are more accurate and stable. In environmental mapping construction based on feature mapping, features—including flat surfaces, corners, and cylinders—are extracted, located, and combined to generate an environmental map. The typical modeling approaches to feature extraction include the RCD algorithm [10], the arc-transversal median (ATM) method, and triangulation-based fusion (TBF). Recently, hybrid integration of feature, topological, and grid mapping has been developed [11,12]. Heinen [13] introduced a Gaussian mixture model to represent the surrounding environment and proposed a new feature-based environment mapping algorithm with the advantages of small memory usage and high processing speed, avoiding discrete errors in fast calculation. Ismail et al. [14] used the data measured by a double ultrasonic sensor and proposed a fusion algorithm for environment feature extraction in which circular-arc feature extraction was combined with Hough transform-based TBF. Lee [15] proposed a new approach to ultrasonic feature extraction modeling, which used an ultrasonic data correlation method to extract circular sets of the outer corners of objects. The integrated circle center of each set was adopted as an EKF-SLAM roadmap. However, the transmitted signals of the 12 ultrasonic sensors were sensitive to crosstalk as they used the same frequencies. Shuai et al. [16] proposed a mapping approach based on ultrasonic line segmentation. The experimental ultrasonic data were measured and then were classified by improved iterative end-point fit (IEPF). The line segments were extracted by searching for transition points, and the parameters of the line segments were estimated by Kalman filter. Finally, the line segments were integrated to generate the environmental mapping. This approach can successfully map the surrounding environment via line segments and improve the efficiency of feature extraction. However, it cannot map complex environmental features such as cylinders and corners.

In summary, although several theoretical studies have investigated ultrasonic-based environmental mapping for mobile robots, the mapping precision and computational efficiency still need to be improved. In particular, the mapping approaches for current indoor service robots focus on laser and visual sensors, which are high in cost and have some inherent negative characteristics. Meanwhile, the ultrasonic approach has many advantages such as low cost, and insensitivity to the object material and to light, making it suitable for widespread application in indoor mobile robots. Currently, the ultrasonic sensors used in mobile robots are capable of obstacle avoidance and angle error estimation [17,18], but the successful construction of mapping for an indoor robot by using ultrasonic signals has rarely been reported. Therefore, it is important to investigate new approaches with low cost and acceptable accuracy for environmental mapping.

Today, various forms of indoor mobile robots, such as automated cleaning robots, accompanying robots, and robotic waiters, are beginning to appear in ordinary life. However, these consumption-type robots face a high cost burden. In this study, we present a low-cost solution to mapping construction by using an ultrasonic array, which has a potential applicability in automated navigation for indoor mobile service robots.

The paper is organized as six sections. Section 1 presents the introduction. In Section 2, a rotating ultrasonic array platform for environmental modeling is established, and its distance data are detected and analyzed. In Section 3, conventional RCD feature extraction is analyzed and its inherent disadvantages are discussed. An MDP algorithm is proposed to optimize and improve the feature extraction. Section 4 establishes the fuzzy model to classify environmental features such as flat surfaces, corners, and cylinders. In Section 5, verification experiments of MDP and the fuzzy classification model based on the ultrasonic sensor array are described. Finally, the conclusion of this paper is presented in Section 6.

2. Experimental Platform Design

In order to attain the distance data of the obstacle object, the experiment platform is built, and fundamental measurements of the distance of objects at different rotary angles and with different object materials are carried out.

2.1. Configuration of Ultrasonic Array

The ultrasonic configuration to detect the distance between the robot and environmental objects, consisting of an ultrasonic sensor module, rotary motor, and control board, is illustrated in Figure 1. Four ultrasonic sensors are mounted on a rotary motor, in a symmetrical layout on the four sides of the motor, forming a rectangular configuration. Each ultrasonic sensor module—which includes a transmitting sensor, a receiving sensor, and the electrical control component—has the following characteristics: 300 cm maximum detection range, 3 cm dead distance, 5 mm distance resolution, and ±35° beam angle. A rotary motor with a model JX-PDI-6221MG [19] steering gear, which can rotate 90 degrees in 0.17 s and is feeding by a power supply that provides 6 V, is chosen to rotate the ultrasonic array. The functions of the control board are to control the rotary speed and direction of the motor, and transmit and receive the ultrasonic waveform by using IntoRobot Atom [20]. IntoRobot Atom is a control system with dual CPU, based on open software and hardware, offering a variety of interfaces and Wi-Fi wireless data transmission. To receive the ultrasonic signal with high precision and avoid signal crosstalk, the rotary speed of the steering gear is set to 25 ms/degree and rotates back and forth in a range of 90°. Thus, the ultrasonic array can scan the full 360° surroundings and detect the distance of environmental objects.

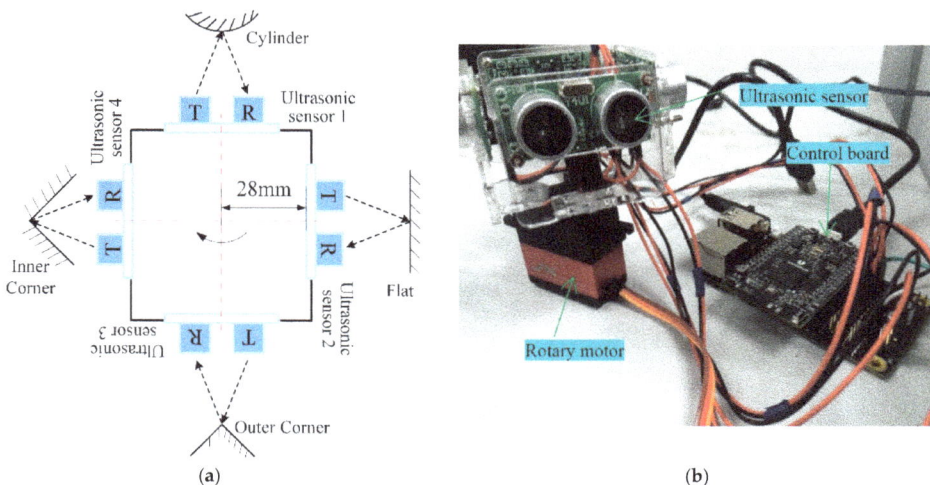

Figure 1. Configuration of ultrasonic array. (**a**) Schematic of ultrasonic sensor array; (**b**) experimental set-up.

2.2. Detected Data Analysis

Figure 2a shows the detected distances, and their errors, from the array to the wall (considered as the object) when the rotary angles are ranged from −30 to +30° at 1° per step. The distance error is defined by subtracting the detected distance from the physically measured value. It is found that the rotary ultrasonic array can detect the object at distances up to 300 cm, which satisfies the detection range requirement of indoor mobile robots. The distance error increases when the rotary angle increases, but can be controlled within 20 mm in the angle range of $\pm 30°$. Figure 2b shows the distance error for different object materials—including metal, glass, concrete, and wood—at a constant physical distance of 100 cm. It can be found that the detected distance errors are consistent for these different object materials, indicating that the ultrasonic signal has adaptability to a wide range of object materials with hard surface and is suitable for object detection in indoor environments containing materials such as glass, wood, concrete, and metal.

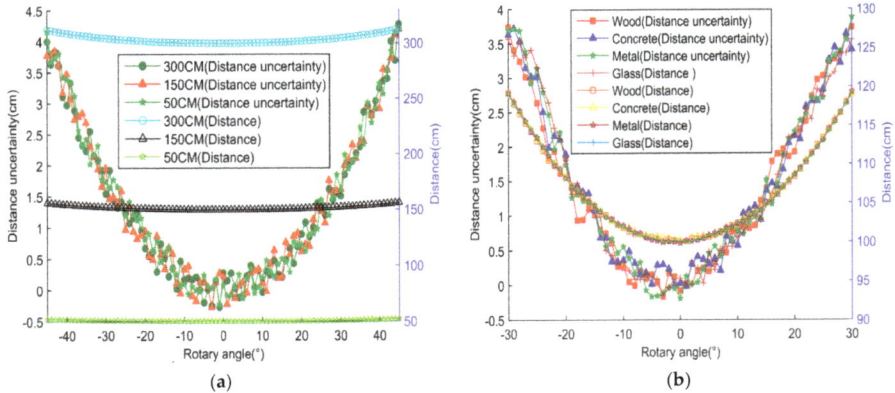

Figure 2. Detected data results. (**a**) Distance and its error at different angles; (**b**) distance and its error with different materials.

3. MDP Optimization Algorithm

In order to filter or decrease the noise and disturbing data in distance measurement, some optimization algorithm is proposed.

3.1. Analysis of Conventional RCD

Range of constant distance (RCD) is a conventional approach to feature recognition in robot navigation. A typical application of RCD is the distance detection of a target object by a laser sensor for automated car driving. The basic principle of distance detection of a target object is shown in Figure 3. The object locations labeled as B, C, and D are considered to be the same within the beam angle θ, which is symmetrical around a center line. RCD detection can achieve highly accurate measurement if using lasers, but RCD based on ultrasound would incur a distance error because of the inherent characteristics of ultrasonic signals. Distance detection by ultrasonic sensors is based on the time of flight (TOF) principle, which implies that the nearest location is detected within the beam angle. If the object is non-circular, non-equidistant, or asymmetric, the location C or h_0 is detected because it is the nearest in distance, while the physical location of the expected detection is D or h_1, which generates a distance bias Δl. The accuracy of feature detection becomes poor if the ultrasonic sensor rotates when using the RCD method, for two reasons: (1) large error in the map outline due to the original distance error, causing the map to be narrowed or widened by RCD; and (2) estimation error when the object features are identified and classified by using the tangent circle principle. Therefore, the RCD method of ultrasonic sensing is not suitable for feature recognition and environmental modeling for mobile robots.

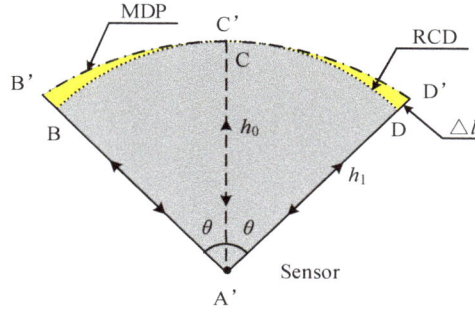

Figure 3. Uncertainty in the conventional RCD model.

To compensate for the distance error Δl of the conventional RCD, an approach termed minimum distance of point (MDP) is proposed. Through the geometric relationship between the object location, object type, and the location of the ultrasonic sensor, a compensated parameter Δl can be calculated to attain a more accurate distance of the detected object. The compensated parameter Δl is a function of the rotary angle, revolving arm length, and the object distance detected.

3.2. MDP Optimization for Feature Extraction

To compensate for the distance error Δl of the conventional RCD, an approach termed minimum distance of point (MDP) is proposed. Through the geometric relationship between the object location, object type, and the location of the ultrasonic sensor, a compensated parameter Δl can be calculated to attain a more accurate distance of the detected object. The compensated parameter Δl is a function of the rotary angle, revolving arm length, and the object distance detected. Figure 4 show the flat and cylinder geometric compensation model. Where, a_1 is the measurement value of point A, and a_2 is the measurement value of point B. b is the distance from the equivalent center of the sensor to the rotation center. h_0 is the final result of point A, and h_1 is the final result of point B. θ is the beam angle. r is the radius of cylinder.

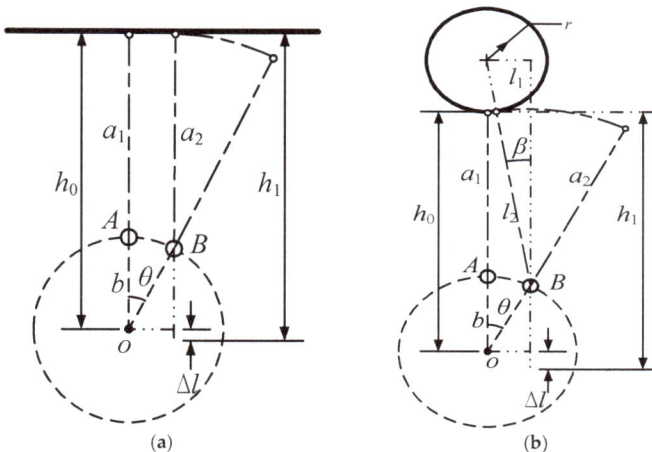

Figure 4. Geometric compensator model. (a) Flat model; (b) cylinder model.

For a flat model, the compensated distance is derived as

$$\Delta l = a_2 - a_1 = b(1 - \cos\theta) \tag{1}$$

For a cylinder model, the compensated distance is derived as

$$l_1 = b\sin\theta \tag{2}$$

$$l_2 = a_1 + r + b(1 - \cos\theta) \tag{3}$$

$$\beta = arc\tan\frac{l_1}{l_2} = arc\tan\frac{b\sin\theta}{a_1 + r + b(1 - \cos\theta)} \tag{4}$$

$$
\begin{aligned}
\Delta l &= \tfrac{l_2}{\cos\beta} - l_2 + b(1 - \cos\theta) \\
&= \frac{a_1 + r + b(1 - \cos\theta)}{\cos\left(arc\tan\frac{b\sin\theta}{a_1 + r + b(1 - \cos\theta)}\right)} - a_1 - r
\end{aligned} \tag{5}
$$

Compared with the conventional RCD, the MDP algorithm not only can obtain a more accurate object distance, but also can extract the other features of the target object for environment mapping. In MDP, the first step to recognize the object features is to attain, from within the original distance dataset, the minimum-distance point that can express the object location. If the minimum distance were directly chosen from the original data, a random error would arise. To improve the accuracy and stability of the minimum-distance point, the least-squares polynomial fitting method is adopted in our study. The m-degree polynomial fitting expression can be given as

$$y = a_0 + a_1 x + a_2 x^2 + \cdots + a_m x^m \tag{6}$$

where y are the detected distance values with compensated parameter Δl at each rotary angle x. m is the index of the detected data. $a_0 \sim a_m$ is the optimal parameter in the fitting model. The quadratic sum of the error e is calculated according to the least-squares method

$$e^2 = \sum_{i=1}^{n}[y_i - (a_0 + a_1 x_i + \cdots + a_m x_i{}^m)]^2 \tag{7}$$

Then, partial differentiation with respect to $\alpha_0 \sim \alpha_m$ yields the optimal estimation, as

$$
\begin{cases}
-2\sum_{i=1}^{n}[y_i - (a_0 + a_1 x_i + \cdots + a_m x_i{}^m)] = 0 \\
-2\sum_{i=1}^{n}[y_i - (a_0 + a_1 x_i + \cdots + a_m x_i{}^m)]x_i = 0 \\
\qquad\qquad\vdots \\
-2\sum_{i=1}^{n}[y_i - (a_0 + a_1 x_i + \cdots + a_m x_i{}^m)]x_i{}^m = 0
\end{cases} \tag{8}
$$

In matrix form, the above can be expressed as

$$
\begin{bmatrix}
n & \sum_{i=1}^{n} x_i & \cdots & \sum_{i=1}^{n} x_i^m \\
\sum_{i=1}^{n} x_i & \sum_{i=1}^{n} x_i^2 & \cdots & \sum_{i=1}^{n} x_i^{m+1} \\
\vdots & \vdots & \ddots & \vdots \\
\sum_{i=1}^{n} x_i^m & \sum_{i=1}^{n} x_i^{m+1} & \cdots & \sum_{i=1}^{n} x_i^{2m}
\end{bmatrix}
\begin{bmatrix}
a_0 \\ a_1 \\ \vdots \\ a_m
\end{bmatrix}
=
\begin{bmatrix}
\sum_{i=1}^{n} y_i \\ \sum_{i=1}^{n} x_i y_i \\ \vdots \\ \sum_{i=1}^{n} x_i^m y_i
\end{bmatrix} \tag{9}
$$

The above Vandermonde matrix is simplified to attain

$$
\begin{bmatrix}
1 & x_1 & \cdots & x_1^m \\
1 & x_2 & \cdots & x_2^m \\
\vdots & \vdots & \ddots & \vdots \\
1 & x_n & \cdots & x_n^m
\end{bmatrix}
\begin{bmatrix}
a_0 \\
a_1 \\
\vdots \\
a_m
\end{bmatrix}
=
\begin{bmatrix}
y_1 \\
y_2 \\
\vdots \\
y_n
\end{bmatrix}
\tag{10}
$$

Equation (10) can be expressed as $X \times A = Y$, so the coefficient matrix with the optimal parameters is attained as

$$
A = (X' \times X)^{-1} \times X' \times Y \tag{11}
$$

where matrix X is the dataset of rotary angles, and matrix Y is the dataset of corresponding distances with compensated parameters Δl. Therefore, when the rotary angles and corresponding distances detected are substituted into the above expression, the fitting expression can be established and the location point of the target object can be determined. Figure 5 shows the result of polynomial fitting to one of the original datasets in our study. Thus, the object location and the ultrasonic distance data can be determined and regarded as the input dataset to further classify the object feature types such as flat surfaces, corners, and cylinders.

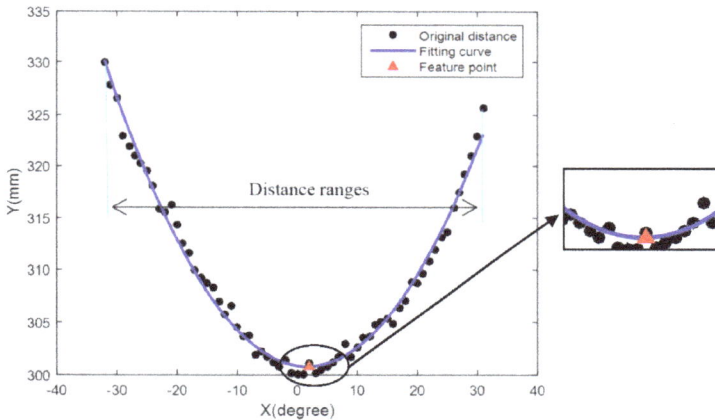

Figure 5. Determination of object location by MDP fitting.

4. Feature Classification Based on Fuzzy Model

Fuzzy theory, which was introduced by Zadeh [21], is a mature solution in motion control, signal processing, and artificial intelligence. The principle of fuzzy theory is to attain an analytical result by applying expert knowledge and practical experience in the form of reasoning rules. Although fuzzy theory is a conventional algorithm, it has many advantages, such as no requirement for a physical expression, less computational time, and strong robustness [22–24]. Generally, building a fuzzy controller consists of the following steps: variable definition, fuzzification, implementation of fuzzy rules, and defuzzification. In this study, a fuzzy model is proposed to recognize and classify object feature types such as corners, flat surfaces, and cylinders. Figure 6 shows the procedure of fuzzy modeling for feature classification. Firstly, the object features are scanned and the distance dataset is attained by the ultrasonic array. The object location is determined by MDP compensation. Also, the ranges of the distance dataset are regarded as the input variables of the fuzzy model. Secondly, the distance data are treated by fuzzification and the reasoning rules relating to boundary state are established. Finally, the types of object feature can be recognized and classified as the outputs of the fuzzy model.

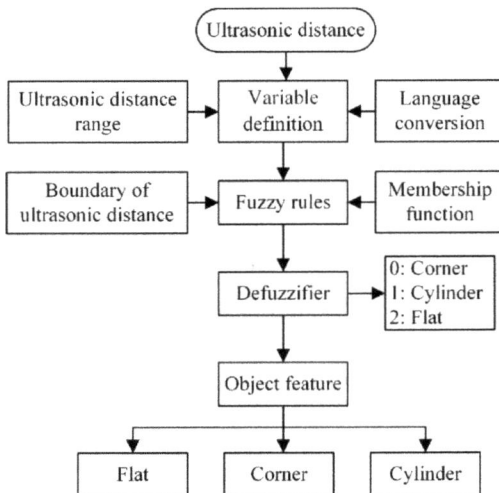

Figure 6. Procedure of fuzzy classification model.

4.1. Variable Definition and Language Conversion

4.1.1. Ultrasonic Distance Range

As implicit in the data measured by the ultrasonic sensor array, the range of the distance dataset has a relationship with the objects' geometric features. In our study, the flat surfaces have the maximum data range, corners the smallest, and the range for cylindrical objects is medium. Thus, the language conversion and data fuzzification can be expressed as in Table 1.

Table 1. Language conversion of ultrasonic distance range

State	Small	Medium	Big
Value	0	1	2

4.1.2. Boundary Definition

The boundary of the fuzzy model can be defined by the covering state (i.e., covered or uncovered) of the distance detected. The range of the distance dataset is reduced if some of the distance detected is covered. Table 2 shows the boundary states of the distance detected in our study, including the non-covering, left-covering, right-covering, and front-covering states, with corresponding setting values of 0, 1, 2 and 3, respectively.

Table 2. Boundary definition

State	Non-Covering	Left Covering	Right Covering	Front Covering
Value	0	1	2	3

4.1.3. Output Feature Types

As shown in Table 3, the classification of features as corner, cylinder, and flat surface, which are the output of the fuzzy model, are defined as 0, 1, and 2, respectively.

State	Corner	Cylinder	Flat
Value	0	1	2

4.2. Data Fuzzification

Data fuzzification is a process to convert the input and output variables into suitable linguistic values. In the fuzzy model, it is essential to fuzzify the ultrasonic distance ranges. Because the detected distance dataset may be fluctuant, the upper limit L_{max} of the distance dataset is defined as the average of the maximum values $L_{max\,1}$, $L_{max\,2}$ of two repeated scans, as

$$L_{max} = \frac{L_{max1} + L_{max2}}{2} \tag{12}$$

Based on the results of our practical testing, the maximum membership degrees of the big, medium, and small states, which can be expressed by L_b, L_m, and L_s, are defined as

$$\begin{cases} L_b = L_{max} \times 0.9 \\ L_m = L_{max} \times 0.55 \\ L_s = L_{max} \times 0.15 \end{cases} \tag{13}$$

The membership function of ultrasonic distance range is attained

$$\mu_0 = \begin{cases} 1, & 0 \leq u \leq L_s \\ \frac{-u}{L_l - L_s} + \frac{-L_l}{L_s - L_l}, & L_s < u \end{cases} \tag{14}$$

$$\mu_1 = \begin{cases} \frac{u}{L_m - L_s} + \frac{L_l}{L_s - L_m}, & L_s < u < L_m \\ \frac{-u}{L_b - L_m} + \frac{-L_b}{L_m - L_b}, & L_m \leq u < L_b \end{cases} \tag{15}$$

$$\mu_2 = \begin{cases} \frac{u}{L_l - L_s} + \frac{L_s}{L_s - L_l}, & L_m < u < L_l \\ 1, & L_l \leq u \end{cases} \tag{16}$$

$$\begin{cases} L_1 = \frac{L_b L_s - L_s^2}{-2L_s + L_b + L_m} \\ L_2 = \frac{L_b^2 - L_s L_m}{2L_b - L_s - L_m} \end{cases} \tag{17}$$

All of the input variables are normalized into [0, 1]. The linguistic values of the inputs and outputs are quantified using the membership functions. Thus, the membership curve can be established as in Figure 7.

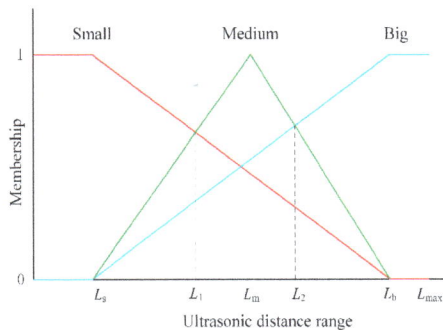

Figure 7. Membership of ultrasound distance range.

4.3. Fuzzy Rules and Defuzzification

Fuzzy rules are used within fuzzy logic systems to deduce an output based on input variables, and defuzzification is the process of producing a fuzzy set into a crisp number.

4.3.1. Fuzzy Rules

The boundaries of ultrasonic distance, and their state (i.e., covered or not), have a significant influence on the ultrasonic distance range. Therefore, the effect of boundary state on ultrasonic distance range must be considered in the reasoning rules of the fuzzy model. It is necessary to redefine the membership of ultrasonic distance range at different boundary states, as shown in Table 4.

Table 4. Membership redefinition of ultrasonic distance range

Membership		Boundary State			
		0	1	2	3
Ultrasonic distance range	0	μ_0 μ_1 μ_2	$\mu_0 - 0.25$ $\mu_1 + 0.25$ μ_2	$\mu_0 - 0.25$ $\mu_1 + 0.25$ μ_2	$\mu_0 - 0.3$ μ_1 $\mu_2 + 0.3$
	1	μ_0 μ_1 μ_2	μ_0 $\mu_1 - 0.2$ $\mu_2 + 0.2$	μ_0 $\mu_1 - 0.2$ $\mu_2 + 0.2$	μ_0 $\mu_1 - 0.3$ $\mu_2 + 0.3$
	2	μ_0 μ_1 μ_2	μ_0 μ_1 μ_2	μ_0 μ_1 μ_2	μ_0 μ_1 μ_2

In Table 4, when the ultrasonic distance range is 2 with the big setting, the memberships of ultrasonic distance keep constant when the boundary is 0, i.e., non-covering. They also keep constant even when the boundary is 1, 2 and 3, because in that case the ultrasonic distance range has a strong stable ability to avoid being disturbed by the covering. When the ultrasonic distance range is 0 with the small setting, the membership is decreased by 0.25 if the boundary is 1 (left covering) or 2 (right covering). The membership is decreased by 0.30 if the boundary is 3 (front covering). Likewise, the membership redefinition of distance range 1 in different covering states can be deduced by similar rules.

4.3.2. Defuzzification

When the maximum membership degree is μ_0, the output of the fuzzy model is defined as 0 for a corner feature. Similarly, the outputs of the fuzzy classification model are 1 for a cylindrical feature, and 2 for a flat surface, when the maximum membership degree is μ_1 and μ_2, respectively.

5. Experiment and Verification

To verify the feasibility of the MDP algorithm and fuzzy model applied in feature classification for mobile robotics, experimental testing and verification analysis are carried out. The environment for classification and mapping is designed as in Figure 8, and includes flat surfaces, corner, and cylinder features. A mobile robot mounted with the ultrasonic sensor array travels by a defined trajectory from A, B, C, . . . , to J. In the experiment, the ultrasonic distance data are detected and preprocessed by some filter method such as K-means cluster [25]. The location and the distance range of the ultrasonic dataset are extracted by the MDP algorithm. Then, the fuzzy model is utilized to classify the features of flat surface, cylinder, and corner. Finally, line fitting, least-squares curve fitting, and integrated combined methods are adopted to generate and construct the environmental map. The flowchart of the feature extraction, feature classification, and mapping is shown in Figure 9.

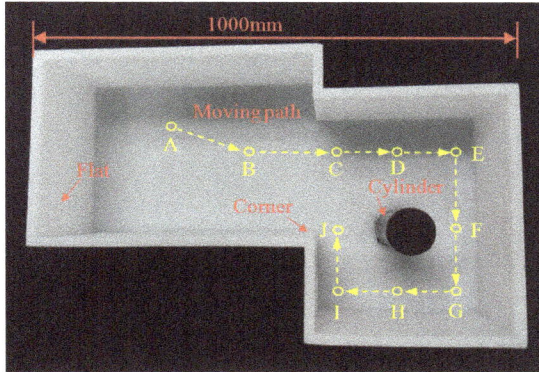

Figure 8. Environment for feature classification and mapping.

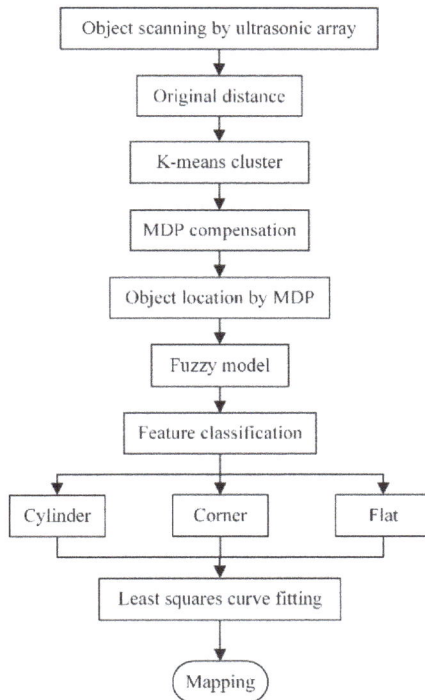

Figure 9. Flowchart of feature extraction, classification, and mapping.

5.1. Distance Data Detection and Preprocessing

In our experiment, 10 groups of ultrasonic distance data are measured and the original data are visualized by MATLAB, as shown in Figure 10a. It is found that the original distance dataset includes some noise and regions of discrete distribution. To reduce the noisy data, a K-means filter is utilized to cluster the original data, and the result is shown in Figure 10b. It can be seen that the noisy and discrete data are cleared and the ultrasonic distance data representing the line and arc features are displayed in outline.

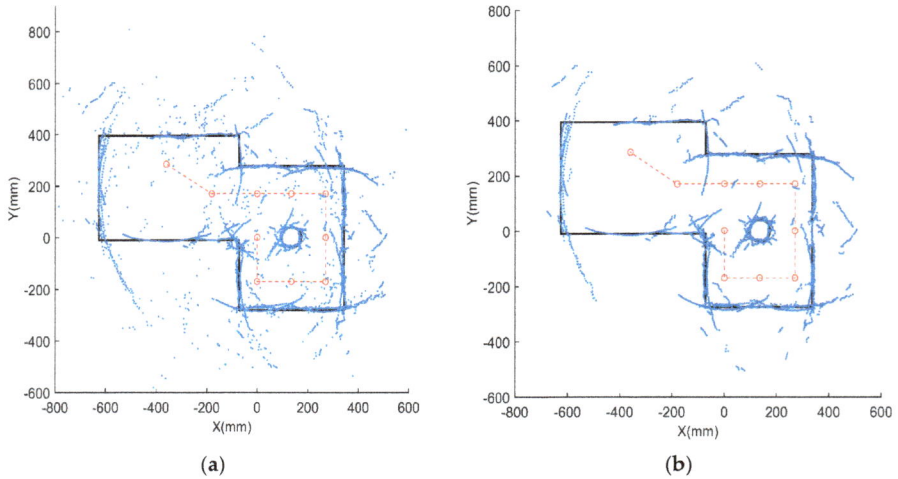

Figure 10. Preprocessing of original data. (**a**) Original distance dataset; (**b**) distance dataset after preprocessing.

5.2. Feature Extraction and Classification

Having attained the ultrasonic distance and its range by data preprocessing, the distance data are compensated further by the MDP method to achieve more accuracy. Then, the MDP algorithm with least-squares polynomial fitting is applied to extract the minimum distance of the distance dataset, allowing the object location to be determined. The result is shown in Figure 11. The preprocessed distance data are clustered again by the differential distance clustering method. It is also confirmed that the ranges of the flat surfaces, cylinder, and corner are big, medium, and small, respectively.

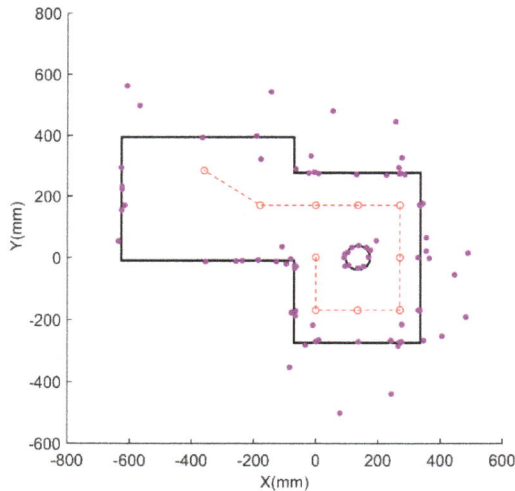

Figure 11. Object location extraction by MDP.

The clustered ultrasonic distance range and the boundary state are input into the fuzzy classification model. After the data fuzzification, reasoning, and defuzzification, the feature types, that is, the flat surfaces, cylinders, and corners, are output as the classification result. In our experiment, eight flat features, one cylindrical feature, and one corner feature are attained by fuzzy classification. The classification results are shown in Figure 12.

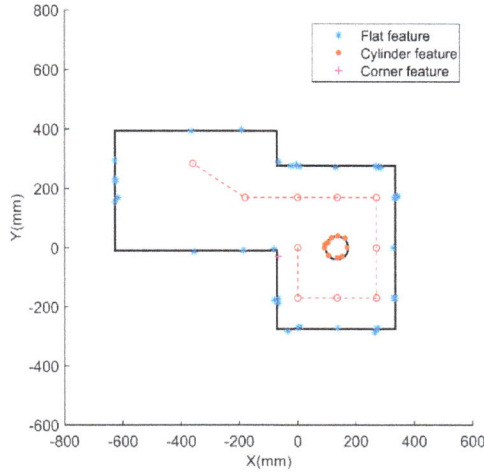

Figure 12. Feature classification by fuzzy model.

5.3. Mapping Construction

The line fitting method is used to connect and form straight lines, an arc, and a corner. In our study, eight straight lines representing the flat features are generated, and the intersections between the straight lines are calculated. Least-squares curve fitting is performed to generate an arc, which represents the cylinder feature. The corner feature is generated by the intersection of two lines. Finally, a comprehensive environment map is constructed successfully, as shown in Figure 13.

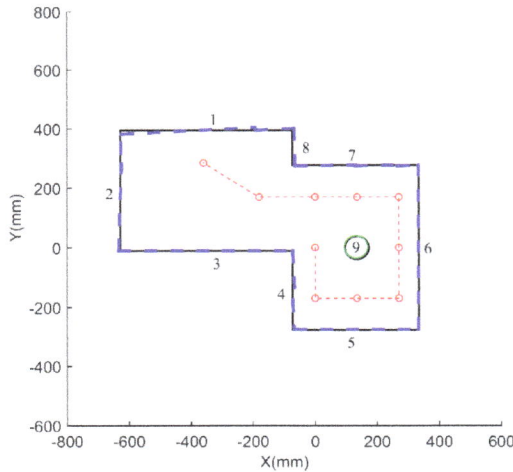

Figure 13. Mapping construction result.

To evaluate the mapping accuracy, the map constructed by MDP and the fuzzy model is compared with the physically measured map. The comparison results are summarized in Table 5. It is shown that, in our designed environment and the setting parameters in fuzzy model, the flat surface with index 1 has the highest distance deviation of 12.8 mm, and its angle deviation is 3°. The flat surface with index 5 has the lowest distance deviation of 0.5 mm and its angle deviation is 0°. In this example studied with basic geometry, it is verified that the MDP and fuzzy model can realize map construction with high accuracy and multiple features by using the rotary ultrasonic sensor array.

Table 5. Accuracy of the mapping result

Index	Distance Uncertainty (mm)	Angle Uncertainty (°)
1	12.8	3
2	6.7	2
3	1.0	0
4	7.8	3
5	0.5	0
6	1.8	0
7	3.8	1
8	7.0	1

6. Conclusions

In this paper, an ultrasonic sensor array with MDP and fuzzy modeling is developed as a feasible, low-cost approach for environment modeling of an indoor mobile robot. A configuration with a rotary ultrasonic sensor array is design to detect the distance of objects in a 360° scope. The MDP algorithm and fuzzy model are proposed to extract and classify environmental features such as flat surfaces, cylinder, and corner. The conclusions are summarized as follows.

(1)   The MDP feature extraction algorithm is proposed and the least-squares polynomial curve fitting is applied to extract the minimum-distance point of the ultrasonic distance dataset. This MDP method can compensate for the feature location error in the conventional RCD algorithm.

(2)   A feature classification model based on fuzzy theory is established. The ultrasonic distance range and distance state detected by the ultrasonic sensor are regarded as the input of the fuzzy model. The rules of data fuzzification, reasoning, and defuzzification are defined according to practical testing experience. With the fuzzy model, the classification results—including flat surfaces, a corner, and a cylindrical feature—can be attained.

(3)   The object feature extraction and the mapping construction by using MDP and the fuzzy model are successfully verified by experiment. A fully accomplished case of mapping with MDP and the fuzzy model is demonstrated in our study using basic regular geometry. Compared with the physical map, the distance error constructed by the rotary ultrasonic array is 1.2 cm and the angle error is 3°, representing high accuracy of environmental mapping for indoor mobile robotics.

Author Contributions: Conceptualization, methodology, and writing—review and editing, Z.L. (Zhili Long); Software, validation, and writing—original draft preparation, R.H.; Formal analysis, investigation, resources and data curation, Y.H.; Visualization and supervision, H.C.; Project administration and funding acquisition, Z.L. (Zuohua Li).

Funding: This work was supported by the following funds: (1) National Natural Science Foundation of China (U1713206), (2) Basic Research Plan of Shenzhen (JCYJ20170413112645981, JCYJ20160427183958817), and (3) Shenzhen Technology Innovation Program (JCYJ20170811160003571).

Conflicts of Interest: The authors declare no conflict of interest.

References

1. Truong, X.; Ngo, T.D. Toward Socially Aware Robot Navigation in Dynamic and Crowded Environments: A Proactive Social Motion Model. *IEEE Trans. Autom. Sci. Eng.* **2017**, *14*, 1743–1760. [CrossRef]
2. Souza Rosa, L.; Bonato, V. A method to convert floating to fixed-point EKF-SLAM for embedded robotics. *J. Braz. Comput. Soc.* **2013**, *19*, 181–192. [CrossRef]
3. Silver, D.; Morales, D.; Rekleitis, I.; Lisien, B.; Choset, H. Arc carving: Obtaining accurate, low latency maps from ultrasonic range sensors. In Proceedings of the 2004 IEEE International Conference on Robotics and Automation (ICRA), New Orleans, LA, USA, 26 April–1 May 2004; IEEE: Piscataway, NJ, USA, 2004; pp. 1554–1561.
4. Hebert, M. Active and passive range sensing for robotics. In Proceedings of the 2000 IEEE International Conference on Robotics and Automation (ICRA), San Francisco, CA, USA, 24–28 April 2000; IEEE: Piscataway, NJ, USA, 2000; pp. 102–110.
5. Benet, G.; Blanes, F.; Simó, J.E.; Pérez, P. Using infrared sensors for distance measurement in mobile robots. *Robot. Auton. Syst.* **2002**, *40*, 255–266. [CrossRef]
6. Blanco, J.; González-Jiménez, J.; Fernández-Madrigal, J. A robust, multi-hypothesis approach to matching occupancy grid maps. *Robotica* **2013**, *31*, 687–701. [CrossRef]
7. Wei, W.; Weidong, C.; Yong, W. Probabilistic Grid Map Based Localizability Estimation for Mobile Robots. *Robots* **2012**, *34*, 485–491.
8. Werner, F.; Sitte, J.; Maire, F. Topological map induction using neighbourhood information of places. *Auton. Robots* **2012**, *32*, 405–418. [CrossRef]
9. Du, Z.; Qu, D.; Xu, F.; Kai, J. Mobile Robot Map Building Based on Ultrasonic Sensors. In Proceedings of the National Conference on Access and Processing of Information, Weihai, China, 6 August 2010; pp. 389–394.
10. Leonard, J.J.; Durrantwhyte, H.F. Mobile Robot Localization by Tracking Geometric Beacons. *IEEE Trans. Robot. Autom.* **2002**, *7*, 376–382. [CrossRef]
11. Lee, S.J.; Dong, W.C.; Wan, K.C.; Lim, J.H.; Kang, C.U. Feature based map building using sparse sonar data. In Proceedings of the International Conference on Intelligent Robots and Systems, Edmonton, AB, Canada, 2–6 August 2005; pp. 492–496.
12. Ismail, H.; Balachandran, B. A Comparison of Feature Extraction Algorithms Based on Sonar Sensor Data. In Proceedings of the ASME 2013 International Mechanical Engineering Congress and Exposition, San Diego, CA, USA, 15–21 November 2013; p. V4A.
13. Heinen, M.R.; Engel, P.M. Incremental feature-based mapping from sonar data using Gaussian mixture models. In Proceedings of the ACM Symposium on Applied Computing, TaiChung, Taiwan, 21–24 March 2011; pp. 1370–1375.
14. Ismail, H.; Balachandran, B. Algorithm Fusion for Feature Extraction and Map Construction from SONAR Data. *IEEE Sens. J.* **2015**, *15*, 6460–6471. [CrossRef]
15. Lee, S.J.; Song, J.B. A new sonar salient feature structure for EKF-based SLAM. In Proceedings of the International Conference on Intelligent Robots and Systems, Taipei, Taiwan, 18–22 October 2010; pp. 5966–5971.
16. Yuan, S.; Huang, L.; Zhang, F.; Sun, Y.; Huang, K. A line extraction algorithm for mobile robot using sonar sensor. In Proceedings of the 2014 11th World Congress on Intelligent Control and Automation (WCICA), Shenyang, China, 29 June–4 July 2014; pp. 3630–3635.
17. Lee, K.; Chung, W.K. Effective Maximum Likelihood Grid Map with Conflict Evaluation Filter Using Sonar Sensors. *IEEE Trans. Robot.* **2009**, *25*, 887–901.
18. Suh, U.; Lee, Y.; Ra, W.; Kim, T.W. Robust Bearing Angle Error Estimation for Mobile Robots with a Gimballed Ultrasonic Seeker. *IEEE Trans. Ind. Electron.* **2018**, *65*, 5785–5795. [CrossRef]
19. Product Introduction. 2014. Available online: http://www.jx-servo.com/English/Product/49513727.html (accessed on 10 October 2018).
20. IntoRobot-Atom_Datasheet. 2015. Available online: http://dl.intoyun.com/terminal/modules/datasheets/en/IntoRobot-Atom_Datasheet_en.pdf (accessed on 10 October 2018).
21. Kinnaird, C.D.; Khotanzad, A. Adaptive fuzzy process control of integrated circuit wire bonding. *IEEE Trans. Electron. Packag. Manuf.* **2002**, *22*, 233–243. [CrossRef]

22. Liu, S.X.; Tong, F.; Luk, B.L.; Liu, K.P. Fuzzy pattern recognition of impact acoustic signals for nondestructive evaluation. *Sens. Actuators A Phys.* **2011**, *167*, 588–593. [CrossRef]

23. Zapata, G.A.; Kawakami, R.; Galvao, H.; Yoneyama, T. Extracting fuzzy control rules from experimental human operator data. *IEEE Trans. Syst. Man Cybern. Part B Cybern.* **1999**, *29*, 398–406. [CrossRef] [PubMed]

24. Meng, J.E.; Mandal, S. A Survey of Adaptive Fuzzy Controllers: Nonlinearities and Classifications. *IEEE Trans. Fuzzy Syst.* **2016**, *24*, 1095–1107.

25. Kanungo, T.; Mount, D.M.; Netanyahu, N.S.; Piatko, C.D.; Silverman, R.; Wu, A.Y. An Efficient k-Means Clustering Algorithm: Analysis and Implementation. *IEEE Trans. Pattern Anal. Mach. Intell.* **2002**, *24*, 881–892. [CrossRef]

sensors

MDPI

Article

Sensor Distribution Optimization for Structural Impact Monitoring Based on NSGA-II and Wavelet Decomposition

Peng Li *, Liuwei Huang and Jiachao Peng

School of Mechatronics & Vehicle Engineering, East China Jiaotong University, Nanchang 330013, China; 18270697178@163.com (L.H.); 13155825909@163.com (J.P.)
* Correspondence: ecjtulipeng@126.com; Tel.: +86-151-7005-2867

Received: 14 October 2018; Accepted: 27 November 2018; Published: 4 December 2018

check for updates

Abstract: Optimal sensor placement is a significant task for structural health monitoring (SHM). In this paper, an SHM system is designed which can recognize the different impact location and impact degree in the composite plate. Firstly, the finite element method is used to simulate the impact, extracting numerical signals of the structure, and the wavelet decomposition is used to extract the band energy. Meanwhile, principal component analysis (PCA) is used to reduce the dimensions of the vibration signal. Following this, the non-dominated sorting genetic algorithm (NSGA-II) is used to optimize the placement of sensors. Finally, the experimental system is established, and the Product-based Neural Network is used to recognize different impact categories. Three sets of experiments are carried out to verify the optimal results. When three sensors are applied, the average accuracy of the impact recognition is 59.14%; when the number of sensors is four, the average accuracy of impact recognition is 76.95%.

Keywords: structural impact monitoring; sensors distribution optimization; NSGA-II; energy analysis of wavelet band; principal component analysis

1. Introduction

With the extensive utilization of load-carrying structures in various engineering applications, there has been increasing interest in methods for predicting and estimating the location and extent of impact damage in structures [1–4]. On the basis of recent research advances, a concept of damage diagnostics for real-time structure monitoring, namely, structural health monitoring (SHM), has been proposed. SHM aims to ensure structural safety by using information provided by the sensor network. Normally, as for a simple structure, the design of a sensor network is based on the engineers' judgment. However, for complex structures, sensor location and the number of sensors in the network become fundamental optimization issues, which cannot be ignored. Under the premise of guaranteed performance, fewer sensors to cover an area will be considered, and the proper placing of these sensors will reduce costs.

In previous studies, sensor distribution optimization is only devoted to the location optimization of sensors, and the number of sensors is limited. This research involved in many modern heuristic algorithms such as genetic algorithm (GA) [5,6], simulated annealing (SA) [7], monkey algorithm (MA) [8], ant colony algorithm (ACO) [9] and differential evolution (DE). Kim established a novel particle swarm optimization framework to achieve robust consensus of decentralized sensors with neighbors rather than through centralized control [10]. An algorithm based on ladder diffusion and ACO is proposed to solve the power consumption and transmission routing problems in wireless sensor networks [9]. Among them, genetic algorithm has been introduced as a promising method for handling single-objective optimization, due to its better convergence, higher calculation precision, lower calculation time and higher robustness.

However, the optimization of the sensors network requires a reduction of the number of sensors and increased monitoring accuracy. It is obviously a multi-objective optimization problem (MOOP). Inherently, sensor location and the number of sensors are two conflicting goals for the design of sensor networks. In recent years, in order to solve the multi-objective optimization problem, many researchers improved the initial heuristic algorithm [11]. Céspedes-Mota uses improved differential evolution algorithm to optimize the distribution of wireless sensor networks according to the distance arranged by sensors [12]. Deb proposes the non-dominated sorting genetic algorithm to solve multi-objective optimization problems [13]. Li also uses this method to optimize the sensors network: he uses the number of sensors and the feature difference among all impact categories as two objective functions, respectively. A set of optimal sensor networks are obtained by the NSGA-II method [14].

In this paper, the NSGA-II method is used to optimize the sensor network. The remaining part of the paper is organized as follows: Section 2 describes the problem. The two objective functions are defined as the following: (i) The number of sensors; (ii) the feature differences among different impact categories. Section 3 describes how to solve the optimization problem by using the idea of MOOP. Moreover, wavelet band energy extraction and PCA are combined to obtain the value of the objective function. The result of sensors distribution optimization by NSGA-II is described in Section 4. Meanwhile, in order to prove the superiority of genetic algorithm, a comparative study is conducted with MA. In Section 5, the performance of the proposed algorithm in optimizing sensor distribution is verified by experiments. Finally, the paper is concluded in Section 6.

2. Description of Problem

In this paper, the composite laminate is used as the object. The study aims to identify structural impact damage, including impact location and impact degrees, by optimizing sensor networks. The adopted material is a [0 deg/90 deg] s-glass/epoxy orthogonal anisotropic laminate. The parameters of composite laminate are shown in Table 1.

Table 1. Parameters of composite laminate.

Equipment	Parameter
thickness and area	15 mm $\times$ 500 mm $\times$ 500 mm
elastic modulus	$E_z = 7.2$ GPa, $E_x = E_y = 6.9$ GPa
Poisson ratio	$V_{xz} = V_{yz} = 0.29$, $V_{xy} = 0.28$
shear elasticity	$G_{xz} = G_{yz} = 7.6$ GPa, $G_{xy} = 4.4$GPa
density	2100 Kg/m^3

As shown in Figure 1, the composite plate is divided into 9×9 grids, which includes 64 grid nodes. The impact load is applied at each node respectively. And each position is subject to two degrees impact, so there are in total 128 impact categories. The sensors are also positioned in some of these 64 nodes. Thus, there are two optimization objectives:

(1) Minimizing the number of sensors
(2) Maximizing the sensor network's optimization performance index based on impact categories

The method of getting the objective function will be introduced in the next section.

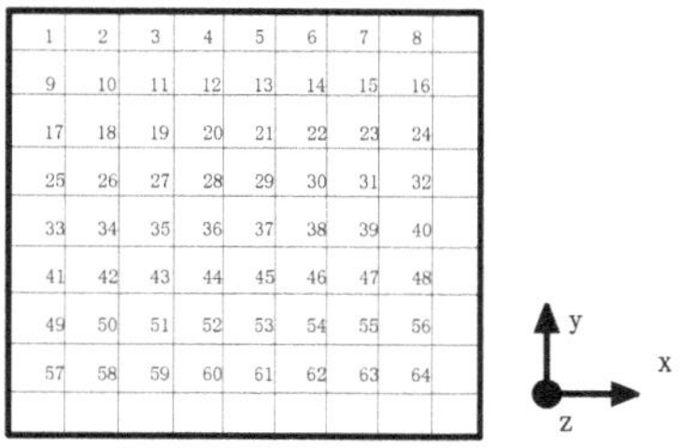

Figure 1. The grid division of composite.

3. Problem Formulation

Because there are two conflicting optimization objectives in the process of sensor network optimization, the optimization problem is a non-dominated multi-objective optimization. The optimization results will include a number of Pareto optimal solutions. Each solution is called non-dominated Pareto optimal, Pareto efficient or non-inferior. Without additional subjective preference information, all Pareto optimal solutions are considered equal. In this paper, the non-dominated sorting genetic algorithm II (NSGA-II) is used to obtain the Pareto optimal solutions of sensor networks.

3.1. Objective Function I

Before solving this problem, the objective function needs to be defined. Reducing the number of sensors can not only reduce the cost of monitoring system, but also accelerate the processing speed of data. The number of sensors is set as the objective function I.

In this paper, the encoding method of the sensor network is binary coding. Because the sensor has 64 alternate locations, each sensor network can be represented by 64 binary digits. When the value of the element is 0 there is no sensor in the position; when the value of the element is 1, the sensor is present in that position. So the sensor network is obtained ($\vec{S} = [s_1, s_2, \cdots, s_{64}]^T$). For example, if the sensor is placed in position 1, the vector $\vec{S}$ is equal to [10000000 00000000 00000000 00000000 00000000 00000000 00000000 00000000]^T.

3.2. Objective Function II

3.2.1. Numerical Simulation

For the purpose of improving the sensor network optimization performance index, the software ANSYS, which performs well with finite element analysis, is used to obtain the shock response of each category impact. It can provide effective data for sensor network optimization. As a rectangular structure is used in this paper, solid units are used. The composite material has a layered structure inside, so layered units are used. Solid-layered-46 is selected as the simulated entity type. Following this, the real constant is defined. The thickness of the composite plate is 15 mm, and the laminate has 10 layers, each with a thickness of 1.5 mm. The internal structure of the laminate is anisotropic orthogonal, that is, the laminate form of the laminate is (0/90) s.

The model is divided into 9 × 9 orthogonal distribution grids. The composite laminate is supported at the four corners, and the finite element analysis model is shown in Figure 2.

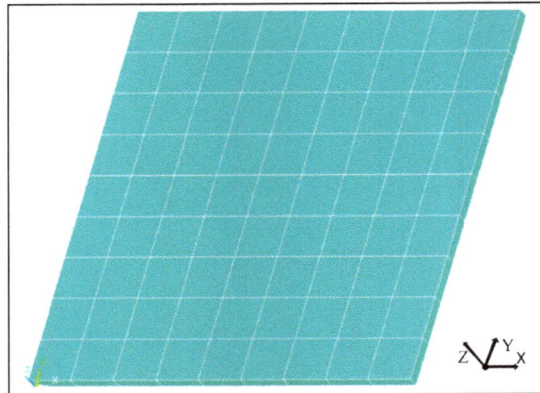

Figure 2. Finite element analysis model of the composite laminate.

As shown in Figure 1, there are 64 nodes on the composite plate. Each node is subjected to two degrees of impact—the full-load impact and half-load impact. The impact process of full-load is divided into two steps. The first load is $30 \times \sin(1744.4 \times t), 0 \leq t \leq 0.0018$ s and is divided into 9 sub-steps. The second load is a zero force, used in order to get the free shock response of the composite laminate. The second load includes 991 sub-steps and the total loading time is 0.1982 s. The total impact process of the simulation analysis is 0.2 s, and the load sub-steps include 1000 steps. The frequency of simulated vibration signal is 5k Hz. The half-load impact steps are similar to the whole load impact steps, the difference between them being impact strength. The half-load of the first step is $15 \times \sin(1744.4 \times t), 0 \leq t \leq 0.0018$ s. Linear analysis is used in this paper because the impact of the experiment is elastic deformation.

The vibration response can be obtained by ANSYS software when the structure is impacted. For example, the response of the 28th downside grid node is obtained as shown in Figure 3. The vibration response curve of the 28th downside grid node is obtained under the full-load impact to the 28th upside grid node.

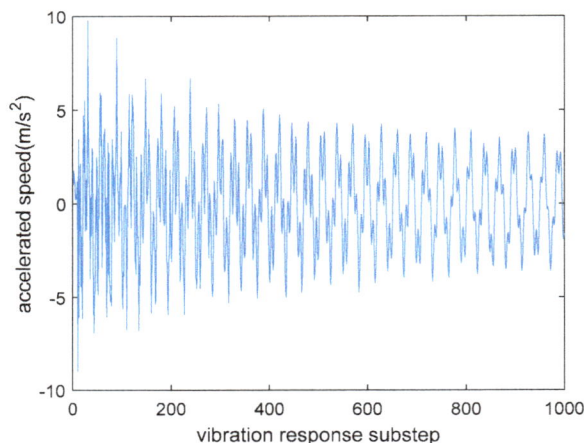

Figure 3. Shock response of 28th downside grid node with load acted on 28th upside grid node.

Since each upside node is subject to two degrees of impact and the upside board of composite has a total 64 nodes, for the *i*th downside node it will receive a total of 128 vibration signals

$\overrightarrow{M}^i_p (p = 1, 2, \ldots 128)$. According to the result of finite element analysis, the shock response matrix at the *i*th downside grid node M^i is defined as

$$M^i = \begin{bmatrix} \overrightarrow{M}^i_1 \\ \vdots \\ \overrightarrow{M}^i_{128} \end{bmatrix} = \begin{bmatrix} M^i_{1,1} & \cdots & M^i_{1,c} \\ \vdots & \ddots & \vdots \\ M^i_{128,1} & \cdots & M^i_{128,c} \end{bmatrix} (i = 1, 2, \ldots, 64) \tag{1}$$

where *i* is the *i*th alternate sensor location and *c* is the length of simulation vibration data. As for the 64 downside grid nodes, the 64 shock response matrices are established by using the result of numerical simulation.

3.2.2. Energy Analysis of Wavelet Band

After obtaining the original simulation data, its features need to be extracted. In recent years, the wavelet transform has been introduced as a promising method in damage identification of structures due to its excellent performance in detecting signal singularity [15–17].

A wavelet packet transform (WPT) is an orthogonal wavelet transform (WT). It inherits the idea of short-time Fourier transform localization [18,19]. Therefore, wavelet feature extraction is often used for impact location. At the same time, the time difference localization method is also a common positioning method [20]. However, our research not only needs to identify the impact location, but also to identify the impact level. Compared with the time difference positioning method, wavelet transform can not only reflect the time difference of the signal, but also reflect the difference in signal strength. Therefore, we chose wavelet decomposition to extract features. Firstly, the principle of wavelet transform needs to be introduced. A function $\psi \in L^2(R)$ is called an orthonormal wavelet while it can be defined by a Hilbert basis. It is a complete orthonormal system for the Hilbert space $L^2(R)$ of square integrable functions. Therefore, the vibration signal can be decomposed by wavelet signal:

$$M^i_p(x) = \sum_{j=1}^{N} \sum_{k \in Z} d^j_k \varphi_{jk}(x) + \sum_{k \in Z} c^N_k \varphi_{Nk}(x) \tag{2}$$

where, *N* is the decomposition layer, d^j_k is the detail coefficient of the *j*th layer, c^N_k is the approximate coefficient of the *n*th layer, and $\varphi_{Nk}(x)$ is the orthogonal scales function. According to Parseval's theorem, the set of orthogonal functions satisfies

$$\int_R \left| M^i_p(x) \right|^2 dx = \sum_{j=1}^{N} \sum_{k \in Z} \left| d^j_k \right|^2 + \sum_{k \in Z} \left| c^N_k \right|^2 \tag{3}$$

where $E_{dN} = \sum_{j=1}^{N} \sum_{k \in Z} \left| d^j_k \right|^2$, $E_{aN} = \sum_{k \in Z} \left| c^N_k \right|^2$, and E_{dN} are known as the total energy of the detail signal of the 1 to *N* layer, E_{aN} is the energy of the approximate signal of the *n*th layer. The total energy of signal $E = E_{dN} + E_{aN}$. The sum of the energy of the signals in each frequency band is consistent with the energy of the original signal, and the vibration signal in each frequency band represents the vibration characteristic information of the original signal in the frequency range. The energy expression of the *j*th layer in the vibration signal E_j can be defined as

$$E_j = \sum_{l=1}^{A} \left| d^j_l \right|^2 \tag{4}$$

where d^j_l is the decomposed signal and *A* is the number of discrete points in the corresponding time period. A set of signals are obtained which correspond to a sequence of energy from low to high frequencies $\{E_j | j = 1, 2, \cdots, 2^N\}$. Finally, row vector $\overrightarrow{E^i_p}$ is obtained by extracting the wavelet frequency band energy:

$$E^i = \begin{bmatrix} \overrightarrow{E}^i_1 \\ \vdots \\ \overrightarrow{E}^i_{128} \end{bmatrix} = \begin{bmatrix} e^i_{1,1} & \cdots & e^i_{1,2^N} \\ \vdots & \ddots & \vdots \\ e^i_{128,1} & \cdots & e^i_{128,2^N} \end{bmatrix} \quad (i = 1, 2, \ldots, 64) \tag{5}$$

The time-frequency energy based on wavelet decomposition reflects the energy E of the original signal over a certain period of time. It can more fully represent the signal characteristics of the data. In order to make computation more rapid, principal component analysis (PCA) is used to reduce the dimensions of the original data.

PCA is a statistical procedure that uses an orthogonal transformation to convert a set of observations of possibly correlated variables into a set of values of linearly uncorrelated variables called principal components. The basic idea is to obtain a set of optimal unit orthogonal vectors based on the linear transformation. The sample data is then rebuilt according to the above orthogonal vector basis, in order to minimize the mean square error between reconstruction samples and original samples [21–23]. The orthogonality between different features is represented by the contribution rate, and then the new data are selected from the original data by setting the cumulative contribution rates R_m. Through the projection of the data from the original 2^N-dimension space to b-dimension space ($2^N > b$), namely dimensionality reduction, the new data after dimension reduction can maximally retain the information of the original data. The steps are as follows:

For the Shock response data matrix of the ith node, the variable coefficient correlation matrix can be written as

$$\left(r_{gh}\right)^i = \frac{\sum_{k=1}^{128} \left(e^i_{kg} - \overline{x_g}\right)\left(e^i_{kh} - \overline{x_h}\right)}{\sqrt{\sum_{k=1}^{128} \left(e^i_{kg} - \overline{x_g}\right)^2 \sum_{k=1}^{128} \left(e^i_{kh} - \overline{x_h}\right)^2}} \tag{6}$$

where $\overline{x_g} = \frac{1}{128} \sum_{k=1}^{128} e^i_{kg}$, $\overline{x_h} = \frac{1}{128} \sum_{k=1}^{128} e^i_{kh}$, and $R^i = (r_{gh})^i$ are the variable coefficient correlation matrix $(2^N, 2^N)$. On the basis of equation $\left|R^i - \lambda I_{(2^N, 2^N)}\right| = 0$, the eigenvalue of matrix R can be obtained $\{\lambda_1, \lambda_2, \ldots, \lambda_{2^N}\}$. The cumulative contribution rate is described as

$$R^i_m = \sum_{j=1}^{m} \lambda_j / \sum_{t=1}^{2^N} \lambda_i \tag{7}$$

where R^i_m is the cumulative contribution rate, λ is the feature value, and m is the number of extracted principal component characteristics. Generally, the selection criteria of m need to satisfy the condition that the cumulative contribution rate is in the 85–95% range. In this study, the cumulative contribution rate is set at 95%. Then the matrix can be modeled as

$$X^i = \begin{bmatrix} \overrightarrow{X}^i_1 \\ \vdots \\ \overrightarrow{X}^i_{128} \end{bmatrix} = \begin{bmatrix} x^i_{1,1} & \cdots & x^i_{1,b} \\ \vdots & \ddots & \vdots \\ x^i_{128,1} & \cdots & x^i_{128,b} \end{bmatrix} \quad (i = 1, 2, \ldots, 64) \tag{8}$$

Through this procedure, we obtained the matrix $X^i(i = 1, 2, \ldots, 64)$ which included the feature vector of all impact categories.

3.2.3. Sensor Location Optimization Performance Index

The objective function II presents the sensor network optimization performance index. For all impact categories, a feature matrix I from a given sensor set S is shown as

$$I = \begin{bmatrix} X^{s_1} & \cdots & X^{s_f} \end{bmatrix} = \begin{bmatrix} \overrightarrow{X}_1^{s_1} & \cdots & \overrightarrow{X}_1^{s_f} \\ \vdots & \ddots & \vdots \\ \overrightarrow{X}_{128}^{s_1} & \cdots & \overrightarrow{X}_{128}^{s_f} \end{bmatrix} \tag{9}$$

The feature differences among all impact categories are expressed according to the distance of the row vectors of the feature matrix I:

$$Z = \begin{bmatrix} z_{1,1} & \cdots & z_{1,128} \\ \vdots & \ddots & \vdots \\ z_{128,1} & \cdots & z_{128,128} \end{bmatrix} \tag{10}$$

where

$$z_{u,v} = \sqrt{[I_u - I_v] \cdot [I_u - I_v]^{\mathrm{T}}} \ (u, v = 1, 2, \ldots, 128) \tag{11}$$

Objective function II, the sensor network optimization performance index, is defined as the minimum value of the matrix element. The greater the value of the performance index, the greater the difference between the different impacts. This makes it easier to identify different impacts. In order to improve the accuracy of impact recognition, objective function II is maximized in the sensor network optimization.

4. Sensor Network Optimization Algorithm

Based on the previously acquired objective functions, NSGA-II is employed to optimize the sensor network. NSGA-II algorithm is proposed by Deb on the basis of NSGA [13]. Its basic process is as follows:

Step 1: Random generation of original population S_t. Each of these individuals $S_t^g (g = 1, 2, \ldots, 200)$ represent a set of sensor arrangements, and each position represents a gene. The number of sensors in each individual is the value of objective function I. Set t = 1;

Step 2: Evaluate objective function II of each sensor set. In this case, each individual corresponds to two objective function values;

Step 3: According to the value of functions, the population is divided into different non-inferior layers $K_1, K_2, \ldots, K_n$. Firstly, find the individual in K_1. If $S_t^g (g = 1, 2, \ldots, 200)$ does not satisfy the following inequality:

$$f_I\left(S_t^g\right) \geq f_I\left(S_t^i\right), \ f_{II}\left(S_t^g\right) < f_{II}\left(S_t^i\right), \exists i = 1, 2, \ldots, 200$$

$S_t^g \in K_1$. After all individuals in K_1 are found, these individuals are labeled, and the individuals in $K_2, \ldots, K_n$ are found in the same way;

Step 4: Establish the optimization pool. Two individuals are selected randomly, and according to the non-inferior layers the better one is chosen and put into the optimization pool;

Step 5: Generating child population S_t' through crossover and mutation operations. Crossover operations—two individuals are randomly selected from the optimization pool and a portion of the genes in the two individuals are randomly exchanged. Figure 4 illustrates this process.

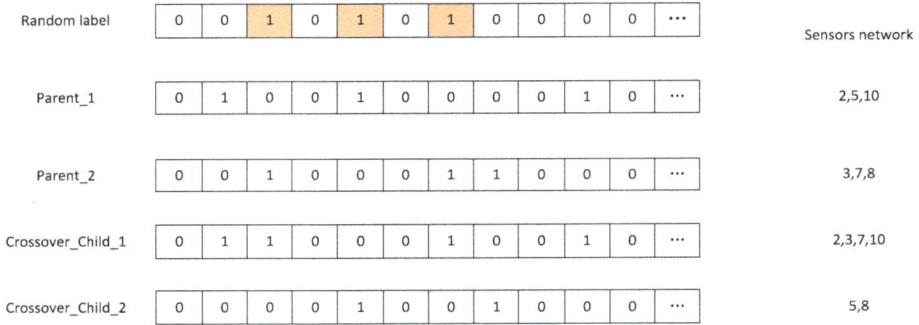

Figure 4. Operation of crossover.

As show in the right side of Figure 4, the number of sensors between the Parent and Crossover-Child may be different. So the method could optimize different numbers of sensor networks at the same time.

Step 6: Mutation operations—an individual is randomly selected from the pool of preferences, and a random exchange exchanges a portion of the genes in the individual. And Figure 5 illustrates this process.

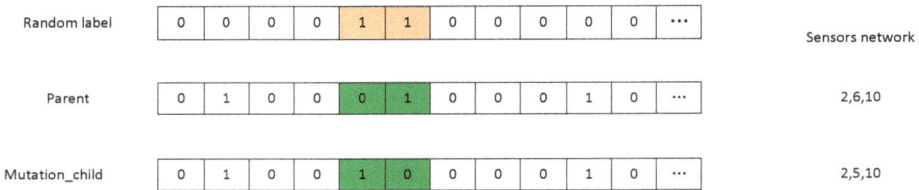

Figure 5. Operation of mutation.

Step 7: Obtain the combined population $S_t'' = S_t' \cup S_t$. According to the value of objective function I and objective function II, the population is divided into different non-inferior layers $K_1, K_2, \ldots, K_m$. The highly non-inferior layers' individuals are selected as the new population S_{t+1};

Step 8: Iteration ends if the end condition is reached or t = 10,000;

Step 9: Set t = t + 1, and go back to step 3.

The parameters of the NSGA-II are shown in Table 2.

Table 2. Parameter set of genetic algorithm (GA).

Parameter	Numeric
initial population size	min(n × 64,200)
crossover probability	0.5
mutation probability	0.16
off-springs population size after mutation operation and crossover operation	min(n × 64 × 1.5300)
termination condition	min(n × 64,200) (selection according to crowding distance)

5. Results and Discussion

The multi-objective optimization monkey algorithm (MOMA), composed of the monkey algorithm and NSGA-II, are used for sensor distribution optimization. The monkey algorithm (MA) is an intelligent optimization algorithm proposed by Zhao et al. [24] to solve large-scale and multi-peak optimization problems. The algorithm simulates the movement of monkeys in the process of climbing in nature, which includes climbing, looking and jumping. This algorithm includes three search processes: the climbing process is mainly used to search for the local optimal solution of the current location; the looking process mainly searches the neighborhood for better solutions than the current position in order to accelerate the search process of the optimal solution; the jump process is to search other areas to avoid the search process in the local area. In this paper, the concept of deep climbing is introduced on the basis of the classical monkey algorithm, which increases the range of optimal solutions and accelerates the convergence rate of the algorithm [25]. As the total number of positions of the sensor is 64, the fast distance is 64. In this paper, the climb process, watch–jump process and deep climb process thresholds are 8, 64 and 64, respectively.

The sensor network optimization results of NSGA-II and MOMA are obtained as shown in Table 3. From Table 3, it is observed that for the different designed thresholds of sensor number, all sensor sets of low threshold are included in sensor sets of high threshold. The NSGA-II and MOMA algorithms get the same optimal sensors network. This proved that the arrangement of the sensor network is optimal. The iteration times of the two algorithms are shown in Figure 6. Since the two algorithms get the same optimal solution, the convergence rate becomes the main index to decide which algorithm is better. NSGA-II is better than the MOMA in terms of iteration time.

Table 3. Sensor sets of Pareto solutions with the different designed thresholds of sensor number.

Designed Threshold	Solution No.	Sensor No. (NSGA-II)						Sensor No. (MOMA)					
	6	4	5	35	36	37	38	4	5	35	36	37	38
	5	18	27	28	30	55		18	27	28	30	55	
6	4	27	30	50	55			27	30	50	55		
	3	11	25	29				11	25	29			
	2	28	30					28	30				
	1	25						25					
	3	11	25	29				11	25	29			
3	2	28	30					28	30				
	1	25						25					
1	1	25						25					

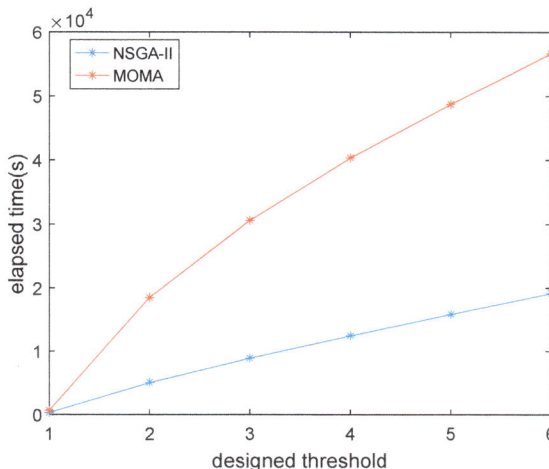

Figure 6. Elapsed time of (non-dominated sorting genetic algorithm) NSGA-II and multi-objective optimization monkey algorithm (MOMA) for the different designed thresholds of sensor number.

6. Method Evaluation

In order to verify the accuracy of impact recognition by sensor networks, an experimental system is designed. In this section, the experimental equipment and classification algorithm are introduced and experimental results are presented.

6.1. Experimental Setup

In this experiment, the identification includes impact degrees and impact locations. We suppose a rubber ball with a diameter of 20 mm dropped from 250 mm is considered a full-load shock, and the height of 125 mm freely dropped is seen as a half-load shock. At the same time, the difference of recognition accuracy and computation time between two, three and four sensors are also considered in the experiment.

The experimental system is shown in Figure 7. Firstly, the vibration signal is collected by the acceleration sensors. Secondly, the signal is introduced into the conditioning circuit to amplify the signal. Thirdly, the signal is imported into the data acquisition module. The parameters of the experimental installations are shown in Table 4.

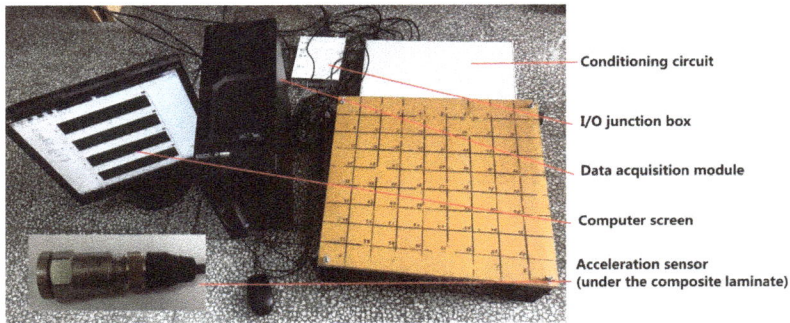

Figure 7. Experimental system.

Table 4. Experimental installation parameters.

Equipment	Model Number	Parameter
acceleration sensors	CA-YD-188T	with a range of -10 g to 10 g, sensitivity is 500 mV/g, frequency response is 0.6~5000
conditioning circuit	YE3826A	12 channels, with a gain of 10, the electric current output is 4 mA
I/O junction box	NI SCB-68A	16 channel analog input channel, custom cable connector kits and mounting accessories
data acquisition module	NI PCI-6251	16 analog inputs at 16 bits, 1.25 MS/s (1 MS/s scanning), Up to 4 analog outputs at 16 bits, 2.8 MS/s (2 µs full-scale settling), Analog and digital triggering, Two 32-bit, 80 MHz counter/timers

6.2. Localization Methodology

In the experiment, 100 samples are taken from each kind of full-load and half-load impact. At the same time, the experiments are carried out when the number of sensors network is 2, 3 and 4, respectively. Taking the number of sensors 4 as an example, the calculation process of impact category recognition is introduced.

Step 1: Dividing the original data. Fifty groups are randomly selected as training samples, and the other 50 groups are selected as test samples. Each category's training sample of a sensor is selected according to the similarity of vibration waveforms. If the average similarity is less than 0.5, this group of data is deleted. From the rest of the data, 30 groups are randomly selected as training samples, so a training sample matrix is obtained.

Step 2: Getting the training sample feature matrix. In the first place, energy analysis of wavelet band is used to extract features from vibration data. Then, principal component analysis (PCA) is used to reduce the dimensions of the training sample matrix. The cumulative contribution rate is 95%. Following this, the data of the 4 sensors are merged together to obtain the feature matrix $X_q^p (p = 1, 2, \ldots, 128, q = 1, 2, \ldots, 30)$.

$$X^p = \begin{bmatrix} \overrightarrow{X}_1^p \\ \vdots \\ \overrightarrow{X}_{30}^p \end{bmatrix} = \begin{bmatrix} x_{1,1}^p & \cdots & x_{1,m\times4}^p \\ \vdots & \ddots & \vdots \\ x_{30,1}^p & \cdots & x_{30,m\times4}^p \end{bmatrix} \quad (p = 1, 2, \ldots, 128) \tag{12}$$

Step 3: Selecting test samples. The average similarity between each test sample of each set of sensors and the shock response of each group's 30 training samples is calculated, respectively. If the similarity between all of the 128 impact classes is less than 0.5, the data of this shock are deleted. Afterwards, 20 groups from qualifying data are selected as test samples.

Step 4: Obtaining the test sample feature matrix. Test samples of the 4 sensors are multiplied by 4 transformation matrixes and the first m data are saved just like the training sample. Then, the matrices of the 4 sensors are combined into the same column, and the characteristics of the test samples for the pth vibration data $B_t^p (p = 1, 2, \ldots, 128, t = 1, 2, \ldots, 20)$ can be expressed as

$$B^p = \begin{bmatrix} \overrightarrow{B}_1^p \\ \vdots \\ \overrightarrow{B}_{20}^p \end{bmatrix} = \begin{bmatrix} b_{1,1}^p & \cdots & b_{1,m\times4}^p \\ \vdots & \ddots & \vdots \\ b_{20,1}^p & \cdots & b_{20,m\times4}^p \end{bmatrix} \quad (p = 1, 2, \ldots, 128) \tag{13}$$

Step 5: Identification of impact category by Product-based Neural Network (PNN). There are many algorithms for solving classification problems, among which PNN is a neural network commonly used in pattern classification [26]. Its training time is short and its classification accuracy is high. No matter how complex the classification problem is, as long as there are enough training data, the optimal solution under the Bayes criterion can be obtained. Therefore, this paper uses PNN algorithm for classification.

When inputting a vector $\overrightarrow{B}_t^p (p = 1, 2, \ldots, 128, t = 1, 2, \ldots, 20)$, the pattern layer computes the distance between the input vector and the training vectors $\overrightarrow{X}_q^p (p = 1, 2, \ldots, 128, q = 1, 2, \ldots, 30)$. The summation layer sums the contribution for each class of inputs and output a probability value $f_p(p = 1, 2, \ldots, 128)$.

Figure 8 shows the basic design of a PNN used for impact recognition. The feature vector $\overrightarrow{B}_q^p$ passes from the input layer through the pattern layer to the output layer. The neurons in the pattern layer enable mapping of the nonlinearity relations between the input and output values, which gives PNN models a better performance over others. The summation f_p is expressed as

$$f_p(B_t^p) = \frac{1}{a(2\pi)^{m\times4/2}\sigma^{m\times4}} \sum_{q=1}^{a} \exp\left(-\sum_{w=1}^{m\times4} \frac{(b_{tw}^P - x_{qw})^2}{2\sigma^2}\right) \tag{14}$$

where a is the training samples number of each category, σ represents the smoothness parameter, and the value of the smoothed parameter is 0.15. b_{tw} represents the wth data of the tth neuron of each sample. The summation layer has 30 neurons in each, with a total 128 categories. The output layer compares the votes for each target predict accumulated in the summing layer. The target category is predicted to the largest vote.

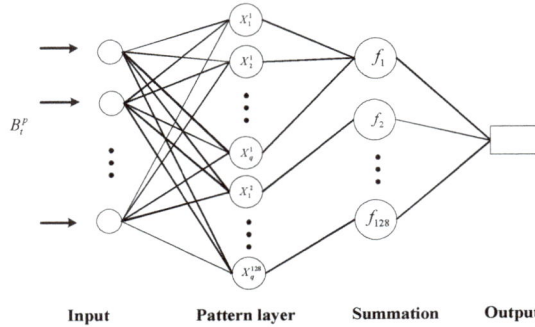

Figure 8. Probabilistic neural network structure for impact recognition.

6.3. Experimental Results and Discussion

Next, in order to confirm the effectiveness of our optimization results, four sets of sensor networks are selected for impact experiments. They are the first non-inferior layer arrangement of 2, 3, and 4 sensors, respectively. And the four sensors in the second non-inferior layer are also arrangement.We analyze the difference in impact identification between the different number of sensors and sensor locations. The impact recognition accuracy obtained by the above method is shown in Figure 7.

As shown in Figure 9, each figure includes 128-type impact recognition accuracy. The first 64 impact categories are half-load shocks and the last 64 are full-load shocks. From Figure 9, the recognition accuracy gradually increases with the number of sensors. Recognition accuracy of the half-load is slightly higher than recognition accuracy of the full-load. Table 5 gives detailed data.

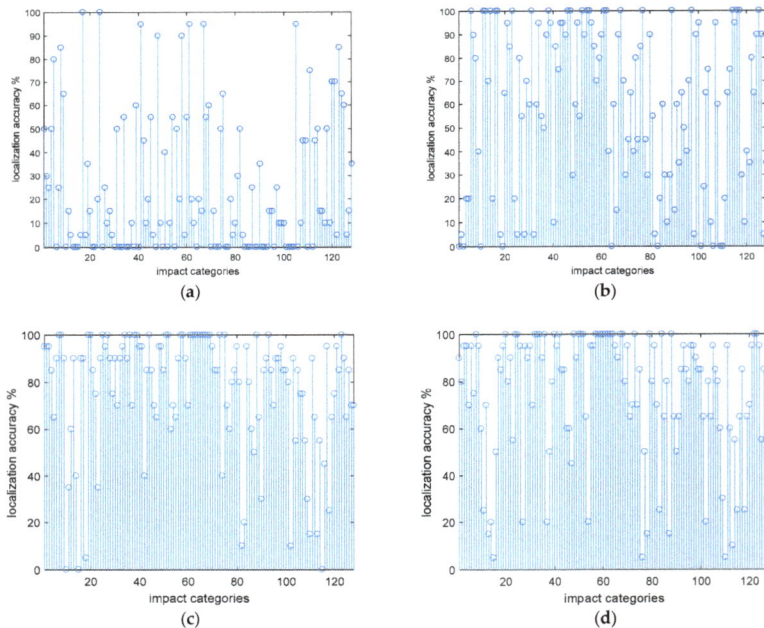

Figure 9. (**a**) The positioning accuracy when installing two sensors; (**b**) the positioning accuracy when installing there sensors; (**c**) the positioning accuracy when installing four sensors; (**d**) the positioning accuracy when installing four sensors, which are second non-inferior layer (10, 15, 35, 38).

Table 5. Different sensors network impact accuracy.

Sensors Network	Average Accuracy	Half-Load Impact Average Accuracy	Full-Load Impact Average Accuracy	Computation Time
(28, 30)	25.27%	26.88%	23.67%	6.01 ms
(11, 25, 29)	59.14%	65.31%	52.97%	8.36 ms
(27, 30, 50, 55)	76.95%	81.72%	72.19%	11.33 ms
(10, 15, 35, 38)	75.55%	80.23%	70.86%	11.33 ms

7. Conclusions

This paper proposes an optimization program of sensor networks for impact identification of composite laminates. The optimization objective includes the optimization of the number of sensors and the sensor network optimization performance index. The number of sensors is defined as objective function I, and the sensor location optimization performance index is defined as objective function II. In order to obtain objective function II, the finite element, energy analysis of wavelet band and PCA methods are used to data processing. Moreover, an experimental system was established to verify whether the impact recognition results were consistent with the assumptions. The final recognition accuracy is obtained using the above mentioned experimental system, and the conclusions are showing as follows:

(a) Comparing NSGA-II and MOMA, the two algorithms can get the same result. However, NSGA-II is better than the MOMA in terms of iteration time, and the gap will become more pronounced with increasing numbers of sensors;

(b) When the number of sensors in the sensor network is 2, 3, and 4, the average recognition accuracy is 25.27%, 59.14% and 76.95%, respectively. The results show that as the number of sensors increases, higher recognition accuracy is obtained;

(c) When there are four sensors, the identification accuracy of the first non-inferior sensor network is 76.95%, and the recognition accuracy of the second non-inferior sensor network is 75.55%. Results show that when the number of sensors is constant, the accuracy of impact recognition will be correlated with the sensor network optimization performance index. The experiment shows that the optimized sensors network can achieve the best impact recognition accuracy;

(d) When there are two sensors in the experiment, the average of full-load impact and half-load impact are 23.67% and 26.88%, respectively; with three sensors, the average of full-load impact and half-load impact are 52.97% and 65.31%, respectively; and with four sensors in the experiment, the average of full-load impact and half-load impact are 72.19% and 81.72%, respectively. The results show that the accuracy of the half-load is slightly higher than full-load under the same conditions;

(e) The calculation time of the experimental results shows that when the number of sensors increases, the recognition accuracy increases. However, computing time also increases. Therefore, in real application, it is necessary to select the appropriate number of sensors according to the real-time requirements.

Author Contributions: Data curation, L.H. and J.P.; Formal analysis, L.H.; Methodology, P.L.; Project administration, P.L.; Software, J.P.; Writing—original draft, L.H.; Writing—review and editing, P.L.

Funding: This research was funded by the National Natural Science Foundation of China, grant number 51365012.

Acknowledgments: This work is supported by the National Natural Science Foundation of China (Grant No. 51365012).

Conflicts of Interest: The authors declare no conflict of interest.

Sensors **2018**, *18*, 4264

References

1. Zhou, G.D.; Yi, T.H.; Li, H.N. Sensor Placement Optimization in Structural Health Monitoring Using Cluster-in-Cluster Firefly Algorithm. *Adv. Struct. Eng.* **2014**, *17*, 1103–1115. [CrossRef]
2. Mallardo, V.; Aliabadi, M.H. Optimal Sensor Placement for Structural, Damage and Impact Identification: A Review. *Sdhm Struct. Durab. Health Monit.* **2013**, *9*, 287–323.
3. Hajializadeh, D.; Obrien, E.J.; Oconnor, A. Virtual Structural Health Monitoring and Remaining Life Prediction of steel bridges. *Can. J. Civil Eng.* **2017**, *44*, 264–273. [CrossRef]
4. Yu, S.; Ou, J. Structural Health Monitoring and Model Updating of Aizhai Suspension Bridge. *J. Aerosp. Eng.* **2017**, *30*, B4016009. [CrossRef]
5. Kuang, K.; Maalej, M.; Quek, S.T. An Application of a Plastic Optical Fiber Sensor and Genetic Algorithm for Structural Health Monitoring. *J. Intell. Mater. Syst. Struct.* **2006**, *17*, 361–379. [CrossRef]
6. Buczak, A.L.; Wang, H.; Darabi, H.; Jafari, M.A. Genetic algorithm convergence study for sensor network optimization. *Inf. Sci. Inform. Comput. Sci. Int. J.* **2001**, *13*, 267–282. [CrossRef]
7. Tong, K.H.; Yassin, A.Y.M.; Bakhary, N.; Kueh, A.B.H. Optimal sensor placement for mode shapes using improved simulated annealing. *Smart Struct. Syst.* **2014**, *13*, 389–406. [CrossRef]
8. Yi, T.H.; Li, H.N.; Gu, M.; Zhang, X.D. Sensor Placement Optimization in Structural Health Monitoring Using Niching Monkey Algorithm. *Int. J. Struct. Stabil. Dyn.* **2014**, *14*, 1440012. [CrossRef]
9. Ho, J.H.; Shih, H.C.; Liao, B.Y.; Chu, S.C. A ladder diffusion algorithm using ant colony optimization for wireless sensor networks. *Inf. Sci. Int. J.* **2012**, *192*, 204–212. [CrossRef]
10. Kim, H.; Chang, S.; Kim, J. Consensus Achievement of Decentralized Sensors Using Adapted Particle Swarm Optimization Algorithm. *Int. J. Distrib. Sens. Netw.* **2014**, *2014*, 1–13. [CrossRef]
11. Li, P.; Liu, Y.; Zou, T.; Huang, J. Optimal design of microvascular networks based on non-dominated sorting genetic algorithm II and fluid simulation. *Adv. Mech. Eng.* **2017**, *9*, 1687814017708175. [CrossRef]
12. Céspedes-Mota, A.; Castañón, G.; Martínez-Herrera, A.F. Optimization of the Distribution and Localization of Wireless Sensor Networks Based on Differential Evolution Approach. *Math. Probl. Eng.* **2016**, *2016*, 1–12. [CrossRef]
13. Deb, K.; Pratap, A.; Agarwal, S.; Meyarivan, T.A.M.T. A fast and elitist multiobjective genetic algorithm: NSGA-II. *IEEE Trans. Evol. Comput.* **2002**, *6*, 182–197. [CrossRef]
14. Li, P.; Wang, Y.; Hu, J.; Zhou, J. Sensors distribution optimization for impact localization using NSGA-II. *Sens. Rev.* **2015**, *35*, 409–418. [CrossRef]
15. Sajid, M.; Ghosh, D. Logarithm of short-time Fourier transform for extending the seismic bandwidth. *Geophys. Prospect.* **2014**, *62*, 1100–1110. [CrossRef]
16. Wang, F.; Ma, X.; Zou, Y.; Zhang, Z. Local power feature extraction method of frequency bands based on wavelet packet decomposition. *Trans. Chin. Soc. Agric. Mach.* **2004**, *3*, 167–170.
17. Bianchi, D.; Mayrhofer, E.; Gröschl, M.; Betz, G.; Vernes, A. Wavelet packet transform for detection of single events in acoustic emission signals. *Mech. Syst. Sig. Process.* **2015**, *64*, 441–451. [CrossRef]
18. Suma, M.N.; Narasimhan, S.V.; Kanmani, B. Interspersed discrete harmonic wavelet packet transform based OFDM—IHWT OFDM. *Int. J. Wavelets Multiresolut. Inf. Process.* **2014**, *12*, 1450034. [CrossRef]
19. Yu, Z.; Xia, H.; Goicolea, J.M.; Xia, C. Bridge Damage Identification from Moving Load Induced Deflection Based on Wavelet Transform and Lipschitz Exponent. *Int. J. Struct. Stab. Dyn.* **2016**, *16*, 1550003. [CrossRef]
20. Mallardo, V.; Aliabadi, M.H.; Khodaei, Z.S. Optimal sensor positioning for impact localization in smart composite panels. *J. Intell. Mater. Syst. Struct.* **2013**, *24*, 559–573. [CrossRef]
21. Li, C.F.; Liu, L.; Lei, Y.M.; Yin, J.Y.; Zhao, J.J.; Sun, X.K. Clustering for HSI hyperspectral image with weighted PCA and ICA. *J. Intell. Fuzzy Syst.* **2017**, *32*, 3729–3737. [CrossRef]
22. Draper, B.A.; Baek, K.; Bartlett, M.S.; Beveridge, J.R. Recognizing faces with PCA and ICA. *Comput. Vis. Image Underst.* **2003**, *91*, 115–137. [CrossRef]
23. Zhou, S.; Mao, M.; Jianhui, S.U. Prediction of Wind Power Based on Principal Component Analysis and Artificial Neural Network. *Power Syst. Technol.* **2011**, *35*, 128–132.
24. Zhao, R.; Tang, W. Monkey algorithm for global numerical optimization. *J. Uncertain Syst.* **2008**, *2*, 165–176.

Sensors **2018**, *18*, 4264

25. Ting-Hua, Y.I.; Zhang, X.D.; Hong-Nan, L.I. Immune monkey algorithm for optimal sensor placement. *Chin. J. Comput. Mech.* **2014**, *30*, 174–179.
26. Sun, Q.; Wu, C.; Li, Y.L. A new probabilistic neural network model based on backpropagation algorithm. *J. Intell. Fuzzy Syst.* **2017**, *32*, 215–227. [CrossRef]

sensors

MDPI

Article

Optimal Placement of Virtual Masses for Structural Damage Identification

Jilin Hou [1,*], **Zhenkun Li [1]**, **Qingxia Zhang [2]**, **Runfang Zhou [1,3]** and **Łukasz Jankowski [4]**

[1] Department of Civil Engineering & State Key Laboratory of Coastal and Offshore Engineering, Dalian University of Technology, Dalian 116024, China; lizhenkunn@gmail.com (Z.L.); zhou0430@126.com (R.Z.)

[2] Department of Civil Engineering, Dalian Minzu University, Dalian 116650, China; zhangqingxia_hit@hotmail.com

[3] College of Urban and Rural Construction, Shanxi Agricultural University, Jinzhong 030801, China

[4] Institute of Fundamental Technological Research, Polish Academy of Sciences, Warsaw 02-106, Poland; ljank@ippt.pan.pl

* Correspondence: houjilin@dlut.edu.cn; Tel.: +86-411-847-06432

Received: 11 December 2018; Accepted: 15 January 2019; Published: 16 January 2019

✓ check for updates

Abstract: Adding virtual masses to a structure is an efficient way to generate a large number of natural frequencies for damage identification. The influence of a virtual mass can be expressed by Virtual Distortion Method (VDM) using the response measured by a sensor at the involved point. The proper placement of the virtual masses can improve the accuracy of damage identification, therefore the problem of their optimal placement is studied in this paper. Firstly, the damage sensitivity matrix of the structure with added virtual masses is built. The Volumetric Maximum Criterion of the sensitivity matrix is established to ensure the mutual independence of measurement points for the optimization of mass placement. Secondly, a method of sensitivity analysis and error analysis is proposed to determine the values of the virtual masses, and then an improved version of the Particle Swarm Optimization (PSO) algorithm is proposed for placement optimization of the virtual masses. Finally, the optimized placement is used to identify the damage of structures. The effectiveness of the proposed method is verified by a numerical simulation of a simply supported beam structure and a truss structure.

Keywords: damage identification; sensor optimization; Virtual Distortion Method (VDM); Particle Swarm Optimization (PSO) algorithm; sensitivity

1. Introduction

Nowadays, structural damage identification becomes a significant field in Structural Health Monitoring (SHM), and many new ideas are proposed in a growing number of studies. Spencer Jr. et al. [1] reviewed recent advances in wireless smart sensors for multi-scale monitoring and control of civil infrastructure. An et al. [2] proposed a novel method for computing the curvature directly from acceleration signals without identifying the modal shapes of the structure. Two examples were adopted to verify the effectiveness of the method, and its robustness to measurement noise. Hu et al. [3] reported on structural health monitoring of a prestressed concrete bridge based on statistical pattern recognition of continuous dynamic measurements over 14 years. Laflamme et al. [4] developed a soft capacitive sensor for structural health monitoring of large-scale systems; the performance of the sensor was then characterized for applications in dynamic vibration-based monitoring [5]. Yang et al. [6] proposed two methods for damage identification of a bridge based on measurements by a test vehicle. Fu [7] used wireless smart sensors to identify modes of structures to monitor sudden events in civil infrastructure. Li et al. [8] monitored fatigue cracks in steel bridges using a large-area strain sensing technology. Structural modes are the most basic characteristics of structures, and the approaches based on modal information are among the most commonly used methods of structural damage identification. Pnevmatikos et al. [9] introduced wavelet analysis for damage detection of a steel frame structure with bolted connections, and the presented experiment showed the effectiveness of the wavelet approach to damage detection of frame structures assembled using bolted connections. Ubertini [10,11] proposed an automated output-only modal identification procedure and utilized carbon nanotube cement-based sensors to identify natural frequencies of a reinforced concrete beam. Xu et al. [12] used embedded piezoceramic transducers to identify damage of a concrete column subject to blast loads. Zhang et al. [13] identified damage of concrete-filled square steel tube (CFSST) column joints under cyclic loading. Ginsberg et al. [14] identified damage parameters of framework by combining sparse solution techniques with an Extended Kalman Filter. The measurement equation was expanded by an additional nonlinear L1-minimizing observation to ensure sparsity of the damage parameter vector. Jiang et al. [15] monitored fatigue damage of modular bridge expansion joints using piezoceramic transducers. Zhang et al. [16] verified a method for concrete strength validation by smart aggregate-based stress monitoring. Most of these approaches are related to modes of structures. However, the modes that can be identified in real application usually do not convey enough information for full characterization of the monitored structure, and they are practically always insensitive to local damage. Researchers have thus proposed methods based on adding components such as mass and stiffness to the structure that can effectively increase the amount of modal information and improve the accuracy of damage identification. Nalitolela et al. [17] proposed a model updating method that adds various physical masses or stiffeners to the structure and utilizes modal information of the updated structures. Then, an improved method was proposed by adding imaginary masses to the preselected degrees of freedom (DOFs) [18]. In 2010, Dems and Mroz [19] further added controllable parameters such as supports, loads and temperature to the original structure, and identified the damage by modal, static and thermodynamic methods. Lu [20] took the beam structure as an example, and comprehensively analyzed the influence of the value, position and the number of the additional masses on damage identification in the additional mass method. Hou et al. [21] derived the virtual mass equation using structural excitation and response based on the VDM, and the effectiveness of the method was verified by an experiment of the frame structure. Therefore, adding virtual masses on structures is an efficient way to obtain more information related to natural frequencies for damage identification. However, there are few studies on the optimal placement of masses and other physical parameters. In fact, the value, positions and the number of the additional virtual masses can greatly affect damage identification results, so the optimal placement of additional virtual masses is the main prerequisite for the accuracy of structural damage identification. Therefore, the problem of optimal placement of virtual masses for the purpose of structural damage identification is studied in this paper.

The problem of optimal placement of virtual masses is similar to the problem of optimal sensor placement, so that similar methods might be applied for this research. In this paper, the optimization criterion and the algorithm for the optimal placement of virtual masses are along the lines provided by the research on the optimal placement of sensors.

There are three criteria that are mainly used: the minimum transmission error criterion [22], the modal kinetic energy criterion [23] and the model reduction criterion [24]. The basic theory behind the minimum transmission error approach is to use the unbiased estimation of the system parameter identification error. When the trace or the determinant of the Fisher information matrix reaches its maximum value, the system parameter identification error reaches its minimum accordingly. Kammer [22] applied the Fisher information matrix to the sensor placement problem for identification of structural modal parameters, and proposed the famous Effective Independence (EI) method, which eliminates in a stepwise manner the DOFs that contribute little to the linear independence of the target mode vectors by maximizing the determinant of the information matrix. Zhan [25] used the modal strain energy method to modify the EI method and applied it to the sensor optimal placement of the truss bridge structure. Yi [26] proposed a new multi-dimensional sensor optimization layout criterion combined with the EI method and the mode assurance criterion, and introduced the Wolf Group algorithm to improve the computational efficiency. Zhang [27] proposed an effective independence–total displacement method to address the problem of optimal sensor placement in hydraulic structures. These sensor optimization studies are in general based on modal observability. Silvers [28] proposed an optimization method, which optimized the sensor arrangement by maximizing the sensitivity of the natural frequency to the damage. Bruggi [29] proposed a method for sensor placement optimization to identify the damage of flexible plates. Li [30] used Non-dominated Sorting Genetic Algorithm II (NSGA-II) and wavelet decomposition to analyze and optimize sensor distribution for structural impact monitoring. The general purpose of the minimum transmission error criterion is to make the modal matrix include as much information as possible, while the purpose of adding masses is to improve the damage identification accuracy. Therefore, this paper draws on the construction of the Fisher information matrix in the transmission error criterion, and it proposes an optimization criterion based on the sensitivity information matrix in damage identification. The aim is to obtain a sensitivity matrix that contains as much information as possible.

After establishing the sensor optimization criterion, the next step is to select the optimization method to find the optimal solution under the corresponding criterion. The current optimization algorithms can be classified as classical optimization algorithms and meta-heuristic algorithms. Classical optimization algorithms utilize classical approaches like the Newton method or the conjugate gradient method to optimize the placement of the measurement points. The optimization efficiency of these methods is relatively high, but they perform an intrinsically local search, so that the globally optimal solution might be difficult to find. The most known meta-heuristic algorithms include genetic, simulated annealing, particle swarm and cross entropy optimization algorithms. They are designed to be global and can thus effectively avoid falling into a locally optimal solution. The Particle Swarm Optimization (PSO) algorithm belongs to the global, meta-heuristic approaches. It was proposed by Kennedy and Eberhart [31] in 1995. The method utilizes a large number of search points treated as particles flying through the search space (particle swarm), which attracted to the optimal solution by changing their velocity based on the individual, local and global experiences. The PSO algorithm has the advantages of a fast convergence, few tunable parameters and an easy implementation. It is widely used in optimization calculations in various fields such as power design, intelligent control, and transportation. He et al. [32] used an improved PSO algorithm to solve the problem of multi-dimensional sensor layout based on information redundancy. The efficiency of the method was verified by taking the Laxiwa arch dam of the upper Yellow River as an example. Zhang [33] proposed an approach for optimal sensor placement based on the PSO algorithm for the structural health monitoring of long-span cable-stayed bridges, and established the fitness function to solve the optimal problem by using the root mean square (RMS) value of the non-diagonal elements in the modal assurance criterion matrix. For applications to discrete optimization variables, Kennedy et al. [34] proposed a binary PSO algorithm for 0–1 programming problems. The particle position was represented by a binary variable, and the velocity of the particle meant the probability of taking 1 as the binary variable.

This paper takes the identification of damage parameters as the ultimate goal and studies the problem of optimal placement of the added virtual masses. It is structured as follows: Firstly, an optimization criterion based on the volumetric maximum of the sensitivity matrix is proposed. Secondly, due to the advantages of a low number of parameters and a small computational cost [35], the PSO algorithm is improved and applied for the optimal placement of virtual masses. Then, the value and the number of virtual masses is optimized. Thirdly, according to the optimization result, the virtual masses are arranged on the considered structure and damage identification is conducted by employing the sensitivity method. Finally, the feasibility and effectiveness of the proposed method are validated by a numerical simulation example of a simply supported beam structure and a truss structure.

2. The Effect of an Added Virtual Mass

In this method, the structure without additional masses is called the original structure, and the structure with an additional virtual mass is called the virtual structure. The notion "virtual" is used to emphasize that the influence of the additional mass is computed based on the recorded responses of the original structure, without mounting a real mass to the system.

Let the excitation be applied and the acceleration be measured in the same structural degree of freedom (DOF), and denote by $h(\omega)$ the corresponding (measured) acceleration frequency response of the original structure. Let a (virtual) mass be added in the same DOF and denote by $H(\omega, m)$ the corresponding acceleration frequency response of the virtual structure. The virtual mass is added just in one DOF, and the other DOFs remain unmodified, therefore the inertia force is generated just in the single involved DOF and it equals $-mH(\omega, m)$.

According to the basic theory of VDM [21], the influence of the additional mass can be equivalently modeled by its inertia force. Therefore, $H(\omega, m)$ can be expressed as the following sum of the original frequency response and the effects of the inertia force:

$$H(\omega, m) = h(\omega) - mH(\omega, m)h(\omega) \tag{1}$$

This formula can be rearranged as:

$$H(\omega, m) = \frac{h(\omega)}{1 + mh(\omega)} \tag{2}$$

In actual engineering projects, the frequency response is usually calculated by the Fourier Transform of time-domain excitations and responses. If the time-domain excitation is denoted by $f(t)$, and the corresponding acceleration response in the same DOF is denoted by $y(t)$, let $A(\omega)$ and $F(\omega)$ denote the corresponding frequency-domain signals obtained by the Fourier Transform. Then, by substituting $h(\omega) = A(\omega)/F(\omega)$ into Equation (2), one obtains the following simple formula for the acceleration frequency response of the virtual structure:

$$H(\omega, m) = \frac{A(\omega)}{F(\omega) + mA(\omega)} \tag{3}$$

Equation (2) can be used to determine the natural frequencies of the virtual structure, which can be then utilized for damage identification. It should be emphasized that the position and the direction of the applied excitation $F(\omega)$ and the measured acceleration response $A(\omega)$ should be the same, and that the virtual mass also must be added in the same position and direction. In other words, the virtual mass is constructed and added in the position where the sensor is.

3. Optimal Sensor Placement for Virtual Masses

In this section, the sensitivity information matrix is constructed by using the natural frequencies of the virtual structure with respect to the damage factor. Then the virtual mass optimization criteria are established based on the sensitivity information matrix. Finally, an optimization method for the virtual mass placement is proposed.

3.1. Sensitivity Information Matrix

It is assumed that there are n substructures in the structure to be identified, and that the damage factor μ_l of the l-th substructure represents its stiffness reduction ratio: it is equal to the stiffness ratio of the l-th substructure after damage to that of l-th substructure before damage. The global structural stiffness matrix after damage is expressed as $\mathbf{K}(\boldsymbol{\mu})$, where:

$$\boldsymbol{\mu} = \{\mu_1, \mu_2, \ldots, \mu_n\} \tag{4}$$

As shown in Figure 1, it is assumed that there are n_m available locations for virtual masses in the structure. And it is supposed that when the mass m is placed at the position i ($i = 1, 2, \ldots, n_m$), then the first k natural frequencies of i-th virtual structure can be identified as $\omega_{1i}(\boldsymbol{\mu}, m), \omega_{2i}(\boldsymbol{\mu}, m), \ldots, \omega_{ki}(\boldsymbol{\mu}, m)$. The j-th natural frequency and the mass normalized mode of the i-th virtual structure are denoted thus by $\omega_{ji}(\boldsymbol{\mu}, m)$ and $\Psi_{ji}(\boldsymbol{\mu}, m)$, respectively.

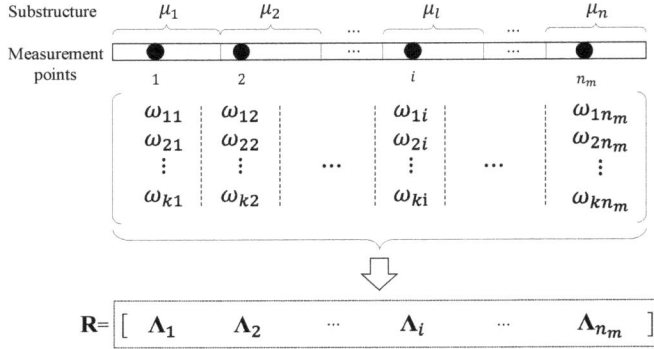

Figure 1. Construction of sensitivity information matrix.

Because the larger order of the natural frequency is, the larger its absolute identification error will be, the relative sensitivity is adopted for analysis. The relative sensitivity $r_{ji,l}$ is the normalized gradient of $\omega_{ji}(\boldsymbol{\mu}, m)$ with respect to the damage factor μ_l:

$$r_{ji,l}(\boldsymbol{\mu}, m) = \frac{1}{\omega_{ji}(\boldsymbol{\mu}, m)} \frac{\partial \omega_{ji}(\boldsymbol{\mu}, m)}{\partial \mu_l} = \frac{\Psi_{ji}^{\mathrm{T}}(\boldsymbol{\mu}, m) \mathbf{K}_l \Psi_{ji}(\boldsymbol{\mu}, m)}{2\omega_{ji}^2(\boldsymbol{\mu}, m)} \tag{5}$$

When the mass m is added in the i-th measuring point, the relative sensitivity of the j-th natural frequency to all n substructure damage factors $\boldsymbol{\mu}$ can be arranged as the following vector:

$$r_{ji}(\boldsymbol{\mu}, m) = \frac{1}{\omega_{ji}(\boldsymbol{\mu}, m)} \frac{\partial \omega_{ji}(\boldsymbol{\mu}, m)}{\partial \boldsymbol{\mu}} = \left\{ r_{ji,1}, r_{ji,2}, \ldots, r_{ji,n} \right\}^{\mathrm{T}} \tag{6}$$

Furthermore, for the i-th measurement point, the relative sensitivity information of all k natural frequencies with respect to all n substructure damage factors can be arranged as a single vector $\Lambda_i = \left\{ r_{1i}^{\mathrm{T}}, r_{2i}^{\mathrm{T}}, \ldots, r_{ki}^{\mathrm{T}} \right\}^{\mathrm{T}}$, which is a column vector with kn elements. The sensitivity information matrix $\mathbf{R}$ of the structure is arranged as shown in Equation (7), and it contains kn rows and n_m columns:

$$\mathbf{R} = \{\Lambda_1, \Lambda_2, \ldots, \Lambda_{n_m}\} \tag{7}$$

In the conventional sensitivity matrix, generally, each column vector represents the sensitivities of all modal information with respect to one considered parameter. In this paper, each column vector of the sensitivity matrix $\mathbf{R}$ represents the sensitivities of all modal information with respect to all considered parameters obtained by adding a virtual mass in one point. This new arrangement is more conducive to the analysis of the correlation between points.

3.2. Optimization Criterion

The optimal placement of virtual masses is to ensure the accuracy of damage identification, so the optimization criterion should assess two conditions: first, the sensitivity for each measurement point should be relatively high; second, sensitivity information for different measurement points should be as irrelevant as possible. The Volumetric Maximum Criterion can guarantee both of the above conditions. The geometric meaning of the optimization criterion based on the volumetric maximum criterion of the sensitivity matrix is described below.

As shown in Figure 2, the vectors $\boldsymbol{\Lambda}_i$, $\boldsymbol{\Lambda}_j$ and $\boldsymbol{\Lambda}_k$ represent the sensitivity information vectors of the *i*-th, *j*-th and *k*-th measurement points in the sensitivity matrix $\mathbf{R}$, respectively. In this figure, the modulus of the *i*-th measurement point sensitivity information vector $\boldsymbol{\Lambda}_i$ is maximum. $\boldsymbol{\Lambda}_i$ can be regarded as the vector of first selected point. To determine the next point, the vector that is the most irrelevant to the *i*-th vector is selected from among the *j*-th and *k*-th vectors. Figure 2 shows that the irrelevance between the vectors $\boldsymbol{\Lambda}_i$ and $\boldsymbol{\Lambda}_j$ is obviously greater than that between the vectors $\boldsymbol{\Lambda}_i$ and $\boldsymbol{\Lambda}_k$.

Moreover, the component λ_{ji} of the vector $\boldsymbol{\Lambda}_j$ in the subspace perpendicular to $\boldsymbol{\Lambda}_i$ is larger than λ_{ki}, which is the component of $\boldsymbol{\Lambda}_k$ in the subspace perpendicular to $\boldsymbol{\Lambda}_j$. Obviously, the area formed by the vectors $\boldsymbol{\Lambda}_i$ and $\boldsymbol{\Lambda}_j$ is larger than that formed by $\boldsymbol{\Lambda}_i$ and $\boldsymbol{\Lambda}_k$. Therefore, the area can be used to describe the irrelevance between two vectors. If this situation is extended to a 3-dimensional or a higher dimensional space, it can be concluded that the greater the irrelevance of the vectors, the larger the volume. Therefore, the volume of the formed parallelogram can be used as the criterion for evaluating the irrelevance of the vectors in the matrix. Consequently, maximization of the volume of the sensitivity matrix can be used as the criterion for the optimal placement of the virtual masses.

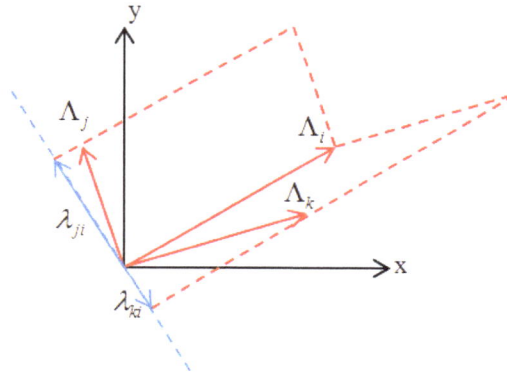

Figure 2. Schematic diagram of the optimization criterion.

To meet the two above conditions, the corresponding objective function based on the volumetric maximum criterion of the sensitivity matrix can be expressed as:

$$f_1(\pi, m) = V(\mathbf{R}(\pi, m)) \tag{8}$$

where π represents the location layout scheme of the virtual masses, $\mathbf{R}(\pi, m)$ represents the structural sensitivity matrix under the corresponding placement scheme of the virtual masses, and V represents the volume formed by the column vectors of measurement points in the sensitivity matrix. When the sensitivity matrix $\mathbf{R}$ contains only one vector, V is the length of that vector; when $\mathbf{R}$ contains 2 vectors, V is the area formed by two vectors; and when $\mathbf{R}$ contains three or more vectors, V can be understood as the volume or the generalized volume formed by the vectors, and its volume can be obtained by Equation (9), where $\det(\mathbf{R}^{\mathrm{T}}\mathbf{R})$ represents the determinant of matrix $\mathbf{R}^{\mathrm{T}}\mathbf{R}$:

$$V(\mathbf{R}) = \sqrt{\det(\mathbf{R}^{\mathrm{T}}\mathbf{R})} \tag{9}$$

In the application of this method, if $\det(\mathbf{R}^{\mathrm{T}}\mathbf{R}) > 0$, then the number n_m of the measurement points should meet the condition $n_m \leq kn$, where k is number of the identified modes and n is the number of the substructures. Given $\mathbf{F} = \mathbf{R}^{\mathrm{T}}\mathbf{R}$, then $\mathbf{F}$ is defined as the Fisher information matrix in this method. Therefore, finding the volume of the sensitivity matrix **R** is the problem of finding the determinant of the Fisher information matrix **F**.

In the process of optimization, it is actually the process of finding the extreme minimum value of the objective function, so Equation (8) can be revised to Equation (10):

$$f_2(\pi, m) = -\sqrt{\det(\mathbf{F})} \tag{10}$$

3.3. Placement Optimization of Virtual Masses

The variables to be optimized in this paper include the value, the number, and the positions of the virtual masses. It is difficult to simultaneously optimize three variables of different characters, so they are optimized separately. Firstly, the sensitivity analysis and error analysis are used to optimize the value of the virtual mass, and then the positions are optimized. Finally, the number of virtual masses is discussed.

3.3.1. Preliminary Optimization of the Value of the Additional Virtual Mass

The purpose of adding virtual masses is to improve the sensitivity of the modal information to local damage. Therefore, the sensitivity analysis of the finite element model is used to determine the value of the additional masses, and Equation (3) shows that the additional virtual masses may cause errors of the frequency response. It may in turn result in errors of the identified natural frequency, thereby reducing the accuracy of the damage identification. Therefore, two factors in the selection of the virtual masses should be considered: the sensitivity and the frequency identification error.

The influence of the additional virtual mass value on the frequency identification error is studied by adding mass to the SDOF (single-degree of freedom) structure. The physical parameters of the structure are assumed as follows: $M = 1$, $K = 1$, $C = 0$. Then the natural frequency of the original structure $\omega_0 = 1$. Measurement error of frequency response are denoted by Δ. When a unit impulse excitation is applied to the original structure, the acceleration frequency response can be expressed as Equation (11):

$$h(\omega) = \frac{\omega^2}{\omega^2 - 1} + \Delta \tag{11}$$

Substituting $h(\omega)$ into Equation (2), the acceleration frequency response $H(\omega, m)$ of the virtual structure with the additional virtual mass m can be easily obtained, and then the corresponding natural frequency ω_e can be estimated by peak-picking, which is shown in Equation (12):

$$\omega_e = \sqrt{\frac{1 + m\Delta}{1 + m + m\Delta}} \tag{12}$$

After adding mass m, the accurate natural frequency ω_a is preliminarily estimated using the stiffness and mass of the structure, which is shown in Equation (13):

$$\omega_a = \sqrt{\frac{1}{1 + m}} \tag{13}$$

The relative error of the estimated natural frequency with respect to the accurate natural frequency can be expressed by substituting Equations (12) and (13) into Equation (14):

$$\delta = \frac{\omega_e - \omega_a}{\omega_a} \tag{14}$$

By considering an example value $\Delta = 0.1$, the curve of the estimated relative error of the natural frequency of the SDOF system with the additional virtual mass m is drawn in Figure 3. It can be seen that the frequency identification error increases with the increase of mass.

In conclusion, from the view of sensitivity analysis, additional masses can improve the sensitivity of the frequency information to local damage. From the view of errors, the greater the added mass is, the greater the error of the estimated natural frequency is. Therefore, the choice of the virtual mass should balance between these two factors, i.e., frequency sensitivity and estimated frequency error.

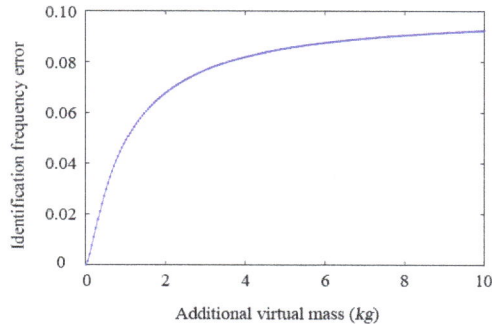

Figure 3. Relationship between the relative frequency identification error and the additional mass.

3.3.2. Optimization Method for Virtual Masses Placement

There are many algorithms that can solve the considered optimization problem. Often, the PSO requires fewer parameters to be tuned and it takes less computational effort in comparison to other meta-heuristic algorithms. The PSO also has the advantages of a fast convergence and an easy implementation. The discrete PSO algorithm can find the global optimal solution in a straightforward procedure, and the calculation results are stable. In this paper, the PSO algorithm is modified and applied for the optimization of virtual masses.

In the PSO algorithm, a search point is treated as a particle that travels through the search space. Each such particle has its own position and velocity, which are modified in the successive optimization steps according to the corresponding fitness value. The PSO algorithm takes fitness function as the criterion to evaluate the quality of the solution in the process of searching for the optimum. Therefore, the selection of the fitness function directly affects the determination of the optimal solution. The optimization criterion based on the volumetric maximum of the sensitivity matrix ensures the maximum irrelevance between the sensitivity information of each measurement point by maximizing the volume of the sensitivity matrix, so that it contains as much information as possible.

The PSO algorithm with a linearly decreasing inertial weight is applied in this paper, and the iterative velocity update equation is as follows:

$$v_{id}^{t+1} = wv_{id}^t + c_1 r_1 \left(p_{id}^t - x_{id}^t \right) + c_2 r_2 \left(p_{gd} - x_{id}^t \right) \tag{15}$$

where w is the inertial weight, v_{id}^t is the velocity in the d-th dimension of the i-th particle during the t-th iteration; c_1 and c_2 are the acceleration coefficients (usually positive constants); r_1 and r_2 are random numbers uniformly distributed on [0,1]; p_{id}^t is the position in the d-th dimension of the past individual best point of the i-th particle during the t-th iteration; p_{gd} is the position in the d-th dimension of the best global extremum point (of the entire particle swarm).

The standard PSO algorithm is mainly applied to the optimization problem of continuous space functions. In this paper, the optimization problem of virtual mass placement is how to choose n_m positions from N possible positions, which is a discrete problem. Therefore, this paper uses the discrete PSO algorithm to optimize the virtual mass placement [31]. The velocity update equation of the discrete binary algorithm is the same as in the original PSO algorithm, but the position update in this method is different and should be studied.

The location of the virtual masses is encoded in the binary code, and x_i^t indicates the position of the i-th particle in the t-th generation. Each x_i^t represents a solution to the optimization problem, $x_i^t = \{x_{i1}^t, x_{i2}^t, \ldots, x_{iN}^t\}$, where N is the number of possible positions. If $x_{ij}^t = 0$, the i-th particle does not arrange the virtual mass at the j-th position in the t-th iteration. Otherwise, when $x_{ij}^t = 1$, it indicates that the j-th position is used to place the virtual mass. The velocity v_{ij}^t represents the probability that the j-th binary bit is 1, therefore it is mapped to the interval [0,1]. The mapping method generally uses the sigmoid function as shown in Equation (16):

$$s(v_{id}) = \frac{1}{1 + exp(-v_{id})} \tag{16}$$

where $s(v_{id})$ is the probability that the position x_{id} equals 1. In the traditional PSO algorithm, the selection of x_{id} is based on Equation (17):

$$x_{id} = \begin{cases} 1 & \text{if } \text{rand}(\ ) \leq s(v_{id}) \\ 0 & \text{otherwise} \end{cases} \tag{17}$$

where rand( ) is a random number uniformly distributed in [0,1]. However, the number of selected positions obtained this way (that is, the number of 1 s) is possibly not equal to the required number n_m. In this paper, the above methods is modified by ranking the difference value between $s(v_{id})$ and the vector of all N random numbers rand( ) from large to small, and assigning 1 to the largest n_m of them, so that always exactly n_m measurement points are selected to place the virtual masses. As shown in Equation (18):

$$x_{id} = \begin{cases} 1 & \text{if } \mathbb{R}(s(v_{id}) - \text{rand}(\ )) \leq n_m \\ 0 & \text{otherwise} \end{cases} \tag{18}$$

where $\mathbb{R}(\cdot)$ represents the position number of the argument in the list of all arguments sorted in the descending order. For example, let $z_i = s(v_{id}) - \text{rand}(\ )$, the variables $z_1, z_2, \ldots, z_N$ are sorted in the descending order, and $\mathbb{R}(z_i)$ is the position of z_i in the sorted list.

The main steps of discrete PSO algorithm is as follows:

(1) Set algorithm parameters;
(2) Initialize the position and the velocity of all particles. The position of each particle x_{ij}^0 is randomly generated to be 0 or 1, where $i = 1, 2, \ldots, \overline{N}$ and $i = 1, 2, \ldots, N$ and $\overline{N}$ is the number of particles in the swarm. The velocity of the particle is generated as a random number between 0 and 1;
(3) Calculate the fitness value of each particle in the population, and compare the particle fitness value with its individual best value P_i. If it is better than P_i, then store the current position as P_i;
(4) Compare the best individual extremum value P_i with the global extremum value P_g. If it is better than P_g, it is stored as an updated value of P_g;
(5) Update the velocity and the positions of the particle. The velocity of the particle can be updated according to Equation (15), and $s(v_{id})$ can be calculated from Equation (16). The positions with the first n_m maximum differences between $s(v_{id})$ and rand( ) are be selected to place the virtual mass (Equation (18)), that is the corresponding bits are set to 1, while the others remain 0;
(6) Stop the operation when the number of iterations reaches a pre-set maximum number of iterations, and output P_g and the corresponding fitness value, otherwise go to step 3.

3.3.3. Determination of the Number of the Virtual Masses

The number of the virtual masses affects the accuracy of damage identification, so it is determined by analyzing and comparing the accuracy of the identified structural damage. The specific method is introduced in the numerical simulation.

Sensors **2019**, *19*, 340

4. Numerical Simulation

4.1. A Simply Supported Beam

A simply supported beam is used to verify the effectiveness of the proposed virtual mass optimization method, see Figure 4. The span of the simply supported beam is 1 m, the width of the section is 0.05 m, the thickness of the section is 0.005 m, the elastic modulus of steel used in the structure is 2.1×10^{11} Pa, and the density of steel is used as 7.85×10^3 kg/m^3. The structure is divided into 20 finite elements. As shown in Figure 4, there are 19 vertical DOFs for candidate position of virtual mass, which are numbered as 1–9.

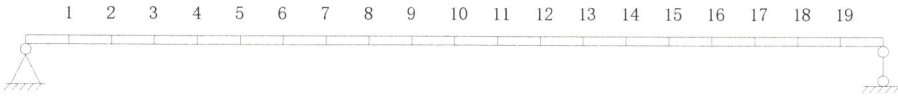

Figure 4. Finite element model of simply supported beam.

The structure is divided into 10 substructures, each substructure contains two finite elements. Structural damage is considered as follows: Substructure 3 and substructure 8 are damaged simultaneously, and their stiffness decreases by 20% and 30%, respectively. This damage scenario can be expressed as $\mu_3 = 0.8$, $\mu_8 = 0.7$.

Firstly, sensitivity analysis and frequency error analysis are used to preliminarily determine the value of the virtual masses. Then, an example of eight virtual masses placement is analyzed, and the improved discrete PSO algorithm is used to search for the optimal positions of virtual masses. Finally, the influence of the number of the masses on the variance of the identified damage factors is studied under 10 groups of noise, and the number of virtual masses is determined.

4.1.1. Determination of Virtual Masses Value

The influence of sensitivity and frequency identification error should be considered in determining the value of virtual masses.

Firstly, relative sensitivity analysis of the simple supported beam model shown in Figure 4 is performed. Different virtual masses are added to an arbitrary node of the structure (the 6th position is taken here as an example). By using Equation (5), the relative sensitivity of the first natural frequency of the structure with different virtual masses is obtained, and the result is shown in Figure 5. The relative sensitivity of the first natural frequency increases with the increase of the additional virtual mass.

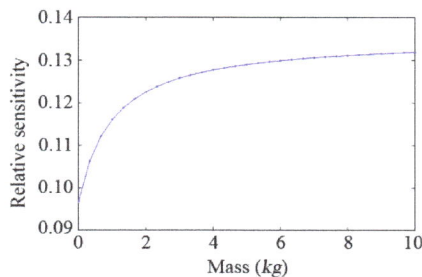

Figure 5. Relationship between the relative sensitivity of the first natural frequency and the additional mass.

The simply supported beam structure used in this paper is a multi-degree freedom (MDOF) system, and the relationship between the frequency identification error of the SDOF system and the additional virtual masses shown in Figure 3 is no longer applicable to the model. For the simply supported beam shown in Figure 4, the virtual masses are added to the 6th node of the structure, and the variance of the identified natural frequency is calculated after applying 20 groups of noise (5% white noise) on the test frequency response. After the structure is attached with different virtual masses, the variance of the identified natural frequency is shown in Figure 6. It shows that the error of the identified natural frequency increases as the virtual mass increases. When the mass is about 3 kg, the variance of the identified natural frequency is still relatively small, and the relative sensitivity is already high, so the virtual mass value is selected to be 3 kg.

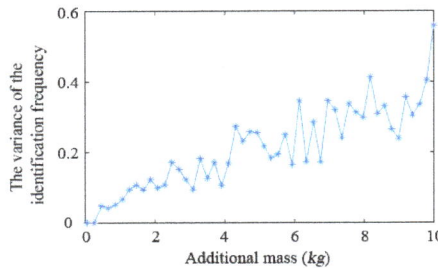

Figure 6. Variance of the identified natural frequency under different virtual mass.

4.1.2. Virtual Masses Placement Optimization Using PSO Algorithm

The following presents the arrangement optimization of the virtual mass locations for 3 kg virtual mass. Firstly, the sensitivity matrix of the structure is calculated by Equation (5). Then the sensitivity values of the first 3 natural frequencies with respect to the damage factors of the 10 substructures are calculated when successively placing 3 kg mass in the 19 vertical DOFs. Finally, the sensitivity matrix is obtained according to Figure 1.

The basic parameters of the PSO algorithm are shown in Table 1. Due to the randomness of the method, the optimization results might be different each time. In this paper, four random trials have been carried out and the results are shown in Table 2:

Table 1. PSO Algorithm Parameter Settings.

Population Size	Particle Dimension	The Maximum Number of Iterations	Inertia Factor		Learning Factor		Random Number		Particle Velocity	
			w_{max}	w_{min}	c_1	c_2	r_1	r_2	v_{max}	v_{min}
40	19	100	0.9	0.4	1.496	1.496	[0–1]	[0–1]	4	4

Table 2. Random optimization results of discrete PSO.

Random Test Number	The Optimal Value	Optimal Fitness Value	Optimization Time
1~4 times	1, 3, 5, 8, 13, 15, 17, 19	−0.022	0.697 s~1 s

It can be seen from Table 2 that the optimized placements of the virtual masses and the best fitness values (Equation (10)) obtained in the 4 tests are exactly the same, but the optimization time differs slightly. To represent the iterative process of a particle, the relationship between the number of iterations and the fitness value for a test particle is drawn in Figure 7. The position distribution diagram of the eight measuring points in the optimized placement is shown in Figure 8. In Figure 7, the fitness function value decreases as the number of iterations increases. Past the 40th iteration, the fitness function value remains stable and no longer changes, which indicates that the improved PSO algorithm has converged. As seen in Table 2, the optimal values and the fitness values obtained in four random trials are identical, which indicates that the algorithm has found the global optimal solution, and thus proves the feasibility and efficiency of the improved PSO algorithm again. It can be seen from Figure 8 that the optimized positions of the measurement points are quite evenly dispersed, and the left and right symmetrical distribution form is centered on the mid-span, indicating that the results obtained by the PSO algorithm based on the maximum of the sensitivity matrix volume are reasonable.

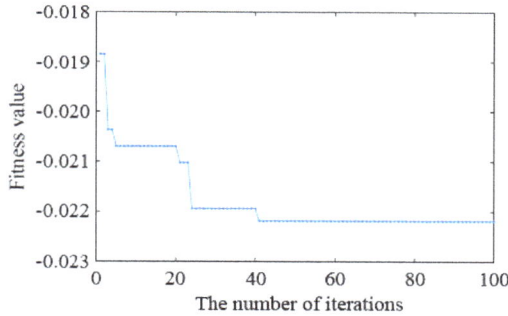

Figure 7. Variation curve of discrete PSO fitness function and iteration times.

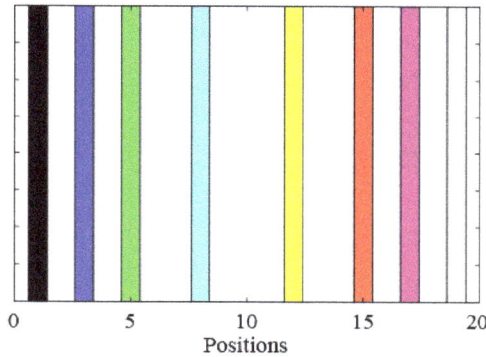

Figure 8. The optimization order of the measuring points.

The proposed algorithm required only 1 s to complete the search on a typical desktop hardware configuration. The discrete PSO algorithm can find the global optimal solution easily, and the calculation results are stable.

4.1.3. Damage Identification

The excitation shown in Figure 9a is used to apply an impulse load. The load duration is 5 ms and the sampling frequency is 5000 Hz.

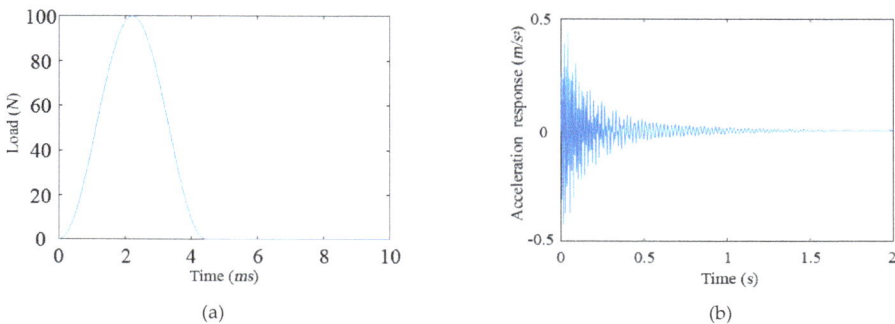

(a)

(b)

Figure 9. (**a**) Pulse excitation. (**b**) Acceleration response.

The influence of 5% white noise is considered. The excitation is applied to the 6-th position, and the corresponding acceleration response is shown in Figure 9b.

(1) Identification of the Natural Frequency with Added Virtual Masses

If the number of the added virtual masses is six, the positions can be optimized by the improved PSO algorithm, which is {1, 3, 8, 12, 17, 19}. The impulse excitation as shown in Figure 9a is applied to these six DOFs, and the acceleration responses in the same DOFs are calculated. The excitation and the response containing 5% noise are substituted into Equation (3), and the amplitude of the frequency response of each DOF is calculated after adding 3 kg of virtual mass. By using the Fourier transform and extracting the peak value, the first three natural frequencies of each DOF with added mass are obtained, as shown in Table 3. The finite element frequencies of the theoretical undamaged model after adding mass are shown in Table 4.

Table 3. First identified three natural frequencies when six virtual masses are arranged on the damaged structure/Hz.

DOF	1	3	8	12	17	19
First natural frequency	3.900	3.899	3.900	3.899	3.899	3.899
Second natural frequency	23.899	24.099	23.900	24.300	24.399	24.099
Third natural frequency	91.298	90.498	91.098	91.098	91.398	90.798

Table 4. First three theoretical frequencies when six virtual masses are added to the undamaged finite element model/Hz.

DOF	1	3	8	12	17	19
First natural frequency	3.899	4.199	4.000	3.999	3.900	3.899
Second natural frequency	23.900	23.999	23.899	23.899	24.399	24.399
Third natural frequency	92.398	92.398	92.398	92.398	91.898	92.498

(2) Damage Identification

The objective function for damage identification is easily built using the natural frequencies identified with added virtual masses and the frequency of the corresponding theoretical finite element model. The result of damage identification is shown in Figure 10, in which the abscissa is the substructure number and the ordinate is the damage factor.

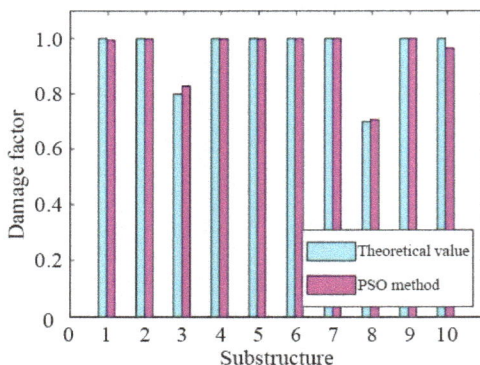

Figure 10. Identified damage factors.

The optimized virtual mass placement obtained by the discrete PSO algorithm can accurately identify the damage location and damage extent. Even in the case of 5% measurement noise, the identification results maintain good accuracy. The optimized arrangement scheme obtained by the discrete PSO algorithm based on the maximum volume of the sensitivity matrix is used for damage identification. Damage identification errors of substructure 3, substructure 8 and substructure 10 are 2.68%, 0.5% and 4.08%, respectively. This method has a high computational efficiency and can find the global optimal solution with a great probability. Moreover, the results fully meet engineering accuracy requirements.

4.1.4. Determination of Virtual Masses Quantity

When different number of virtual masses is placed on the beam, the identified results are different, as shown in Figure 11. It can be seen from Figure 11 that the larger the number of the virtual masses is, the closer the identified damage factor approaches the theoretical value. The identification error is the largest when four masses are arranged, and it decreases gradually with the increase of the number of virtual masses.

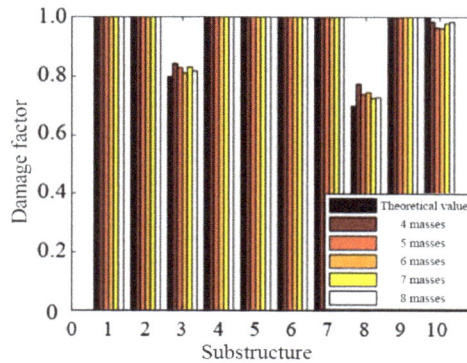

Figure 11. Damage identification results when arranging different numbers of masses.

Then, in the case of applying 10 groups of noise (5% white noise), when 4–10 virtual masses are optimally placed, the variance results of the damage factors are shown in Figure 12.

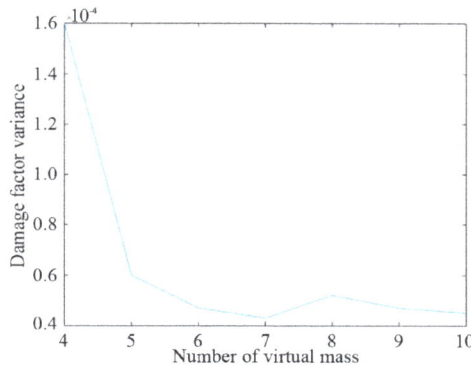

Figure 12. Variance of the damage factor when using different numbers of masses.

The variance of the damage factors decreases with the increasing number of the virtual masses. When there are more than six masses, the variance of the damage factors tends to be stable and fluctuates only slightly. Overall, the number of the virtual masses is selected as six, the mass value is 3 kg, and the optimized placement is {1, 3, 8, 12, 17, 19}.

4.2. A Truss Structure

Figure 13 shows a truss structure model consisting of 15 members and nine nodes. The length of every member is 3 m, the height of the truss is 2.6 m, the elastic modulus of the rod is 2.0×10^{11} Pa, the density is 7.8×10^3 kg/m^3; the members are round steel tubes, the diameter of the steel tubes is 0.1 m, and the wall thickness is 0.05 m. Each member of the truss structure is a single element, so there are 15 elements and 15 DOFs.

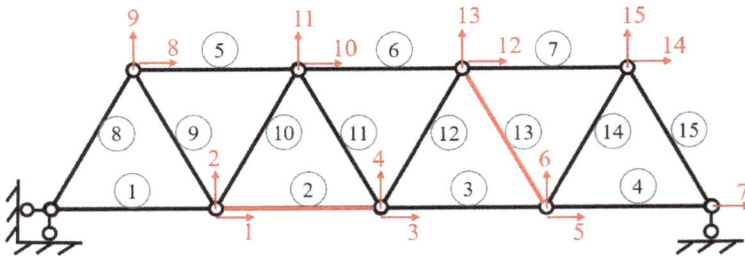

Figure 13. Truss structure model.

Figure 13 shows the numbers of elements and DOFs. It is assumed that the members numbered 2 and 13 are damaged, and the damage factors are 0.6 and 0.7, respectively. The first four natural frequencies of structure with added masses are used as the basis for damage identification. The natural frequencies of the original structure and the damaged structure are shown in Table 5:

Table 5. The first four natural frequencies of the undamaged and the damaged structure/Hz.

Frequency Order	1	2	3	4
Undamaged structure	37.7299	65.6261	120.2154	192.8448
Damaged structure	34.9979	63.8086	119.3401	188.4917

Firstly, the additional virtual mass of this truss structure is determined to be 200 kg based on the structural sensitivity analysis. The truss structure has a total of 15 DOFs, and they are all used as candidate positions for additional masses. Seven cases are considered by adding 4–10 virtual masses in the structure. The improved PSO algorithm is used to determine the optimum placements of the virtual masses, which are shown in Table 6.

Table 6. Number and optimized locations of additional virtual masses.

Case	Number	Locations
1	4	7,8,9,13
2	5	7,9,10,13,15
3	6	7,8,9,13,14,15
4	7	7,9,10,11,13,14,15
5	8	7,8,9,10,11,13,14,15
6	9	1,7,8,9,10,11,13,14,15
7	10	1,5,7,8,9,10,11,13,14,15

The basic parameters in the PSO are the same as in Section 4.1.2. From the optimized results, all cases include lateral DOFs and longitudinal DOFs, and the positional arrangement is relatively scattered, which seems reasonable as assessed using the engineering common sense.

Numerical simulations are performed for each of the seven cases listed in Table 6, and 5% white noise is considered in dynamic simulation. The natural frequencies of the damaged structure with added virtual masses are used for damage identification, and the identified damage factors are shown in Figure 14.

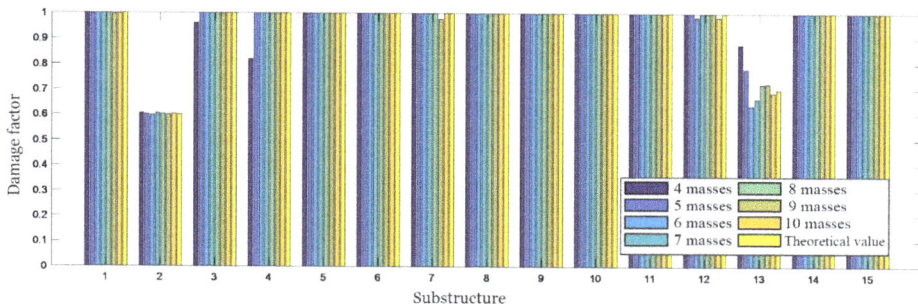

Figure 14. Damage identification results obtained by employing different numbers of masses.

It can be seen from Figure 14, that when only four masses are added to the structure, there are some relatively large errors in the results of damage identification. This is because once a mass is added, four natural frequencies can be obtained. Therefore, four masses correspond to 16 natural frequencies used for damage identification. However, the structure has a total of 15 damage factors to be identified, and 15 parameters cannot be identified accurately by using only 16 natural frequencies due to the influence of measurement noise. When five masses are added to the structure, the damage can be relatively accurately localized. As the number of the added masses increases, more frequency information can be obtained, and the accuracy of damage identification is relatively high. When the number of the added masses is greater than seven, the identified damage factors have a slight fluctuation, but tend to be stable. The identified damage factors are very close to the actual values, which is enough to meet the engineering accuracy.

5. Conclusions

This paper proposes an optimal placement method of virtual masses for the purpose of damage identification. Firstly, a specific form of the sensitivity matrix is established, which is more conducive to the analysis of the correlation between points. Then, an optimization criterion called Volumetric Maximum Criterion is proposed, which is based on the volumetric maximum of the sensitivity matrix. This criterion promotes sensitivity matrices (and the corresponding placements) that contain more information that can be used to identify structural damage. Thereupon, an improved version of the PSO algorithm is proposed for optimization. Finally, the optimal placement of virtual masses is verified in numerical examples of a simply supported beam and a truss structure. It can be concluded that:

1. The Volumetric Maximum Criterion uses volume to quantify the information contained in the sensitivity matrix. It can ensure that the sensitivity matrix contains as much sensitivity information as possible, thereby ensuring the accuracy of damage identification;
2. The improved PSO algorithm is accurate and efficient for optimization of virtual mass placements, and it can find the global optimal solution;
3. A method of sensitivity analysis and error analysis is proposed and discussed for the purpose of determination of virtual mass value. The method guarantees that natural frequencies used for damage identification have high sensitivities to damage and a small identification error;
4. The problem of optimal sensor placement is similar to the optimal arrangement of virtual masses, so the basic theory of the proposed methods in this paper, including the arrangement of sensitivity matrix, Volumetric Maximum Criterion and improved PSO algorithm, can be studied further for optimal sensor placement of modal identification.

Author Contributions: Conceptualization, J.H., Q.Z. and Ł.J.; methodology, J.H. and Q.Z.; software, Z.L. and R.Z.; validation, Z.L. and R.Z.; data curation, Z.L. and R.Z.; writing—original draft preparation, Z.L.; writing—review and editing, Z.L., J.H., Q.Z. and Ł.J.; supervision, J.H.

Funding: This research was funded by National Key Research and Development Program of China (2018YFC0705604), National Natural Science Foundation of China (NSFC) (51878118), Liaoning Provincial Natural Science Foundation of China (20180551205), and the project DEC- 2017/25/B/ST8/01800 of the National Science Centre, Poland.

Conflicts of Interest: The authors declare no conflict of interest.

References

1. Spencer, B.F., Jr.; Jo, H.; Mechitov, K.A.; Li, J.; Sim, S.H.; Kim, R.E.; Cho, S.; Linderman, L.E.; Moinzadeh, P.; Giles, R.K.; et al. Recent Advances in Wireless Smart Sensors for Multi-scale Monitoring and Control of Civil Infrastructure. *J. Civ. Struct. Health Monit.* **2015**, *6*, 17–41. [CrossRef]

2. An, Y.H.; Blachowski, B.; Ou, J.P. A degree of dispersion-based damage localization method. *Struct. Control Health Monit.* **2016**, *23*, 176–192. [CrossRef]

3. Hu, W.H.; Tang, D.H.; Teng, J.; Said, S.; Rohrmann, R.G. Structural Health Monitoring of a Prestressed Concrete Bridge Based on Statistical Pattern Recognition of Continuous Dynamic Measurements Over 14 Years. *Sensors* **2018**, *18*, 4117. [CrossRef] [PubMed]

4. Laflamme, S.; Kollosche, M.; Connor, J.J.; Kofod, G. Soft Capacitive Sensor for Structural Health Monitoring of Large-Scale Systems. *Struct. Control Health Monit.* **2012**, *19*, 70–81. [CrossRef]

5. Laflamme, S.; Ubertini, F.; Saleem, H.; D'Alessandro, A.; Downey, A.; Ceylan, H.; Materazzi, L. Dynamic Characterization of a Soft Elastomeric Capacitor for Structural Health Monitoring. *ASCE J. Struct. Eng.* **2014**, *141*, 04014186. [CrossRef]

6. Yang, Y.; Zhu, Y.H.; Wang, L.L.; Jia, B.Y.; Jin, R.Y. Structural Damage Identification of Bridges from Passing Test Vehicles. *Sensors* **2018**, *18*, 4035. [CrossRef]

7. Fu, Y.G.; Hoang, T.; Mechiow, K.; Kim, J.R.; Zhang, D.C.; Spencer, B.F., Jr. Sudden Event Monitoring of Civil Infrastructure Using Demand-Based Wireless Smart Sensors. *Sensors* **2018**, *18*, 4480. [CrossRef]

8. Kong, X.; Li, J.; Collins, W.; Bennett, C.; Laflamme, S.; Jo, H. A Large-area Strain Sensing Technology for Monitoring Fatigue Cracks in Steel Bridges. *Smart Mater. Struct.* **2017**, *26*, 085024. [CrossRef]

9. Pnevmatikos, N.G.; Blachowski, B.; Hatzigeorgiou, G.D.; Swiercz, A. Wavelet analysis based damage localization in steel frames with bolted connections. *Smart Struct. Syst.* **2016**, *18*, 1189–1202. [CrossRef]

10. Ubertini, F.; Gentile, C.; Materazzi, A.L. Automated modal identification in operational conditions and its application to bridges. *Eng. Struct.* **2013**, *46*, 264–278. [CrossRef]

11. Ubertini, F.; Materazzi, A.L.; D'Alessandro, A.; Laflamme, S. Natural frequencies identification of a reinforced concrete beam using carbon nanotube cement-based sensors. *Eng. Struct.* **2013**, *60*, 265–275. [CrossRef]

12. Xu, K.; Deng, Q.S.; Cai, L.J.; Ho, S.C.; Song, G.B. Damage Detection of a Concrete Column Subject to Blast Loads Using Embedded Piezoceramic Transducers. *Sensors* **2018**, *18*, 1377. [CrossRef]

13. Zhang, J.; Xu, J.D.; Guan, W.Q.; Du, G.F. Damage Detection of Concrete-Filled Square Steel Tube (CFSST) Column Joints under Cyclic Loading Using Piezoceramic Transducers. *Sensors* **2018**, *18*, 3266. [CrossRef] [PubMed]

14. Ginsberg, D.; Fritzen, C.P.; Loffeld, O. Sparsity-constrained extended Kalman filter concept for damage localization and identification in mechanical structures. *Smart Struct. Syst.* **2018**, *21*, 741–749. [CrossRef]

15. Jiang, T.Y.; Zhang, Y.W.; Wang, L.; Zhang, L.; Song, G.B. Monitoring Fatigue Damage of Modular Bridge Expansion Joints Using Piezoceramic Transducers. *Sensors* **2018**, *18*, 3973. [CrossRef]

16. Zhang, H.B.; Hou, S.; Ou, J.P. Validation of Finite Element Model by Smart Aggregate-Based Stress Monitoring. *Sensors* **2018**, *18*, 4062. [CrossRef]

17. Nalitolela, N.G.; Penny, J.E.T.; Friswell, M.I. A mass or stiffness addition technique for structural parameter updating. *Int. J. Anal. Exp. Modal Anal.* **1992**, *7*, 157–168.

18. Nalitolela, N.G.; Penny, J.E.T.; Friswell, M.I. Updating model parameters by adding an imagined stiffness to the structure. *Mech. Syst. Signal Process.* **1993**, *7*, 161–172. [CrossRef]

19. Dems, K.; Mróz, Z. Damage identification using modal, static and thermographic analysis with additional control parameters. *Comput. Struct.* **2010**, *88*, 1254–1264. [CrossRef]

20. Lu, P.; Wang, L.; Duan, J.; Zhang, Z.Y.; Fang, L. Influencing factors of beam structure damage identification based on additional mass. *J. PLA Univ. Sci. Technol.* **2017**, *18*, 295–301. [CrossRef]

21. Hou, J.L.; An, Y.H.; Wang, S.J.; Wang, Z.Z.; Jankowski, Ł.; Ou, J.P. Structural Damage Localization and Quantification Based on Additional Virtual Masses and Bayesian Theory. *J. Eng. Mech.* **2018**, *144*, 04018097. [CrossRef]

22. Kammer, D.C. Sensor placement for on-orbit modal identification and correlation of large space structures. *J. Guid. Control Dyn.* **1991**, *14*, 251–259. [CrossRef]

23. Papadopoulos, M.; Garcia, E. Sensor Placement Methodologies for Dynamic Testing. *AIAA J.* **1998**, *36*, 256–263. [CrossRef]

24. GUYAN, R.J. Reduction of Stiffness and Mass Matrices. *AIAA J.* **1965**, *3*, 380. [CrossRef]

25. Zhan, J.Z.; Yu, L. An Effective Independence-Improved Modal Strain Energy Method for Optimal Sensor Placement of Bridge Structures. *Appl. Mech. Mater.* **2014**, *670–671*, 1252–1255. [CrossRef]

26. Yi, T.H.; Li, H.N.; Wang, C.W. Multiaxial sensor placement optimization in structural health monitoring using distributed wolf algorithm. *Struct. Control Health Monit.* **2016**, *23*, 719–734. [CrossRef]

27. Zhang, J.W.; Liu, X.R.; Zhao, Y.; Wu, G.; Liu, X.L. Optimal sensor placement for hydraulic structures based on effective independence-total displacement method. *J. Vib. Shock* **2016**, *35*, 148–153. [CrossRef]

28. Silvers, J.E. Frequency Response Sensitivity Analysis to Determine Sensor Placement for Vibration-Based Damage Detection in Structural Elements. Ph.D. Thesis, Purdue University, West Lafayette, IN, USA, 2013.

29. Bruggi, M.; Mariani, S. Optimization of sensor placement to detect damage in flexible plates. *Eng. Optim.* **2013**, *45*, 659–676. [CrossRef]

30. Li, P.; Huang, L.W.; Peng, J.C. Sensor Distribution Optimization for Structural Impact Monitoring Based on NSGA-II and Wavelet Decomposition. *Sensors* **2018**, *18*, 4264. [CrossRef]

31. Eberhart, R.; Kennedy, J. *A New Optimizer Using Particle Swarm Theory*; Micro Machine and Human Science: Nagoya, Japan, 1995; pp. 39–43.

32. He, L.J.; Lian, J.J.; Ma, B.; Wang, H.J. Optimal multiaxial sensor placement for modal identification of large structures. *Struct. Control Health Monit.* **2014**, *21*, 61–79. [CrossRef]

33. Zhang, X.; Wang, P.; Xing, J.C.; Yang, Q.L. *Optimal Sensor Placement of Long-Span Cable-Stayed Bridges Based on Particle Swarm Optimization Algorithm*; Practical Applications of Intelligent Systems; Springer: Berlin/Heidelberg, Germany, 2014; pp. 207–217.

34. Kennedy, J.; Eberhart, R.C. A discrete binary version of the particle swarm optimization algorithm. Computational cybernatics and simulation. In Proceedings of the 1997 IEEE International Conference on Systems, Man, and Cybernetics—Computational Cybernetics and Simulation (SMC 97), Orlando, FL, USA, 12–15 October 1997.

35. Huang, Y.J.; Wang, H.; Khajepour, A.; Li, B.; Ji, J.; Zhao, K.G.; Hu, C. A review of power management strategies and component sizing methods for hybrid vehicles. *Renew. Sustain. Energy Rev.* **2018**, *96*, 132–144. [CrossRef]

sensors

MDPI

Article

A Method for Settlement Detection of the Transmission Line Tower under Wind Force

Xinbo Huang [1,*] , Yu Zhao [1], Long Zhao [2] and Luya Yang [1]

[1]   School of Electronics and Information, Xi'an Polytechnic University, Xi'an 710048, China;
      zhaoyu@xpu.edu.cn (Y.Z.); yangluya@xpu.edu.cn (L.Y.)
[2]   School of Electro-Mechanical Engineering, Xidian University, Xi'an 710070, China; zhaolong@xpu.edu.cn
*   Correspondence: huangxinbo@xpu.edu.cn; Tel.: +86-186-0296-5050

Received: 28 October 2018; Accepted: 6 December 2018; Published: 10 December 2018

check for updates

Abstract: In view of the settlement problem of transmission tower foundation, the vibration characteristics of transmission towers under wind force are measured experimentally. In this paper, the 110 kV cat head transmission tower of Xi'an Polytechnic University is measured and analyzed. Firstly, the acceleration sensor and meteorological sensor are installed on the tower to collect the vibration response and environment parameters of the tower in real time. Then, an experiment platform is built to simulate the tower settlement, and the vibration response of the tower after settlement is measured in time. Finally, the low-order modal frequencies of the transmission tower before and after settlement under wind force load are extracted by stochastic subspace identification (SSI), and the relationship between modal frequencies of different modes is analyzed via temperature correction. By comparison and analysis, it is obvious that the X-direction modal frequencies before and after settlement under natural wind load are changed, and the change rate increases with the increase of settlement displacement, which can be used as effective evidence for judging the settlement of transmission tower foundation.

Keywords: transmission tower; settlement; wind force; acceleration; modal frequencies

---

1. Introduction

In electrical power systems, the transmission tower is an integral component of power transmission, and its safe operation is an important factor. Terrain conditions of the transmission tower are complex, where often exist phenomena such as displacement, inclination, cracking and subsidence [1]. In particular, the settlement tilt of a tower can be caused by severe natural environmental conditions (such as rain, snow, wind and so on) or man-made destruction and other factors, which may even cause the tower collapse. Because the stress of tower foundations will change when settlement occurs [2], many scholars have tried to solve this problem with experimental simulation, and a significant amount of theoretical research has been undertaken. Some researchers have taken the tower line system as the research object and analyzed the vibration characteristics of the tower line system under the action of wind [3,4]. The finite element analysis method has often been used to analyze the dynamic characteristics and stability in wind conditions in the case of the complex structure of the tower [5,6]. Some scholars have conducted nonlinear buckling analysis of the tower line system, determined the critical wind load and analyzed the dynamic characteristics of the tower line system under different wind loads. It has been found that low modal frequency is a useful indicator for predicting the occurrence of structural instability [7]. In addition, some scholars have carried out wind tunnel tests on the EHV (Extra-high Voltage) transmission tower line system and studied the dynamic characteristics of the system under different wind speeds [8]. However, the method of mechanical analysis is usually used to study the response of the tower to wind, and there are few methods to identify the abnormal structure of the tower.

Structural health monitoring technologies are widely used in structural damage detection, including for electric power tower structures. The method of measuring tower inclination with an inclination sensor has been widely used [9]. However, this method can only indirectly reflect the stress of the large deformation of a tower, and parameters of load balance, stealth faults (such as the small deformation of the tower) or the yield failure of the local rod cannot be found in time. Some researchers realized the remote monitoring of tower structure damage by installing strain sensors on the transmission tower [10]. Although the resistance strain gauge can identify the force change of the rod, there are still some problems, such as complex attachment processes, susceptibility to rust and electromagnetic interference on the measuring circuit. Therefore, some researchers use grating fiber strain sensors to improve the range and accuracy of the measurement [11,12]; however, the development of this method is limited by installation quantity, installation location and the direction of the sensor.

The technology of modal identification has been studied in the field of structural health monitoring. In references [13,14], passing vehicles on a reinforced concrete bridge vibrated the bridge and the damage to the bridge was studied by modal identification. In reference [15], a wind turbine blade is vibrated by an excitation device and the damage to the wind turbine blade is identified by the modal technique, and the position of the damage or length of the crack is determined. Under the action of wind load, the transmission conductor will generate vertical vibration due to the presence of the Karman vortex. In reference [16], the modal analysis of the transmission line is carried out to realize the detection of the broken conductor of the transmission line. However, there is almost no modal analysis study for transmission towers. Transmission tower structures are completely different from reinforced concrete bridges, wind turbines and transmission lines. Their vibrations are multi-directional, so existing technologies cannot be applied directly to iron towers. However, iron towers prone to accidents are often in unattended areas such as in mountains and hills. The environmental impact in these areas are small, which can greatly reduce the workload of research. Therefore, a method for settlement detection of the transmission line tower under wind force is proposed in this paper. By installing acceleration sensors and meteorological sensors on the transmission tower, the system can collect vibration acceleration signals of the tower and related parameters (wind speed, wind direction, temperature) of the surrounding environment in real time. The modal frequency of the tower can be identified through modal analysis under wind force, by which it can be judged whether tower settlement occurs. Finally, the system is tested on the 110 kV cat head transmission tower of Xi'an Polytechnic University. The measured vibration data are analyzed by stochastic subspace identification (SSI) and modal parameters are modified via the ambient temperature. The result shows that the system can accurately obtain the structural information of the tower by identifying its modal parameters, which can be used as effective evidence for judging the settlement of the transmission tower foundation.

2. Principle and Experiments

2.1. Principle

There are many types of transmission towers with different complex structures and stiffness. Moreover, even the same tower has different degrees of stiffness due to the installation process. However, if we can find a single parameter to reflect the structure of the tower system and use it as the reference to the normal structure, the structure can be considered to have a fault when the parameter changes significantly. The transmission tower vibration has multiple degrees of freedom under wind load. According to the dynamic equation of a linear system with N degrees of freedom, the vibration equation of the tower can be expressed as follow:

$$M\ddot{x}(t) + C\dot{x}(t) + Kx(t) = f(t) \tag{1}$$

where M, C, K are the mass, damping and stiffness matrices of the tower structure; $x(t)$ is the displacement response time history vector; and $f(t)$ is the excitation time history vector, which is also the wind load action. In this operation, other factors can also influence the vibration of power transmission towers. From the motion equation, although the excitation on the right side of the equation changes, the stiffness and mass of the tower do not change. Different excitations have different frequencies, which can excite different modal. However, these have little effect on the modal parameters of the system. The experimental environment of this paper is the 110 kV three-tower and two-gear transmission lines of Xi'an Polytechnic University. It mainly considers the vibration response of the transmission tower under wind excitation. When transmission tower settlement occurs, especially uneven settlement, the stiffness matrix K and damping matrix C will change and the modal frequency of the corresponding tower system will also change. According to this characteristic, we can monitor whether tower settlement occurs by the modal frequency of the tower system.

2.2. Experiments

2.2.1. Experimental Tower

The transmission tower studied in this paper is a ZM-110 kV cat head tower which is built in Xi'an Polytechnic University. The LGJ-95/15 conductors are installed on the tower and the direction of the lines are north-south. The cat head tower's total height is 19 m and the root is 3.103 m, as shown in Figure 1a.

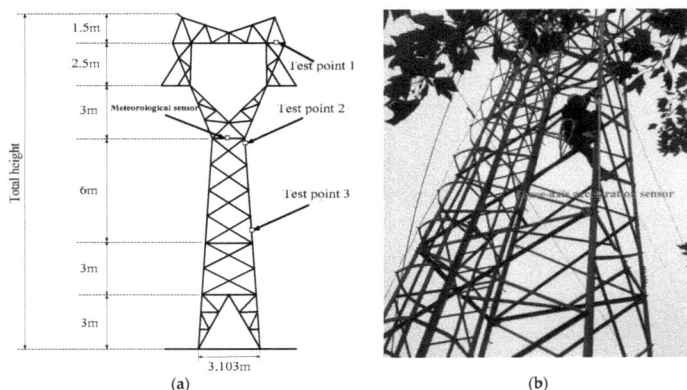

Figure 1. ZM-110 kV cat head tower: (**a**) transmission tower model; (**b**) installation of the monitoring device.

2.2.2. Sensor Installation

One meteorological sensor and three three-axis acceleration sensors were installed on the transmission tower so as to measure the meteorological parameters around the tower and the vibration signals at three measuring points. The layout of the points is shown in Figure 1a. The tower will vibrate when the natural wind exerts load excitation on the tower. Because the stiffness of the tower is high, the natural wind load is not very strong in general. Moreover, the degrees of freedom of the tower legs are constrained and vibration near the tower legs is not obvious. On the contrary, vibration near the tower top is easier to identify. Thus, the measuring points are selected to be the junction of the tower cross arm and the ground support, the tower mouth, and the tower body main material at the height of 7.5 m from the ground, respectively [17]. Selecting these positions as the measuring points is helpful to precisely extract the vibration acceleration of the tower. The three-axis acceleration sensor has a magnetic force on one side, allowing it to be firmly attached to the angled steel of the tower, as shown in Figure 1b.

2.2.3. Real-Time Measurement

In order to prove that the modal frequencies of the transmission tower system can be well extracted by modal identification, the transmission tower in this paper was measured and analyzed at 14:00 on 22 September, 2018. At the same time, the tower settlement experiment platform was built to simulate the settlement process of single tower foundation, and the low-order modal frequencies of the transmission tower system before and after settlement were extracted and analyzed. The weather was sunny, with a south-east wind and wind level of 3–4. The sampling frequency of the three-axis acceleration sensor was set to 200 Hz and the sampling time was 30 min because the low-order modal frequency of the tower is within 1–15 Hz [18]. Figure 2 shows the wind speed history curve measured by the meteorological sensor during the test period.

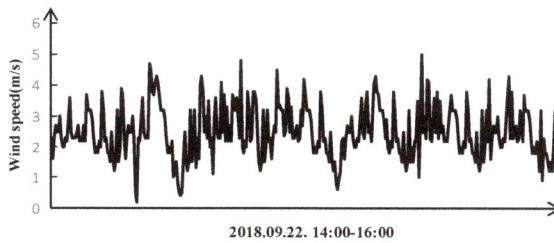

Figure 2. Wind speed time history curve during the test period.

Considering that the tower vibration is not a free vibration in single direction, the three-axis acceleration sensor was used to monitor the acceleration signals in the X, Y and Z directions of the tower in this paper; these signals were processed and analyzed separately. Figure 3 shows the acceleration time history curves for a 5 s interval in the X, Y and Z directions of test point 1 under normal conditions.

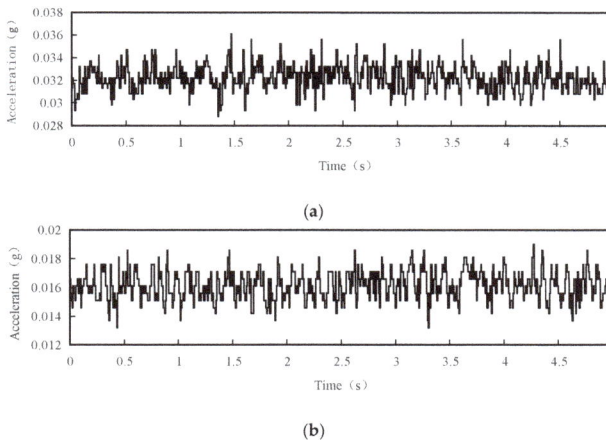

Figure 3. *Cont.*

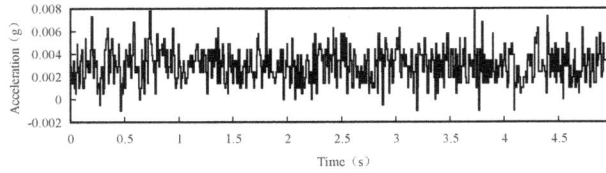

(c)

Figure 3. Acceleration time history curves with 5 s interval of the test point 1 under normal conditions: (**a**) acceleration time history curve in the X direction; (**b**) acceleration time history curve in the Y direction; (**c**) acceleration time history curve in the Z direction.

The collected vibration signals needed to be pre-processed before subsequent analysis. A low-pass filter with cut-off frequency of 15 Hz is used to identify the low-order vibration frequency characteristics of the tower, and high-order vibration frequencies with small amplitudes are neglected. Figure 4 shows the acceleration time history curves for a 5 s interval after filtering of the test point 1 under normal conditions.

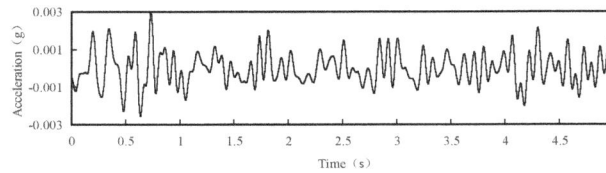

(a)

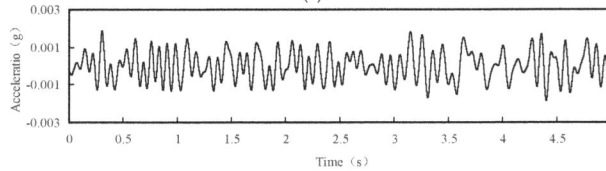

(b)

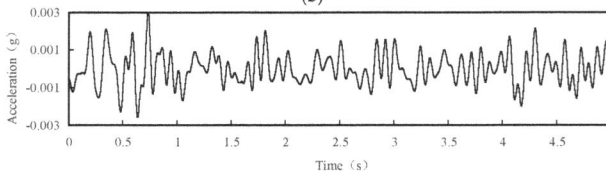

(c)

Figure 4. Acceleration time history curves with interval of 5 s of the test point 1 after filtering under normal conditions: (**a**) acceleration time history curve in the X direction after filtering; (**b**) acceleration time history curve in the Y direction after filtering; (**c**) acceleration time history curve in the Z direction after filtering.

2.3. Vibration Analysis

Stochastic subspace identification (SSI) is a new method for linear system identification developed in recent years. It can extract model parameters directly from the corresponding output signals of the environmental excitation rather than from artificial excitation. The key of this method is to determine the order of the system. A stabilization diagram is a novel approach which can be used for modal identification in the case of strong noise [19]. In this paper, stochastic subspace identification is used to analyze the acceleration signal after filtering, and the stabilization diagram is used for modal identification of the system.

Figure 5 shows the stabilization diagram in the X direction of test point 1 under normal conditions. Its abscissa represents the modal frequency and V1–V8 represent the eight modes of the tested tower system, respectively. According to the requirements of the stability axis, $\left| \frac{f_i - f_{i+1}}{f_i} \right| < [\Delta f_e]$, where i is the modal order, f is the modal frequency and $[\bullet]$ is the modal frequency limit parameter. The stability of the system can be determined by the intensity of the points on the stable axis. As shown in Figure 5, even if the power spectral density (PSD) is not prominent, the eight modes in the X direction can be well analyzed and the modal frequency is also stable.

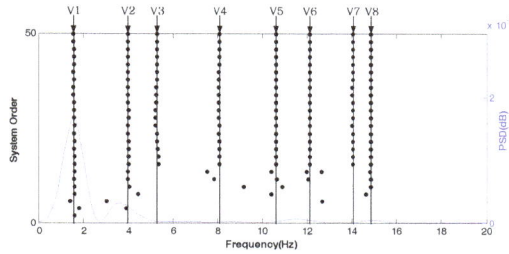

Figure 5. Stabilization diagram in the X direction of test point 1 under normal conditions.

Table 1 lists the modal frequencies in the X direction of test points under normal conditions calculated by SSI. Because of the influence of the installation position and the wind condition of the acceleration sensor, the modal frequencies in some orders of the three test points change slightly. However, the overall frequency identification results are relatively good. In the following analysis, we use the average value of the three test points for analysis.

Table 1. Modal frequency in the X direction of the test points under normal conditions.

	1st Mode	2nd Mode	3rd Mode	4th Mode	5th Mode	6th Mode	7th Mode	8th Mode
Test point 1	1.538	3.987	5.227	8.006	10.590	12.100	14.030	14.830
Test point 2	1.538	3.987	5.224	8.000	10.611	12.101	14.030	14.830
Test point 3	1.538	3.987	5.227	8.003	10.612	12.008	14.029	14.761
Average	1.538	3.987	5.226	8.003	10.611	12.100	14.030	14.807

Figure 6 shows the stabilization diagram in the Y direction of test point 1 under normal conditions. It can be seen that the modal frequencies of each order are relatively stable.

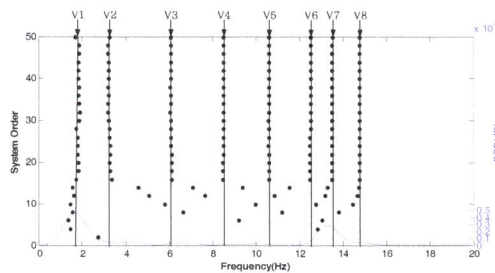

Figure 6. Stabilization diagram in the Y direction of test point 1 under normal conditions.

Table 2 shows the modal frequencies in the Y direction of the test points under normal conditions calculated by SSI. The modal frequencies of three test points are less consistent than those in the X direction, but the modal frequencies are stable.

Table 2. Modal frequency in the Y direction of the test points under normal conditions.

	1st Mode	2nd Mode	3rd Mode	4th Mode	5th Mode	6th Mode	7th Mode	8th Mode
Test point 1	1.698	3.260	6.076	8.503	10.601	12.510	13.503	14.761
Test point2	1.694	3.262	6.075	8.497	10.611	12.501	13.500	14.760
Test point 3	1.698	3.264	6.078	8.504	10.612	12.511	13.499	14.761
Average	1.697	3.262	6.076	8.501	10.608	12.507	13.501	14.761

Figure 7 shows the stabilization diagram in the Z direction of test point 1 under normal conditions. The modal frequency of the Z direction is very unstable. If the modal parameters of the Z direction are used as evidence for tower settlement, error would be introduced.

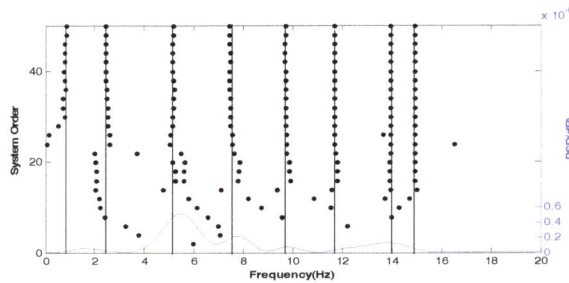

Figure 7. Stabilization diagram in the Z direction of the test point 1 under normal conditions.

3. Field Test and Discussion

3.1. Experiment and Analysis of Tower Foundation Settlement

Transmission tower settlement is caused by its foundation and underground soil, which is uniform or uneven. Under uneven settlement, the height difference of the four tower foundations may cause a large change in the force of the transmission tower, which also causes the stiffness and the modal frequency of the tower system to change. Since a damage test cannot be carried out, the process of tower foundation settlement cannot be readily simulated. Therefore, a hydraulic jack was used to raise the tower foundation, which alters the height of the tower foundation and verifies the feasibility of the proposed method.

3.1.1. Tower Foundation Settlement

The anchor bolts of the ZM-110 kV cat tower in Xi'an Polytechnic University are outlying, which is convenient for building the settlement experimental platform. Figure 8 is a schematic diagram of the experimental platform. Before tests, the tower foundation must be cleaned to ensure the cleanliness of the tower foundation as much as possible, so as not to introduce unnecessary errors. Firstly, the ultra-thin hydraulic jack can be placed at the bottom of the tower foundation by releasing the anchor bolts. Then, the tower foundation can be lifted by raising the ultra-thin hydraulic jack, and recording the height of the tower foundation from the groove bottom as h_x. When the tower is installed and designed, the initial height of the tower foundation and the groove bottom is recorded as h_0. Finally, the longitudinal displacement $\Delta h = h_x - h_0$ can be obtained. Because the displacement of the ultra-thin hydraulic jack is limited, 2-mm stainless-steel plates can be added to the groove bottom without withdrawing the last measuring jack. The longitudinal displacement can be increased by applying longitudinal displacement with another hydraulic jack.

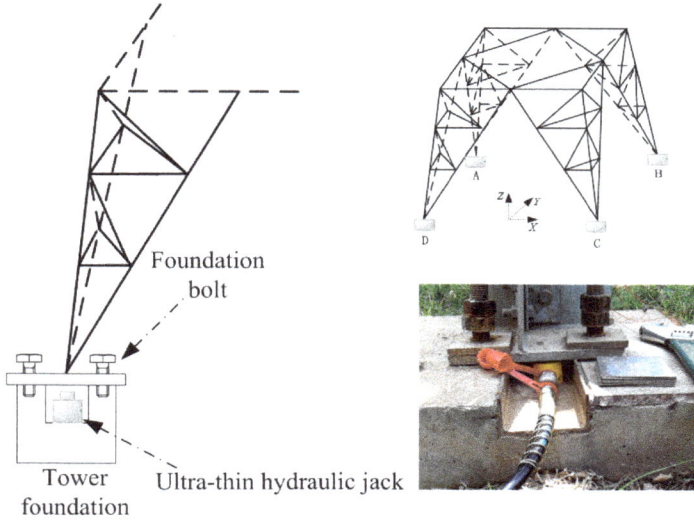

Figure 8. Tower foundation settlement experiment platform.

3.1.2. Settlement Experiment Analysis

In order to determine which direction of modal frequency under natural load can provide better evidence for settlement monitoring, we use the settlement experimental platform to settle 4 mm and 6 mm of tower foot A, and the recorded settlement data are processed and analyzed.

In Table 3, the average modal frequencies and the change rate in the X direction of tower foot A before and after settlement are recorded. ω_0, ω_4 and ω_6 represents the modal frequencies in settlements of 0 mm, 4 mm and 6 mm, respectively. $\frac{|\omega_4-\omega_0|}{\omega_0}$% and $\frac{|\omega_6-\omega_0|}{\omega_0}$% represent the change rate of the modal frequencies of the tower system in settlements of 4 mm and 6 mm, respectively. It can be seen that the modal frequency of the tower experienced an obvious change after 4 mm settlement; the change rates of the modal frequency of six modes exceeded 3% and the change rate of the third mode exceeded 10%. When settling 6 mm, the change rate is further increased, with the change rates of five modes exceeding 5%. The repeated tests show that the results are stable. It can be seen that the X direction low-order modal frequencies of the tower system can be used as evidence for tower foundation settlement.

Table 3. Average modal frequency and change rate in the X direction of tower foot A before and after settlement.

	Settling 0 mm (ω_0)	Settling 4 mm (ω_4)	$\frac{|\omega_4-\omega_0|}{\omega_0}$%	Settling 6 mm (ω_6)	$\frac{|\omega_6-\omega_0|}{\omega_0}$%
1st mode	1.538	1.593	3.576	1.601	4.096
2nd mode	3.987	3.859	3.217	3.724	6.596
3rd mode	5.226	5.842	11.787	5.936	13.596
4th mode	8.003	7.722	3.511	7.596	5.086
5th mode	10.611	10.205	3.826	9.912	6.588
6th mode	12.100	11.950	1.240	11.720	3.140
7th mode	14.030	13.560	3.350	12.987	7.434
8th mode	14.807	14.790	0.115	14.760	0.317

Figure 9 is a graph of the modal frequency change rate and order in the Y direction of tower foot A before and after settlement. After settlement, the modal frequency of the first mode changes clearly, while changes for other modes are less obvious. Thus, the modal frequency in the Y direction is not suitable as evidence of tower foundation settlement under natural load.

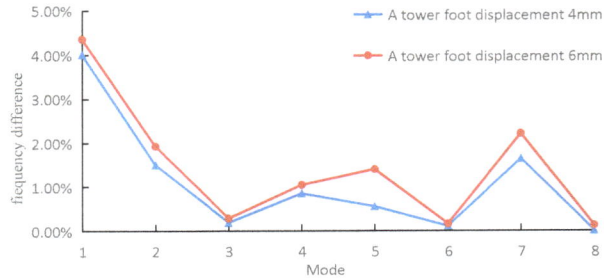

Figure 9. Change rate of modal frequency in the Y direction of the tower foot A before and after settlement.

In order to ensure that the low-order modal frequency in the X direction can be used as evidence for tower foundation settlement, we carried out the settlement experiments of 4 mm and 6 mm on tower foot C. Figure 10 is the change rate of the modal frequency in the X direction before and after the settlement of tower foot A and C. The variation of the modal frequencies of each mode of tower foot C after settlement is consistent with that of tower foot A after settlement, which verifies the conclusion that the low-order modal frequency in the X direction can be used to indicate tower foundation settlement. In addition, it can be seen from Figure 10 that 3rd, 5th, and 7th mode frequencies in the X direction of the tower system change clearly, and the change rate of the modal parameters are effective to judge whether settlement occurs during the actual monitoring process.

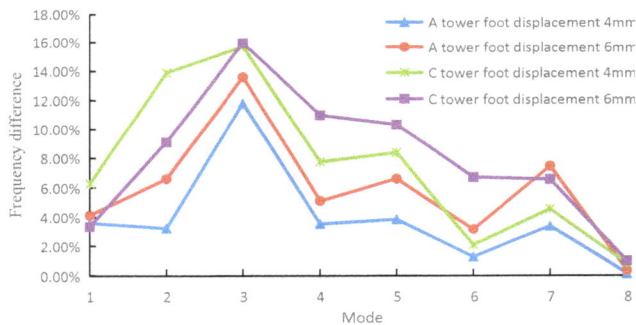

Figure 10. Change of modal frequency in the X direction of the A and C tower feet before and after settlement.

3.2. Influence of Temperature on Modal Frequency

A large number of studies have shown that the ambient temperature has a great impact on the dynamic characteristics of the structure, which is the most important factor. The modal frequency fluctuations caused by changes of normal temperature can obscure the frequency changes caused by minor structural damage, and even cause misunderstanding of the structural dynamic properties [20]. Considering the effect of the temperature on the modal frequency of the tower system, a one-week test was carried out from 10 October 2018. The ambient temperature and modal frequencies of the 3rd, 5th, and 7th in the X direction of the tower were recorded every 30 min from 9:00 am to 18:00 pm. Figure 11 shows the relationship between the modal frequency and the temperature change over the seven days. It can be seen that the modal frequency is negatively correlated with the ambient temperature.

In order to reduce the influence of ambient temperature on measurements, we fitted the ambient temperature and the modal frequencies of the 3rd, 5th and 7th mode of the tower. The obtained correction coefficient is used to reduce the error caused by the ambient temperature. Figure 12 shows the fitted curves of the ambient temperature and the modal frequency. The variation of the modal frequency is linear with respect to the variation of the ambient temperature.

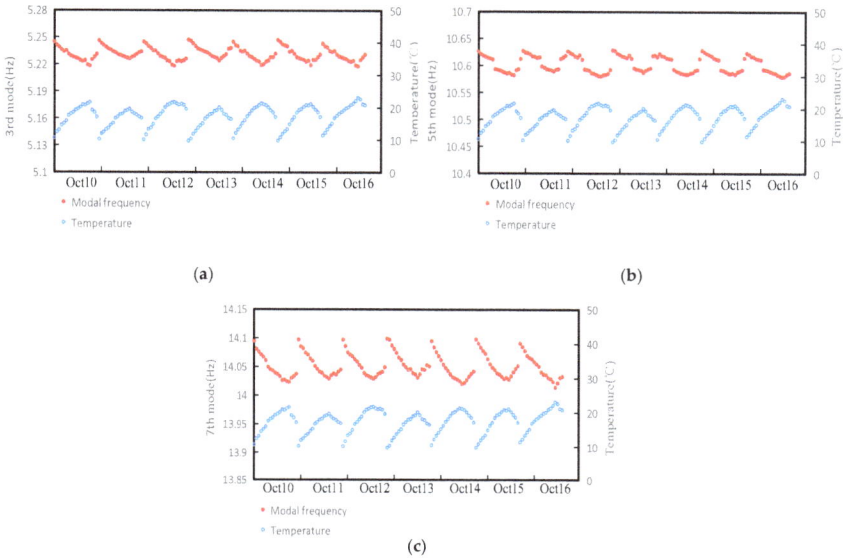

Figure 11. Curves of ambient temperature and modal frequency: (**a**) curve of ambient temperature and 3rd modal frequency; (**b**) curve of ambient temperature and 5th modal frequency; (**c**) curve of ambient temperature and 7th modal frequency.

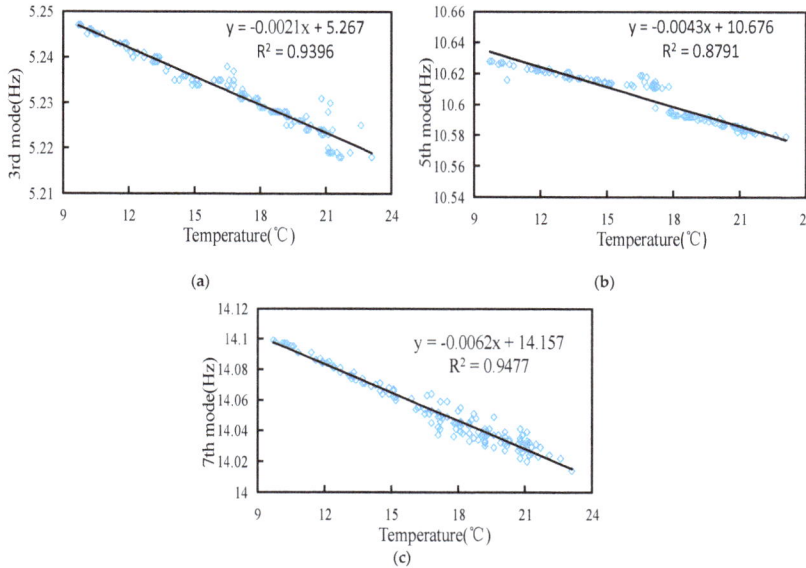

Figure 12. Fitted curves of ambient temperature and modal frequency: (**a**) fitted curve of ambient temperature and 3rd modal frequency; (**b**) fitted curve of ambient temperature and 5th modal frequency; (**c**) fitted curve of ambient temperature and 7th modal frequency.

4. Monitoring Technology Realization

4.1. System Structure

In order to realize the on-line monitoring of the tower settlement, a monitoring system for transmission tower settlement is designed. The system shown in Figure 13 mainly includes five parts: three-axis acceleration sensors, monitoring device, meteorological sensors, power supply module and monitoring center. The outer shell of the monitoring device is a stainless-steel rigid box, which is installed on the cross arm of the tower to obtain the vibration signal of the tower, the wind speed and direction, and the temperature around the tower in real time. The monitoring center communicates with the monitoring device by 4G wireless communication, which can receive the vibration data of the tower in real time or regular time. The filtering algorithm, stochastic subspace identification (SSI) modal identification model and tower settlement fault diagnosis model are embedded via the expert software of the monitoring center. Using the filtering algorithm to process the measured data is helpful to improve the vibration frequency identification accuracy of the transmission tower. The SSI identification model can extract the low-order frequencies of the transmission tower system from the received vibration data. The tower settlement fault diagnosis model determines whether a tower settlement accident has occurred by comparing the modal parameters with the tower without settlement and gives an alarm in case of an accident. At the same time, the operator can obtain the accident information at initiation and take timely corrective measures to prevent further development of the accident. In field maintenance or testing, the operator can obtain the monitoring device data through Bluetooth by using a handheld device or laptop. Solar power and battery are used to provide electricity for the whole system. The best installation angle of solar panels is determined by actual operating conditions, and the battery is located in the stainless-steel rigid box.

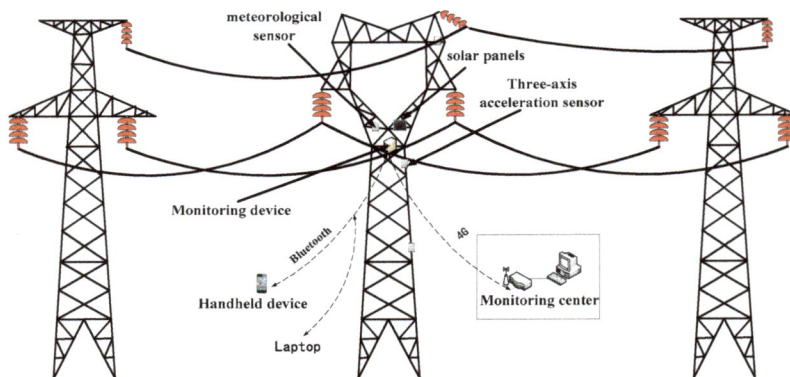

Figure 13. System structure.

4.2. Monitoring Device

The principle diagram of the monitoring device is shown in Figure 14a, including the main control unit (MCU), power module, communication unit (4G/Bluetooth/RS485) and sensors. The power module is composed of solar panels, batteries and controllers, which provide power for the whole system. It has low power consumption and can last for more than 30 days in rainy weather. Sensors include the three-axis accelerometer sensor and the meteorological sensor. The three-axis acceleration sensor is used to collect the vibration response of the tower and the meteorological sensor is used to measure the environmental parameters around the tower. The monitoring device sends the measurement data to the monitoring center and receives the commands from the monitoring center, which include time setting, sampling interval, sampling duration, monitoring device initialization, etc. Figure 14b shows the main control and power supply control panels.

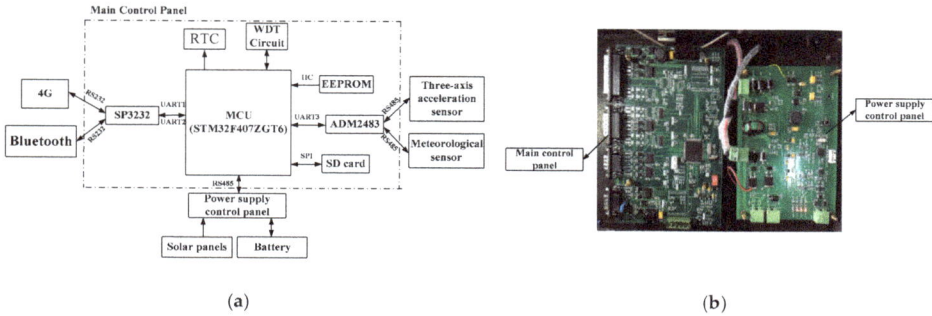

Figure 14. Monitoring device: (**a**) principle diagram of the monitoring device, RTC(Real-Time Clock),WDT(Watchdog Timer); (**b**) main control and power supply control panels.

4.3. Implementation Process

In reference [21], the minimum critical displacement of the 110 kV cat head tower is 7.5 mm when the settlement is simulated by finite element analysis, and stress distribution is verified by the method of affixing the grating fiber strain gauge and artificially simulating transmission tower sink. Based on above reasons, the threshold value should be less than the critical displacement value of a single tower and considered as a certain margin. In this paper, when the settlement was 4 mm and 6 mm, the maximum change rate of the modal frequency was 11.787% and 13.596% respectively. Considering the error of the artificial simulation settlement and the actual settlement, for the 110 KV cat head tower studied in this paper, the maximum modal change rate is 10%, which is used as the threshold of the monitoring device.

Figure 15 shows a flow chart of the system. After the installation of the device, ensuring that the tower settlement does not occur within one day, the low-order modal frequency matrix ω_0 of the tower system measured on the first day is used as a parameter when the tower structure is intact. The device can collect the environmental parameters around the acceleration signal of the tower system in real time or at a fixed time. The acquired signal is filtered to remove DC (direct current) components and trend items. The monitoring center analyzes the received signal and calculates the correlation function matrix of the acceleration and the PSD. The stochastic subspace is combined with the stable graph method to obtain the modal frequency matrix ω_i of the system. Considering the effect of the temperature on the system, the obtained modal frequency is corrected by the temperature, which is $\hat{\omega}_i$. The modal frequency change matrix is calculated as $\delta_i = \frac{|\hat{\omega}_i - \omega_0|}{\omega_0}\%$, and the largest value δ_{Max} in the modal frequency matrix δ_i is obtained, $\delta_{Max} = Max\{\delta_i\}$, to determine if the δ_{Max} is exceeded. If it is exceeded, then an alarm is issued.

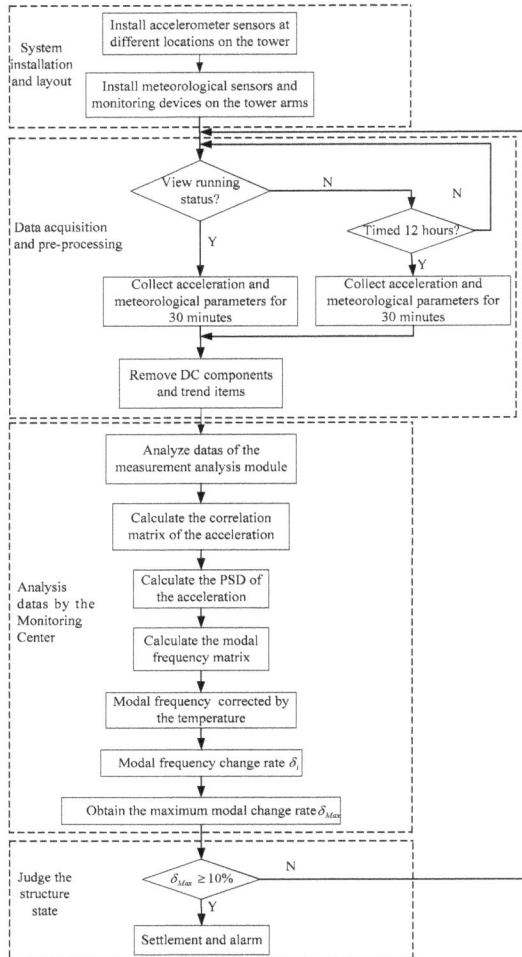

Figure 15. Flow chart of the system.

5. Conclusions

In this paper, a method for settlement detection of the transmission line tower under wind force is proposed. Through the measurement and experimental analysis of the ZM-110 kV cat tower of Xi'an Polytechnic University, the following conclusions are drawn.

The low-order modal frequencies in the X and Y directions of the transmission towers under natural load can be extracted by stochastic subspace identification (SSI) and stabilization diagram.

The artificial simulation of the tower settlement shows that the low-order modal frequencies of the different tower feet in the X direction before and after settlement are obvious, and the change rate increases with the increase of the settlement displacement. Therefore, under the action of the natural load, the change of the low-order modal frequency in the X direction can be used as a basis to detect whether the transmission tower is settled.

The experiment found that the ambient temperature has a certain influence on the modal frequency of the tower system, and the modal frequency has a negative correlation with the temperature change. The temperature and modal frequency curves were fitted, and it was found that the change of the modal frequency is linear with the change of the ambient temperature. The coefficients of the modal frequency and temperature change were determined.

This paper studies the dynamic response of transmission towers under wind excitation and considers the effects of the ambient temperature. However, there are other sources of vibration in some areas, such as falling rocks, which interact with the wind to complicate the vibration response of the tower. Next, we will study modal parameter identification under the influence of multiple excitation.

Author Contributions: Conceptualization, L.Z., Y.Z. and X.H.; Funding acquisition, X.H.; Methodology, Y.Z. and L.Z.; Project administration, L.Z. and X.H.; Software, L.Z. and X.H.; Writing—original draft, Y.Z. and X.H.; Writing—review & editing, L.Y.

Funding: This work was supported by the Xi'an Municipal Science and Technology Bureau (grant 201805023YD1CG7 (3)), Shaanxi Province (grant 2018JQ5049 and 2018ZDXM-GY-040) and Shaanxi Provincial Education Department (grant 17JK0322).

Conflicts of Interest: The authors declare no conflict of interest.

References

1. Huang, X.B.; Chen, R.G.; Wang, X.J. Online monitoring of tower inclination of transmission line. In *Transmission-Line Monitoring and Fault Diagnosis*, 2nd ed.; China Electric Power Press: Beijing, China, 2014; pp. 361–370.
2. Huang, X.B.; Chen, Z.L.; Zhao, L.; Zhu, Y.C.; Xu, G.H.; Si, W.J. Stress simulation and experiment for tower foundation settlement of 110 kV transmission line. *Electr. Power Autom. Equip.* **2017**, *37*, 361–370.
3. Zhang, M.; Zhao, G.; Wang, L.; Li, J. Wind-Induced Coupling Vibration Effects of High-Voltage Transmission Tower-Line Systems. *Shock Vib.* **2017**, *2017*, 1205976. [CrossRef]
4. Battista, R.C.; Rodrigues, R.S.; Pfeil, M.S. Dynamic behavior and stability of transmission line towers under wind forces. *J. Wind Eng. Ind. Aerodyn.* **2003**, *91*, 1051–1067. [CrossRef]
5. Wang, Z.Q.; Wang, J.; Huang, Z.H.; Wu, X.Q.; Meng, X.B. Strain Modal Analysis of Lattice Transmission Tower. *J. North China Electr. Power Univ.* **2017**, *44*, 62–70.
6. Wang, F.; Han, J.K.; Wang, C.C.; Wang, C.Z.; Wong, X.L.; Zhou, Y.Y.; Fang, H. Strong wind simulation of transmission tower structures. *Build. Struct.* **2018**, *48*, 39–44.
7. Fei, Q.G.; Zhou, H.G.; Han, X.L.; Wang, J. Structural health monitoring oriented stability and dynamic analysis of a long-span transmission tower-line system. *Eng. Fail. Anal.* **2012**, *20*, 80–87. [CrossRef]
8. Li, Z.L.; Xiao, Z.Z.; Han, F. Aeroelastic model design and wind tunnel tests of 1000 kV Hanjiang long span transmission line system. *Power Syst. Technol.* **2008**, *32*, 1–5.
9. Malhara, S.; Vittal, V. Mechanical state estimation of overhead transmission lines using tilt sensors. *IEEE Trans. Power Syst.* **2010**, *25*, 1282–1290. [CrossRef]
10. Xia, Y.; Zhang, P.; Ni, Y.Q.; Zhu, H.P. Deformation monitoring of a super-tall structure using real-time strain data. *Eng. Struct.* **2014**, *67*, 29–38. [CrossRef]
11. Huang, X.B.; Liao, M.J.; Xu, G.H. Stress monitoring method applying FBG sensor for transmission line towers. *Electr. Power Autom. Equip.* **2016**, *36*, 68–72.
12. Bang, H.; Kim, H.; Lee, K. Measurement of strain and bending deflection of a wind turbine tower using arrayed FBG sensors. *Int. J. Prec. Eng. Manuf.* **2012**, *13*, 2121–2126. [CrossRef]
13. Ding, K. Application of Wavelet Analysis of Curvature Modal to Damage Detection of Bridges. *Noise Vib. Control* **2013**, *33*, 131–135.
14. Huynh, H.C.; Park, J.H.; Kim, J.T. Structural identification of cable-stayed bridge under back-to-back typhoons by wireless vibration monitoring. *Measurement* **2016**, *88*, 385–401. [CrossRef]
15. Ou, Y.W.; Eleni, N.C.; Vasilis, K.D.; Minas, D.S. Vibration-based experimental damage detection of a small-scale wind turbine blade. *Struct. Health Monit.* **2017**, *16*, 79–96. [CrossRef]
16. Zhao, L.; Huang, X.B.; Jia, J.Y.; Zhu, Y.C.; Cao, W. Detection of Broken Strands of Transmission Line Conductors Using Fiber Bragg Grating Sensors. *Sensors* **2018**, *18*, 2397. [CrossRef] [PubMed]
17. Wang, Z.Q.; Wang, J. Optimal placement of biaxial acceleration sensor for transmission towers. *Chin. J. Sci. Instrum.* **2017**, *38*, 2200–2209.

18. Li, H.N.; Bai, H.F.; Yi, Y.H.; Ren, L. Field measurement and quasi-static studies on gust response of transmission tower. *J. Vib. Shock* **2010**, *29*, 17–21.

19. Chang, J.; Zhangi, Q.W.; Sun, L.M. Application of stabilization diagram for modal parameter identification using stochastic subspace method. *Eng. Mech.* **2007**, *24*, 39–44.

20. Sohn, H.; Dzwonczyk, M.; Straser, E.G.; Kiremidjian, A.S.; Law, K.H.; Meng, T. An experimental study of temperature effect on modal parameters of the Alamosa Canyon Bridge. *Earthq. Eng. Struct. Dyn.* **1999**, *28*, 879–897. [CrossRef]

21. Huang, X.; Zhao, L.; Chen, Z.; Cheng, L. An online monitoring technology of tower foundation deformation of transmission lines. *Struct. Health Monit.* **2018**. [CrossRef]

![sensors](sensors logo)

MDPI

Article

Influence of Uniaxial Stress on the Shear-Wave Spectrum Propagating in Steel Members

Zuohua Li [†], Jingbo He [†], Diankun Liu, Nanxi Liu [ID], Zhili Long and Jun Teng *

Harbin Institute of Technology Shenzhen, Shenzhen 518055, China; lizuohua@hit.edu.cn (Z.L.);
hejingbo@stu.hit.edu.cn (J.H.); liudiankun2010@163.com (D.L.); liu_nanxi@foxmail.com (N.L.);
longzhili@hit.edu.cn (Z.L.)
* Correspondence: tengj@hit.edu.cn; Tel.: +86-755-2603-3806
† These authors contributed equally to this work.

Received: 24 December 2018; Accepted: 23 January 2019; Published: 25 January 2019

check for
updates

Abstract: Structural health monitoring technologies have provided extensive methods to sense the stress of steel structures. However, monitored stress is a relative value rather than an absolute value in the structure's current state. Among all the stress measurement methods, ultrasonic methods have shown great promise. The shear-wave amplitude spectrum and phase spectrum contain stress information along the propagation path. In this study, the influence of uniaxial stress on the amplitude and phase spectra of a shear wave propagating in steel members was investigated. Furthermore, the shear-wave amplitude spectrum and phase spectrum were compared in terms of characteristic frequency (CF) collection, parametric calibration, and absolute stress measurement principles. Specifically, the theoretical expressions of the shear-wave amplitude and phase spectra were derived. Three steel members were used to investigate the effect of the uniaxial stress on the shear-wave amplitude and phase spectra. CFs were extracted and used to calibrate the parameters in the stress measurement formula. A linear relationship was established between the inverse of the CF and its corresponding stress value. The test results show that both the shear-wave amplitude and phase spectra can be used to evaluate uniaxial stress in structural steel members.

Keywords: uniaxial stress measurement; structural steel members; amplitude spectrum; phase spectrum; shear-wave birefringence; acoustoelastic effect

---

1. Introduction

1.1. Absolute Stress in Structural Steel Members

Many large-scale steel structures have been built worldwide due to their high degree of industrialization [1,2]. Fully understanding the performance degradation of steel structures during their entire life cycle has become a significant topic [3], which has received increasing attention in academic and engineering fields [4,5]. Structural health monitoring [6,7] is one of the most effective technologies to sense the real response of the monitored objects. Many excellent monitoring technologies [8,9], systems [10], and advanced intelligent algorithms [11] have been developed and applied to solve engineering problems. A stress monitoring system [12,13], which plays an important role in structural health monitoring technologies, has been regarded as a mature way to obtain structural stress information from the macroscale stress distribution of a whole structure [14] to the microscale stress concentration of a local member [15]. However, the monitored stress value using a stress monitoring system is a relative value rather than an absolute value. The absolute stress, which represents the current state of structures, is a significant indicator for judging the safety of structures [16].

Existing stress measurement methods, such as diffraction [17,18] and magnetic methods [19], can be used to detect the absolute stress of materials. However, these methods are unable to adequately test large-scale steel members and are unsuitable for field applications because a strict testing environment is required during the testing process. In addition, the testing equipment is complex, and the testing process is time consuming. Generally, absolute stress measurements of structural steel members using structural health monitoring technologies remains a challenging task [20,21].

1.2. Ultrasonic Stress Measurement Methods

In recent years, ultrasonic methods, which are based on acoustoelastic effects, have been studied to evaluate the internal and initial stress in complex structures [22–25]. Compared with other stress measurement methods, such as X-ray diffraction [17], neutron diffraction [18], and magnetic [19] methods, ultrasonic methods have shown great prospects for use in in-site stress measurements [26]. Essentially, ultrasonic methods establish a linear relationship between the stress and ultrasonic wave velocities, that is, a time-of-flight (TOF) measurement [27]. Compared with other ultrasonic waves, a longitudinal critically refracted (Lcr) wave exhibits the greatest sensitivity to stress [28]. Hence, these waves have been widely used to evaluate welding residual stress [29], rail stress [30], steam turbine disk stress [31], and steel member stress [22]. To improve the signal-to-noise ratio, the laser-generated Lcr wave method was presented to evaluate the stress in a noncontact manner [32], and the piezoelectric effect-generated Lcr wave was investigated to detect the stress in an immersion manner [33]. Combining the experimental and the numerical analysis results, the colored stress distribution nephogram of a tested member can be sketched [34]. Because the Lcr wave energy is relatively small and rapidly decays, guided ultrasonic wave methods have been proposed and used to monitor the stress in steel strands [35] and aluminum plates [36,37], which is a further application of the acoustoelastic effect. Recently, the influence of a uniaxial load on the electromechanical impedance of embedded piezoceramic transducers in steel fiber concrete was investigated [38]. A normalized root-mean-square deviation index was developed to analyze the electromechanical impedance information, and the experimental results showed that the index increases with the uniaxial load, thus providing a potential method to evaluate the uniaxial stress of steel fiber concrete.

In addition to the methods described above, the shear wave [39] can also be used to evaluate stress. The effect of birefringence [40] describes a phenomenon in which the velocity of a shear wave varies when the shear wave vibrates in different directions, which endows the shear wave with unique advantages to evaluate the stress in materials. If two individual stress values in plate-like components need to be detected, then the combination of longitudinal and polarized shear waves is advantageous [41]. In addition, by measuring the velocities of the shear wave in two different polarization directions, the influence of texture during the stress evaluation can be separated [42]. Note that the aforementioned shear-wave methods are based on TOF measurements. The accuracy of the ultrasonic stress evaluation results is influenced by the TOF data collection. In fact, many uncertain factors, such as microcracks [43], inhomogeneous materials [44], coupling conditions [45], and temperature [46], may lead to a distortion of the waveform, which limits the industrial application of ultrasonic methods. It is critical to distinguish the influence of the uncertainty factors from that of stress [47]. To date, only a few systems have been used in practical engineering [48].

In addition to the above methods, shear-wave frequency domain signals have received attention in recent years. Shear-wave spectrum analysis methods are based on acoustoelastic theory and the shear-wave birefringent effect. When a beam of a shear wave is perpendicularly incident to a stressed solid, it separates into two modes. The two separated shear-wave modes travel with different velocities, which produces interference effects. The received shear-wave spectrum contains the interference information, which can be used to evaluate the absolute stress in solids [49]. Recently, the shear-wave amplitude spectrum method was proposed to measure the absolute stress in steel members [24]. The experimental results showed that the inverse of the CF linearly changed with the applied uniaxial stress, and then the mechanically applied stresses of the structural steel members were evaluated. The ultrasonic shear-wave amplitude spectrum method makes use of the amplitude spectrum to establish the relationship between the stress and the CF. In fact, the phase spectrum also contains the stress information along the shear-wave propagation path, which may provide a new method to detect the absolute stress in structural steel members. However, the effect of stress on the phase spectrum is not as well understood, which is the focus of this study.

1.3. Goals and Objectives of This Study

In light of the challenges described above, here we investigate the influence of uniaxial stress on shear-wave spectrum propagation in steel members. Compared to our previous work [24], which aimed to measure the absolute stress using the shear-wave amplitude spectrum, this paper further studies the phase spectrum of a shear wave propagating in steel members. Moreover, the shear-wave amplitude spectrum and phase spectrum are compared in terms of CF collection, parametric calibration and absolute stress measurement principles, which represents an expansion of our previous method [24]. For this purpose, the theoretical formulas of the shear-wave pulse echo phase spectrum are derived. Accordingly, the relationship between the uniaxial stress and the CF is established. Three structural steel members are tested to investigate the effect of the applied uniaxial stress on the shear-wave amplitude and phase spectra. The parameters representing the quantitative relationship between the stress and the CF are calibrated using the experimental data. The results show that the amplitude and phase spectra have the potential to be used for stress monitoring of in-service structures.

2. Theory

2.1. Theoretical Derivation of the Shear-Wave Pulse Echo Spectrum

The theoretical expression of the shear-wave pulse echo spectrum is derived on the assumption that the steel member interface exerts no effect on the propagation of the shear wave. In addition, the steel member material is assumed to be isotropic and homogeneous as well as be elastic in its range. When a beam of ultrasonic shear-waves is perpendicularly incident on a steel member, the motion equation of the shear-wave propagating is

$$u_0 = y(t) = \sum_{i=0}^{n} A_i \cos(w_i t + j_i), (i = 0, 1, 2, ..., n),$$ (1)

where the shear-wave contains various components of the harmonic vibration, u_0 and A_i are the amplitude of the vibration source and the amplitude of the ith component, respectively; w_i and φ_i are the angular frequency and initial phase of the ith component, respectively; and t is the vibration time.

Uniaxial stress in steel members can cause an acoustic anisotropy of the material; that is, stress causes the ultrasonic velocity to change when vibrating in different directions. When an ultrasonic shear-wave is perpendicularly incident on a steel member under a uniaxial stressed state, it separates into two shear-wave modes with one polarization direction parallel to the stress direction and the other mode perpendicular to the stress direction [49]. The motion equations of the two separated shear waves are [40]

$$u_1(x_3, t) = y(t - \frac{x_3}{v_{31}}) \cos \theta,$$ (2)

$$u_2(x_3, t) = y(t - \frac{x_3}{v_{32}}) \sin \theta,$$ (3)

where x_1, x_2 and x_3 are axes of the Cartesian coordinate; the two separated shear waves propagate in the positive direction of x_3; v_{31} and v_{32} are the velocities of the two shear-waves traveling in the direction of x_3 with a particle vibration parallel to x_1 and x_2, respectively; and θ is the angle between the incident shear-wave direction and the x_1 direction.

When the two separated shear waves travel from the starting point on one side of the steel member to the rear side, they will be reflected and travel back to the starting point. The synthesis of the two reflected shear waves is

$$u_r(t) = y(t - \frac{x_3}{v_{31}}) \cdot \cos^2\theta + y(t - \frac{x_3}{v_{32}}) \cdot \sin^2\theta, \tag{4}$$

where $y(t - x_3/v_{31})$ and $y(t - x_3/v_{32})$ contain the information of the two separated shear waves' TOF delay and $\cos^2\theta$ and $\sin^2\theta$ contain the amplitude information of the two wave components that synthesize pulse echo in the incident direction. Let $M = \cos^2\theta$ and $N = \sin^2\theta$; then Equation (4) can be simplified as the following form:

$$u_r(t) = y(t - \frac{x_3}{v_{31}}) \cdot M + y(t - \frac{x_3}{v_{32}}) \cdot N. \tag{5}$$

$U_0(f)$ and $U_r(f)$ are defined as the Fourier transforms of $y(t)$ and $u_r(t)$, respectively. The synthesis of the two reflected shear waves in the frequency domain is [24]

$$U_r(f) = U_0(f) \cdot L(\theta, f), \tag{6}$$

$$L(\theta, f) = \begin{array}{l} \cos(2\pi f \frac{x_3}{v_{31}}) \cdot M + \cos(2\pi f \frac{x_3}{v_{32}}) \cdot N \\ -i\left[\sin(2\pi f \frac{x_3}{v_{31}}) \cdot M + \sin(2\pi f \frac{x_3}{v_{31}}) \cdot N\right] \end{array} \tag{7}$$

where $L(\theta, f)$ is defined as the interference factor (IF). Equation (6) is the theoretical expression of the shear-wave pulse echo spectrum propagating in steel members. Note that Equation (6) is equivalent to Equation (4) and contains the interference information of the two separated shear waves.

2.2. Theoretical Derivation of the Shear-Wave Amplitude Spectrum

Equation (6) contains information on the amplitude spectrum and phase spectrum. By taking the modular operation on both sides of Equation (6), the theoretical formula of the shear-wave amplitude spectrum can be obtained, which is shown in the following formula.

$$|U_r(f)| = |U_0(f)| \cdot |L(\theta, f)|, \tag{8}$$

$$|L(\theta, f)| = \sqrt{1 + 2MN(\cos(2\pi P f) - 1)}, \tag{9}$$

where $|L(\theta, f)|$ is the amplitude of the interference factor (AIF) with a value ranging from 0 to 1; P equals $(2l/v_{31} - 2l/v_{32})$, which is the TOF difference of the two separated shear waves.

The AIF is a periodic function of the frequency and polarized angle. When the AIF reaches a minimum, the frequency and the polarized angle can be solved.

$$\begin{cases} f^* = \frac{2N_1 - 1}{2P}, (N_1 = 1, 2, 3, ...) \\ \theta = \frac{N_2\pi}{4}, (N_2 = 1, 3, 5, ...) \end{cases} . \tag{10}$$

In Equation (10), f^* is defined as the CF. The CFs are defined as the first CF ($f_1{}^*$), the second CF ($f_2{}^*$), the third CF ($f_3{}^*$), . . . , when N_1 equals 1, 2, 3, . . . , respectively.

Equation (8) shows that the amplitude spectrum is a product of the incident shear-wave amplitude spectrum and the AIF. The periodic values for the frequency and the polarized angle are $1/P$ and $\pi/2$, respectively. Particularly, when the shear-wave polarized angle is an odd multiple of $\pi/4$, the AIF reaches 0 at the minimum point. This finding indicates that the energy of the harmonic component with the frequency of $(2N_1 - 1)/2P$ decreases to 0. Correspondingly, the amplitude value in the amplitude spectrum with a frequency of $(2N_1 - 1)/2P$ decreases to 0. In Equation (9), P is the TOF difference of the two separated shear waves. Because the TOF difference of the two separated shear waves is determined by the uniaxial stress, the CF in Equation (10) is related to the uniaxial stress in the steel member. This effect establishes the foundation for detecting the uniaxial stress in steel members from the shear-wave amplitude spectrum.

2.3. Theoretical Derivation of the Shear-Wave Phase Spectrum

$\varphi_0(f)$ and $\varphi_r(f)$ are defined as the phase spectra of U_0 (f) and U_r (f), respectively. The spectra U_0 (f) and U_r (f) are complex functions. By combination with Equation (6), the following expression can be obtained.

$$\varphi_r(f) = \varphi_0(f) + \varphi_L(f), \tag{11}$$

where $\varphi_L(f)$ is defined as the phase of the interference factor (PIF).

From Equation (7), the following formula can be obtained:

$$\varphi_L(f) = -\pi f \left(\frac{2l}{v_{31}} + \frac{2l}{v_{32}} \right) - \arctan((M - N) \cdot \tan(\pi f P)). \tag{12}$$

By substituting Equation (12) into Equation (11), the theoretical expression of the shear-wave pulse echo phase spectrum can be obtained.

$$\varphi_r(f) = \varphi_0(f) - \pi f \left(\frac{2l}{v_{31}} + \frac{2l}{v_{32}} \right) - \arctan((M - N) \cdot \tan(\pi f P)). \tag{13}$$

From Equation (13), the shear-wave pulse echo phase spectrum contains three parts. The first part, $\varphi_0(f)$, is the phase spectrum of the incident shear wave. The second part, $-\pi f(2l/v_{31} + 2l/v_{32})$, is the delayed phase values of the shear-wave propagating a length of $2l$. The third part equals $-\arctan((M - N) \cdot \tan(\pi f P))$, which is the phase value caused by the TOF difference for the two separated shear waves. When the amplitude of the separated shear waves is identical, that is, N equals M, the third part in Equation (13) is 0, and the corresponding shear wave polarized angle is 45°. In the following theoretical derivation, as the third part in Equation (13) is significant for CF collection, the shear wave polarized angle should not be 45°. Comparing Equations (8) and (11), the interference factor plays different roles in the amplitude and phase spectra. The AIF ($|L(\theta,f)|$) and the PIF ($\varphi_r(f)$) indicate the amplitude change and the phase change in the synthesis shear wave, respectively.

When the shear-wave pulse echo amplitude spectrum reaches a minimum value, the phase difference of the two separated shear-waves should be $(2N_3 - 1)\pi$. Hence, the phase difference of the two separated shear-waves corresponds to the TOF difference, P, should be $(2N_3 - 1)\pi$.

$$2\pi f P = (2N_3 - 1)\pi, \ (N_3 = 1, 2, 3, ...). \tag{14}$$

The solution of Equation (14) is

$$f^* = \frac{2N_3 - 1}{2P}, \ (N_2 = 1, 2, 3, ...), \tag{15}$$

which is identical to the CF derived from the amplitude spectrum.

Particularly, when the polarized angle of the shear-wave is 45°, the third part in Equation (13) is 0. Then, the shear-wave pulse echo phase spectrum is

$$\varphi_{45°}(f) = \varphi_0(f) - \pi f \left(\frac{2l}{v_{31}} + \frac{2l}{v_{32}} \right). \tag{16}$$

The phase difference between an arbitrary polarized angle θ, and the polarized angle of 45° is

$$\Delta \varphi_r = -\arctan((M - N) \cdot \tan(\pi f P)). \tag{17}$$

where $\Delta\varphi_r$ is defined as the phase difference (PD). A typical illustration of the PD is shown in Figure 1, in which P is taken as equal to 100 ns as an example. The CF in the curve of the PD corresponds to an inflection point. The inflection point in the PD curve can be used to identify the CF.

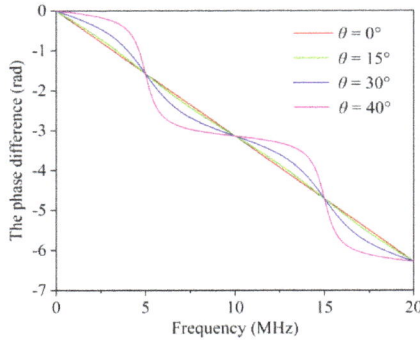

Figure 1. Functional image of the phase difference.

A method for obtaining the inflection point is to draw an image of the derivation of the PD (DPD), in which the maximum point corresponds to the CF. An illustration of the DPD curve when $P = 100$ ns is shown in Figure 2. The maximum values in the curve correspond to the CFs. Both Equations (10) and (15) show that the CF is a key indicator because it is directly related to the TOF difference of the two separated shear waves. Hence, the CF can be collected by determining the maximum value in the curve of the DPD function.

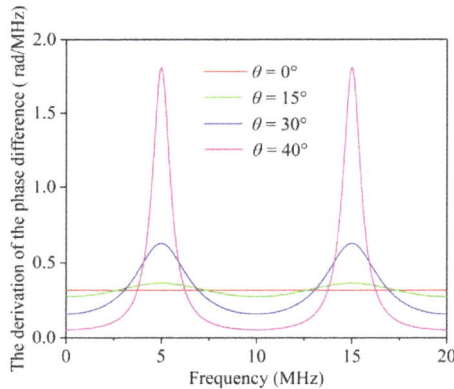

Figure 2. An illustration of the derivation of the phase difference.

2.4. Uniaxial Stress Measurement Using the Shear-Wave Pulse Echo Spectrum

According to the acoustoelastic effect, the velocities of the shear waves are different when their particle vibration directions are perpendicular and parallel to the stress direction, respectively. The velocities of the shear waves can be related to the uniaxial stress, which theoretical formulas can be found in references [24,50]. By further combining Equations (10) and (15), we obtain the following formula [24].

$$\sigma = \frac{\kappa}{f^*} - \gamma, \ (N_2 = 1),$$

(18)

$$\kappa = \frac{2N_1 - 1}{2t_0} \cdot \frac{-8\mu^2}{4\mu + n}, \ (N_1 = N_3 = 1, 2, 3, ...),$$

(19)

$$\gamma = \frac{-8\mu^2}{4\mu + n}\alpha,\tag{20}$$

where σ is the uniaxial stress in the direction of x_2; λ and μ are the second-order elastic constants; l, m, and n are the third-order elastic constants; t_0 is the shear-wave TOF in the free-stressed state and equals $2l/v_0$; α is a factor to indicate the initial anisotropy of the materials; and κ and γ should be fitted using the uniaxial compressive test.

3. Experimental Studies

3.1. Equipment and Sample

The devices and the measurement schematic diagram for identifying the uniaxial stress effect on the spectrum of the shear wave are shown in Figures 3 and 4, respectively. The probe used in the experiments is a normal incidence shear-wave transducer (V156-5/25"; central frequency: 5 MHz; Olympus NDT, Waltham, MA, USA), which can introduce shear waves directly into the steel member without the use of refraction. The shear-wave transceiver probe is excited by the ultrasonic generator (5072PR; Olympus NDT, Waltham, MA, USA) and a pure shear wave is generated. The generated shear wave is perpendicularly incident on the steel member loaded by a universal testing machine (SHT4605; MTS Systems (Shenzhen, China) Co., LTD). After being reflected from the rear side of the steel member, the shear wave travels back to the steel member surface and is received by the transceiver probe. The shear-wave pulse echo signals travel back to the ultrasonic generator and are finally collected by the oscilloscope (MDO3024; Tektronix, Beaverton, OR, USA). A personal computer (PC) is used to process the received signals. The polarized angle of the shear wave can be determined by rotating the transceiver probe that is imbedded in a card slot. More details of collecting the pulse echo shear waves can be found in paper [24].

Three steel members, made of Q235 steel, are designed and used as the test specimens. The dimensions of the three steel members are 80 mm × 45 mm × 24 mm (sample C1), 80 mm × 45 mm × 30 mm (sample C2), and 80 mm × 45 mm × 36 mm (sample C3). GW-type-III ultrasound coupler is used as the couplant to couple the probe and the specimens.

As the shear-wave length is small enough (approximately 0.64 mm) and the shear-wave travel length is short enough (24 mm to 36 mm), we do not consider the influence of steel member boundaries on shear-wave propagation. Therefore, guided waves are not formed when the shear wave propagates in steel members. The experiment was conducted at room temperature (25 °C), and the temperature was considered constant. Therefore, variations in the operational temperature were not considered.

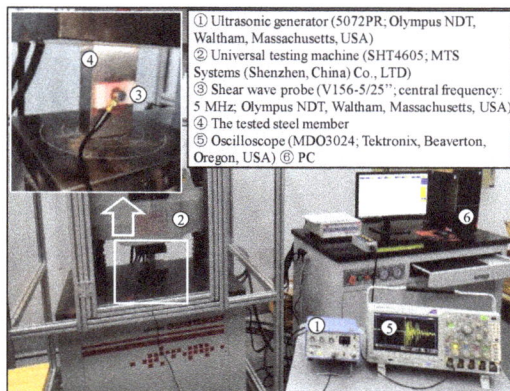

① Ultrasonic generator (5072PR; Olympus NDT, Waltham, Massachusetts, USA)
② Universal testing machine (SHT4605; MTS Systems (Shenzhen, China) Co., LTD)
③ Shear wave probe (V156-5/25"; central frequency: 5 MHz; Olympus NDT, Waltham, Massachusetts, USA)
④ The tested steel member
⑤ Oscilloscope (MDO3024; Tektronix, Beaverton, Oregon, USA) ⑥ PC

Figure 3. Measurement devices: photographs.

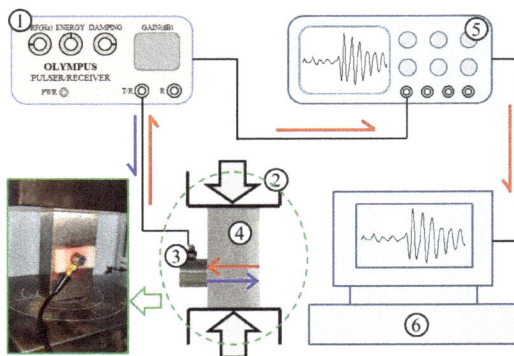

Figure 4. Measurement system: schematic diagram.

3.2. Influence of the Uniaxial Stress on the Shear-Wave Amplitude Spectrum

A universal testing machine was used as the loading device to apply compressive stress along the vertical axis of the three specimens. The increasing step load history applied to the three specimens is shown in Figure 5. The step load increased from 20 MPa to 230 MPa with a step amplitude of 10 MPa. The transceiver probe is attached to the specimens' surface. During each loading stabilization, the shear-wave pulse echo signals are collected using the oscilloscope with a sampling rate of 100 MSa/s. The typical time-domain signals of the received shear waves (sample C3, σ = 200 MPa) are shown in Figure 6. Using the Fourier transform method, the pulse echo signal can be converted into frequency domain signals [24].

The shear wave is perpendicularly incident on the steel member with a polarized angle of 45°. The second pulse echo signals are extracted, and the corresponding applied uniaxial stresses are recorded. The Fourier transform method is used to transform the time domain signal to the amplitude spectrum. Hence, the change in the amplitude spectra affected by the stresses for the three specimens can be obtained. Figure 7 shows the normalized amplitude spectra under different compressive stress states in samples C1, C2, and C3.

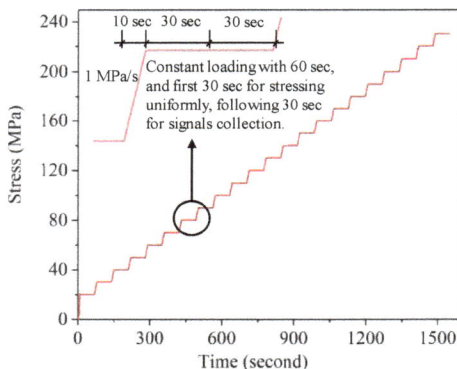

Figure 5. The increasing step load history applied on the three specimens.

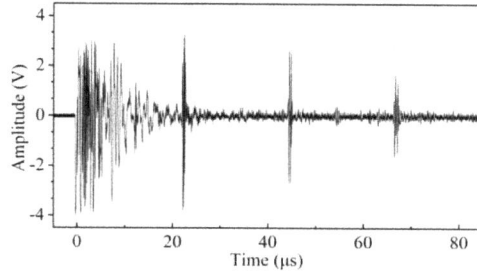

Figure 6. The typical time-domain signals of the received shear waves (sample C3, $\sigma = 200$ MPa).

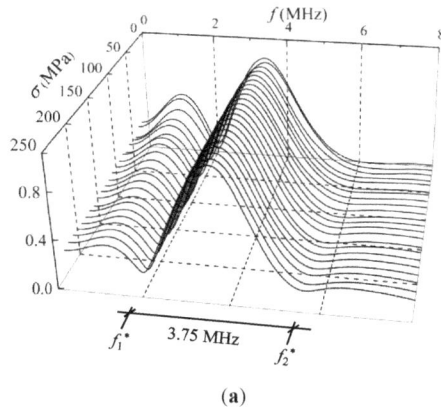

(**a**)

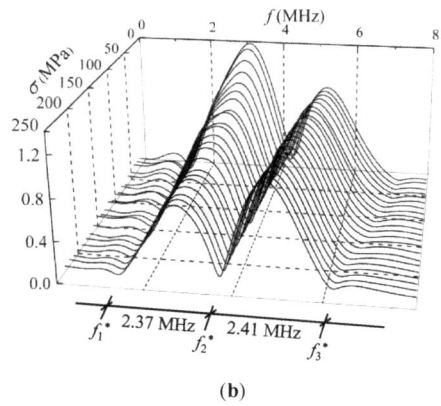

(**b**)

Figure 7. *Cont.*

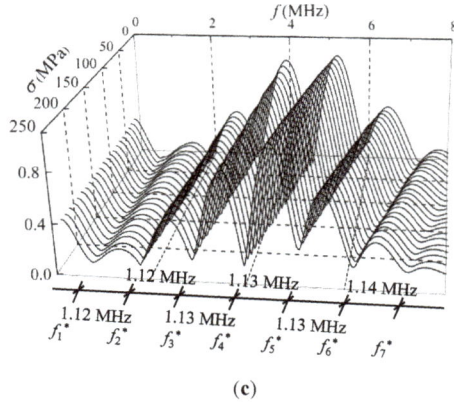

(c)

Figure 7. The normalized amplitude spectra under different compressive stress states in (**a**) sample C1; (**b**) sample C2; and (**c**) sample C3.

3.3. Influence of Uniaxial Stress on the Shear-Wave Phase Spectrum

The polarized angles of 40° and 45° were selected during the experiments. Using the Fourier transform method, the shear-wave phase spectrum could be obtained from the collected time-domain signals. The phase spectrum corresponding to different uniaxial compressive stresses could be obtained for the three specimens. A typical illustration of the change in the phase spectrum affected by the stress state is shown in Figure 8.

The PD describes the difference of the two phase spectra for an arbitrary polarized angle and a polarized angle of 45°. With the phase spectra of polarized angles at 40° and 45°, the PD under different stress states could be sketched. Further, the DPD curves was successfully obtained. The influence of stress on the normalized DPD curves for the three specimens is shown in Figure 9.

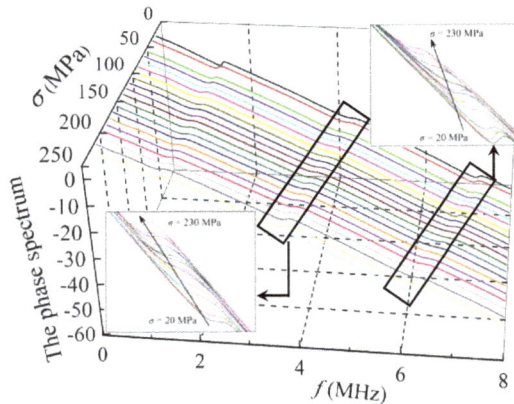

Figure 8. The change in the phase difference curves affected by the stress ($\theta = 40°$).

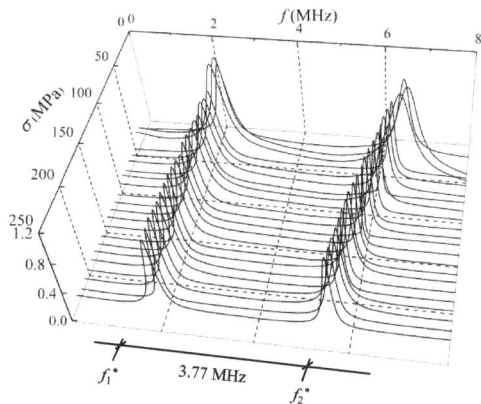

(**a**)

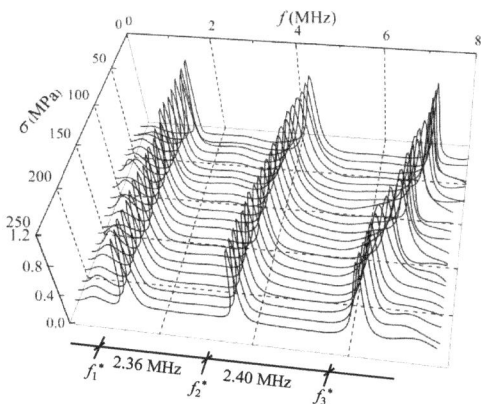

(**b**)

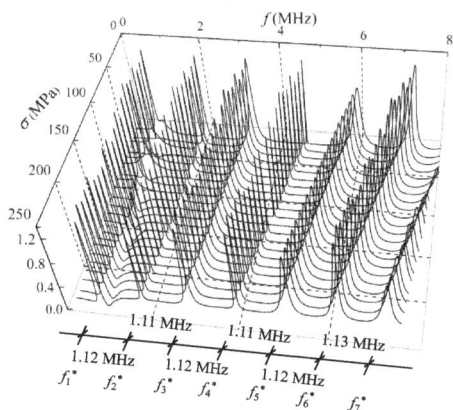

(**c**)

Figure 9. The normalized derivation of the phase difference curves under different compressive stress states in (**a**) sample C1; (**b**) sample C2; and (**c**) sample C3.

3.4. Parameter Calibration of the Stress Measurement Formula

The aim of sketching the amplitude spectra and the DPD curves is to collect the CFs in each stress state. For the convenience of making a comparison between the amplitude spectra and the DPD curves, the second CFs for sample C1, the third CFs for sample C2, and the fifth CFs for sample C3 are extracted from the amplitude spectra and DPD curves under identical applied uniaxial stresses. The comparison of the CFs in the amplitude spectra and the DPD curves for the three samples are shown in Figure 10. Using the least squares method listed in reference [24], the parameters in Equation (18) can be obtained from a linear fitting of the stress and the inverse of the CF, which is shown in Figures 11 and 12. The coefficients of the fitting line are listed in Table 1.

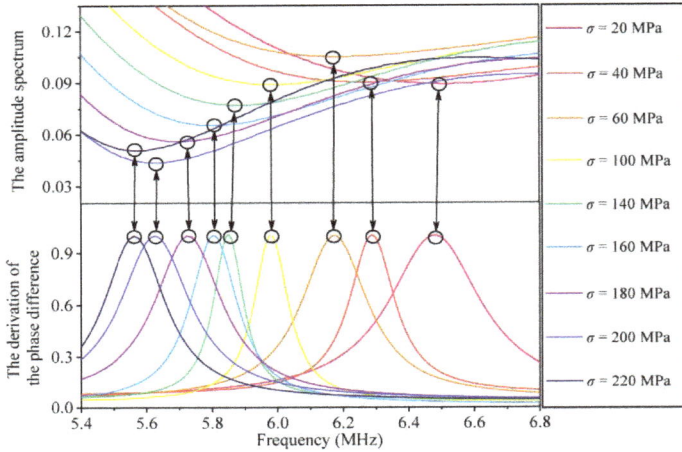

(a)

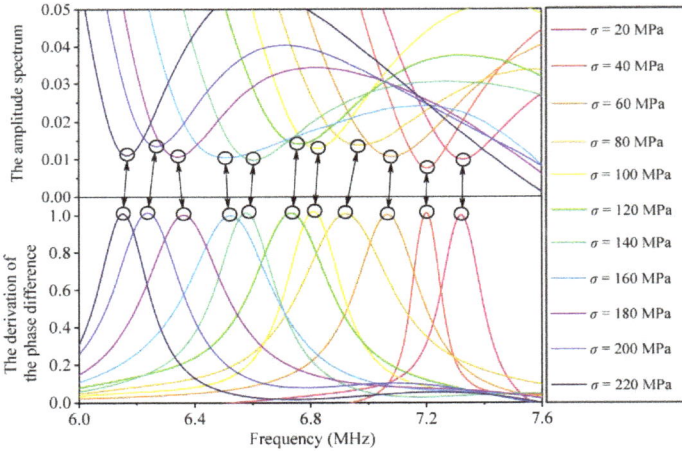

(b)

Figure 10. *Cont.*

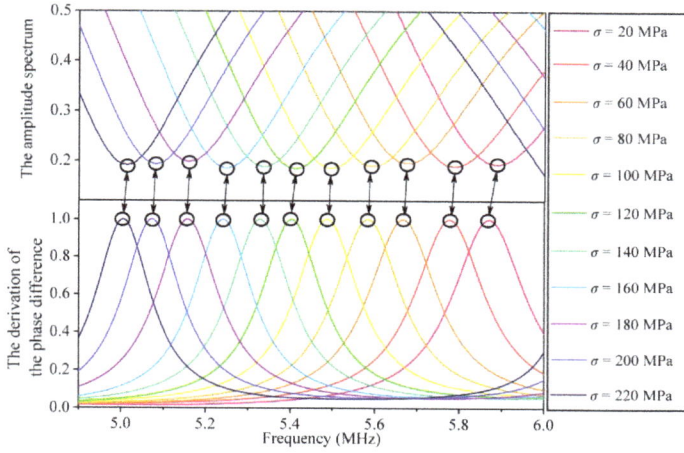

(c)

Figure 10. Comparison of the amplitude spectra and the derivation of the phase difference curves: (a) sample C1; (b) sample C2; and (c) sample C3.

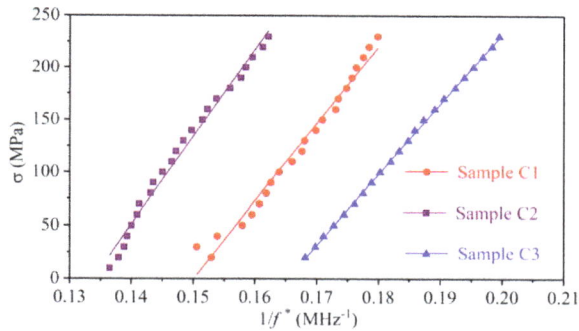

Figure 11. Fitting lines between the stress and the inverse of the characteristic frequency extracted from the amplitude spectra.

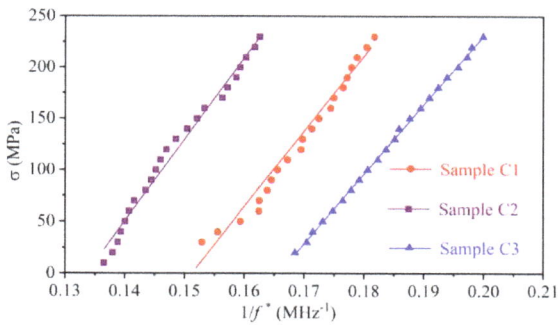

Figure 12. Fitting lines between the stress and the inverse of the characteristic frequency extracted from the derivation of the phase difference curves.

Table 1. Coefficients of the fitting line using the amplitude spectrum and the phase spectrum.

Parameters	Methods	Sample C1		Sample C2		Sample C3	
		Values	Difference	Values	Difference	Values	Difference
κ (MPa·MHz)	Amplitude Spectrum	7356.73	0.23%	8308.07	0.16%	6691.01	0.01%
	Phase Spectrum	7339.74		8321.41		6755.27	
γ (MPa)	Amplitude Spectrum	1103.99	5.83	1111.95	6.06	1106.05	13.75
	Phase Spectrum	1109.82		1118.01		1119.80	
R^2	Amplitude Spectrum	0.9901	/	0.9937	/	0.9998	/
	Phase Spectrum	0.9795	/	0.9828	/	0.9996	/

4. Results and Discussion

4.1. Influence of Uniaxial Stress on Shear-Wave Amplitude Spectrum

The amplitude spectrum of a shear wave traveling in a steel member under different stress state is definitely different, as shown in Figure 7. The main reason is that the interference effect between the two separated shear waves lead to an energy loss of the harmonic components. In particular, the energy of the harmonic component corresponding to the CF decreased to 0 when the shear-wave polarized angle reached 45°. In addition, the amplitude spectra did not change with stress when the shear-wave polarized angle was 0°. The reason for this phenomenon is that the velocities of the two separated shear waves are identical. Therefore, no interference occurs between the two separated shear waves. Another explanation of this phenomenon can be found in Equations (8) and (9), in which the AIF equals 1 when the shear-wave polarized angle is 0°. Hence, the pulse echo amplitude spectrum does not change with stress.

As shown in Figure 7, the minimum point is periodically presented in the amplitude spectra with a repetition period of 3.75 MHz, 2.39 MHz, and 1.25 MHz for samples C1, C2, and C3, respectively. According to Equation (10), the periodic values for the CF are $1/P$. Because the TOF difference (P) of the two separated shear waves depends on the dimensions of the sample, sample C3 with the thickest size shows the least periodic value, while sample C1 with the thinnest size presents the maximal period. The minimum point in an amplitude spectrum corresponds to the CF. The CF shows a tendency to move left with increasing stress, which lays the foundation for stress evaluation. This is consistent with the theoretical analysis result in Equation (10): the CF decreases when the stress-induced TOF difference P increases.

4.2. Influence of Uniaxial Stress on Shear-Wave Phase Spectrum

Figure 8 shows the typical phase spectra when the applied compressive stress increases from 20 MPa to 230 MPa. All the phase spectra have a tendency to decrease with increasing frequency. The inflection points periodically appear on the phase spectra curves. With increasing compressive stress, the inflection points tend to move left. However, it is difficult to observe any quantitative relationship between the inflection points and the stresses.

The DPD curves of the three samples in Figure 9 show an obvious variation tendency. Compared with the phase spectrum, the influence of stress on the DPD curves is obvious because a peak point appears, which corresponds to a specific inflection point. The peak points periodically appeared in each DPD curve with nearly identical periods to the amplitude spectra for the three samples. As the peak point in the DPD curve corresponds to the CF, the CF can be obtained from the DPD curve. In Figure 9, the DPD curves of three samples show that the CF decreases with stress, which is consistent with the results obtained from the amplitude spectrum.

4.3. Comparison of the Amplitude and Phase Spectra

Equations (10) and (15) indicate that the abscissas corresponding to the minimum point in the amplitude spectrum and the maximum point in the DPD curve should be identical, which was verified by the experimental results in Figure 10. The CFs obtained from the amplitude spectra and the DPD curves in Figure 10 are nearly identical. Therefore, both the amplitude spectrum and the phase spectrum can be used to extract the CF. The difference between the two methods is that the stress exerts a direct influence on the amplitude spectrum, while the effect of stress on the phase spectrum is difficult to observe, and the CF is extracted from the DPD curve.

For the amplitude spectrum, the CF can be obtained with only one shear-wave pulse echo signal with a polarized angle of 45°. For the phase spectrum, two shear-wave pulse echo signals are required: one is the signal with a polarized angle of 45°, and the other is the signal with a polarized angle of close to 45°. Therefore, the shear-wave pulse echo signal with a polarized angle of 45° is required for both the amplitude and phase spectra.

Notably, the essence of the stress effect on the two spectra is identical, that is, the interference effect of the two separated shear waves. The velocities of the two separated shear waves propagating in a stressed steel member are different; thus, the interference effect occurs for the two separated shear waves. For the amplitude spectra, when the shear-wave polarized angle is 45°, the interference effect induces the amplitude of the CF to decrease to 0; thus, a minimum point appears, and the CF can be obtained. For the phase spectra, the maximum point in the DPD curve corresponds to the CF, and it changes with stress. Although the method of collecting the CF is different, the value of the CF corresponding to a certain stress state is identical. Therefore, the calibrated parameters using the two sets of data should be identical.

Note that the CF corresponding to a peak point in the DPD curve is easily observed, while the minimum point in the amplitude spectrum is not always obvious, as shown in Figures 7 and 9. For instance, the second CFs (f_2^*) for sample C1 and the third CFs (f_3^*) for sample C2, which are shown in Figure 7a,b, are difficult to collect. The main reason is that the amplitude spectrum indicates the energy amplitude of a certain frequency, which is directly related to the transceiver probe. In this work, the central frequency of the probe is approximately 5 MHz, and the amplitude spectrum energy is centered on the range of 3–7 MHz. When the CFs are beyond the range of 3–7 MHz, the change in the amplitude spectrum energy is not dramatic and is difficult to observe. Therefore, from the aspect of CF extraction, the phase spectrum method is more advantageous.

4.4. Parametric Calibration for the Stress Measurement Formula

Figures 11 and 12 show that the inverse of the CFs obtained from the amplitude spectra and the DPD curves linearly change with the stress. The calibrated parameters in Table 1 indicate that the correlation coefficients (R^2) of all the lines are larger than 0.95, which verifies the correctness of Equation (18).

The difference in the calibrated parameter κ using the amplitude spectrum and the phase spectrum is less than 1% for the three samples. This error may come from the ambient effect and can be ignored. The parameter γ is related to the material and the initial acoustic anisotropy. Because the three samples were cut from one steel plate, the calibrated parameter γ is nearly identical for the three samples using the two types of spectra, in which the maximum error is 13.75 MPa. With the calibrated parameters, the uniaxial stress in a steel member can be evaluated by collecting the shear-wave pulse echo signals and extracting the CFs from the amplitude spectra or the DPD curves. The results in this work provide a potential way to detect the uniaxial absolute stress in structural steel members using the amplitude spectrum method and the phase spectrum method.

The experiments of this work were implemented on the laboratory scale, and a perfect linear relationship between the stress and the inverse of CFs was obtained. However, some necessary factors need to be considered for the absolute stress evaluation of realistic structures. For instance, the ambient temperature of realistic structures is uncontrollable, the surface roughness of the tested steel members will be notable due to the development of corrosion, and the coupling state between the probe and steel member surface is difficult to maintain in a constant state. All of these factors may exert a direct influence on the absolute stress measurement. Another limitation of the two methods is that structural steel members are usually non-removable after installation. It is hard to calibrate the parameters on the original tested member. A possible solution is to replicate a steel member with the same material and dimensions as the tested steel member. However, the replicated steel member is not the original in-service steel member, and the potential differences between the tested and replicated steel members may lead to errors in parametric calibration. To realize the practical engineering application of the shear-wave spectrum method, further research efforts should focus on the influence of temperature, surface roughness and coupling state on parametric calibration and the absolute stress measurement.

5. Conclusions

In this paper, the influence of uniaxial stress on the shear-wave spectrum propagating in steel members was investigated. Three steel members were used to study the effect of the applied uniaxial stress on the amplitude spectrum and phase spectrum. The conclusions are summarized as follows:

(1) The theoretical expressions of the shear-wave pulse echo phase spectrum were derived. The essence of the stress effect on the shear-wave phase spectrum is the interference effect of the two separated shear waves, which follows the same principles as the shear-wave amplitude spectrum.

(2) The CFs can be obtained from both the amplitude spectrum and the DPD curve. The extracted CFs are identical when the steel member is under the same stress state. To collect a CF, one shear-wave signal is required for the amplitude spectrum, while two shear-wave signals are needed for the phase spectrum.

(3) The inverse of the CF showed a linear relationship with the corresponding uniaxial stress, thus establishing a basis for the uniaxial stress evaluation. The calibrated parameters obtained from the two methods are nearly identical. Using the calibrated parameters, the uniaxial stress in a steel member can be evaluated by extracting the CF from the shear-wave pulse echo signal.

Author Contributions: Conceptualization, Z.L.; Data curation, J.H.; Formal analysis, J.H.; Funding acquisition, Z.L. and J.T.; Investigation, J.H.; Methodology, J.H.; Project administration, Z.L. and J.T.; Resources, Z.L. and J.T.; Software, D.L.; Supervision, J.T.; Validation, D.L. and N.L.; Visualization, D.L.; Writing—original draft, J.H.; Writing—review & editing, N.L. and Z.L.

Funding: This work was financially supported by the National Key Research and Development Program of China under Grant 2016YFC0701102, the National Natural Science Foundation of China under Grant 51538003, the National Major Scientific Research Instrument Development Program of China under Grant 51827811, and the Shenzhen Technology Innovation Program under Grant JCYJ20170811160003571. The authors are thankful for this financial support.

Conflicts of Interest: The author(s) declared no potential conflicts of interest with respect to the research, authorship, and/or publication of this article.

References

1. Nichols, J.M. Structural health monitoring of offshore structures using ambient excitation. *Appl. Ocean Res.* **2003**, *25*, 101–114. [CrossRef]
2. Li, J.; Hao, H.; Fan, K.; Brownjohn, J. Development and application of a relative displacement sensor for structural health monitoring of composite bridges. *Struct. Control Health Monit.* **2015**, *22*, 726–742. [CrossRef]
3. Biondini, F.; Frangopol, D.M. Life-cycle performance of deteriorating structural systems under uncertainty: Review. *J. Struct. Eng.* **2016**, *142*, F4016001. [CrossRef]
4. Wang, H.; Li, G.; Huang, X. Behavior of coupled shear walls with buckling-restrained steel plates in high-rise buildings under lateral actions. *Struct. Des. Tall Spec.* **2016**, *25*, 22–44. [CrossRef]
5. Wang, J.; Zhao, H. High performance damage-resistant seismic resistant structural systems for sustainable and resilient city: A review. *Shock Vib.* **2018**, 8703697. [CrossRef]
6. Sousa, H.; Wang, Y. Sparse representation approach to data compression for strain-based traffic load monitoring: A comparative study. *Measurement* **2018**, *122*, 630–637. [CrossRef]

7. Asadollahi, P.; Huang, Y.; Li, J. Bayesian finite element model updating and assessment of cable-stayed bridges using wireless sensor data. *Sensors* **2018**, *18*, 3057. [CrossRef]

8. Luo, M.Z.; Li, W.J.; Wang, B.; Fu, Q.Q.; Song, G.B. Measurement of the length of installed rock bolt based on stress wave reflection by using a giant magnetostrictive (GMS) actuator and a PZT sensor. *Sensors* **2017**, *17*, 444. [CrossRef]

9. Laflamme, S.; Kollosche, M.; Connor, J.J.; Kofod, G. Soft capacitive sensor for structural health monitoring of large-scale systems. *Struct. Control Health Monit.* **2012**, *19*, 70–81. [CrossRef]

10. Yi, T.H.; Li, H.N.; Gu, M. Recent research and applications of GPS-based monitoring technology for high-rise structures. *Struct. Control Health Monit.* **2013**, *20*, 649–670. [CrossRef]

11. Ay, A.M.; Wang, Y. Structural damage identification based on self-fitting ARMAX model and multi-sensor data fusion. *Struct. Health Monit.* **2014**, *13*, 445–460. [CrossRef]

12. Teng, J.; Lu, W.; Wen, R.F.; Zhang, T. Instrumentation on structural health monitoring systems to real world structures. *Smart Struct. Syst.* **2015**, *15*, 151–167. [CrossRef]

13. Meoni, A.; D'Alessandro, A.; Downey, A.; García-Macías, E.; Rallini, M.; Materazzi, A.L.; Torre, L.; Laflamme, S.; Castro-Triguero, R.; Ubertini, F. An experimental study on static and dynamic strain sensitivity of embeddable smart concrete sensors doped with carbon nanotubes for SHM of large structures. *Sensors* **2018**, *18*, 831. [CrossRef] [PubMed]

14. Shen, Y.B.; Yang, P.C.; Zhang, P.F.; Luo, Y.Z.; Mei, Y.J.; Cheng, H.Q.; Jin, L.; Liang, C.Y.; Wang, Q.Q.; Zhong, Z.N. Development of a multitype wireless sensor network for the large-scale structure of the National Stadium in China. *Int. J. Distrib. Sens. N.* **2013**, *9*, 709724. [CrossRef]

15. Li, G.W.; Pei, H.F.; Yin, J.H.; Lu, X.C.; Teng, J. Monitoring and analysis of PHC pipe piles under hydraulic jacking using FBG sensing technology. *Measurement* **2014**, *49*, 358–367. [CrossRef]

16. Teng, J.; Lu, W.; Cui, Y.; Zhang, R.G. Temperature and displacement monitoring to steel roof construction of Shenzhen Bay Stadium. *Int. J. Struct. Stab. Dyn.* **2016**, *16*, 1640020. [CrossRef]

17. Suzuki, K. Proposal for a direct-method for stress measurement using an X-ray area detector. *NDT E Int.* **2017**, *92*, 104–110. [CrossRef]

18. Hemmesi, K.; Farajian, M.; Boin, M. Numerical studies of welding residual stresses in tubular joints and experimental validations by means of x-ray and neutron diffraction analysis. *Mater. Des.* **2017**, *126*, 339–350. [CrossRef]

19. Zhou, D.; Pan, M.; He, Y.; Du, B. Stress detection and measurement in ferromagnetic metals using pulse electromagnetic method with U-shaped sensor. *Measurement* **2017**, *105*, 136–145. [CrossRef]

20. Rossini, N.S.; Dassisti, M.; Benyounis, K.Y.; Olabi, A.G. Methods of measuring residual stresses in components. *Mater. Des.* **2012**, *35*, 572–588. [CrossRef]

21. Liu, T.J.; Zou, D.J.; Du, C.C.; Wang, Y. Influence of axial loads on the health monitoring of concrete structures using embedded piezoelectric transducers. *Struct. Health Monit.* **2017**, *16*, 202–214. [CrossRef]

22. Li, Z.H.; He, J.B.; Teng, J.; Wang, Y. Internal stress monitoring of in-service structural steel members with ultrasonic method. *Materials* **2016**, *9*, 234. [CrossRef] [PubMed]

23. He, J.B.; Li, Z.H.; Teng, J.; Wang, Y. Absolute stress field measurement in structural steel members using the Lcr wave method. *Measurement* **2018**, *122*, 679–687. [CrossRef]

24. Li, Z.H.; He, J.B.; Teng, J.; Huang, Q.; Wang, Y. Absolute stress measurement of structural steel members with ultrasonic shear-wave spectral analysis method. *Struct. Health. Monit.* **2017**. [CrossRef]

25. Li, Z.H.; He, J.B.; Teng, J.; Wang, Y. Cross-correlation-based algorithm for absolute stress evaluation in steel members using the longitudinal critically refracted wave. *Int. J. Distrib. Sens. Netw.* **2018**, *14*. [CrossRef]

26. Withers, P.J.; Turski, M.; Edwards, L.; Bouchard, P.J.; Buttle, D.J. Recent advances in residual stress measurement. *Int. J. Press. Vessel. Pip.* **2008**, *85*, 118–127. [CrossRef]

27. Guz', A.N.; Makhort, F.G. The physical fundamentals of the ultrasonic nondestructive stress analysis of solids. *Int. J. Appl. Mech.* **2000**, *36*, 1119–1149. [CrossRef]

28. Bray, D.E.; Tang, W. Subsurface stress evaluation in steel plates and bars using the L-CR ultrasonic wave. *Nucl. Eng. Des.* **2001**, *207*, 231–240. [CrossRef]

29. Karabutov, A.; Devichensky, A.; Ivochkin, A.; Lyamshevb, M.; Pelivanova, I.; Rohadgic, U.; Solomatina, V.; Subudhic, M. Laser ultrasonic diagnostics of residual stress. *Ultrasonics* **2008**, *48*, 631–635. [CrossRef]

30. Egle, D.M.; Bray, D.E. Measurement of acoustoelastic and 3rd-order elastic-constants for rail steel. *J. Acoust. Soc. Am.* **1976**, *60*, 741–744. [CrossRef]

31. Lee, H.Y.; Nikbin, K.M.; O'Dowd, N.P. A generic approach for a linear elastic fracture mechanics analysis of components containing residual stress. *Int. J. Press. Vessel. Pip.* **2005**, *82*, 797–806. [CrossRef]

32. Sanderson, R.M.; Shen, Y.C. Measurement of residual stress using laser-generated ultrasound. *Int. J. Press. Vessel. Pip.* **2010**, *87*, 762–765. [CrossRef]

33. Javadi, Y.; Najafabadi, M.A. Comparison between contact and immersion ultrasonic method to evaluate welding residual stresses of dissimilar joints. *Mater. Des.* **2013**, *47*, 473–482. [CrossRef]

34. Sadeghi, S.; Najafabadi, M.A.; Javadi, Y.; Mohammadisefat, M. Using ultrasonic waves and finite element method to evaluate through-thickness residual stresses distribution in the friction stir welding of aluminum plates. *Mater. Des.* **2013**, *52*, 870–880. [CrossRef]

35. Chaki, S.; Bourse, G. Stress level measurement in prestressed steel strands using acoustoelastic effect. *Exp. Mech.* **2009**, *49*, 673–681. [CrossRef]

36. Gandhi, N.; Michaels, J.E.; Lee, S.J. Acoustoelastic Lamb wave propagation in biaxially stressed plates. *J. Acoust. Soc. Am.* **2012**, *132*, 1284–1293. [CrossRef]

37. Wali, Y.; Njeh, A.; Wieder, T.; Ghozlen, M.B. The effect of depth-dependent residual stresses on the propagation of surface acoustic waves in thin Ag films on Si. *NDT E Int.* **2007**, *40*, 545–551. [CrossRef]

38. Wang, Z.J.; Chen, D.D.; Zheng, L.Q.; Huo, L.S.; Song, G.B. Influence of axial load on electromechanical impedance (emi) of embedded piezoceramic transducers in steel fiber concrete. *Sensors* **2018**, *18*, 1782. [CrossRef]

39. Allen, D.R.; Sayers, C.M. The measurement of residual-stress in textured steel using an ultrasonic velocity combinations technique. *Ultrasonics* **1984**, *22*, 179–188. [CrossRef]

40. Lipeles, R.; Kivelson, D. Theory of ultrasonically induced birefringence. *J. Chem. Phys.* **1977**, *67*, 4564–4570. [CrossRef]

41. Crecraft, D.I. The measurement of applied and residual stresses in metals using ultrasonic waves. *J. Sound Vib.* **1967**, *5*, 173. [CrossRef]

42. Herzer, H.R.; Becker, M.M.; Schneider, E. The acousto-elastic effect and its use in NDE. In *Handbook of Advanced Non-Destructive Evaluation*; Ida, N., Meyendorf, N., Eds.; Springer: Cham, Switzerland, 2018; pp. 1–17.

43. Djerir, W.; Ourak, M.; Boutkedjirt, T. Characterization of the critically refracted longitudinal (L-CR) waves and their use in defect detection. *Res. Nondestruct. Eval.* **2014**, *25*, 203–217. [CrossRef]

44. Palanichamy, P.; Joseph, A.; Jayakumar, T.; Raj, B. Ultrasonic velocity measurements for estimation of grain size in austenitic stainless steel. *NDT E Int.* **1995**, *28*, 179–185. [CrossRef]

45. Lhémery, A.; Calmon, P.; Chatillon, S.; Gengembre, N. Modeling of ultrasonic fields radiated by contact transducer in a component of irregular surface. *Ultrasonics* **2002**, *40*, 231–236. [CrossRef]

46. Zou, D.J.; Liu, T.J.; Liang, C.F.; Huang, Y.C.; Zhang, F.Y.; Du, C.C. An experimental investigation on the health monitoring of concrete structures using piezoelectric transducers at various environmental temperatures. *J. Intell. Mater. Syst. Struct.* **2015**, *26*, 1028–1034. [CrossRef]

47. Liu, H.B.; Li, Y.P.; Li, T.; Zhang, X.; Liu, Y.K.; Liu, K.; Wang, Y.Q. Influence factors analysis and accuracy improvement for stress measurement using ultrasonic longitudinal critically refracted (LCR) wave. *Appl. Acoust.* **2018**, *141*, 178–187. [CrossRef]

48. Vangi, D.; Virga, A. A practical application of ultrasonic thermal stress monitoring in continuous welded rails. *Exp. Mech.* **2007**, *47*, 617–623. [CrossRef]

49. Blinka, J.; Sachse, W. Application of ultrasonic-pulse-spectroscopy measurements to experimental stress analysis. *Exp. Mech.* **1976**, *16*, 448–453. [CrossRef]

50. Javadi, Y.; Azari, K.; Ghalehbandi, S.M.; Roy, M.J. Comparison between using longitudinal and shear waves in ultrasonic stress measurement to investigate the effect of post-weld heat-treatment on welding residual stresses. *Res. Nondestruct. Eval.* **2017**, *28*, 101–122. [CrossRef]

MDPI

Article

Robust Identification of Strain Waves due to Low-Velocity Impact with Different Impactor Stiffness

Alessio Beligni * , Claudio Sbarufatti, Andrea Gilioli, Francesco Cadini and Marco Giglio

Politecnico di Milano, Mechanical Engineering Department, via La Masa 1, 20156 Milano, Italy;
claudio.sbarufatti@polimi.it (C.S.); andrea.gilioli@polimi.it (A.G.); francesco.cadini@polimi.it (F.C.);
marco.giglio@polimi.it (M.G.)
* Correspondence: alessio.beligni@polimi.it; Tel.: +39-02-2399-8213

Received: 12 February 2019; Accepted: 9 March 2019; Published: 14 March 2019

✓ check for updates

Abstract: Low-velocity impacts represent a major concern for aeronautical structures, sometimes producing barely detectable damage that could severely hamper the aircraft safety, even with regards to metallic structures. For this reason, the development of an automated impact monitoring system is desired. From a passive monitoring perspective, any impact generates a strain wave that can be acquired using sensor networks; signal processing techniques allow for extracting features useful for impact identification, possibly in an automatic way. However, impact wave characteristics are related to the impactor stiffness; this presents a problem for the evaluation of an impact-related feature and for the development of an automatic approach to impact identification. This work discusses the problem of reducing the influence of the impactor stiffness on one of the features typically characterizing the impact event, i.e., the time of arrival (TOA). Two passive sensor networks composed of accelerometers and piezoelectric sensors are installed on two metallic specimens, consisting of an aluminum skin and a sandwich panel, with aluminum skins and NOMEX™ honeycomb core. The effect of different impactor stiffnesses is investigated by resorting to an impact hammer, equipped with different tips. Subsequently, a method for data processing is defined to obtain a feature insensitive to the impactor stiffness, and this method is applied to multiple impact signals for feature uncertainty evaluation.

Keywords: low-velocity impacts; strain wave; impactor stiffness; data processing; feature selection; impact identification

---

1. Introduction

During the last few decades, structural health monitoring (SHM) systems have become a crucial research topic for the enhancement of the material and structural potential and the reduction of the maintenance and operation costs. In the aeronautical field, a complete SHM system is composed of different sub-systems, with different tasks including the detection of anomalous structural behavior and usage monitoring, comprehensively referred to as Integrated Health and Usage Monitoring Systems (IHUMS). In a damage-tolerant scenario, structures are designed to withstand both service loads, such as those related to a typical operation, and impulsive loads, due to impacts, where typical examples are bird-strikes [1,2], hailstones [1] and debris impacts [3,4].

Focusing on impacts, theoretical knowledge states that impulsive loads generate elastic waves in solid materials [5,6]; those waves are dispersive and the possibility of understanding their behavior has long been of primary interest [7,8]. This aspect is of fundamental importance today, not only due to the difficulty of obtaining simple analytical solutions, but also due to the rapidly increasing use of composite materials in the aeronautical field [9–11], even if metallic structures still play a major role.

Experimental results, combined with analytical and Finite Element (FE) models, have led to a better understanding of the behavior of dispersive waves [12,13]; those waves are the basis for impact and damage monitoring. Waves can be generated using an actuator or by a foreign object impact event and recorded using a sensor. The former approach is typically referred to as *active monitoring*, targeting the identification of potential damage after an impact event [14], and has been widely applied to various structures, including metallic, honeycomb- [15–17] and fiber-reinforced [18] composite components under different sources of damage, such as fatigue, wear, overloads, low- and high-velocity impacts, etc. The latter defines the field of *passive impact monitoring* [11], aiming to identify the occurrence of an impact event, possibly estimating its location [19] and inversely reconstructing the impact force [20] based on signals acquired from passive sensors. In a model-based SHM framework, passive impact monitoring can be used as input to numerical models for damage estimation [14,21], thus triggering the active damage monitoring system for potential real-damage identification.

For passive impact monitoring, the ability to acquire the strain wave is thus of primary importance; different sensor technologies can be adopted: in [7] strain gauges were used to acquire the bending wave generated by an impact of a steel ball on a beam, while in [22,23] the capability and limits of strain gauges for dynamic event acquisition were explained. In [22], the authors used an apparatus to produce strain waves up to 300 kHz in frequency and 2000 $\mu\varepsilon$ in amplitude and acquired the dynamic signal with strain gauges and laser interferometer, for comparison; gauge length influence, response lag, signal attenuation, static gauge factor variation and other phenomena were analyzed. In [23], the author studied the effect of gauge length on the strain gauge cutoff frequency. In [24], strain-wave-induced accelerations were generated by an impact on a Hopkinson bar and measured by means of accelerometers. The latter was used in [25] to identify external forces and structural damage parameters by means of a sparsity-based reconstruction method. In [26] the strain waves induced by dropping a steel ball on an aluminum plate were recorded using piezoelectric films and acoustic emission sensors; piezoelectric technology for strain acquisition was reviewed in [27]. Finally, in [28] methods for detecting dynamic strain signals with optical fiber (OF) sensors were reviewed, while in [29] an OF-based strain measurement system was tested on-board with its working principle based on a stimulated Brillouin scattering and birefringence phenomena of OF sensors to separate the strain and the temperature effect for wide-area monitoring, i.e., 8 m long full-scale airplane tail-plane. Even microphones were used to identify impact events on CFRP plates [30,31], with a frequency range from 20 Hz to 20 kHz.

However, the ability to acquire the dynamic strain waves is not typically sufficient, in fact, signal processing techniques are necessary to extract the desired information. Several impact identification methods have been developed during the years, mainly based on analytical techniques, model-based techniques and machine learning-based approaches [11,32–36] for impact detection, localization and energy estimation.

To verify the effectiveness of a sensing technology or of an impact identification method, experimental tests are required, and a lot of aerospace materials have been tested: in [37,38] and [39] low-velocity impacts on sandwich structures, fiber-metal laminates and composite materials were reviewed, respectively. During impact tests, not only the structure response, but also the shape of the impacting object [40,41] and the impactor stiffness play key roles in impact and damage identification [8]. This last parameter, together with the energy of the impact, is responsible for the strain wave generation in different frequency ranges, which translate into different wave velocities due to the dispersive nature of wave propagation in panel-like structures, such those considered in this study. Despite this, only a few works in the literature have dealt with the problem of adopting different impactor stiffnesses during the tests: in [42], impact location performance is assessed using a ping pong ball and a rubber ball. In [43], two composite panels were impacted with a hammer equipped with three different tips; in that paper the authors calibrate an inverse method for impact location and force reconstruction using different hammer tips. In [44], an aluminum plate was impacted with six impactors made of different materials, aiming at developing a technique for recognizing the impacting object stiffness. Finally, in [45] an aluminum plate was impacted with a hammer equipped with three different tips, in order to compare the performances of three impact identification techniques, but the training database used for one type of tip did not work for all the tips due to the impact signal frequency content. Some of these examples demonstrate the possibility of performing impact identification in generic scenarios, with different impacting materials leveraging on post-processing algorithms that are mostly insensitive to wave velocity, or training some techniques to cope with related uncertainty. However, at the same time, they show the difficulties encountered when different impactor stiffness and wave velocities are involved, the latter severely hampering the impact identification. Conversely, the focus of the present work is to reduce the direct influence of the different impactor stiffnesses on the impact-related feature of interest, i.e., the time of arrival (TOA) of an elastic wave at the sensor position, before the application of any impact identification technique, the latter being left to future work by the same authors.

To this aim, we propose experimentally investigating the effects of different impactor stiffnesses, reproducing low-velocity impacts and evaluating the velocity of the generated strain waves in correspondence to different impact positions. In order to do so, first, a simple metallic plate and a sandwich panel were chosen as specimens, representative of typical aeronautical structures; this choice then led to the investigation of the influence of different structures on the feature extracted (i.e., the strain wave velocity). Subsequently, different classical sensor technologies were adopted to acquire the dynamic strain signal and evaluate the possibility of extracting features from each sensor network. Finally, the strain wave velocity was assessed by processing the acquired signals; the possibility of extracting a unique feature representative of all the impacts for each specimen tested was also evaluated.

The paper is organized as followed. In Section 2, the experimental setup is described first; then, the preliminary signal analysis and feature selection are presented. Finally, a detailed description of the methods adopted for evaluating the chosen feature and its statistical characterization are presented. In Section 3, the results of the feature evaluation are presented and discussed. Finally, conclusions are drawn in Section 4.

2. Materials and Methods

In this section, the experimental setup and the techniques used to obtain the desired results are examined. First, the acquisition system and the equipment are described, then the acquired signals are analyzed and the desired feature is defined. Finally, methods of signal processing and feature extraction are discussed.

2.1. Experimental Setup

Two metallic specimens are considered: a simple aluminum plate and a sandwich panel, with aluminum skins and a NOMEX^{TM} (DuPont^{TM}, Wilmington, DE, USA) honeycomb core. Their dimensions and mechanical characteristics are listed in Tables 1 and 2, for the metallic plate and sandwich panel, respectively.

Table 1. Dimensions and mechanical characteristics of the metallic plate.

Material	Al2024-T3
Dimensions	$400 \times 400 \times 1.5$ mm
Density	2780 Kg/m^3
Young Modulus	73.1 GPa
Poisson coeff.	0.33

Table 2. Dimensions and mechanical characteristics of the sandwich panel.

Dimensions	$400 \times 400 \times 22.05$ mm
Skin material	Al2024-T3
Skin thickness	1.5 mm ($\times 2$ skins)
Core material	HRH10-3/16-2
Core thickness	19.05 mm
Core density	32 Kg/m^3

In order to reproduce low-velocity impacts, a PCB mod. 086C03 impact hammer with a steel extender was adopted. In order to reproduce different impactor stiffness, the hammer was equipped with four different tips: a steel tip (PCB 084B03), a Teflon (The Chemours Company, Wilmington, DE, USA) tip (PCB 084B04) and two tips made of two different types of rubber, one softer and one stiffer (PCB 084C11, PCB 084C05). Each specimen was rigidly supported at the four corners and was impacted in 27 different positions with each tip, for a total of 108 impacts, by a single operator.

The strain waves generated by the impacts were acquired using two different sensing technologies: accelerometers and piezoelectric sensors. The accelerometers are monoaxial Brüel & Kjær (Nærum, Denmark) DeltaTron type 4508 while the piezoelectric sensors are PIC255 disks with a diameter of 10 mm and thickness 1 mm, produced by PI ceramic (Lederhose, Thuringia, Germany). A total of four sensors for each type are glued at the edges of a 200×200 mm square, which is also the limit of the impact area, as shown in Figure 1a.

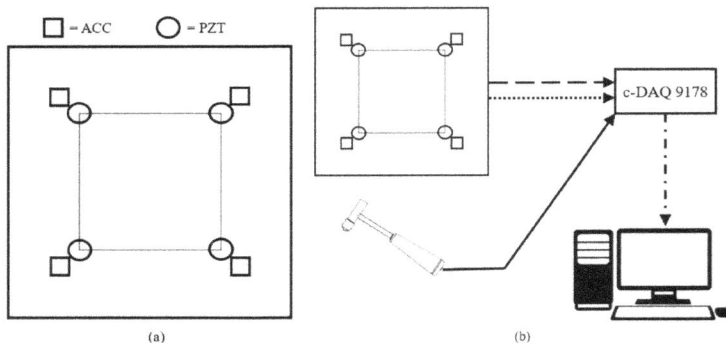

Figure 1. Impact area and sensor positions (**a**) and experimental setup (**b**).

The hammer signal was acquired by a NI-9234 acquisition card, as well as for all the accelerometer signals. The piezoelectric signals required the adoption of a NI-9229 acquisition card able to acquire signals up to 60 V, due to the capability of the piezoelectric disks to reach high voltage values. All the NI acquisition cards were gathered with a NI c-DAQ-9178 chassis, to simultaneously acquire all the signals using the SignalExpress software (National Instruments, Austin, TX, USA). A trigger, based on the hammer signal, was used and the acquisition frequency was 51.2 kHz. All connections were BNC cables, except for the laptop, which was connected by USB to the c-DAQ chassis. An overview of the entire setup is shown in Figure 1b.

2.2. Preliminary Signal Analysis and Feature Extraction

Both chosen sensor technologies were able to acquire signals that are representative of the occurring impact phenomenon, however differences in the signal shapes are expected due to the different sensor working principles. In Figure 2a, the comparison of accelerometer (ACC) and piezoelectric (PZT) signal in response of a representative impact shows that both sensor typologies are sensitive to the incoming impact wave, but they also offer different responses.

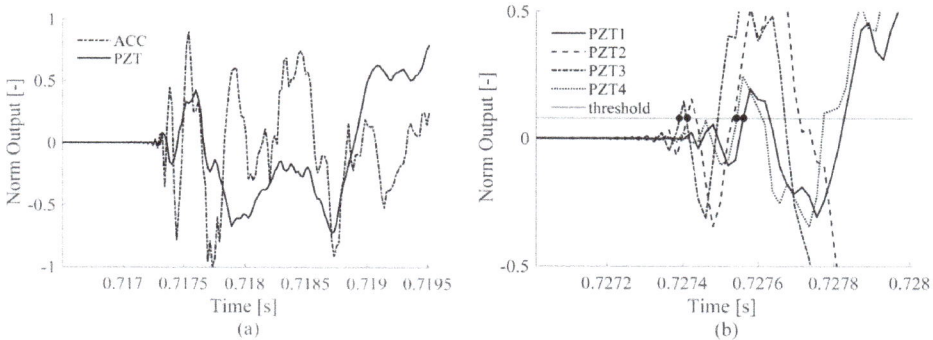

Figure 2. Original signals during an impact event for ACC and PZT sensors, on the aluminum plate (**a**) and the TOA measurement with multiple PZTs on the same aluminum plate (**b**).

It can be noted in Figure 2a that both sensors produced a very low baseline signal, including the environmental and sensors uncertainties before the occurrence of an impact, thus facilitating the identification of the impact event.

With this aim, the feature chosen to represent the impact event is the TOA of the elastic wave to the sensors, defined as the time at which the sensor signal exceeds a properly defined threshold value for the first time. Figure 2b shows the TOAs identified by the PZT sensors (black dots) for a representative impact on the aluminum plate, using non-filtered signals.

However, although a stable baseline is visible in Figure 2a, the TOA could not be identified regardless of the properties of the impacted structure, the impactor, the impact energy and the sensor dynamic behavior, which influence the frequency content of the signal. In fact, the strain waves generated by impacts are known from the theory to be multiple flexural dispersive waves [5]. Different strain wave components are generated after the impact, associated to different frequencies, different amplitudes and travelling at different speeds. Thus, the same wave front should be identified by each sensor for a correct selection of the TOA and an efficient passive impact identification. In fact, the distance of the impact position from a sensor determines the TOA of the wave with respect to the sensor itself; the sensor closest to the impact location has the lowest TOA for the same wave front, while the most distant sensor has the largest TOA. The sensors can be listed in ascending TOA order, creating a sensor sequence. The performances of the passive impact identification methodology are strongly related to the capability of defining the correct TOA and thus the correct sensor sequence. Errors in TOA evaluation could severely hamper the methodology's accuracy. For these reasons, it is necessary to process the original data to obtain a more regular signal, before impact identification.

2.3. Frequency Analysis

Considering the different physical quantities measured by the PZT and ACC, a frequency analysis was performed to compare the sensors and to identify the frequency range conveying significant information. First, the one-sided power spectral density (PSD) was computed for all the sensors, for each impact position and each specimen and its maximum peak was set to zero; then, the maximum impact frequency was evaluated, imposing a limit equal to -10 dB and selecting the highest frequency that crosses this limit [5]. The -10 dB value was chosen to obtain a frequency for which 90% of the signal energy is considered, thus most of the impact energy was contained in the frequencies from zero to the maximum impact frequency estimated. One example is reported in Figure 3a,b, where the -10 dB limit was extracted from the ACC and PZT sensor signal during a representative impact, respectively.

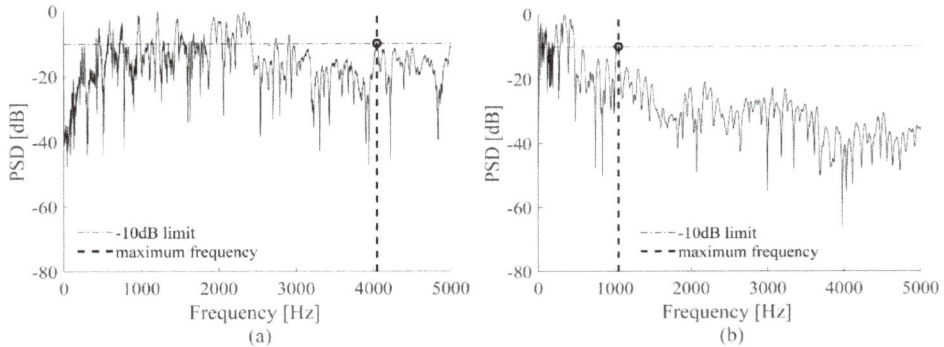

Figure 3. −10 dB extraction procedure for ACC (**a**) and PZT (**b**) sensors.

After evaluating the significant impact frequency for each sensor on each specimen and for each impact, a normality test, i.e., the Lilliefors Test [46], was executed in Matlab using the function *lillietest*. Table 3 summarizes the Lilliefors normality test results; the result $H = 1$ indicates that the null hypothesis of normal distribution was rejected at the 5% level of significance.

Table 3. Lilliefors normality test results.

	ACC—Al Plate	PZT—Al Plate	ACC—Sandwich	PZT—Sandwich
H	1	1	1	1

The results show that the distribution of the maximum impact frequencies for each sensor on each specimen cannot be assumed to be Gaussian with sufficient confidence. Thus, the cumulative distribution function (CDF) was used to define the cutoff sensor frequency for the acquired signal on each specimen, specifically considering the 95% of the CDF as the limit frequency. Figure 4a,b shows the procedure results for the ACC sensor, while Figure 4c,d shows the results for the PZT sensor, on the aluminum plate and the sandwich panel, respectively.

For the ACC case, a unique frequency can be considered since the results were very similar for both the specimens (Figure 4a,b), i.e., 4912 Hz. On the other hand, two different characteristic frequencies were evaluated for the PZT case on the two specimens (Figure 4c,d), i.e., 619 Hz for the skin plate and 1689 Hz for the sandwich panel.

At the end of the frequency analysis, the two main results are: (i) the limit sensor signal frequency reached during the tests for the ACC sensor is 4912 Hz, valid for both the specimens, while (ii) a limit sensor frequency of 619 Hz and 1689 Hz is reached by the PZT sensor on the aluminum plate and the sandwich panel, respectively.

In addition to these values, 8000 Hz and 2000 Hz were also considered significant threshold frequencies. At the maximum frequency of 8000 Hz the ACC sensor still showed a linear response, as reported in the producer certification. For both the ACC and PZT sensors the 2000 Hz value was chosen as the theoretical maximum frequency reachable using an impact hammer identical to the one used during the test [47]. All significant cutoff frequencies for both sensors on each specimen are summarized in Table 4.

Finally, assuming all the frequencies previously discussed, a low pass Infinite Impulse Response (IIR) filter was applied to the signals using Matlab; considering that IIR filters induce some time delay, the Matlab function *filtfilt* was used to compensate for this effect and avoid any filter influence on the TOA assessment.

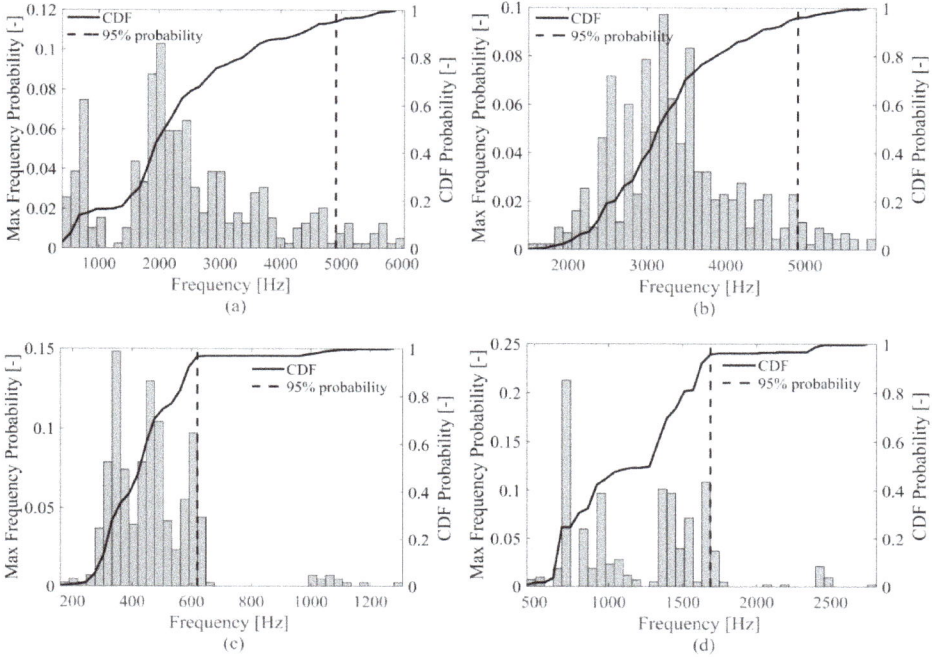

Figure 4. CDF and 95% limit for the ACC on the aluminum plate (**a**) and the sandwich panel (**b**) and for the PZT on the aluminum plate (**c**) and the sandwich panel (**d**).

Table 4. Identified significant cutoff frequencies for the ACC and PZT sensor, on both specimens.

	ACC—Al Plate	PZT—Al Plate	ACC—Sandwich	PZT—Sandwich
Frequency [Hz]	8000	-	8000	-
	4912	4912	4912	4912
	2000	2000	2000	2000
	619	619	1689	1689

2.4. Threshold Selection

Considering the problems of selecting the baseline threshold discussed in Section 2.2, two different approaches were chosen for the extraction of the TOA. The first one was a visual approach, in which the TOA was identified visually for each impact, manually selecting the baseline threshold. This was done in order to obtain a reference result that is as unaffected as possible by the unavoidable errors related to the dispersive nature of the strain waves. In fact, if a visual identification of the TOA was performed, the signal was observed and the threshold exceedance point was selected in a consistent manner, referring to the occurrence of the same wave front at different sensors, thus guaranteeing the correctness of the sequence at which the strain waves reach the sensors. This second point is of great importance in the impact identification procedure, as is highlighted in the following. However, this visual approach is highly time consuming, the results could be affected by human errors and the methodology does not permit the definition of a fixed threshold value for an automatic impact detection algorithm, which is a desired result in any structural health monitoring application.

For this reason, a second approach is proposed, namely an automatic TOA evaluation, for which the threshold was selected based on the percentage of correct sensor sequences, considering all the impact positions on each specimen. First, the signals were gathered considering the sensor technology and the specimens, then they were normalized, dividing the sensor's output by the maximum peak reached during each impact. Finally, the threshold was varied from the minimum value (−1) to the maximum value (+1) and, in correspondence to each threshold value, the sequence accuracy was evaluated as:

$$sequence\ accuracy = \frac{number\ of\ correct\ sequences}{number\ of\ impacts}. \tag{1}$$

The sequence accuracy resulting for each threshold value is drawn in Figure 5 for the accelerometers and in Figure 6 for the piezoelectric sensors. The main outcome of this procedure is the threshold value selection to be used in the automatic approach, for each sensor technology on each specimen. Using the selected threshold value, guarantees the higher sequence accuracy for the impact database considered, for the automatic TOA evaluation.

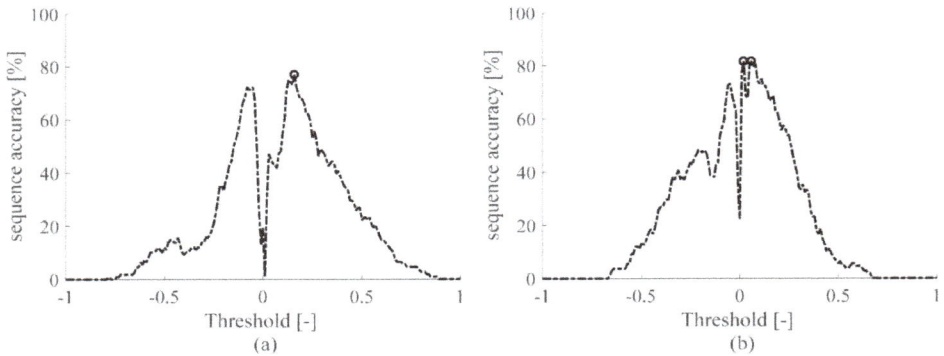

Figure 5. Sequence accuracy of the ACC signals on the aluminum plate (**a**) and the sandwich panel (**b**), when the 0–619 Hz or 0–1689 Hz filter is applied, respectively.

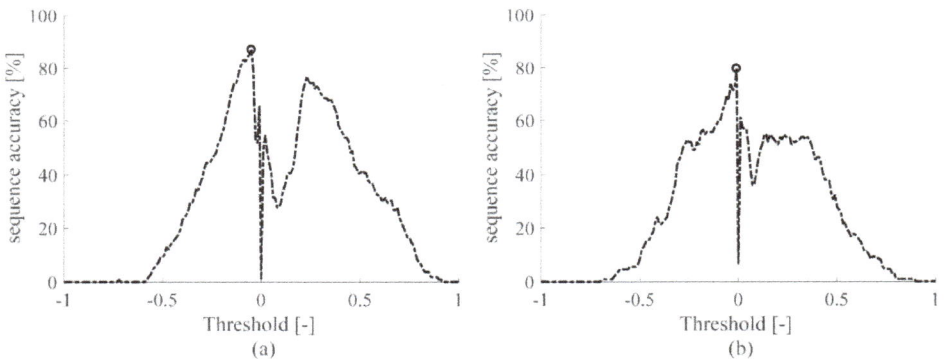

Figure 6. Sequence accuracy for the PZT signals on the aluminum plate (**a**) and the sandwich panel (**b**), when the 0–619 Hz or 0–1689 Hz filter is applied, respectively.

Note that Figures 5 and 6 show only the results obtained for the signals filtered with the lowest cutoff frequency among those available in Table 4, specifically with low-pass ranges 0–619 Hz and 0–1689 Hz for the aluminum plate and sandwich panel, respectively. By considering only the lowest cutoff frequencies, the best sequence accuracy results were obtained. The interested reader can refer to Table A1 in the Appendix A, where the performances obtained with all the low-pass filters adopted are reported for both specimens.

2.5. TOA Data Processing

In order to simulate a real impact monitoring scenario, where no trigger of the impact instant is available, a relative *TOA* is presented hereafter, defined as the difference (ΔT) between the *TOA* of the *i*-th sensor (TOA_i) and the *TOA* of the first sensor reached by the impact wave (TOA_0).

$$\Delta T_i = TOA_i - TOA_0 \quad i = 1, 2, \ldots, N-1, \tag{2}$$

where N is the number of sensors for impact identification.

Similar considerations apply for the impact distance. Assuming the impact location as a known variable, thus neglecting the error in impact location due to the operator, a relative distance (ΔL) was computed as the difference between the *i*-th sensor distance from the impact location (D_i) and the distance from the same impact location of the first sensor reached by the impact wave (D_0).

$$\Delta L_i = D_i - D_0 \quad i = 1, 2, \ldots, N-1 \tag{3}$$

The $N-1$ pairs of values (ΔL–ΔT) are available after each impact and, after repeating the impact K times for K different locations, $K(N-1)$ (ΔL–ΔT) couples were collected in a graph as shown in Figure 7. The linearity between the relative distance and the relative *TOA* is evident and a linear regression was performed. The choice of representing the results within a unique, coherent framework was made to show that different impactor stiffnesses excite different wave propagation modes, which, in turn, are associated to different frequencies, thus inducing different wave velocities. In fact, the slope of the linear regression in the (ΔL–ΔT) graph was used to represent the wave velocity.

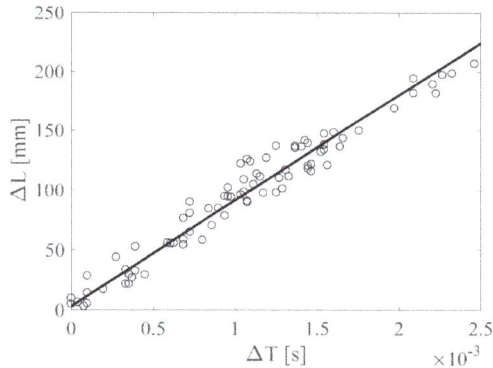

Figure 7. (ΔL–ΔT) graph example.

Regarding the linear regression, first the correlation between the ΔL and ΔT data was checked and the Pearson coefficient was evaluated for all the datasets; Table A2 in the Appendix A reports the results obtained using the manual approach, for the ACC and PZT sensors, respectively. All of the Pearson coefficients were close to 1 and all the associated *p*-values were approximately zero, suggesting that a linear correlation was present for the (ΔL–ΔT) data, thus justifying the use of the linear regression for the data representation. The same conclusions apply for the data collected by the automatic approach with the results listed in Table A2 in the Appendix A.

An outlier analysis was then used to eliminate data that were not representative of the studied phenomenon, due to errors in the experimental tests' execution. The presence of outliers is related to the modality chosen for the test execution and for the *TOA* evaluation. In fact, all the impacts were executed manually, resulting in a potentially different real impact position from the target impact location, inducing errors on the ΔL parameter. The same considerations were made for the ΔT

parameter; even when it was evaluated based on the visual approach, signal misinterpretation could happen due to the difficulty of distinguishing the various strain wave contributions associated to different frequencies and velocities within the time series. Also, sensors or acquisition system failures could give results not representative of the impact phenomenon. These facts induce errors in the ΔL–ΔT data that may give rise to outliers.

As a consequence, the (ΔL–ΔT) data displayed a certain amount of dispersion and, even though the linear regression visibly appeared to be a good model for the (ΔL–ΔT) dataset representation, it required verification that the three hypotheses for the simple linear regression are valid, namely (i) the normality of the residuals, (ii) the homoscedasticity and (iii) the independence of the residuals. For brevity's sake, the hypotheses were verified for the majority of the analyzed datasets, thus assuming that the simple linear regression is adequate for modeling the (ΔL–ΔT) data:

$$\Delta L = m\Delta T + q, \tag{4}$$

where ΔL and ΔT were evaluated from the experimental dataset, m is the slope and q the intercept of the linear regression line. Finally, the uncertainties related to the linear regression parameters, i.e., slope m and intercept q, could be evaluated as follows:

$$\sigma_m = \sqrt{\frac{n}{n\sum \Delta T_i^2 - \left(\sum \Delta T_i\right)^2}} \tag{5}$$

$$\sigma_q = \sqrt{\frac{\sum \Delta T_i^2}{n\sum \Delta T_i^2 - \left(\sum \Delta T_i\right)^2}}, \tag{6}$$

where σ_m and σ_q represent the slope and intercept standard deviations, respectively, and n is the total number of samples included in the analysis.

3. Results and Discussion

In this section, the results are presented for the metallic plate first, then for the sandwich panel; a comparison between the ACC and PZT results is shown for all the low-pass filters adopted. Then the automatic approach results are presented. The linear regression parameter uncertainties were evaluated for all cases.

3.1. Aluminum Plate Results

For the first part of the work, the Aluminum plate signals were analyzed using the visual approach. In this case the feature TOA and thus the parameter ΔT, were evaluated by visually determining the instant at which the impact wave arrived at the sensor.

Figure 8 shows the results for the ACC sensors, while Figure 9 shows the results for the PZT sensors; both figures show the 0–4912 Hz low-pass filter results for all the hammer tips, in position (a). Position (b) refers to the 0–2000 Hz low-pass filter results, while position (c) to the 0–619 Hz low-pass filter results. Finally, Figure 10 depicts the results for the 0–8000 Hz range, for the ACC sensor only. For simplicity of interpretation, the linear regressions are also reported and used as references; different colors are used to identify the different tips used.

Figures 9a and 10 show the results for ACC and PZT, respectively, for which the widest frequency range was adopted for each sensor, i.e., 0–8000 Hz for ACC and 0–4912 Hz for PZT. In both images different lines have different slopes; considering that each colored line represents a different hammer tip, it is possible to state that the impactor stiffness significantly influences the (ΔT–ΔL) relationship.

In Figure 8a, the cutoff frequency estimated for the ACC with the approach introduced in Section 2.3 was used; in this case, a slope similarity was visible for all the lines. The impactor stiffness failed to influence the results if the estimated cutoff frequency for the ACC sensors was adopted.

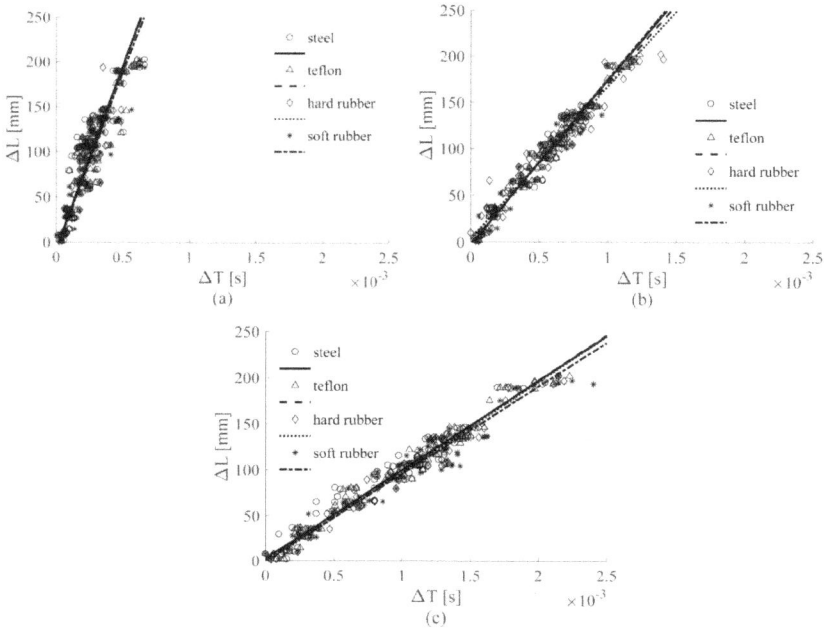

Figure 8. (ΔL–ΔT) visual approach results for the ACC sensors, Al plate, all tips, filtered with the 0–4912 Hz filter (**a**), 0–2000 Hz filter (**b**) and 0–619 Hz filter (**c**).

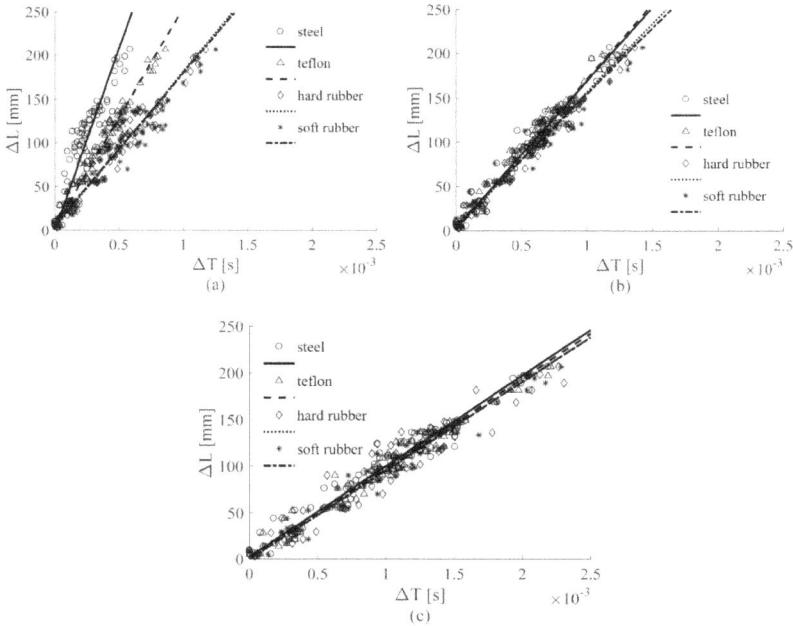

Figure 9. (ΔL-ΔT) visual approach results for the PZT sensors, Al plate, all tips, filtered with the 0–4912 Hz filter (**a**), 0–2000 Hz filter (**b**) and 0–619 Hz filter (**c**).

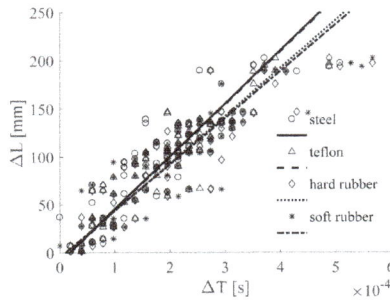

Figure 10. (ΔL–ΔT) visual approach results for ACC sensors, Al plate, all tips, filtered with the 0–8000 Hz filter.

Figures 8b and 9b show the results for the 0–2000 Hz range for the ACC and PZT sensors, respectively; lower cutoff frequencies resulted in slower waves—in fact, the line slopes were lower in both the cases if compared to those ones obtained with higher cutoff frequencies. Moreover, the difference between the lines in the PZT case decreased, indicating that a reduction of the filtering frequency causes a reduction in the impactor stiffness influence on the wave velocity.

Finally, Figures 8c and 9c show the results for the 0–619 Hz range for the ACC and PZT sensors, respectively. The adopted frequency is the cutoff frequency estimated with the approach in Section 2.3 for analyzing PZT signals; thus from Figure 9c, a significant impactor stiffness influence reduction was obtained for the PZT results. The same conclusion was reached from Figure 9a, in which the ACC results were obtained by adopting the cutoff frequency estimated analyzing the ACC signals. Moreover, the results obtained with the ACC and the PZT sensors showed very similar slopes.

3.2. Sandwich Panel Results

Figure 11 shows the visual approach results for the ACC sensor, while Figure 12 gives those for the PZT sensor; in these figures, position (a) refers to the 0–4912 Hz low-pass filter, position (b) to the 0–2000 Hz low-pass filter, and position (c) to the 0–1689 Hz low-pass filter results. Figure 13 depicts the results for the 0–8000 Hz range, for the ACC sensor only; linear regression lines were used as the references for all the conclusions.

Figure 13 also shows that, considering the sandwich panel case, in the 0–8000 Hz filtering range, different hammer tips correspond to lines with different slopes; the same is visible in Figure 12a, which refers to the results obtained filtering the PZT signals with the wider range, i.e., 0–4912 Hz. Thus, also for the sandwich panel, different hammer tips generated strain waves with different velocities, as well as for the aluminum plate case.

In Figure 11a, the cutoff frequency defined for the ACC was used, resulting in similar slopes for all the lines; thus, the impactor stiffness has no influence on the ACC results if the estimated cutoff frequency for the ACC case was adopted.

Figures 11b and 12b illustrate the results for the 0–2000 Hz range for the ACC and the PZT sensors, respectively; in the sandwich panel case, as well as for the aluminum plate case, the cutoff frequency reduction caused a decrease in the wave velocity, i.e., the slope coefficient m, as shown in Table 5 below for the ACC case.

Finally, Figures 11c and 12c show the results for the 0–1689 Hz range for the ACC and the PZT sensors, respectively. For the sandwich panel case, as well as for the aluminum plate case, the lowest cutoff frequency was the value estimated with the approach in Section 2.3, for the PZT sensors. For the PZT results in Figure 12c, a strong impactor stiffness influence reduction was observed, similar to that obtained in Figure 11a with ACC, adopting the cutoff frequency value estimated by analyzing the ACC signals. Moreover, results obtained with the ACC and the PZT sensors, for the 0–1689 Hz range, showed similar slopes, as also seen in the aluminum plate case.

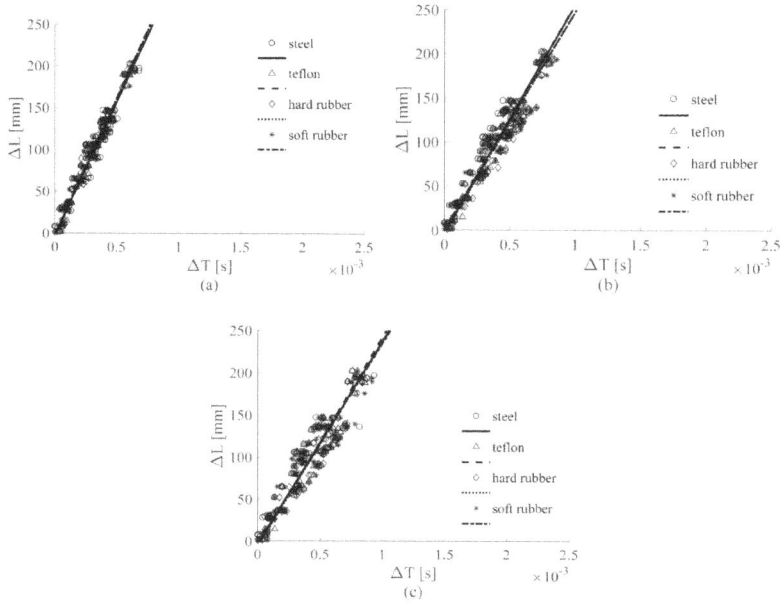

Figure 11. (Δ*L*–Δ*T*) visual approach results for the ACC sensors, sandwich panel, all tips, filtered with the 0–4912 Hz filter (**a**), 0–2000 Hz filter (**b**) and 0–1689 Hz filter (**c**).

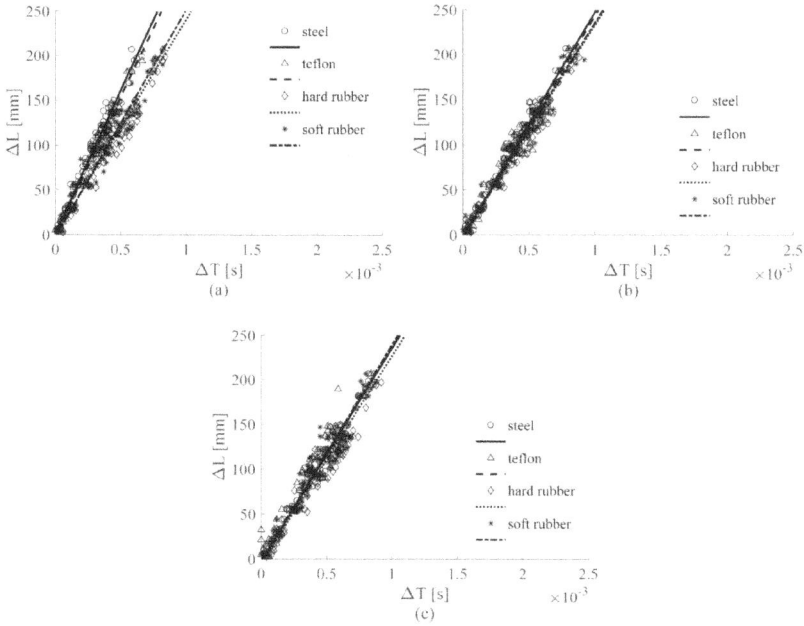

Figure 12. (Δ*L*–Δ*T*) visual approach results for the PZT sensors, sandwich panel, all tips, filtered with the 0–4912 Hz filter (**a**), 0–2000 Hz filter (**b**) and 0–1689 Hz filter (**c**).

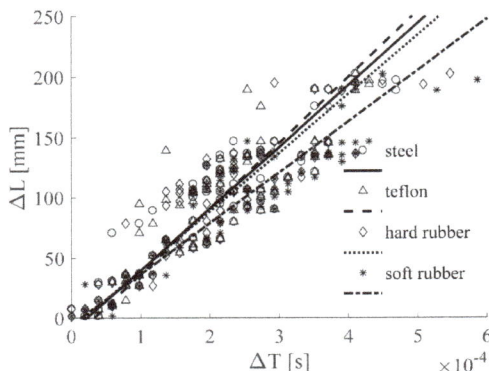

Figure 13. (ΔL–ΔT) visual approach results for the ACC sensors, sandwich panel, all tips, filtered with the 0–8000 Hz filter.

Table 5. Linear regression slope coefficient m for the ACC sensor, 0–4912 Hz and 0–2000 Hz cases.

Frequency [Hz]	Linear Regression Slope Coefficient m—ACC Sensors—[m/s]			
	Steel	**Teflon**	**Hard Rubber**	**Soft Rubber**
0–4912	320.1	331.1	326.5	323.9
0–2000	258.4	261.6	251.1	247.0

Note that for the sandwich panel the 2000 Hz limit was very close to the lowest cutoff frequency, i.e., 1689 Hz; for this reason, the impactor stiffness influence reduction was already observed in the PZT case in Figure 12b for the 0–2000 Hz range.

For this first part of the work, a low-pass filter that adopts the cutoff frequency evaluated with the approach introduced in Section 2.3 was applied to reduce the influence of the impactor stiffness.

Another fundamental factor that significantly affects the results is related to the stiffness of the impacted structure; in fact, structures with different stiffness produce strain waves with different velocities. Considering the results obtained with the cutoff frequency 0–619 Hz for the aluminum plate and 0–1689 Hz for the sandwich panel for the ACC case (see Figures 8c and 11c), the line groups showed different slopes. Similar conclusions can be drawn by observing the results for the PZT case (see Figures 9c and 12c). Thus, it can be concluded that by processing the signal with a filter built according to the approach introduced in Section 2.3, the influence of the impactor stiffness was reduced, while the influence of the impacted structure stiffness remained unaltered.

Finally, for the results obtained in Sections 3.1 and 3.2 it is possible to conclude that:

- Different impactor stiffnesses produce different impact wave velocities, represented as line slopes in the plots.
- Frequencies identified with the approach introduced in Section 2.3 are suitable for the reduction of the impactor stiffness influence on the impact wave velocity evaluation, for both sensors considered.
- Frequencies identified with the approach introduced in Section 2.3 are unsuitable for the reduction of the impacted structure stiffness influence on the impact wave velocity evaluation, for both sensors considered.

Due to the impactor stiffness influence reduction, unique datasets were created gathering all the tip results, for the aluminum plate 0–619 Hz filtered, and for the sandwich panel, 0–1689 Hz filtered. Thus, a unique linear regression representative of all the impactor stiffness cases could be calculated and the linear regression parameter uncertainties could be evaluated. The results are listed in Table A3 in the Appendix A.

3.3. Automatic Approach Results

The automatic approach was considered in the second part of the work. In this case the feature TOA, and consequently also the parameter ΔT, were evaluated choosing a threshold value according to the procedure explained in Section 2.4 and then by selecting the correct points using an automated algorithm. Only the cases 0–619 Hz and 0–1689 Hz were considered, not only because of the impactor stiffness influence reduction, as just shown, but also for the higher level of sequence accuracy that could be obtained using those frequencies, as shown in Section 2.4.

Figure 14 shows the results obtained on the aluminum plate, for the ACC and PZT sensors respectively. The results are consistent with the results obtained visually shown in Figures 8c and 9c.

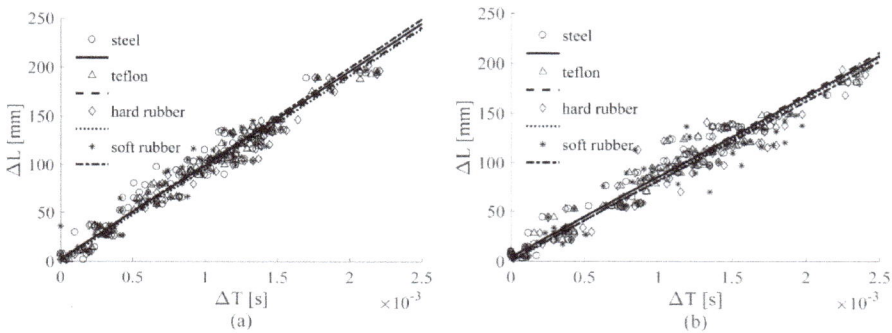

Figure 14. (ΔL–ΔT) automatic approach results for the Al plate, all tips, the ACC sensors (**a**) and the PZT sensors (**b**), both filtered with the 0–619 Hz filter.

Figure 15 shows the results obtained on the sandwich panel, for the ACC and PZT sensors, respectively. The results are very similar to those obtained manually in Figures 11c and 12c. For both the visual and the automated approach, the specimen stiffness influence could not be removed from the results.

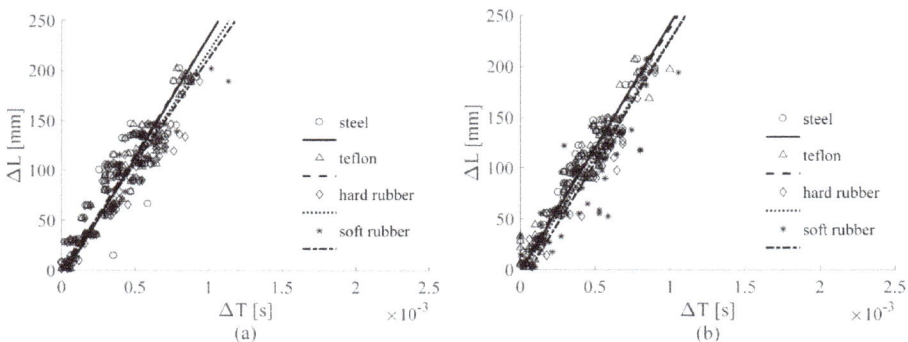

Figure 15. (ΔL–ΔT) automatic approach results for sandwich panel, all tips, ACC sensors (**a**) and PZT sensors (**b**), both filtered with the 0–619 Hz filter.

Thus, conclusions identical to those gleaned from the visual approach of Sections 3.1 and 3.2 were drawn.

As before, in this case a unique dataset was created to calculate a unique linear regression and then to evaluate the linear regression parameter uncertainties. The results are listed in Table A4 in the Appendix A. The comparison of the results in Tables A3 and A4 in the Appendix A shows that the automatic approach allowed for obtaining standard deviation values similar to those obtained with the visual approach. In fact, for the aluminum plate the automatic approach showed an increment in the slope standard deviation equal to 9% and 4%, for the ACC sensors and the PZT sensors, respectively, whereas the sandwich panel showed an increment equal to 21% and 33%, for the ACC sensors and the PZT sensors, respectively.

4. Conclusions

In this work, a methodology for the reduction of the impactor stiffness influence on the evaluation of the impact strain wave time of arrival (TOA) and velocities has been proposed. The procedure can be summarized as follows:

- First, a test setup was developed to acquire impact strain waves using two typical sensor technologies for the aeronautical field, i.e., accelerometers and piezoelectric sensors. Different impactor stiffnesses were reproduced using an impact hammer equipped with different tips.
- The procedure cutoff frequency was evaluated analyzing the acquired signals and was subsequently applied by post-processing the signals.
- The TOA feature was evaluated to estimate the strain wave velocities, using the post-processed signals. Two approaches were used, a visual and an automatic approach; in the second case the threshold had to be carefully chosen to develop the automatic algorithm for feature extraction.

Finally, the results were interpreted using a linear regression model and the linear regression parameter uncertainties were evaluated. The first part of the work confirmed that different impactor stiffnesses induce different impact wave velocities and that a cutoff frequency could therefore be defined and used for signal processing to obtain results that are not affected by the impactor stiffness. On the contrary, the impact wave velocity dependence was not reduced due to different specimen stiffnesses.

In the second part of the work an automatic approach was developed; it showed good capability to determine the correct sensor sequence and good accordance with the visual approach for evaluating the impact wave velocity. This approach could be useful for defining a procedure for the development of an automated impact monitoring system able to neglect the impactor stiffness in identifying the impact event. However, further work is required on the automatic TOA evaluation procedure to enhance its performance, especially considering the sequence accuracy.

Finally, all the results were represented using a linear regression and its parameter uncertainties, i.e., slope m and intercept q standard deviations. Those uncertainties are essential to evaluate the error propagation related to impact wave velocity evaluation errors, if we consider the impact monitoring system as part of a wider structural health monitoring system.

Author Contributions: Conceptualization, C.S. and A.G.; Data curation, A.B., C.S. and A.G.; Formal analysis, C.S., A.G. and F.C.; Funding acquisition, M.G.; Investigation, A.B., C.S., A.G. and F.C.; Methodology, C.S.; Supervision, C.S. and M.G.; Writing—original draft, A.B.; Writing—review & editing, F.C.

Funding: This research received no external funding.

Conflicts of Interest: The authors declare no conflict of interest.

Appendix A

Table A1. Sequence accuracy overview.

	ACC [%]	PZT [%]	Frequency [Hz]
Aluminum Plate	59.26	-	8000
	57.41	75.00	4912
	62.04	76.85	2000
	76.85	87.04	619
Sandwich Panel	55.56	-	8000
	67.59	79.63	4912
	81.48	82.41	2000
	82.41	83.33	1689

Table A2. Pearson coefficients for the datasets obtained with the ACC and PZT sensor, for the visual and the automatic approach.

4912 Hz—Visual	ACC	PZT	Hammer Tip
Aluminum Plate	0.8856	0.9011	Steel
	0.8989	0.9863	Teflon
	0.8956	0.9699	Hard rubber
	0.8910	0.9527	Soft rubber
Sandwich Panel	0.9715	0.9884	Steel
	0.9857	0.9894	Teflon
	0.9783	0.9706	Hard rubber
	0.9807	0.9659	Soft rubber
2000 Hz—Visual	**ACC**	**PZT**	**Hammer Tip**
Aluminum Plate	0.9804	0.9779	Steel
	0.9921	0.9874	Teflon
	0.9635	0.9818	Hard rubber
	0.9650	0.9768	Soft rubber
Sandwich Panel	0.9574	0.9876	Steel
	0.9647	0.9701	Teflon
	0.9601	0.9770	Hard rubber
	0.9379	0.9802	Soft rubber
619 Hz/1689 Hz—Visual	**ACC**	**PZT**	**Hammer Tip**
Aluminum Plate	0.9766	0.9736	Steel
	0.9828	0.9765	Teflon
	0.9837	0.9709	Hard rubber
	0.9684	0.9638	Soft rubber
Sandwich Panel	0.9339	0.9782	Steel
	0.9445	0.9463	Teflon
	0.9425	0.9642	Hard rubber
	0.9366	0.9667	Soft rubber
619 Hz/1689 Hz—Automatic	**ACC**	**PZT**	**Hammer Tip**
Aluminum Plate	0.9630	0.9800	Steel
	0.9850	0.9798	Teflon
	0.9804	0.9588	Hard rubber
	0.9701	0.9479	Soft rubber
Sandwich Panel	0.9148	0.9726	Steel
	0.9394	0.9585	Teflon
	0.9047	0.9549	Hard rubber
	0.9457	0.8860	Soft rubber

Table A3. Linear regression coefficient uncertainties; visual approach.

	ACC		PZT		Frequency [Hz]
	σ_m [V]	σ_q [V]	σ_m [V]	σ_q [V]	
Aluminum Plate	1.260×10^7	1.386×10^5	1.812×10^7	1.989×10^5	0–619
Sandwich Panel	8.511×10^8	9.367×10^6	6.917×10^8	7.547×10^6	0–1689

Table A4. Linear regression coefficient uncertainties; automatic approach.

	ACC		PZT		Frequency [Hz]
	σ_m [V]	σ_q [V]	σ_m [V]	σ_q [V]	
Aluminum Plate	1.378×10^7	1.465×10^5	1.892×10^7	1.976×10^5	0–619
Sandwich Panel	1.066×10^7	1.130×10^5	9.246×10^8	1.010×10^5	0–1689

References

1. Abrate, S. Soft impacts on aerospace structures. *Prog. Aerosp. Sci.* **2016**, *81*, 1–17. [CrossRef]
2. Allaeys, F.; Luyckx, G.; Van Paepegem, W.; Degrieck, J. Characterization of real and substitute birds through experimental and numerical analysis of momentum, average impact force and residual energy in bird strike on three rigid targets: A flat plate, a wedge and a splitter. *Int. J. Impact Eng.* **2017**, *99*, 1–13. [CrossRef]
3. Ohkami, Y.; Tanaka, H. Estimation of the Force and Location of an Impact Exerted on a Spacecraft. *JSME Int. J.* **1998**, *41*, 829–835. [CrossRef]
4. Nguyen, S.N.; Greenhalgh, E.S.; Graham, J.M.R.; Francis, A.; Olsson, R. Runway debris impact threat maps for transport aircraft. *Aeronaut. J.* **2014**, *118*, 229–266. [CrossRef]
5. Nilsson, A.; Liu, B. *Vibro-Acoustics*, 2nd ed.; Science Press: Beijing, China; Springer: Berlin/Heidelberg, Germany, 2015; Volume 1.
6. Worden, K. Rayleigh and Lamb waves—Basic principles. *Strain* **2001**, *37*, 167–172. [CrossRef]
7. Inoue, H.; Kishimoto, K.; Shibuya, T. Experimental wavelet analysis of flexural waves in beams. *Exp. Mech.* **1996**, *36*, 212–217. [CrossRef]
8. Christoforou, A.P.; Yigit, A.S. Effect of Flexibility on Low Velocity Impact Response. *J. Sound Vib.* **1998**, *217*, 563–578. [CrossRef]
9. Hellard, G. A Long Story of Innovations and Experiences Composites in Airbus. Available online: https://docplayer.net/25342669-Composites-in-airbus-a-long-story-of-innovations-and-experiences-presented-by-guy-hellard.html (accessed on 12 March 2019).
10. Smith, F. *The Use of Composites in Aerospace: Past, Present and Future Challenges*; Avalon Consultancy Services LTD.: Newbury, UK, 2013; p. 140.
11. Staszewski, W.J.; Mahzan, S.; Traynor, R. Health monitoring of aerospace composite structures—Active and passive approach. *Compos. Sci. Technol.* **2009**, *69*, 1678–1685. [CrossRef]
12. Hosseini, S.M.H.; Gabbert, U. Numerical simulation of the Lamb wave propagation in honeycomb sandwich panels: A parametric study. *Compos. Struct.* **2013**, *97*, 189–201. [CrossRef]
13. Smelyanskiy, V.N.; Hafiychuk, V.; Luchinsky, D.G.; Tyson, R.; Miller, J.; Banks, C. Modeling wave propagation in Sandwich Composite Plates for Structural Health Monitoring. In Proceedings of the Annual Conference of the Prognostics and Health Management Society, Montreal, QC, Canada, 25–29 September 2011; Volume 2, pp. 1–10.
14. Luchinsky, D.G.; Hafiychuk, V.; Smelyanskiy, V.N.; Kessler, S.; Walker, J.; Miller, J.; Watson, M. Modeling wave propagation and scattering from impact damage for structural health monitoring of composite sandwich plates. *Struct. Health Monit.* **2013**, *12*, 296–308. [CrossRef]
15. Zhao, J.; Li, F.; Cao, X.; Li, H. Wave propagation in aluminum honeycomb plate and debonding detection using scanning laser vibrometer. *Sensors* **2018**, *18*, 1669. [CrossRef]
16. Fiesler Saxena, I.; Guzman, N.; Hui, K.; Mal, A.K. Disbond detection in a composite honeycomb structure of an aircraft vertical stabilizer by fiber Bragg gratings detecting guided ultrasound waves. *Proc. Institution Mech. Eng. Part C J. Mech. Eng. Sci.* **2017**, *231*, 3001–3010. [CrossRef]

17. Sikdar, S.; Banerjee, S. Identification of disbond and high density core region in a honeycomb composite sandwich structure using ultrasonic guided waves. *Compos. Struct.* **2016**, *152*, 568–578. [CrossRef]

18. Dziendzikowski, M.; Kurnyta, A.; Dragan, K.; Klysz, S.; Leski, A. In situ Barely Visible Impact Damage detection and localization for composite structures using surface mounted and embedded PZT transducers: A comparative study. *Mech. Syst. Signal Process.* **2016**, *78*, 91–106. [CrossRef]

19. Jung, H.K.; Park, G. Integrating passive- and active-sensing techniques using an L-shaped sensor array for impact and damage localization. *J. Intell. Mater. Syst. Struct.* **2018**, *29*, 3436–3443. [CrossRef]

20. Sanchez, J.; Benaroya, H. Review of force reconstruction techniques. *J. Sound Vib.* **2014**, *333*, 2999–3018. [CrossRef]

21. Manes, A.; Gilioli, A.; Sbarufatti, C.; Giglio, M. Experimental and numerical investigations of low velocity impact on sandwich panels. *Compos. Struct.* **2013**, *99*, 8–18. [CrossRef]

22. Ueda, K.; Umeda, A. Dynamic response of strain gages up to 300 kHz. *Exp. Mech.* **1998**, *38*, 93–98. [CrossRef]

23. Othman, R. Cut-off frequencies induced by the length of strain gauges measuring impact events. *Strain* **2012**, *48*, 16–20. [CrossRef]

24. Knapp, J.; Altmann, E.; Niemann, J.; Werner, K.-D. Measurement of shock events by means of strain gauges and accelerometers. *Measurement* **1998**, *24*, 87–96. [CrossRef]

25. Ginsberg, D.; Fritzen, C. Sparsity-constrained identification of external forces and structural damage. In Proceedings of the European Workshop on Structural Health Monitoring Series, Manchester, UK, 10–13 July 2018; pp. 1–9.

26. Ciampa, F.; Meo, M. Acoustic emission source localization and velocity determination of the fundamental mode A_0 using wavelet analysis and a Newton-based optimization technique. *Smart Mater. Struct.* **2010**, *19*. [CrossRef]

27. Chopra, I. Review of the state of the art of smart structures and integrated systems. *AIAA J.* **2002**, *40*, 2145–2187. [CrossRef]

28. Wild, G.; Hinckley, S. Acousto-ultrasonic optical fiber sensors: Overview and state-of-the-art. *IEEE Sens. J.* **2008**, *8*, 1184–1193. [CrossRef]

29. Saito, N.; Yari, T.; Enomoto, K. Development Overview of a Distributed Strain Sensing Technology Using Optical Fiber Sensors for Aircraft Structures. In Proceedings of the European Workshop on Structural Health Monitoring Series, Bilbao, Spain, 5–8 July 2016; pp. 1–9.

30. Atobe, S.; Kobayashi, H.; Hu, N.; Fukunaga, H.; City, C. Real-Time Impact Force Identification of Cfrp Laminated Plates Using Sound Waves. In Proceedings of the 8th International Conference on Composite Materials, Jeju Island, Korea, 21–26 August 2011; pp. 1–6.

31. Atobe, S.; Nonami, S.; Hu, N.; Fukunaga, H. Identification of impact force acting on composite laminated plates using the radiated sound measured with microphones. *J. Sound Vib.* **2017**, *405*, 51–268. [CrossRef]

32. Li, Q.; Lu, Q. Impact localization and identification under a constrained optimization scheme. *J. Sound Vib.* **2016**, *366*, 133–148. [CrossRef]

33. Samagassi, S.; Khamlichi, A.; Driouach, A.; Jacquelin, E. Reconstruction of multiple impact forces by wavelet relevance vector machine approach. *J. Sound Vib.* **2015**, *359*, 56–67. [CrossRef]

34. Theodosiou, T.C.; Rekatsinas, C.S.; Saravanos, D.A. Estimation of impact location and characteristics in laminated composite plates. In Proceedings of the European Workshop on Structural Health Monitoring Series, Manchester, UK, 10–13 July 2018; pp. 1–10.

35. Worden, K.; Staszewski, W.J. Impact Location and Quantification on a Composite Panel using Neural Networks and a Genetic Algorithm. *Strain* **2000**, *36*, 61–68. [CrossRef]

36. De Stefano, M.; Gherlone, M.; Mattone, M.; Di Sciuva, M.; Worden, K. Optimum sensor placement for impact location using trilateration. *Strain* **2015**, *51*, 89–100. [CrossRef]

37. Chai, G.B.; Zhu, S. A review of low-velocity impact on sandwich structures. *Proc. Inst. Mech. Eng. Part L J. Mater. Des. Appl.* **2011**, *225*, 207–230. [CrossRef]

38. Chai, G.B.; Manikandan, P. Low velocity impact response of fibre-metal laminates—A review. *Compos. Struct.* **2014**, *107*, 363–381. [CrossRef]

39. Richardson, M.O.W.; Wisheart, M.J. Review of low-velocity impact properties of composite materials. *Compos. Part A Appl. Sci. Manuf.* **1996**, *27*, 1123–1131. [CrossRef]

40. Kurşun, A.; Şenel, M.; Enginsoy, H.M.; Bayraktar, E. Effect of impactor shapes on the low velocity impact damage of sandwich composite plate: Experimental study and modelling. *Compos. Part B Eng.* **2016**, *86*, 143–151. [CrossRef]

41. Iqbal, M.A.; Gupta, N.K. Ballistic limit of single and layered aluminium plates. *Strain* **2011**, *47*, 205–219. [CrossRef]

42. Kundu, T.; Das, S.; Jata, K.V. An improved technique for locating the point of impact from the acoustic emission data. *Proc. SPIE-Int. Soc. Opt. Eng.* **2007**, *6532*, 1–12.

43. Park, J.; Ha, S.; Chang, F.-K. Monitoring Impact Events Using a System-Identification Method. *AIAA J.* **2009**, *47*, 2011–2021. [CrossRef]

44. Agbasi, C.; Banerjee, S. Classification of Low Velocity Impact Using Spiral Sensing Technique. In *Experimental and Applied Mechanics*; Conference Proceedings of the Society for Experimental Mechanics Series; Springer: Cham, Switzerland, 2014; Volume 6, pp. 79–87.

45. Sanchez, N.; Meruane, V.; Ortiz-Bernardin, A. A novel impact identification algorithm based on a linear approximation with maximum entropy. *Smart Mater. Struct.* **2016**, *25*, 095050. [CrossRef]

46. Lilliefors, H.W. On the Kolmogorov-Smirnov Test for Normality with Mean and Variance Unknown. *J. Am. Stat. Assoc.* **1967**, *62*, 399–402. [CrossRef]

47. PCB Piezotronics, Inc. *Model 086E80 ICP®Impact Hammer Installation and Operating Manual*; PCB Piezotronics, Inc.: Depew, NY, USA, 2015.

sensors

MDPI

Article

Structural Damage Identification of Bridges from Passing Test Vehicles

Yang Yang [1,*], Yuanhao Zhu [1], Li Lei Wang [1], Bao Yulong Jia [2] and Ruoyu Jin [3]

1   MOE Key Laboratory of New Technology for Construction of Cities in Mountain Area, and School of Civil Engineering, Chongqing University, Chongqing 400045, China; m18723232480@163.com (Y.Z.); leeleiwang@163.com (L.L.W.)
2   Horoy Property Group (Shenzhen) Co., Ltd., Shenzhen 518000, China; aijiangmini@163.com
3   Subject of Built Environment, School of Environment and Technology, University of Brighton, Brighton BN2 4GJ, UK; R.Jin@brighton.ac.uk
*   Correspondence: yangyangcqu@cqu.edu.cn

Received: 1 September 2018; Accepted: 9 November 2018; Published: 19 November 2018

check for
updates

Abstract: This paper presents two approaches for the structural damage identification of a bridge from the dynamic response recorded from a test vehicle during its passage over the bridge. Using the acceleration response recorded by the vibration sensors mounted on a test vehicle during its passage over the bridge, along with the computed displacement response, the bending stiffness of the bridge can be determined using either: (1) the frequency-domain method based on the improved directed stiffness method with the identified frequency and corresponding mode shape, or (2) the time-domain method based on the residual vector of the least squares method with a fourth-order displacement moment. By comparing the bending stiffness values identified from the vehicle-collected data for the bridge under the undamaged and damaged states that are monitored regularly by the test vehicle, the bridge damage location and severity can be identified. Through numerical simulations and field tests, the present approaches are shown to be effective and feasible.

Keywords: bending stiffness; damage identification; environmental noise; bridge; test vehicle

1. Introduction

The physical properties of a structure such as stiffness and mass are important for structural health monitoring, because variations in these properties indicate the direct occurrence of damage. In most of the damage detection schemes, the mass of a structure is usually assumed to remain unchanged before and after the occurrence of damage. Accordingly, the change in stiffness of a structure is the most crucial dynamic property for damage identification.

Hou et al. [1] presented comprehensive reviews for the literature on the damage detection of structures. Amezquita-Sanchez and Adeli [2] presented a state-of-the-art review of recent articles on signal processing techniques for vibration-based SHM. Considering the bending stiffness index identification, Maeck [3,4] proposed the bending stiffness estimation approach for structures using the frequencies, mode shapes, and their derivatives, which is called the direct stiffness calculation (DSC) technique. Xu et al. [5,6] proposed the method of statistical moment-based damage detection (SMBDD) for inversely calculating the stiffness of steel-framed structures, which is sensitive to local structural damage, but insensitive to measurement noise. By integrating the generalized pattern search algorithm with the indirect identification technique using a passing vehicle, Li et al. [7] calculated the bending stiffness of a bridge, and pointed out that parameters such as the penalty values and mesh features should be further studied. Considering the difficulty of choosing the appropriate penalty factors for use in the DSC technique, Yang et al. [8–10] calculated the stiffness through an improved DSC technique for application to practical structures, which was verified both theoretically and experimentally.

Most of the identification techniques for bridges are referred to as the direct identification method, since they rely on the data collected by the vibration sensors that are directly mounted on the bridge. Blachowski et al. [11] proposed the axial strain accelerations degree of dispersion method with PCB piezoelectric accelerometers arranged directly in a truss structure. Kim et al. [12] studied Nair's damage indicator and its statistical pattern with a field experiment of a real continuous steel Gerber-truss bridge by the acceleration response of the bridge. Sevillano et al. [13] used a modal interval analysis method to address the uncertainty in vibration-based damage detection of a concrete frame. Yang et al. [14] proposed the deterministic and stochastic approaches for damage identification of experimental benchmark Reinforced Concrete (RC) frame model based on the fusing damage index by combining two types of statistical moment. Mao and Wang et al. [15,16] investigated the relationships between the dynamic properties and the environmental factors, especially the temperature based on the one-year monitoring data under normal operating conditions and one typhoon monitoring data by the sensors directly arranged on a Sutong Cable-Stayed Bridge. The indirect identification technique differs from the conventional direct method for measuring the bridge dynamic properties in that no vibration sensors need to be installed on the bridge. Rather, only one or a few vibration sensors need to be mounted on an instrumented test vehicle to record its response when passing over the bridge, from which the dynamic properties of the bridge are identified. The indirect identification technique, using a test vehicle to extract the first few bridge frequencies, was first proposed in 2004 by Yang et al. [17,18], and subsequently validated experimentally by Yang and Lin [19]. Originally, the main focus of the indirect identification technique is to extract the frequencies of the bridge, which is the most basic parameter related to the health status of a bridge. This technique is based on the transformation of the recorded data for the test vehicle from the time domain to the frequency domain using fast Fourier transform (FFT) [18,19], empirical mode decomposition [20], or other techniques [21–24]. Along these lines, Feng and Feng [21] proposed a bridge damage detection procedure that utilizes the vehicle-induced displacement response of the bridge, particularly, the curvature of the first mode shape, for simulated damage cases. OBrien and Keenahan [22] used a vehicle equipped with traffic speed deflectometers (TSDs) for determining the apparent profile of a bridge by an optimization algorithm, and showed that the time-shifted difference in the apparent profile can be probably used as a damage indicator of the bridge in the presence of noise by simulation. Behroozinia and Khaleghian et al. [23] presented a finite element model of the intelligent tire by using implicit dynamic analysis for defect tire detection. McGetrick et al. [24] used the test vehicle to identify the frequency and damping of a bridge, considering both smooth and rough bridge surfaces, and various vehicle speeds. It is noted that the application of the indirect method has been mainly focused on the frequency, damping, and indirect parameters with relation to the damage of the bridge in previous studies. More significantly, other properties of the bridge—particularly those for directly identifying stiffness, which reveals that the health status of a bridge—have not been evaluated using the indirect technique.

In this paper, it is assumed that the test vehicle is allowed to regularly monitor the bridge termly. The response of the test vehicle recorded during the *current* travel is assumed to be the *damaged* state and that of the *previous* travel is assumed to be the *undamaged* state. If no damage is detected by comparison of the two states, the current state is reset as the undamaged one, and another monitoring continues. By comparing the bending stiffness values identified from the vehicle-collected data for the bridge under the undamaged and damaged states monitored regularly by the test vehicle, the bridge damage location and severity can be identified based on the undamaged state. Only the *acceleration response* of the test vehicle is measured, and the *displacement response* is calculated by integration. Compared with previous studies, the bending stiffness estimation approach for each element of the bridge for damage identification is the main object of this paper, and the more prominent advantage of the indirect technique. The technique was developed by Yang et al. [25], and is used to obtain the mode shape of the monitored bridge by the test vehicle response. This mode shape is subsequently utilized to calculate the bending stiffness of the bridge, which is referred to as the *frequency domain method.* Using this method, a reliance on assumed penalty factors is necessary. On the other hand, making use of the relationship between the displacement response of the test vehicle and the bending stiffness of the bridge [17], the fourth-order statistical moment (the fourth-order statistical moment of structural response is expressed in terms of a probability density function (PDF) $p(x)$ as $M_4 = \int_{-\infty}^{+\infty} (x - \bar{x})^4 p(x) dx$, where x is the structural response with $\bar{x}$ as its mean value.) of the displacement response of the bridge is computed using the procedure documented in Xu et al. [5,6]. Subsequently, the bending stiffness of the bridge is acquired for damage detection, which is referred to as the *time-domain method.*

The adopted frequency domain method is a fast, initial evaluation technique for detecting the structural condition, since no optimization is required. In contrast, the adopted time domain method is a time-consuming, meticulous evaluation technique for damage detection, since all of the relevant parameters have to be optimized. In this paper, the used response data of the test vehicle are generated by simulation and field experimental tests, where the paper is focused on the *feasibility* of the indirect approach for damage detection, making use of such simulated and recorded data, which can be used for updating a real-time identification of structural damage in a timely manner.

2. Theoretical Background and Formulations

2.1. Frequency Domain Method

Figure 1 shows the mathematical model for a test vehicle moving on a bridge. In this model, the vehicle is simplified as a moving mass m_v, supported by a spring of stiffness k_v; the bridge is a simply-supported beam of span L, uniform mass density m^* per unit length, and uniform bending rigidity EI. To focus on the physical behavior of the vehicle, the following assumptions are adopted without a loss of generality for the problem. (1) Road surface roughness is ignored in the derivation, but is included in one of the studied numerical cases and field tests to evaluate the influence of this assumption. (2) Vehicle mass is negligibly small in comparison with the bridge mass. (3) Prior to the arrival of the test vehicle, the bridge remains at rest, i.e., zero initial conditions are assumed for the bridge, which is acceptable because the bridge vibrations caused by ambient excitations are small compared to those caused by moving vehicular loads. (4) Damping is neglected for both the vehicle and the bridge, which is acceptable, because the vibrations of both the vehicle and the bridge under moving loads are forced vibrations where damping is usually insignificant. (5) The test vehicle travels at a constant speed, v, during its passage over the bridge.

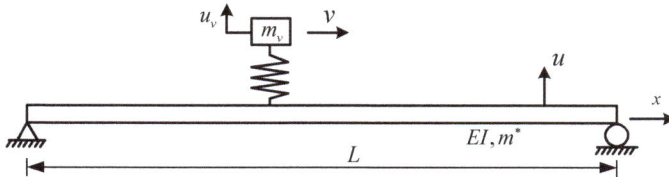

Figure 1. Moving test vehicle over a bridge.

The equations of motion can be written for the vehicle and bridge as follows:

$$m_v \ddot{u}_v(t) + k_v(u_v(t) - u(x,t)|_{x=vt}) = 0 \tag{1}$$

$$m^* \ddot{u}(x,t) + EI u''''(x,t) = f_c(t)\delta(x - vt) \tag{2}$$

where $u(x,t)$ is the vertical displacement of the bridge, $u_v(t)$ is the vertical displacement of the vehicle, measured from its static equilibrium position, $\delta(x - vt)$ is the Dirac delta function, and the superposed dot and prime denote derivatives with respect to time t and coordinate x, respectively. The contact force $f_c(t)$ is expressed as follows:

$$f_c(t) = -m_v g + k_v(u_v(t) - u(x,t)|_{x=vt}) \tag{3}$$

where g is the acceleration of gravity.

Using the modal superposition method, one can obtain the solution for the acceleration response of the test vehicle as follows [18,20]:

$$\ddot{u}_v(t) = \sum_{n=1}^{\infty} \left\{ A_{1,n} \cos\left(\frac{(n-1)\pi v}{L}\right)t + A_{2,n} \cos\left(\frac{(n+1)\pi v}{L}\right)t + A_{3,n} \cos(\omega_v t) \right. \\ \left. + A_{4,n} \cos\left(\omega_{b,n} - \frac{n\pi v}{L}\right)t + A_{5,n} \cos\left(\omega_{b,n} + \frac{n\pi v}{L}\right)t \right\} \tag{4}$$

where n is the counter for the bridge mode, $n\pi v/L$ is the driving frequency, ω_v is the vehicle frequency, as shown in Equation (5), and $\omega_{b,n}$ is the bridge n-th mode frequency identified by FFT, as shown in Equation (6):

$$\omega_v = \sqrt{k_v/m_v} \tag{5}$$

$$\omega_{b,n} = \frac{n^2\pi^2}{L^2}\sqrt{\frac{EI}{m^*}} \tag{6}$$

The coefficients in Equation (4) are given as follows:

$$A_{1,n} = -\left(\frac{(n-1)\pi v}{L}\right)^2 \times \frac{\Delta_{st,n}\omega_v^2}{2(1-S_n^2)(\omega_v + \frac{(n-1)\pi v}{L})(\omega_v - \frac{(n-1)\pi v}{L})} \tag{7}$$

$$A_{2,n} = -\left(\frac{(n+1)\pi v}{L}\right)^2 \times \frac{-\Delta_{st,n}\omega_v^2}{2(1-S_n^2)(\omega_v + \frac{(n-1)\pi v}{L})(\omega_v - \frac{(n-1)\pi v}{L})} \tag{8}$$

$$A_{3,n} = -\omega_v^2 \times \left\{ \frac{2\Delta_{st,n}\omega_v^2\left(\frac{\pi v}{L}\right)^2 n}{2(1-S_n^2)(\omega_v + \frac{(n-1)\pi v}{L})(\omega_v - \frac{(n-1)\pi v}{L})(\omega_v + \frac{(n+1)\pi v}{L})(\omega_v - \frac{(n+1)\pi v}{L})} \right. \\ \left. - \frac{2\Delta_{st,n}S_n\omega_v^2\left(\frac{n\pi v}{L}\right)\omega_{b,n}}{(\omega_v - \omega_{b,n} + \frac{n\pi v}{L})(\omega_v + \omega_{b,n} - \frac{n\pi v}{L})(\omega_v + \omega_{b,n} + \frac{n\pi v}{L})(\omega_v - \omega_{b,n} - \frac{n\pi v}{L})} \right\} \tag{9}$$

$$A_{4,n} = \left(\omega_{b,n} - \frac{n\pi v}{L}\right)^2 \times \frac{-S_n\Delta_{st,n}\omega_v^2}{2(1-S_n^2)(\omega_v - \omega_{b,n} + \frac{n\pi v}{L})(\omega_v + \omega_{b,n} - \frac{n\pi v}{L})} \tag{10}$$

$$A_{5,n} = -\left(\omega_{b,n} + \frac{n\pi v}{L}\right)^2 \times \frac{S_n\Delta_{st,n}\omega_v^2}{2(1-S_n^2)(\omega_v + \omega_{b,n} + \frac{n\pi v}{L})(\omega_v - \omega_{b,n} - \frac{n\pi v}{L})} \tag{11}$$

where the vehicle-induced static deflection $\Delta_{st,n}$ of the bridge and the speed parameter S_n of the n-th mode of the bridge are defined as follows:

$$\Delta_{st,n} = \frac{-2m_v g L^3}{n^4\pi^4 EI} \tag{12}$$

$$S_n = \frac{n\pi v}{L\omega_{b,n}} \tag{13}$$

To extract the mode shapes of the bridge [25], the component response corresponding to the bridge frequency of the n-th mode should be singled out from the vehicle response by a feasible filtering technique based on Hilbert transform [25]:

$$z(t) = R_b(t) + i\hat{R}_b(t) = \left[\left|\frac{\omega_{b,n}^2 S_n\Delta_{st,n}\omega_v^2}{(1-S_n^2)\left(\omega_v^2 - \omega_{b,n}^2\right)}\right| \sin\frac{n\pi vt}{L}\right] e^{i(\omega_{b,n}t - \frac{\pi}{2})} \tag{14}$$

$$R_b = A_{4,n} \cos\left(\omega_{b,n} - \frac{n\pi v}{L}\right)t + A_{5,n} \cos\left(\omega_{b,n} + \frac{n\pi v}{L}\right)t \tag{15}$$

$$\hat{R}_b(t) = H[R_b(t)] = A_{4,n} \sin\left(\omega_{b,n} - \frac{n\pi v}{L}\right)t + A_{5,n} \sin\left(\omega_{b,n} + \frac{n\pi v}{L}\right)t \tag{16}$$

where the coefficients $A_{4,n}$ and $A_{5,n}$ are defined in equations (10) and (11), respectively. Equation (14) indicates that, in the dynamic response of the test vehicle during its passage over the bridge, the component response of the n-th bridge frequency, $\omega_{b,n}$, oscillates with a varying amplitude, but with a shape identical to the n-th mode shape of the bridge in a sinusoidal form. In other words, the bridge component response oscillates within the envelope formed by the mode shape of the bridge, as implied by the instantaneous amplitude of the vehicle response.

With the n-th frequency and corresponding mode shape of the bridge made available by the procedure presented above, the bending stiffness of each element of the bridge can be calculated by the improved DSC method [8–10]. The improvement to the original DSC technique [3,4] is based on the fundamental mechanics of beams, where the bending stiffness EI of each cross-section is equal to the modal bending moment M at the same cross-section divided by the corresponding modal curvature, namely:

$$EI = \frac{M}{d^2\phi/dx^2} = \frac{M}{\kappa} \tag{17}$$

where x is the axis of the beam, ϕ is the mode shape function, and κ is the modal curvature. Equation (17) is valid for each mode of the beam if the effects of damping and shear deformation are ignored. This elementary beam theory can be approximately applied for the damage identification of beam structures, along with the indirect identification technique, as discussed in this paper.

According to the D'Alembert's principle [26], the cross-section of a beam should be in dynamic equilibrium in the presence of inertia force. With the improved DSC method [8–10], the internal force at each cross-section can be calculated for the n-th frequency and corresponding mode shape. In this study, the modal curvature of the n-th mode shape is calculated using the central difference method [27]. With the modal curvature and internal forces made available for each cross-section, the bending stiffness can be calculated for the n-th mode of the beam. Based on this approach, the bending stiffness at each node of the bridge model, e.g., using the finite element method (FEM), can be obtained from the frequency-domain method.

In this study, it is assumed that the bridge under investigation is monitored regularly by the test vehicle termly, and the acceleration response is recorded from the test vehicle during each passage. Such a procedure of comparison is repeated for the bridge throughout its service. The following is a summary of the analysis procedure:

(1) It is assumed that a *previous* monitoring of the bridge of concern has been completed using the procedure stated below, which is regarded as the *undamaged* state.

(2) The acceleration response is recorded for the test vehicle during its passage over the bridge for the *current* monitoring, which is suspected as the *damaged* state.

(3) Identify the n-th frequency of vibration of the bridge from the recorded vehicle response in the previous and current runs of monitoring.

(4) Recover the n-th mode shape of the bridge from the instantaneous amplitude of the component response corresponding to the n-th frequency.

(5) Calculate the stiffness EI using the n-th frequency and corresponding mode shape for the bridge, based on which the structural damage is detected.

(6) If no damage is detected, then the current monitoring is regarded as the *undamaged* state, and the same procedure of damage detection is repeated for the next monitoring.

Therefore, the corresponding flowchart is presented in Figure 2.

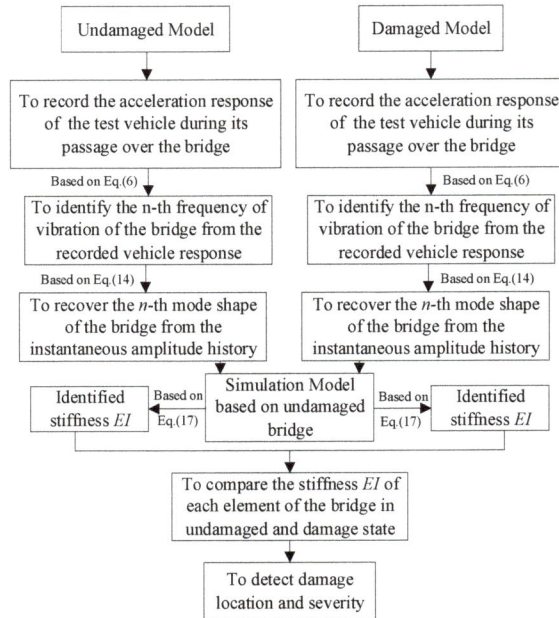

Figure 2. Flowchart of the frequency domain method.

2.2. Time-Domain Method

The statistical moment-based damage detection method is proposed by Xu et al. [5,6] for identifying the stiffness properties of a shear building before and after the occurrence of damages using the measured building story responses. Subsequently, for determining the damage location and severity in the structure, the stiffness properties identified for the two states are compared. It is demonstrated [5,6] that the fourth-order moment, rather than the second-order or the sixth-order moments, of the displacement story response is more suitable for identifying the stiffness properties, as a tradeoff between the sensitivity of the index to structural damage and the stability to random excitation. Such a technique was experimentally verified using shaking table tests for three shear building models [5].

Unlike previous studies [5,6], the fourth-order moment of displacement is adopted herein for the bridge structure using the response data collected by the passing test vehicle. The acceleration response of the test vehicle in Equation (4) can be integrated to yield the displacement response as follows:

$$
\begin{aligned}
u_v(t) = \sum_{n=1}^{\infty} \Big\{ &\overline{\overline{A}}_{1,n} \cos\Big(\frac{(n-1)\pi v}{L}\Big)t + \overline{\overline{A}}_{2,n} \cos\Big(\frac{(n+1)\pi v}{L}\Big)t + \overline{\overline{A}}_{3,n} \cos(\omega_v t) \\
&+ \overline{\overline{A}}_{4,n} \cos\Big(\omega_{b,n} - \frac{n\pi v}{L}\Big)t + \overline{\overline{A}}_{5,n} \cos\Big(\omega_{b,n} + \frac{n\pi v}{L}\Big)t \Big\}
\end{aligned}
\tag{18}
$$

where the coefficients in the above equation are listed below:

$$
\begin{aligned}
&\overline{\overline{A}}_{1,n} = A_{1,n} / - \Big(\frac{(n-1)\pi v}{L}\Big)^2, \overline{\overline{A}}_{2,n} = A_{2,n} / - \Big(\frac{(n+1)\pi v}{L}\Big)^2, \overline{\overline{A}}_{3,n} = A_{3,n} / - \omega_v^2 \\
&\overline{\overline{A}}_{4,n} = A_{4,n} / - \Big(\omega_{b,n} - \frac{n\pi v}{L}\Big)^2, \overline{\overline{A}}_{5,n} = A_{5,n} / - \Big(\omega_{b,n} + \frac{n\pi v}{L}\Big)^2
\end{aligned}
\tag{19}
$$

For the case where the parameters v, L, ω_v, m_v are constants, the displacement response of the test vehicle is only related to the frequency and bending stiffness of the bridge. In practice, it is assumed that the structural mass remains unchanged before and after damage [3–10]. Thus, the displacement response of the test vehicle is indicative of the bending stiffness of the bridge, which is the property exploited in the following discussion.

In this paper, we assume that a bridge is divided into N elements, and that each element has N_s sampling points. For the i-th element of the bridge, the displacement of the test vehicle can be given as $u_{v_i}(i) = [u_{v_i}(1), u_{v_i}(2), \ldots, u_{v_i}(N_s)]$ Thus, the average displacement response of the test vehicle at the i-th element can be computed as follows:

$$\overline{u}_{v_i} = \frac{1}{N_s} \sum_{j=1}^{N_s} u_{v_i}(j) \tag{20}$$

Accordingly, the fourth-order moment vector at each element of the bridge can be computed from the displacement response of the test vehicle as follows:

$$\hat{M}_4 = [\hat{M}_{41}, \hat{M}_{42}, \ldots, \hat{M}_{4N}] \tag{21}$$

where the entry for the i-th element is expressed as follows:

$$\hat{M}_{4i} = \int_{-\infty}^{+\infty} \left(u_{v_i} - \overline{u}_{v_i}\right)^4 p(u_{v_i}) du_{v_i} \tag{22}$$

where $p(u_{v_i})$ is the PDF of the structural response u_{v_i}. Thus, it can be calculated by using summation-type relationships as follows [5,6]:

$$\hat{M}_{4i} = \frac{1}{N_s} \sum_{j=1}^{N_s} u_{v_i}(j)^4 - \frac{4}{N_s} \overline{u}_{v_i} \sum_{j=1}^{N_s} u_{v_i}(j)^3 + \frac{6}{N_s} \overline{u}_{v_i}^2 \sum_{j=1}^{N_s} u_{v_i}(j)^2 - 3\overline{u}_{v_i}^4 \tag{23}$$

First, an initial value is assign to the stiffness EI of the bridge using the value obtained from the *previous* monitoring. With this value, the vehicle response can be solved from Equations (1) and (2). Then, the fourth-order moment vector corresponding to the *previous* monitoring can be computed from Equations (18), (21), and (23), which is considered as the theoretical statistical moment vector, $M_4 = [M_{41}, M_{42}, \ldots, M_{4N}]$. Simultaneously, the fourth-order moment vector, $\hat{M}_4$, can be computed using the test vehicle response recorded during the *current* monitoring. Therefore, the residual vector between M_4 and $\hat{M}_4$ is calculated as follows:

$$F(EI) = M_i(EI) - \hat{M}_i \tag{24}$$

Ideally, if the given vector of the stiffness values of all of the elements EI is equal to the actual values, the two norms of the residual vector, $\|F(EI)\|$, becomes zero. Practically, the vector of the optimal stiffness values can be identified by the least-squares method. Giving the EI of an element an initial value EI_0 from a *previous* monitoring, compute the corresponding fourth-order member moment, compare the actual value of the fourth-order member moment established from the *current* monitoring to that computed from the initial value EI_0, and finally, minimize $\|F(EI)\|$ to assess the damage condition of the bridge. Based on this approach, the bending stiffness at each element can be obtained.

The time-domain method in the indirect identification technique can be evaluated according to the analysis procedure of the following steps:

(1) Measure the displacement responses of the test vehicle, or calculate the displacement from the acceleration response of the test vehicle during its passage over the bridge for the undamaged and damaged states.

(2) The actual statistical moments of the measured displacement responses of the test vehicle with the undamaged and damaged states, $\hat{M}_{4i}$, are estimated using Equation (23).

(3)   Given the vector that collects all of the stiffness parameters for all of the elements representing the bridge FE model using initial values based on the calculated stiffness, e.g., from the frequency-domain method, the theoretical statistical moments of the displacement responses of the test vehicle, M_4, are calculated based on the FE model of the bridge and also making use of Equation (23).

(4)   Substituting $\hat{M}_4$ and M_4 into Equation (24), the vector collecting the structural stiffness values of all of the elements of the FE model of the bridge can be identified by the constrained nonlinear least-squares method for the undamaged and damaged states.

(5)   All of the attributes of the structural damage of the bridge, including the existence, location, and severity, can be detected by comparing the identified vector of the stiffness values of the undamaged bridge, $\hat{E}I^u$, to that of the damaged bridge, $\hat{E}I^d$.

Therefore, the corresponding flowchart is presented in Figure 3.

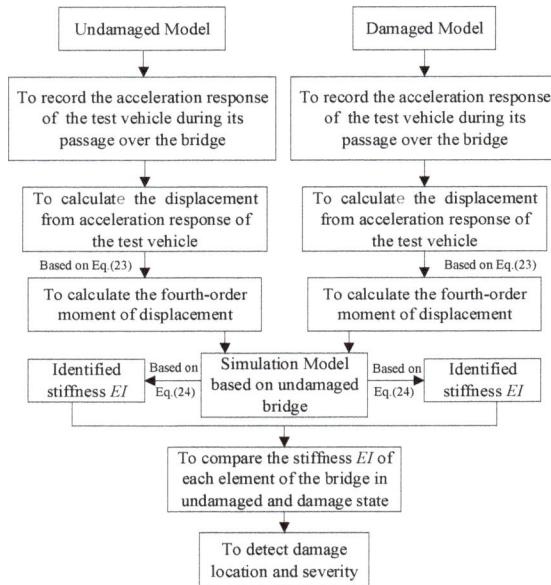

Figure 3. Flowchart of the time-domain method.

3. Parameters of the Test Vehicle Numerical Study

In order to investigate the feasibility and limitations of the presented approaches from the dynamic response of the passing test vehicle, several numerical cases are studied herein using the FEM, based on a well-developed simulation algorithm for vehicle–bridge interaction [17,25]. For the considered numerical simulation, the simply supported bridge is one span of the Da-Wu-Lun bridge [19], which is a part of the Taiwan Provincial Highway 2 near the northern coast of Taiwan. The considered bridge unit is composed of six prestressed I girders, placed at a center-to-center distance of 2.8 m, and has a span length of 30 m, as shown in Figure 4. The cross-section of the bridge has a total width of 16.5 m with a 20-cm thick concrete deck slab and a five-cm thick Asphalt Concrete(AC) pavement layer. The cross-sectional area and moment of inertia of each I girder are 0.64 m^2 and 0.2422 m^4, respectively. The concrete of the bridge has an elastic modulus of 29 GPa and a material density of 2400 kg/m^3. Figure 5 shows the FE model of the considered bridge span with 10 beam elements (i.e., 11 nodes) where the numbers in circles are the element numbers, and the others are the node numbers.

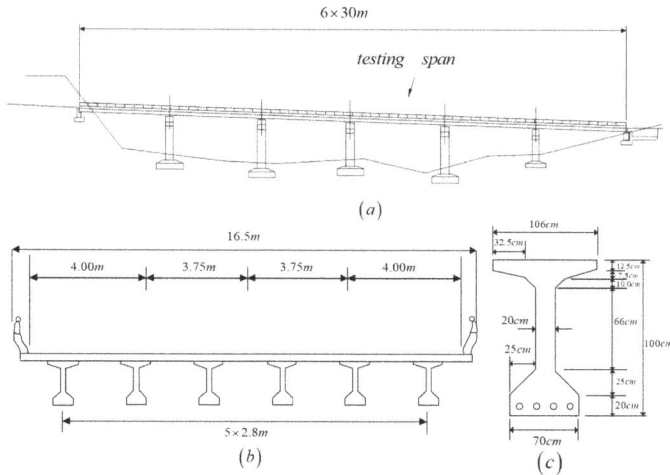

Figure 4. Bridge considered for the simulation, (**a**) bridge elevation, (**b**) bridge cross-section, (**c**) girder cross-section.

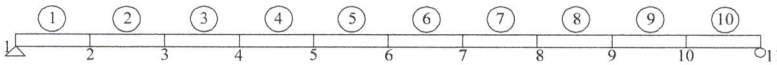

Figure 5. Finite element (FE) model of one span of the bridge with 10 elements.

The accuracy of the single-mode closed-form solution obtained for the vehicle–bridge couple system, and the vehicle response in particular, will be verified by the three-dimensional elements and two-dimensional elements for a typical example. As for the above bridge modal, the following data are adopted for the test vehicle: mass m_v = 500 kg, stiffness k_v = 90 kg/m, v = 1 m/s, and zero damping. For this vehicle, the vehicle to bridge mass ratio is 1:100. The vertical displacement of the vehicle obtained by the three-dimensional element and two-dimensional element approaches have been plotted in Figure 6a,b, respectively. As can be seen from Figure 6 and all of the analyzed results, the solutions obtained by the two approaches show a high degree of coincidence for the vehicle response; however, the analytical results are considered acceptable for the purpose of identifying the key parameters involved. Therefore, studying the frequency-domain method and time-domain method of the indirect measurement technique as the key point, the simulation model below are all based on the two-dimensional element approach for simplicity.

With the test vehicle acceleration and displacement responses discussed above, several test vehicle parameters are required for the indirect technique of bridge damage identification. The test vehicle parameters are frequency ω_v, mass m_v, stiffness k_v, and speed v. Considering the frequency-domain method and based on previous studies [17,25], the ratio of the bridge fundamental frequency ω_b to ω_v, i.e., $r = \omega_b/\omega_v$, is an important design parameter of the field test for the intended purpose of stiffness identification. In this study, a test vehicle with $m_v = 500$ kg and $v = 3$ m/s, k_v is adjusted for the different values of r, and the *EI* of the bridge is computed. Figure 7 shows results corresponding to $r = 0.7$ to 1.4 (for nodal point numbers, refer to Figure 5). It is noted that the nodes located in the neighborhood of the abutments (nodes 2 and 10) did not correspond to accurate results because of the unsuitable combination of the identified mode shape from the test vehicle [25] and the improved DSC method [8–10] in the frequency-domain method. This is attributed to the higher errors of the identified mode shapes near the boundaries compared to near the mid-span. Except for these nodes, it is shown that when $r = 0.7$ and 1.4, the calculated *EI* is closed to the specified *EI* where the difference is below 5%. With r approaching 1.0 from above or below, the calculated *EI* becomes coarser due to resonance where the bridge vibration includes significant vehicle vibration in the same frequency band. Accordingly, the calculated *EI* is inaccurate compared with the specified *EI*; refer to the result for $r = 0.9$ in Figure 7. An important point follows from this discussion, namely, if the natural frequency of the test vehicle is close to the natural frequency of the bridge, it is difficult to use the frequency-domain method for damage identification, because the collected data include a similar frequency signal for the test vehicle and the bridge. Therefore, for the ratio $r \leq 0.7$ or $r \geq 1.4$, the identification of *EI* is suitable for damage identification.

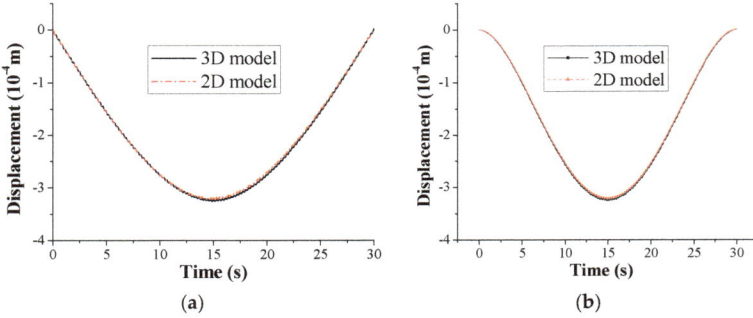

Figure 6. Vertical displacement response of (**a**) bridge midpoint and (**b**) test vehicle.

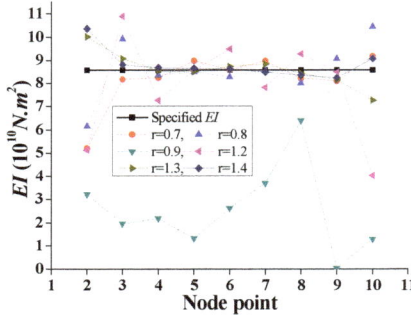

Figure 7. Simulation results of the calculated *EI* for elements along the bridge for different *r* values.

Another considered factor is the constant speed of the test vehicle, v. The calculated *EI* using the frequency-domain method is shown in Figure 8 for speeds ranging from 2 m/s to 7 m/s. It is shown that the calculated *EI* should be based on the vehicle speed not exceeding 6 m/s to obtain suitable stiffness identification, not including boundary elements, i.e., nodes 2 and 10.

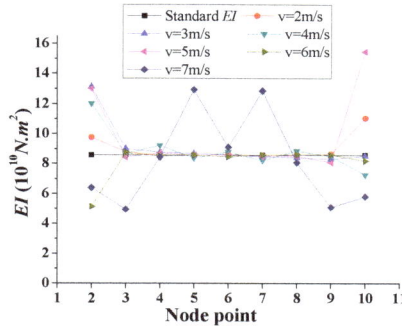

Figure 8. Simulation results of calculated *EI* with different test vehicle speeds.

4. Considered Scenarios in the Numerical Study

Based on the previous section, the following values are considered for the main parameters of the test vehicle in this section: mass m_v = 500 kg, stiffness k_v = 90 kg/m, and v = 3 m/s. The finite element (FE) model of the bridge in Figure 5 is adopted with a time step of 0.01 s to demonstrate the sensitivity of the detection methods. The acceleration and displacement responses that are numerically generated for the test vehicle during its passage over the bridge are processed using the procedures outlined above to identify the bridge bending stiffness (*EI*). The considered damage scenarios are listed in Table 1 and identified by D, with a subscript indicating the damaged element number. The severity of the damage is denoted by the percentage of reduction from the original (undamaged) sectional bending stiffness.

Table 1. Description of damage scenarios in the numerical simulations.

Scenario	Group	Damaged Element(s)	Reduction in Element Stiffness (%)				
1	D_6	6	$D_6 = 0$	$D_6 = 10$	$D_6 = 20$	$D_6 = 30$	$D_6 = 40$
2	D_3, D_6	3 & 6	$D_3 = 10$ $D_6 = 30$	$D_3 = 20$ $D_6 = 30$	$D_3 = 30$ $D_6 = 30$	$D_3 = 30$ $D_6 = 40$	$D_3 = 20$ $D_6 = 40$
3	D_1, D_{10}	1 & 10	$D_1 = 10$ $D_{10} = 10$	$D_3 = 20$ $D_6 = 20$	$D_3 = 20$ $D_6 = 30$	$D_3 = 20$ $D_6 = 40$	$D_3 = 30$ $D_6 = 40$

4.1. Frequency-Domain Method

The fundamental frequency and corresponding mode shape are accurate and convenient regarding extraction from the acceleration responses of the test vehicle [25]. When the improved DSC method [8–10] is used, only the measurements in one mode are sufficient to identify the damage. Using the indirect identification technique, only the fundamental frequency and the corresponding mode shape are used to calculate the bending stiffness at the nodes herein.

To demonstrate the damage detection for a single damage location using the frequency-domain method, it is assumed that the bridge in Figure 9 experienced damage Scenario 1 (Table 1). From Figure 9, it is clear that the stiffness values at nodes 6 and 7, corresponding to D_6, are the lowest. Figure 10 shows the *EI* variation ratio, i.e., stiffness degradation level, at the different nodes. For the undamaged state, the corresponding stiffness values at nodes can be calculated from the indirect identification technique of from the original design documents. Figure 10a,b shows the variation ratio considering the undamaged *EI* similar to the on-site test immediately after a newly constructed bridge using the indirect identification technique, and the specified undamaged *EI* similar to the original design document, respectively. It is indicated that the *EI* variation ratios at nodes 6 and 7 are close to the true values, i.e., the mean values of the damage percentages of the adjacent elements $(D_6 + D_5)/2$ for node 6 and $(D_6 + D_7)/2$ for node 7. Thus, the distribution of *EI* and the corresponding variation ratio along the bridge are satisfactory for detecting the damage location and severity, except in boundary elements 1 and 10.

Figure 11 shows the calculated *EI* in the case of multiple damage locations (Scenario 2 of Table 1) corresponding to the damaged element 3 (nodes 3 and 4) and element 6 (nodes 6 and 7). Moreover, Figure 12a,b indicates the variation ratio with respect to the calculated and specified undamaged *EI*, respectively. The simulation results indicate that the magnitudes of the variation ratio increase with the increase of damage severity.

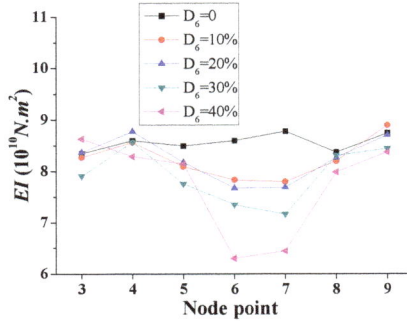

Figure 9. Simulation results of calculated *EI* for a single damage (Scenario 1).

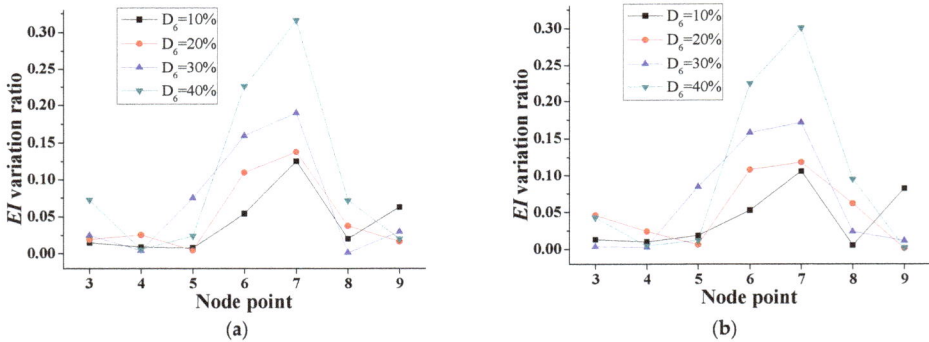

Figure 10. *EI* variation ratios for a single damage (Scenario 1). (**a**) Ratio with respect to calculated undamaged *EI*. (**b**) Ratio with respect to specified undamaged *EI*.

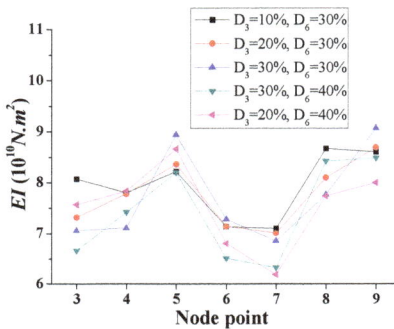

Figure 11. Simulation results of calculated *EI* for double interior damages (Scenario 2).

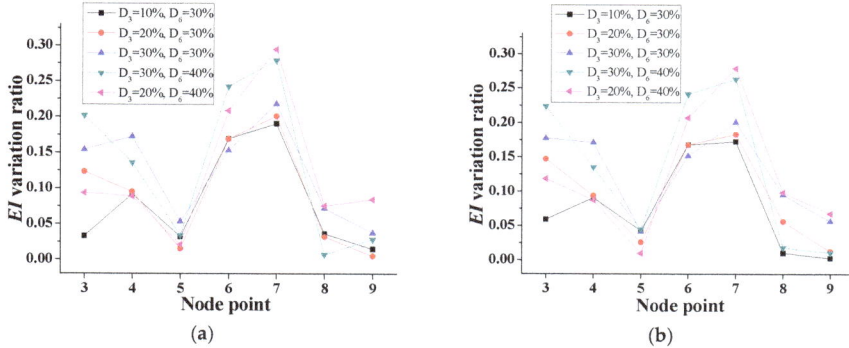

Figure 12. *EI* variation ratio for double interior damages (Scenario 2). (**a**) Ratio of calculated to undamaged *EI*. (**b**) Ratio of specified to undamaged *EI*.

Similar observations can be made for scenarios 1 and 2, as shown in Table 1. Element 6 is damaged in both scenarios, with only one damaged element in Scenario 1, and two damaged elements in Scenario 2. The calculated *EI* and corresponding variation ratio exhibit almost no change for element 6 in these two scenarios with the same assumed damaged case. This important feature of the ability to detect damage locations and severity without the influence of damage of other locations is essential for practical applications of structural damage identification.

4.2. Time-Domain Method

Using Equation (21), the fourth-order moment vectors of the "measured" displacement response of the test vehicle can be estimated for the previously discussed bridge model and test vehicle parameters considering different damage scenarios. Thus, the consequent *EI* of each element, which may differ from the calculated *EI* at each node using the frequency-domain method, can be calculated. The identified *EI* of each element is represented in scenarios, as shown in Figures 13–15.

According to the time-domain method, the distribution of stiffness is determined for each element as shown in Figure 13, showing the lowest stiffness for the damaged element 6. The computed ratios of reduction in stiffness for this element of 0.093, 0.194, 0.295, and 0.396 are very close to the damaged cases, i.e., D_6 = 10%, 20%, 30%, and 40%, respectively. On the other hand, the variations of stiffness in undamaged elements of Figure 13 are very small, below 5%.

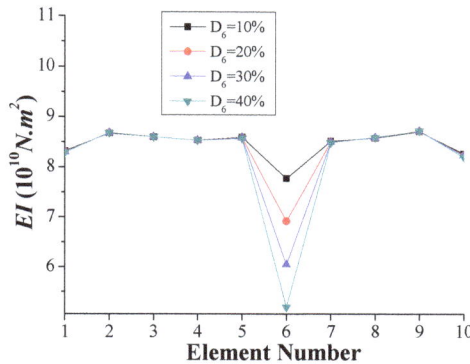

Figure 13. Simulation results of calculated *EI* for a single damage (Scenario 1).

When two damaged inner elements, 3 and 6, are simulated (Scenario 2), the identified stiffness at each element, as shown in Figure 14, is accurate compared to the specified distribution of stiffness. The differences of the identified stiffness values and those of the specified ones for the damaged and undamaged elements are below 1% and 5%, respectively. As stated previously, the boundary elements cannot be identified properly using the frequency-domain method. However, the time-domain method permits the *EI* of the boundary elements to be accurately calculated, as shown in Figure 15 for the damaged stiffness of boundary elements 1 and 10. The variations of the stiffness between the identified stiffness values and the specified ones are below 1%.

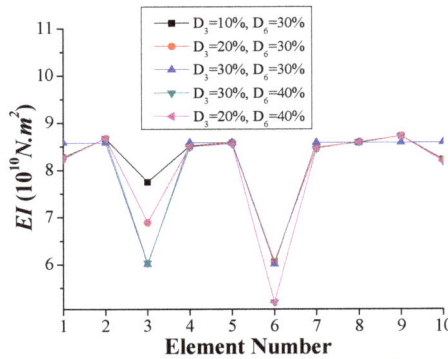

Figure 14. Simulation results of calculated *EI* for two interior damages (Scenario 2).

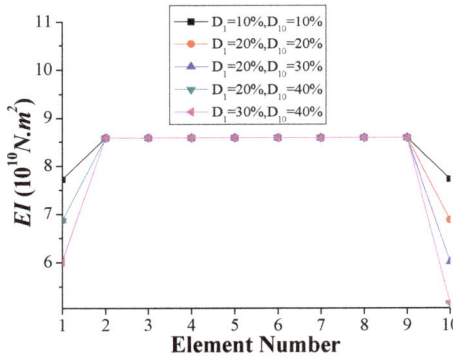

Figure 15. Simulation results of calculated *EI* for boundary damages (Scenario 3).

Based on the above results, the time-domain method can be used for boundary damage identification where the frequency-domain method cannot. However, the frequency-domain method is more efficiently computationally compared to the time-domain method, which requires solving an optimization problem. For the discussed bridge simulation, calculating *EI* using the time-domain method starting with specified undamaged stiffness for the initial values at each element may require extensive computations. However, if the initial stiffness values are identified from the application of the frequency-domain method, the computing time that is required to solve the optimization problem by applying the time-domain method can be significantly reduced.

5. Measurement Error Study

For a reliable damage detection method, a significant challenge is posed by environmental noise in practical applications, e.g., thermal conditions or the effects of the roughness of the road surface.

In this study, it is assumed that the influence of the environmental effects on the dynamic response of the passing test vehicle is represented by *white noise*. The displacement response of the test vehicle with random noise is expressed as follows [28,29]:

$$y_m = y_{calculated} + E_p P \sigma(y_{calculated}) \tag{25}$$

where $y_{calculated}$ is the calculated displacement and acceleration responses of the test vehicle from the FE model, E_p is the noise level, P is an independent random variable of Gaussian distribution with zero mean and unit standard deviation, and $\sigma(y_{calculated})$ is the specified standard deviations of the calculated displacement responses of the test vehicle.

5.1. Considering Noise in the Frequency-Domain Method

To numerically demonstrate the sensitivity of the identified bending stiffness using the frequency-domain method, it is assumed that the discussed bridge simulation experienced the scenarios that are summarized in Table 1, but with a consideration of different noise levels, as shown in Figures 16–18. Using Equation (25), the simulations with added Gaussian random white noise for each level of the bridge are repeated 10 times in order to reduce the effects of the random errors (similar to 10 *in situ* measurements). The average of the noisy data is used for the subsequent damage detection to estimate the bending stiffness at nodes by the frequency-domain method. Figure 16 shows the identified stiffness values of the undamaged case from the calculated results of the 10 random realizations corresponding to each considered noise level. As expected, the bending stiffness identified from the first mode shape with comparatively low noise level is more accurate than that with higher noise level. This indicates that the variation ratios between the identified and the specified *EI* values are below 3%, even if noisy data are used with up to a 20% noise level. Moreover, the stiffness can be reasonably identified, even for up to 30% noise levels, except for node 3, where the errors are over 20%.

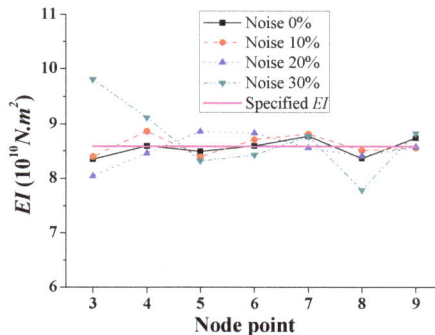

Figure 16. Simulation results of calculated *EI* with different noise levels.

To evaluate the proposed approach for calculating the bending stiffness under different damage scenarios (similar to the above case of the undamaged bridge) with environmental noise, the numerical simulations considering the artificial noise prescribed with different levels are performed under these damaged scenarios. Figure 17 shows the calculated *EI* from the data with the noise at different levels for damaged element 6 with a 10% and 40% reduction in stiffness. From Figure 17a, the results of the damage identification at a 10% noise level are quite satisfactory, and the lowest *EI* values are observed at nodes 6 and 7 (the end nodes of damaged element 6), identifying the damaged location accurately. At the higher noise levels, the proposed approach also revealed the stiffness reduction at node 6. However, the *EI* values at the boundary nodes are even lower than those in the specified damaged regions at the noise level of 30%, which would pose an inevitable impediment for the efficient damage identification. Furthermore, when the bending stiffness of element 6 is presumed to have a 40% reduction in stiffness, all of the results at different noise levels meet the requirements of determining the damage location based on the lowest identified *EI* values, as shown in Figure 17b, where the lower *EI* values at nodes 6 and 7 are apparent. Moreover, as expected, due to the increased noise, larger random errors in the calculated *EI* are observed compared with the specified *EI*. For the scenario of high environmental noise, the curvature that fluctuates κ in Equation 17 would fluctuate much more due to the amplification of differential effects with the $d^2\varphi/dx^2$ [8–10]. Accordingly, the calculated *EI* based on the classical beam theory, Equation (17), may lead to an unreasonable stiffness value, especially in the vicinity of damaged element(s) with the highest chance of a much higher curvature variation [8–10], and near support element(s) with the near-zero curvature [8–10]. Due to the limited number of repeated numerical simulations, the results are non-ergodic [30], and the averaging techniques can hardly eliminate the interference effects of the Gaussian random noise. Therefore, the "real" signal becomes largely contaminated, resulting in increased or decreased values of the measured data at some positions. Fortunately, these results indicate that the environmental noise would exert smaller influences on the damage identification results when the bending stiffness of the damaged element is significantly reduced.

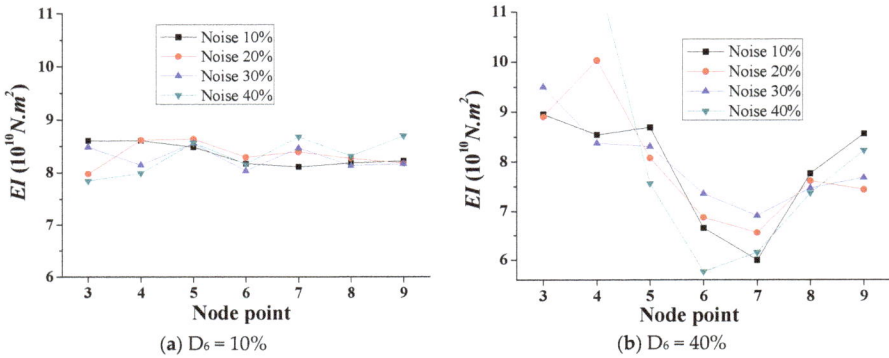

Figure 17. Simulation results of calculated *EI* with different noise levels (Scenario 1).

Figure 18 shows the calculated *EI* with different noise levels for double-interior damages (D_3 = 20% and D_6 = 30% versus D_3 = 30% and D_6 = 40%). The reduced stiffness of the two damaged elements are generally identified for different noise levels. In addition, the results demonstrate the higher damage of element 6 for most of the noise scenarios. However, the calculated results for the 40% noise level that is shown in Figure 18b illustrate the lower *EI* values at nodes 3 and 4 compared with those at nodes 6 and 7. Therefore, the high environment noise levels would deteriorate the efficiency and accuracy of the proposed frequency domain approach to some extent, and make the damage identification more complex and less reliable.

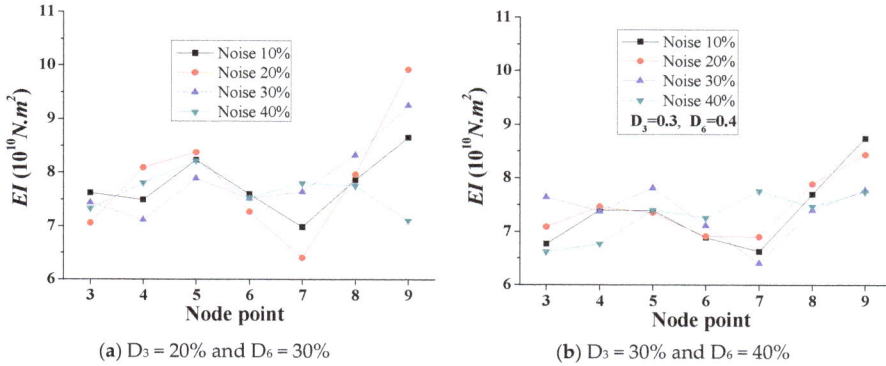

Figure 18. Simulation results of calculated *EI* with different noise levels (Scenario 2).

5.2. Considering Noise in the Time-Domain Method

This section presents the effects of the environmental noise on damage identification using the time-domain method. Besides the single and double-element damage scenarios, the scenario of damaged boundary elements is also included. As discussed in the previous section, the calculation following Equation (25) is repeated 10 times, and the averaged results are attained for the analysis. Figure 19 shows the calculated *EI* at different noise levels for the above-mentioned damage scenarios using the time-domain method. It is observed that the differences between the *EI* values of the damaged element(s) are negligible for the different noise levels. Therefore, the environmental noise has insignificant effects on the calculated stiffness when using the time-domain method. In addition, from Figure 19a,b, the damage of element 6 is clear, and the higher reduction of the bending stiffness is accurately reproduced. However, the undamaged boundary elements 1 and 10 can be mistaken for "damaged" elements due to the calculated stiffness reduction at high noise levels. This may lead to unnecessary inspection fieldwork in the case of slightly damaged elements, as shown in Figure 19a. For the scenario of double-element damage, Figure 19c, d reveals the accurate damage locations and damage severity. The effects of the environmental noise and the interference between the two damaged elements are negligible. Unlike the frequency-domain approach, the damages at the boundary elements are well recognized by the time-domain method, as shown in Figure 19e,f. Consequently, the time-domain method is advantageous, with higher accuracy, robustness, and reliability than the frequency-domain method.

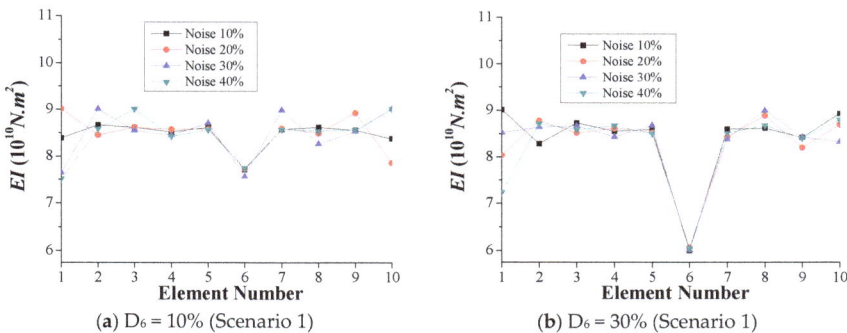

Figure 19. *Cont.*

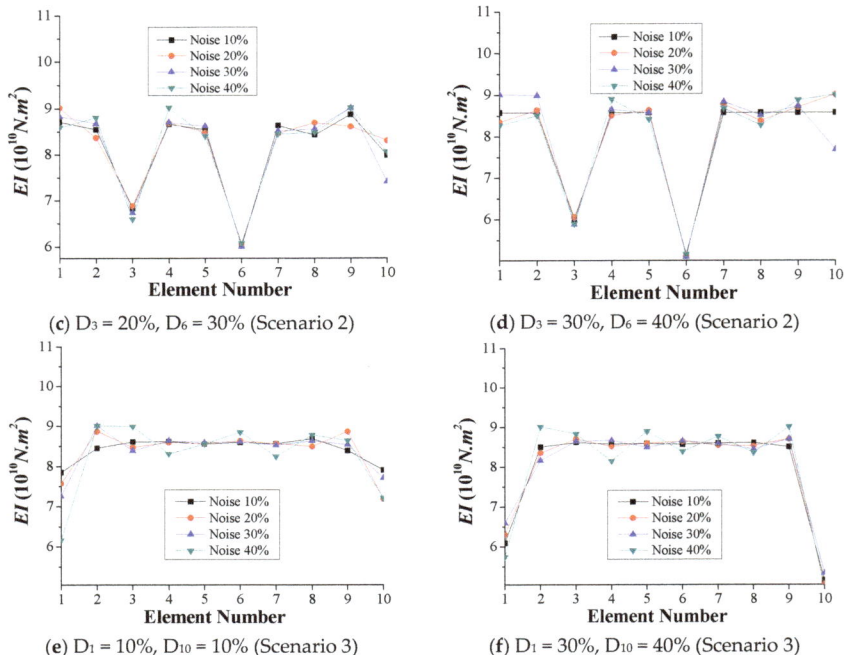

Figure 19. Simulation results of calculated *EI* with different noise levels.

6. Field Test Study

The Hongxing bridge, located in Fuling District of Chongqing City, is a simply supported three-spanned bridge with each span's length at 20 m, as shown in Figure 20a. The cross-sectional moment of inertia is 0.38 m^4, and the elastic modulus is 3.0×10^{10} N/m^2. The bridge was recently built in 2018, and has not been officially open to the public. Therefore, it had little traffic flow, and noise interference was relatively weak. According to field investigations, the road roughness is shown in Figure 20b; it is suitable for the actual experimental study of the indirect measurement technique.

Considering the second span as the test beam bridge, researchers kept the speed of the test vehicle–car (tractor) system at 1 m/s, as shown in Figure 21. According to the test vehicle going across the test beam bridge, the acceleration response of the test vehicle with an acceleration sensor installed in the center of the test vehicle could be recorded. For reducing the effect of the surface road surface, two different weights of the test vehicle, namely a big vehicle (1100 kg) and small vehicle (1050 kg) with the same vehicle frequency, could pass the test beam bridge, respectively. The difference between the responses of the two test vehicles could be regarded as the initial acceleration response signals. Researchers let the big vehicle and small vehicle pass the second span of the bridge three times, respectively, and then were able to use displacement response-measuring technology based on the double integral of the recorded acceleration response with zero initial conditions at each time. Therefore, there are three displacement responses each for the big vehicle and small vehicle. After averaging the displacement responses of the big vehicle and the small vehicle for reducing random noise, which is shown in Figure 22, the difference between the displacement response of the big vehicle and the small vehicle can be regarded as the initial displacement response signal for the analysis procedure of the time-domain method in Section 2.2. The corresponding acceleration response calculated by the initial displacement response differential twice can be regarded as the initial acceleration signals for an analysis procedure of the frequency-domain method in Section 2.1.

(a)   (b)

Figure 20. Hong Xing Bridge of Chongqing City.

Figure 21. Field tests on site.

Based on the analysis procedure of the frequency-domain method in Section 2.1 and time-domain method in Section 2.2, considering the same element and node numbers shown in Figure 5, the identified stiffness *EI* at the element nodal points calculated by the frequency-domain method is shown in Figure 23. Compared to the original *EI*, the maximum relative error in the identified stiffness *EI* occurs in the element node point number 9 with a value of approximately 16%; however, the identified results are acceptable within an engineering acceptance range.

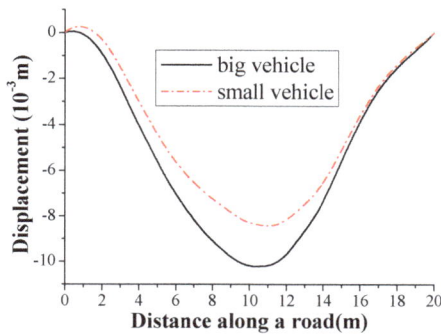

Figure 22. The average of three displacement responses on site.

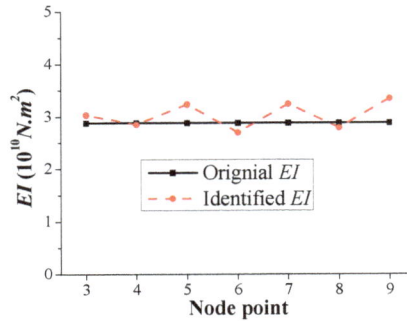

Figure 23. Identified *EI* results calculated by the frequency-domain method.

It can also be seen that the identified stiffness *EI* at the element number calculated by the time-domain method is shown in Figure 24, which is obviously better than the results of Figure 23. Compared with the original *EI*, the maximum relative error in the identified stiffness *EI* occurs in element number 10 with a value of approximately 5%, and the remaining *EI* results are all below 1%. It indicated again that the time-domain method is advantageous with higher accuracy, robustness, and reliability than the frequency domain method.

This is a preliminary verification for an indirect measurement technique. It is noted that the applicability of the frequency-domain method and time-domain method to practical bridges with recorded data from field tests should be further promoted, and details will be presented in future publications.

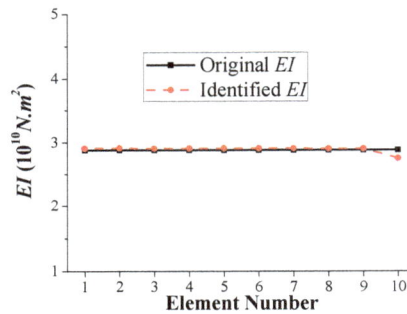

Figure 24. Identified *EI* results calculated by time-domain method.

7. Results Discussion

In the simulation numerical, the results indicated the ability to detect damage locations and severity without the influence of damage at other locations in single and double-damage location(s). Based on the sensitivity of different environment noise levels to damage identification, it is noted that higher environment noise would deteriorate the efficiency and accuracy of the proposed frequency-domain approach to some extent, and make the damage identification more complex and less reliable. However, the time-domain method is advantageous with higher accuracy, robustness, and reliability than the frequency-domain method. A filtering method to reduce the measurement noise should be studied in the future.

In the field test, performing multiple passes on the bridge and then averaging the signals was considered to reduce the measurement noise. Two different weight test vehicles with the same vehicle frequency were used for reducing the effect of road surface roughness. The results show that the identification can be accepted as an engineering requirement. More technologies and an updated designed test vehicle to eliminate these disturbance factors will be promoted in future work.

8. Conclusions

This study presents an indirect approach for identifying the structural damage of a bridge from a passing test vehicle. Both the frequency-domain and time-domain methods have been embedded into the proposed indirect approach, which is numerically examined in the single, double, and boundary-damage scenarios considering different noise levels. During the passage of the test vehicle over a bridge, the fundamental frequency and corresponding mode shape of the bridge can be extracted from the field measurements recorded by the vibration sensors mounted on the test vehicle. Subsequently, the stiffness for structural damage identification can be calculated from the improved direct stiffness calculation technique, which is referred to as the frequency-domain method. For the displacement response measured by the test vehicle, or the twice integration of the acceleration response, the fourth-order moment vectors can be calculated from the statistical moment-based damage detection method combined into the indirect approach, which is referred to as the time-domain method. Through a numerical case study, the main conclusions are as follows:

1. The proposed indirect approach, including the frequency-domain and time-domain methods, requires no parametric inputs, which is more general compared to other structural damage identification, e.g., wavelet-based methods. Therefore, the proposed approach can be directly adopted for the structural damage identification of in-service bridge structures without additional and cumbersome calibration.

2. The frequency-domain method is advantageous with its high cost efficiency, since it can estimate the initial stiffness of the bridge based on the first mode of vibration, and is sufficient for identifying damage location(s) apart from the end regions of the bridge. However, this method requires that the speed of the passing vehicle should be lower than 6 m/s during the measurement, and it is not applicable for damage identification in the boundary nodes.

3. Although the time-domain method is computationally intensive due to the additional optimization steps, it has the advantages of high accuracy, reliability, and robustness, and is feasible for use especially in the end regions of the bridge, which is suitable for identifying damage location(s) and damage severities.

4. The field test study shows that the identified results errors from using the frequency-domain method and time-domain method are below 16% and 5% respectively; this indicates that the two methods are useful for assessment bending stiffness with a practical bridge. Moreover, it indicated that the time-domain method is advantageous with higher accuracy, robustness, and reliability as compared to the frequency-domain method.

5. In the practical assessment of the bridge health conditions, the frequency-domain approach is suitable in the preliminary phase to estimate the initial damage conditions of the bridge on site. Subsequently, in the final phase of the investigation, the time-domain approach can provide more detailed and comprehensive results with high accuracy and reliability.

Since the conclusions are drawn from the analytical analysis, numerical simulations, and initial field-test verification, as in the practical applications of the proposed damage detection approach for damage identification based on old damaged bridge test, are not included in this paper, and will be presented in future publications. It is noted that the modified bending stiffness results by using the frequency-domain method, especially for the end regions of the bridge, and high measurement noise should be further promoted, and will be presented in future publications. Finally, future research should focus on developing techniques and equipment for designing a test vehicle and considering it without road closure, and more corresponding parameters with a vehicle–bridge couple system associated with practical challenges should be studied.

Sensors **2018**, *18*, 4035

Author Contributions: Y.Y. took part in the entire researching process. Y.Z. and L.L.W. contributed to the experiment design. B.Y.J. held on the simulation process. R.J. revised the manuscript.

Funding: This study was supported in part by the Fundamental Research Fund for the Central University (Project No. 2018CDQYTM0044), National Natural Science Foundation of China (Grant No. 51778090), the 111 Project of China (Grants No. B18062), and the National Natural Science Foundation of China (Grants No. 51708062).

Conflicts of Interest: The authors declare no conflict of interest.

References

1. Hou, L.Q.; Zhao, X.F.; Ou, J.P.; Liu, C.C. A Review of Nondeterministic Methods for Structural Damage Diagnosis. *J. Vib. Shock* **2014**, *33*, 50–58.

2. Amezquita-Sanchez, J.P.; Adele, H. Signal Processing Techniques for Vibration-Based Health Monitoring of Smart Structures. *Arch. Comput. Methods Eng.* **2016**, *23*, 1–15. [CrossRef]

3. Maeck, J. Damage Assessment of Civil Engineering Structure by Vibration Monitoring. Ph.D. Thesis, Department of Civil Engineering, Katholieke Universiteit Leuven, Leuven, Belgium, 2003.

4. Huth, O.; Feltrin, G.; Maeck, J.; Kilic, N.; Motavalli, M. Damage Identification Using Modal Data: Experiences on a Prestressed Concrete Bridge. *ASCE J. Struct. Eng.* **2005**, *131*, 1898–1910. [CrossRef]

5. Xu, Y.L.; Zhang, J.; Li, J.C. Experimental Investigation on Statistical Moment-Based Structural Damage Detection Method. *J. Struct. Health Monit.* **2009**, *8*, 555–571. [CrossRef]

6. Xu, Y.L.; Zhang, J.; Li, J.; Wang, X.M. Stochastic Damage Detection Method for Building Structures with Parametric Uncertainties. *J. Sound Vib.* **2011**, *330*, 4725–4737. [CrossRef]

7. Li, W.M.; Jiang, Z.H.; Wang, T.L.; Zhu, H.P. Optimization Method Based on Generalized Pattern Search Algorithm to Identify Bridge Parameters Indirectly by a Passing Vehicle. *J. Sound Vib.* **2014**, *333*, 364–380. [CrossRef]

8. Yang, Y.; Mosalam, K.M.; Liu, H.; Huang, X.N. An Improved Direct Stiffness Calculation Method for Damage Detection of Beam Structures. *Struct. Control Health Monit.* **2013**, *20*, 835–851. [CrossRef]

9. Yang, Y.; Mosalam, K.M.; Liu, G.; Wang, X.L. Damage Detection Using Improved Direct Stiffness Calculations-A Case Study. *Int. J. Struct. Stab. Dyn.* **2016**, *16*, 1640002. [CrossRef]

10. Yang, Y.; Yang, Y.B.; Chen, Z.X. Seismic Damage Assessment of RC Structures under Shaking Table Tests Using The Modified Direct Stiffness Calculation Method. *Eng. Struct.* **2017**, *131*, 574–585. [CrossRef]

11. Blachowski, B.; An, Y.H.; Spencer, B.F.; Ou, J.P. Axial Strain Accelerations Approach for Damage Localization in Statically Determinate Truss Structures. *Comput.-Aided Civ. Infrastruct. Eng.* **2017**, *32*, 304–314. [CrossRef]

12. Kim, C.W.; Chang, K.C.; Kitauchi, S.; McGetrick, P.J. A Field Experiment on a Steel Gerber-truss Bridge for Damage Detection Utilizing Vehicle-induced Vibrations. *Struct. Health Monit.* **2016**, *15*, 421–429. [CrossRef]

13. Sevillano, E.; Sun, R.; Perera, R. Damage Evaluation of Structures with Uncertain Parameters via Interval Analysis and FE Model Updating Methods. *Struct. Control Health Monit.* **2016**, *24*, 1–22. [CrossRef]

14. Yang, Y.; Li, J.L.; Zhou, C.H.; Law, S.S.; Lv, L. Damage Detection of Structures with Parametric Uncertainties Based on Fusion of Statistical Moments. *J. Sound Vib.* **2018**. [CrossRef]

15. Mao, J.X.; Wang, H.; Feng, D.M.; Tao, T.Y.; Zheng, W.Z. Investigation of Dynamic Properties of Long-span Cable-stayed Bridges based on One-year Monitoring Data under Normal Operating Condition. *Struct. Control Health Monit.* **2018**, *25*, 1–19. [CrossRef]

16. Wang, H.; Mao, J.X.; Huang, J.H.; Li, A.Q. Modal Identification of Sutong Cable-Stayed Bridge during Typhoon Haikui Using Wavelet Transform Method. *J. Perform. Constr. Facil.* **2016**, *30*, 04016001-1-11. [CrossRef]

17. Yang, Y.B.; Lin, C.W.; Yan, J.D. Extracting Bridge Frequencies from the Dynamic Response of a Passing Vehicle. *J. Sound Vib.* **2004**, *272*, 471–493. [CrossRef]

18. Yang, Y.B.; Lin, C.W. Vehicle-Bridge Interaction Dynamics and Potential Applications. *J. Sound Vib.* **2005**, *284*, 205–226. [CrossRef]

19. Yang, Y.B.; Lin, C.W. Use of a Passing Vehicle to Scan the Fundamental Bridge Frequencies: An Experimental Verification. *Eng. Struct.* **2005**, *27*, 1865–1878.

20. Yang, Y.B.; Chang, K.C. Extraction of Bridge Frequencies from the Dynamic Response of a Passing Vehicle Enhanced by the EMD Technique. *J. Sound Vib.* **2009**, *322*, 718–739. [CrossRef]

21. Feng, D.M.; Feng, M.Q. Output-Only Damage Detection Using Vehicle-induced Displacement Response and Mode Shape Curvature Index. *Struct. Control Health Monit.* **2016**, *23*, 1088–1107. [CrossRef]

22. Obrien, E.J.; Keenahan, J. Drive-by Damage Detection in Bridges Using the Apparent Profile. *Struct. Control Health Monit.* **2015**, *22*, 813–825. [CrossRef]

23. Behroozinia, P.; Khaleghian, S.; Taheri, S.; Mirzaeifar, R. Damage Diagnosis in Intelligent Tires Using Time-domain and Frequency-domain Analysis. *Mech. Based Des. Struct. Mach.* **2018**. [CrossRef]

24. McGetrick, P.J.; Gonazalez, A.; O'Brien, E.J. Theoretical Investigation of the Use of a Moving Vehicle to Identify Bridge Dynamic Parameters. *Non-Destruct. Test. Cond. Monit.* **2009**, *51*, 433–438. [CrossRef]

25. Yang, Y.B.; Li, Y.C.; Chang, K.C. Constructing the Mode Shapes of a Bridge from a Passing Vehicle: A Theoretical Study. *Smart Struct. Syst.* **2014**, *13*, 797–819. [CrossRef]

26. Vujanovic, B. Conservation Laws of Dynamical Systems via D'Alembert's Principle. *Int. J. Non-Linear Mech.* **1978**, *13*, 185–197. [CrossRef]

27. Clough, R.W.; Penzien, J. *Dynamics of Structures*; McGraw-Hill: New York, NY, USA, 1993.

28. Bu, J.Q.; Law, S.S.; Zhu, X.Q. Innovative Bridge Condition Assessment from Dynamic Response of a Passing Vehicle. *ASCE J. Eng. Mech.* **2006**, *132*, 1372–1379. [CrossRef]

29. Xiang, Z.H.; Dai, X.W.; Zhang, Y.; Lu, Q.H. The Tap-Scan Method for Damage Detection of Bridge Structures. *Interact. Multiscale Mech.* **2010**, *3*, 173–191. [CrossRef]

30. Kiureghian, A.D. Non-ergodicity and PEER's framework formula. *Earthq. Eng. Struct. Dyn.* **2005**, *34*, 1643–1652. [CrossRef]

![sensors logo] *sensors*

MDPI

Article

Principal Component Analysis Method with Space and Time Windows for Damage Detection

Ge Zhang, Liqun Tang, Licheng Zhou *, Zejia Liu *, Yiping Liu and Zhenyu Jiang

School of Civil Engineering and Transportation, State Key Laboratory of Subtropical Building Science, South China University of Technology, Guangzhou 510640, China; zhangge13756010981@163.com (G.Z.); lqtang@scut.edu.cn (L.T.); tcypliu@scut.edu.cn (Y.L.); zhenyujiang@scut.edu.cn (Z.J.)
* Correspondence: ctlczhou@scut.edu.cn (L.Z.); zjliu@scut.edu.cn (Z.L.); Tel.: +86-20-87111030-3304 (L.Z.)

Received: 8 April 2019; Accepted: 29 May 2019; Published: 2 June 2019

✔ check for updates

Abstract: Long-term structural health monitoring (SHM) has become an important tool to ensure the safety of infrastructures. However, determining methods to extract valuable information from large amounts of data from SHM systems for effective identification of damage still remains a major challenge. This paper provides a novel effective method for structural damage detection by introduction of space and time windows in the traditional principal component analysis (PCA) technique. Numerical results with a planar beam model demonstrate that, due to the presence of space and time windows, the proposed double-window PCA method (DWPCA) has a higher sensitivity for damage identification than the previous method moving PCA (MPCA), which combines only time windows with PCA. Further studies indicate that the developed approach, as compared to the MPCA method, has a higher resolution in localizing damage by space windows and also in quantitative evaluation of damage severity. Finally, a finite-element model of a practical bridge is used to prove that the proposed DWPCA method has greater sensitivity for damage detection than traditional methods and potential for applications in practical engineering.

Keywords: principal component analysis; space window; time window; damage detection

1. Introduction

The safety of infrastructures such as bridges and high-rise buildings is of the utmost concern to the public. During operation, civil structures are subjected to various kinds of external loads, such as traffic, wind, temperature, etc. In fact, these evolving loads may be much more complicated than those considered in the design phase. Therefore, it is of importance to monitor the structural responses, such as strain, displacement, and acceleration with the aim of assessment of their real-time states. Nowadays, long-term structural health monitoring (SHM) systems are widely used to acquire data of structural responses, as well as external loads to monitor the states of civil structures. However, how to process and analyze these data for identifying possible structural changes has been a great challenge [1,2]. In general, structural responses may not change evidently when only a small amount of damage is imparted. Moreover, response variations may be masked by the uncertainties in the structural parameters of practical structures, as well as by the presence of noise. All of these factors result in the raw data being uninformative regarding the occurrence of structural changes, therefore, resulting in the need for feature extraction of measurement data [3]. In order to detect damage effectively, the extracted features are required to be sensitive to damage, while insensitive to parametric uncertainties or noise.

Damage detection methods can be generally classified into two categories, namely model-based [4] and model-free methods [5]. Model-based methods require an accurate finite-element model as well as a model-updating process for damage identification [6]. They have the ability not only to identify the presence and location of damage but also to quantify it in meaningful engineering units. However, computational complexity and model updating of these methods, especially for large-scale structures, have been a challenge in SHM [5]. As an alternative, model-free methods have drawn much attention for the sake that they have been demonstrated applicable to damage identification [7]. These methods utilize time series of measurement data for analysis without the need for geometrical and material information. Due to this reason, they are more inexpensive and efficient compared to model-based methods.

During the past few decades, various kinds of model-free data-interpretation methods for damage detection have been developed, including the autoregressive (AR) model, autoregressive moving average (ARMA) model, autoregressive integrated moving average (ARIMA) model, correlation analysis (CA), instance-based method (IBM), wavelet-based (WB) methods, neural network (NN) model, robust regression algorithm (RRA), principal component analysis (PCA), etc. AR establishes a time-series model to predict future values based on the past measured data. Residual errors or AR parameters are usually used as sensitive features for damage detection [8,9]. ARMA and ARIMA, which are improved methods compared to AR, also take advantage of coefficients as indices for identifying damage [10,11]. CA detects damage through variations of correlation coefficients for measurement datasets since the correlation coefficients will change when damage occurs. This method has been demonstrated to have good performance with regard to identifying and localizing damage [12]. However, it fails to identify damage when the measurement noise is at high levels [13]. IBM computes the minimum distance of a cluster of sensor data (generally for three or four sensors) at each time step [14]. The occurrence of damage is determined if the phase of the minimum distance exceeds a threshold [15]. WB methods are also effective tools for on-line and off-line damage detection [16]. These methods firstly decompose original signals in different time domains and scales. Then, mode shapes, wavelet spectra, wavelet component energy, and the tendency of wavelet coefficients are selected as sensitive features to detect damage [17–20]. The NN model has been widely utilized to identify anomalous structural behavior by using static and dynamic responses [21,22]. The number of hidden layers, the number of neurons in each layer, the neuron activation function and error criteria should be carefully considered in the NN method [23]. Some investigators also verified that incorporating other methods into a traditional NN model significantly enhances the effectiveness of damage detection [24,25]. As for RRA, it is focused on the correlation between a pair of sensors and construction of a robust regression relationship for measurement data [26]. An anomaly is identified when correlation coefficients exceed threshold bounds. This method has demonstrated the ability to identify and localize damage in simple as well as complex structures [27].

PCA is another popular method used for damage identification in long-term SHM. It exhibits reliable and effective performance in modal analysis, reduced-order modelling, feature extraction, and structural damage detection [28–32]. In addition, it proves to be an effective tool to improve the training efficiency and enhance the classification accuracy for other machine learning algorithms, such as unsupervised learning methods [33–37]. Since the total historical dataset including responses of both healthy and damaged states is involved in the analysis process, PCA is not sensitive to the occurrence of damage in real time in SHM. Moreover, large amounts of historical data may cause computational complexity. Posenato et al. then proposed the moving PCA (MPCA) method to enhance discrimination features between undamaged and damaged structural responses [13,27,38]. This method essentially uses a sliding fixed-size time window for time-series data instead of handling the total historical dataset. An eigenvector time series will be obtained as the time window moves forward. The components of the most important eigenvector are utilized as sensitive features for damage detection. Due to the moving temporal window, MPCA enhances the detection effectiveness compared to that of the traditional PCA method through monitoring the evolution of eigenvector components between undamaged and damage states. In other words, MPCA is used to monitor the components of the eigenvector variance (CEVs) between a healthy state and damaged state for damage identification. It was demonstrated that the sensitivity of MPCA for damage identification was significantly improved compared to other methods such as ARIMA, CWT, RRA and IBM [12,13,39]. However, in the data-interpretation process of both PCA and MPCA, data from all sensors should be used to calculate the eigenvalues and eigenvectors. It makes sense that responses located far from the damaged area are insensitive to damage. In other words, part of the data includes information insensitive to damage, consequently reducing the sensitivity regarding damage detection. If a space window is applied to exclude the data from those sensors located far from the damage, it is possible to improve the damage detectability. As a consequence, if both space and time windows are applied in the traditional PCA method, this is expected to further improve the damage detectability. In fact, Posenato et al. have also proposed a sensor clustering overlapping algorithm for MPCA when there exists a large number of sensors [13]. The clustering process is essential to implement space windows for the installed sensors. However, the authors aimed to deal with measurements from fewer sensors for computational efficiency. They did not carry out further investigation on the damage detectability.

According to the above discussion, both PCA and MPCA methods use all sensors for analysis and may decrease the detection performance. This paper will provide a double-window PCA (DWPCA) method for structural damage identification. The primary idea is to combine space and time windows with the traditional PCA method. It is found that discrimination of the eigenvectors between damaged and healthy states is enhanced due to the introduction of space and time windows. Numerical results show that the proposed method, in contrast with MPCA, improves the sensitivity for damage identification and is also quicker to detect damage after its occurrence. Further investigations indicate that the novel approach exhibits a better performance regarding damage localization and quantitative evaluation. Finally, the proposed DWPCA is shown to be robust in the presence of noise and shows potential for applications in practical engineering.

This paper is organized as follows: Section 2 describes PCA, MPCA and the proposed DWPCA method. Section 3 presents a detailed description of the planar beam model for simulations, as well as the methodology to determine the space window. In Section 4, comparative studies with MPCA are conducted to verify the advantages of the proposed method. In Section 5, application of the proposed DWPCA to a full-scale structure is presented. In Section 6, valuable conclusions are drawn according to the numerical results.

2. The Proposed Double-Window Principal Component Analysis Method

In the following, the descriptions of PCA, MPCA, and the proposed DWPCA method will be presented in sequence. It should be noted that MPCA introduces a moving time window in the traditional PCA method, while the proposed DWPCA introduces both space and time windows.

2.1. PCA

PCA is a useful tool for reducing data dimensionality while retaining essential information for manipulated datasets. The main objective is to transform original data to a smaller set of uncorrelated variables [40]. For damage detection, PCA can be used to eliminate noise and simultaneously derive damage-sensitive features such as eigenvectors. The data-processing steps of PCA are detailed as below. The first step of PCA is the construction of a matrix, $\mathbf{U}(t)$, that contains the time histories of all measured data:

$$
\mathbf{U}(t) = \begin{bmatrix} u_1(t_1) & u_2(t_1) & \cdots & u_M(t_1) \\ u_1(t_2) & u_2(t_2) & \cdots & u_M(t_2) \\ \vdots & \vdots & \ddots & \vdots \\ u_1(t_N) & u_2(t_N) & \cdots & u_M(t_N) \end{bmatrix},
\tag{1}
$$

where t represents time, $u_i(i = 1, 2, \cdots, M)$ denotes the response from the i-th sensor installed in the monitored structure, M is the total sensor number, $t_j(j = 1, 2, \cdots, N)$ denotes the j-th time step of measurements, and N is the total number of time observations during monitoring. Note that the data of each column are the time series of measurement events from each individual sensor.

Subsequently, time series of each column or each sensor should be normalized by subtracting the mean value given by:

$$\bar{u}_i = \frac{1}{N}\sum_{j=1}^{N} u_i(t_j). \tag{2}$$

The normalized matrix can then be written as:

$$\mathbf{U}'(t) = \begin{bmatrix} u_1(t_1)-\bar{u}_1 & u_2(t_1)-\bar{u}_2 & \cdots & u_M(t_1)-\bar{u}_M \\ u_1(t_2)-\bar{u}_1 & u_2(t_2)-\bar{u}_2 & \cdots & u_M(t_2)-\bar{u}_M \\ \vdots & \vdots & \ddots & \vdots \\ u_1(t_N)-\bar{u}_1 & u_2(t_N)-\bar{u}_2 & \cdots & u_M(t_N)-\bar{u}_M \end{bmatrix}. \tag{3}$$

The next step is to construct the $M \times M$ covariance matrix, which is defined as:

$$\mathbf{C} = \frac{1}{M}\mathbf{U}'^{\mathrm{T}}\mathbf{U}'. \tag{4}$$

Finally, the eigenvalue λ_i and the corresponding eigenvector $\boldsymbol{\psi}_i$ of the covariance matrix can be obtained by solving the following equation:

$$(\mathbf{C}-\lambda_i\mathbf{I})\boldsymbol{\psi}_i = 0, \tag{5}$$

where $\mathbf{I}$ denotes the $M \times M$ identity matrix, $\boldsymbol{\psi}_i = \begin{bmatrix} \psi_{i,1} & \psi_{i,2} & \cdots & \psi_{i,M} \end{bmatrix}^{\mathrm{T}}$ in which $\psi_{i,j}(j = 1, 2, \cdots, M)$ is the component corresponding to the jth sensor.

Generally, one would sort the eigenvalues into decreasing order, namely $\lambda_1 > \lambda_2 > \cdots > \lambda_M$. Then, the first eigenvector $\boldsymbol{\psi}_1$ related to λ_1 contains the largest variance and thereby retains essential information for the original matrix U. In fact, most of the variance is contained in the first few principal components, while the remaining less important components involve the measurement of noise. For this reason, the first few eigenvectors are always used as sensitive features to detect and localize damage. It can be seen that neither a space window nor time window is applied in PCA. The total historical dataset including responses of healthy and damaged states is used for analysis, thereby leading to low damage detectability.

2.2. MPCA

MPCA is an improved method based on PCA which involves applying a moving time window of fixed size. Only the time series of observations inside the moving time window are used to construct the covariance matrix for the derivation of eigenvalues and eigenvectors. Previous studies have proven that the introduction of the moving time window enhances the discrimination between features of undamaged and damaged structures, and thereby renders better performance for damage detection [14,26]. Additionally, the sensitivity of MPCA for damage identification has proven to be significantly improved as compared with other methods such as PCA, ARIMA, DWT, RRA and IBM [12,13,39]. The proper choice of the window size T is also important in the first step. If the response time series have periodic characteristics, the temporal window size should be equivalent to the longest period. Once the time window size or the number of consecutive measurements for each sensor inside the window is fixed, the matrix $\mathbf{U}$ in Equation (1) at the k-th time step can be rewritten as:

$$\mathbf{U}(k) = \begin{bmatrix} u_1(t_k) & u_2(t_k) & \cdots & u_M(t_k) \\ u_1(t_{k+1}) & u_2(t_{k+1}) & \cdots & u_M(t_{k+1}) \\ \cdots & \cdots & \ddots & \cdots \\ u_1(t_{k+T-1}) & u_2(t_{k+T-1}) & \cdots & u_M(t_{k+T-1}) \end{bmatrix}, \tag{6}$$

where $k = 1, 2, \cdots, N - T + 1$. Note that the mean value of each column of $\mathbf{U}(k)$ at the k-th time step would become,

$$\overline{u}_i(k) = \frac{1}{T} \sum_{j=k}^{k+T-1} u_i(t_j). \tag{7}$$

Next, repeating the steps of PCA, one is able to obtain the eigenvalue $\lambda_i(k)$ and eigenvector $\psi_i(k)$. It should be noted that $\lambda_i(k)$ and $\psi_i(k)$ are time series and vary with the time step.

During continuous monitoring, responses are divided into two phases: training and monitoring phases. In the training phase, the structure is assumed to behave normally (no damage). Then, eigenvector variance between the training phase and the monitoring phase at the kth time step can be determined by the following equation:

$$\Delta\psi_i(k) = \psi_i(k) - \overline{\psi}_i, \tag{8}$$

where $\overline{\psi}_i$ denotes the mean value of the ith eigenvector in training phase, while $\Delta\psi_i(k)$ represents the eigenvector variance between $\psi_i(k)$ and $\overline{\psi}_i$ at the kth time step of the monitoring phase, and $\Delta\psi_i(k) = [\Delta\psi_{i,1}(k)\Delta\psi_{i,2}(k)\cdots\Delta\psi_{i,M}(k)]^T$ where $\Delta\psi_{i,j}(k)$ is the component of the eigenvector variance (CEV) corresponding to the jth sensor. It should be noted that $\Delta\psi_{i,j}(k)$ is generally utilized as the feature for anomaly detection in MPCA. CEV $\Delta\psi_{i,j}(k)$ by MPCA can be expressed in terms of eigenvector components as follows:

$$\Delta\psi_{i,j}(k) = \psi_{i,j}(k) - \overline{\psi}_{i,j}, \tag{9}$$

where $\overline{\psi}_{i,j}$ denotes the mean value of the eigenvector component in a healthy state, and $\Delta\psi_{i,j}(k)$ can be considered as the variation between $\psi_{i,j}(k)$ in monitoring phase and $\overline{\psi}_{i,j}$ in healthy state. If $\Delta\psi_{i,j}(k)$ exceeds a threshold, alarm will be flagged.

When damage occurs, structural responses may change, consequently causing variations in eigenvectors and CEVs. Thus, one may follow the $\Delta\psi_{i,j}(k)$ over time to examine whether damage exists. As indicated in Equation (6), MPCA uses only the latest T observations instead of the whole time series. Once damage occurs, fewer data that are irrelevant to the damage, as compared with PCA, are considered for the calculation, resulting in a better sensitivity for damage detection. However, MPCA generally takes into account responses from all sensors. Some of the sensors may be insensitive to damage located at a certain position. If a space window is used to group sensitive sensors, it is possible to enhance damage detectability.

2.3. The Proposed DWPCA

When damage occurs, data from sensors close to damage location change significantly while data from sensors away from damage may be unchanged. Hence, a novel DWPCA method is proposed herein to combine space and time windows with PCA. It can also be treated as an improved method for MPCA by the introduction of a space window. The application of the space window, in the aim of enhancing damage detectability, is to group sensors sensitive to damage and to exclude those that are insensitive. The key step for the choice of the space window is to determine the damage-sensitive area (DSA), where measurement data change significantly when damage occurs. It should be noted that the DSA varies with damage location as well as damage level. For example, the damage location commonly decides the position of the DSA and a high damage level reasonably leads to a large DSA.

For the space window, a criterion as shown below is used for determination of the damage-sensitive sensors which fall within the DSA:

$$\left| \left[u_i^d(t) - u_i^h(t) \right] / u_i^h(t) \right| \geq \eta, \tag{10}$$

where the superscripts d and h denote damaged and healthy states, respectively. For each measurement time step, $u_i^d(t)$ represents the data from the i-th sensor in a damaged case, and $u_i^h(t)$ represents the response in a healthy state. η is the lowest limit of a detectable relative change in response. Note that $\Delta u_i(t) = u_i^d(t) - u_i^h(t)$ represents the variation of response under a damage condition. If $\left| \Delta u_i(t)/u_i^h(t) \right|$ is lower than the sensor sensitivity, it is impossible to detect the damage. Therefore, η should be chosen as the sensor sensitivity. It should be noted that Equation (10) only applies to responses which are sensitive to local damage. And in this paper, strains are used in the analysis. However, Equation (10) may not be applicable to vibration monitoring, due to the fact that vibration responses are integral structural effects and may not be sensitive to local damage.

A sensor is defined as damage-sensitive if the responses it acquires in the case of damage satisfy the following formula:

$$n_d/n \geq p_0, \tag{11}$$

where n represents the total number of observations over time, n_d represents the total number of observations satisfying Equation (10), p_0 is a given constant parameter that determines the lowest possibility to define a damage-sensitive sensor. In fact, it is difficult to determine the exact value for p_0 because it depends on the specific structures. In general, an approximate range for p_0 can be given as from 50% to 100%. This means if more than half of all observations at a certain scenario satisfy Equation (10), then a sensor can be treated as damage-sensitive. Once an accurate FE model is given, an accurate method for determining DSA according to Equation (11) can be provided based on numerical simulations of various damaged cases (various combinations of damage locations and severities are considered). Consequently, the space window can be defined as the set of sensors installed in DSA. However, it is not possible to provide an accurate method to determine the DSA without an accurate FE model, because the determination of DSA depends on specific structures including materials, types of structures and boundary conditions. In such case, only empirical experience is available to determine the DSA and the space window. For example, one can use diverse space windows of which each involves several neighboring sensors. Because damage in general has more effects on those sensors which are nearby, the space window that is nearest to the location of the damage is most likely to group the sensors that are more sensitive to damage.

The space window is presented in the form of $\begin{bmatrix} i_1 & i_2 & \cdots & i_S \end{bmatrix}$, where i_1, i_2, $\cdots$, i_S are the sensor numbers, and S represents the total number of sensors in the space window, in other words, within the DSA. For example, $\begin{bmatrix} 1 & 3 & 5 & 7 \end{bmatrix}$ means Sensor 1, Sensor 3, Sensor 5, and Sensor 7 are grouped inside the space window. Once the window is determined, one can conduct PCA with a moving time window for measurement values to detect damage. In consideration of space and time windows, Equation (6) can be rewritten as:

$$\mathbf{U}(k) = \begin{bmatrix} u_{i_1}(t_k) & u_{i_2}(t_k) & \cdots & u_{i_S}(t_k) \\ u_{i_1}(t_{k+1}) & u_{i_2}(t_{k+1}) & \cdots & u_{i_S}(t_{k+1}) \\ \cdots & \cdots & \cdots & \cdots \\ u_{i_1}(t_{k+T-1}) & u_{i_2}(t_{k+T-1})) & \cdots & u_{i_S}(t_{k+T-1}) \end{bmatrix}. \tag{12}$$

Then, repeating the steps of PCA, one is able to obtain the time-variant eigenvector $\psi_i(k) = \begin{bmatrix} \psi_{i,1} & \psi_{i,2} & \cdots & \psi_{i,S} \end{bmatrix}^T$. Similarly, with Section 2.2, herein $\Delta\psi_{i,j}(k)$ by DWPCA in a spatial window can be obtained from Equation (9). By following the CEV $\Delta\psi_{i,j}(k)$ at each time step, damage can be detected if $\Delta\psi_{i,j}(k)$ exceeds a certain threshold value. Meanwhile, it is possible to localize damage through the space window by observing rapidly changing components. It can be seen in Equation (12) that only sensitive responses are considered in the analysis for DWPCA. Thus, it is expected to improve the performance of damage identification.

3. Validation of DWPCA with a Planar Beam

3.1. Model for Simulation

To evaluate the effectiveness and efficiency of the proposed DWPCA method, large datasets including responses from a structure under various damaged scenarios are needed. In practice, it is not possible to acquire such datasets from a real civil structure because intentionally imparting damage to the structure is not allowed for safety reasons. As a result, this study adopts a finite element (FE) model of a planar beam established by ANSYS for calculations to obtain necessary datasets, as shown in Figure 1. In the simulations, strain responses of the FE model under seasonal temperature variations are computed under different damaged scenarios. Simulations for various damage severities at different locations are achieved by exerting certain stiffness reductions in certain finite elements of the model.

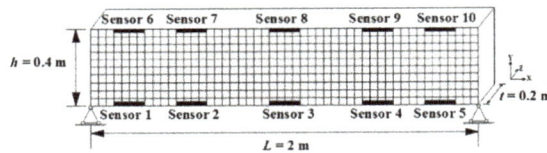

Figure 1. Finite element model of a simply supported beam for simulations with ten strain sensors installed.

The planar beam in Figure 1 is assumed to be 2.0 m in length (L) with a rectangular cross-section of 0.4 m in height (h) and 0.2 m in width (t). The FE model is evenly discretized into 500 quadrilateral elements with 50 elements for each row in the x-direction (beam length) and 10 elements for each column in the y-direction (beam height). Note that each element of the meshes has a length of 0.04 m and a height of 0.04 m. It is also assumed that the beam is composed of concrete with a Young's modulus of 34.5 GPa, a Poisson's ratio of 0.2, and a thermal expansion coefficient of $1 \times 10^{-5}/°C$. For the thermal loads, seasonal temperature variations are applied on the bottom and top surfaces. The temperature on the bottom surface is set to be $T_b = 20 + 10\sin(\pi t/730)(°C)$, while that on the top surface is $T_t = T_b + 10(°C)$. Note that the sinusoidal function in T_b has a period of one year, which is consistent with the period of seasonal temperature variations. In addition, a linear temperature distribution along the beam height is taken into consideration. During simulations, equivalent forces caused by temperature variations are exerted on the nodes of each finite element to obtain strain responses. In this paper, thermal excitations of four years are applied. Figure 2 illustrates the evolution of temperatures at the top and bottom of the beam over four years.

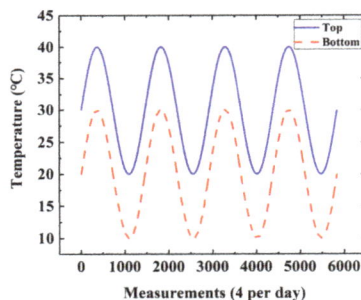

Figure 2. Evolution of temperatures at the top and bottom of the beam over four years.

Additionally, a virtual SHM system is installed in the beam structure. As shown in Figure 1, the system is assumed to be composed of ten strain sensors, of which five are installed on the top surface and the other five on the bottom. For each damaged scenario, strain histories are computed for four years with four measurements per day, i.e., 5840 measurement events in total for each sensor. In addition, permanent damage is introduced at the beginning of the third year (after 2920 measurements) by stiffness reductions at certain finite elements in the model.

In this paper, the following damaged scenarios, as illustrated in Figure 3, are considered for comparative studies between MPCA and the proposed DWPCA:

(1). Scenario A: Damage in four finite elements at Sensor 1 as shown in Figure 3a;
(2). Scenario B: Damage in four finite elements at Sensor 3 as shown in Figure 3b;
(3). Scenario C: Damage in four finite elements at Sensor 6 as shown in Figure 3c;
(4). Scenario D: Damage in four finite elements at Sensor 8 as shown in Figure 3d;
(5). Scenario E: Damage in four finite elements near Sensor 1 as shown in Figure 3e;
(6). Scenario F: Damage in four finite elements near Sensor 3 as shown in Figure 3f;
(7). Scenario G: Damage in four finite elements near Sensor 6 as shown in Figure 3g;
(8). Scenario H: Damage in four finite elements near Sensor 8 as shown in Figure 3h.

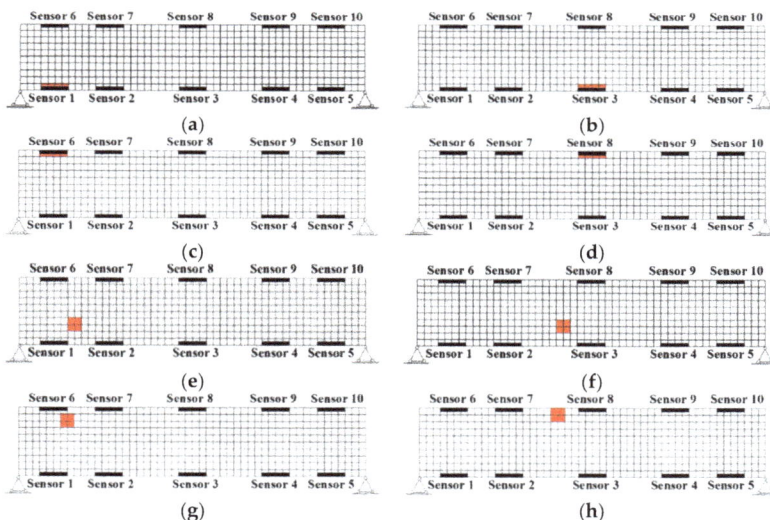

Figure 3. Damaged scenarios for the evaluation of damage detection algorithms: (**a**) damage in four finite elements at Sensor 1; (**b**) damage in four finite elements at Sensor 3; (**c**) damage in four finite elements at Sensor 6; (**d**) damage in four finite elements at Sensor 8; (**e**) damage in four finite elements near Sensor 1; (**f**) damage in four finite elements near Sensor 3; (**g**) damage in four finite elements near Sensor 6; (**h**) damage in four finite elements near Sensor 8.

3.2. Determination of the DSA

To determine the DSA in this study, the value of the parameter η is chosen as 5% because the sensor sensitivity is assumed to be 5% in the simulation. As for p_0, it is chosen as 60%, because it is observed that if more than 60% of all observations in a certain scenario satisfy Equation (10), the sensor is found to be damage-sensitive according to the simulation results. Figures 4 and 5 illustrate the calculated strain variation, as well as the DSA, for Scenario A and Scenario B with a stiffness reduction of 80%, respectively. It is seen that damage-sensitive elements are indeed more likely to lie in the vicinity of the damage location. However, as shown in Figures 4b and 5b, some finite elements are sensitive to damage even though they are located far from the damage location. As a result, the determination of the DSA should not be directly based on damage location. Numerical simulations can help this problem.

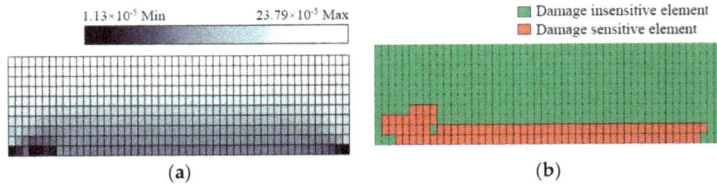

Figure 4. Simulation results for Scenario A with a stiffness reduction of 80%: (**a**) contour for strain variation; (**b**) damage-sensitive area indicated by finite elements in red.

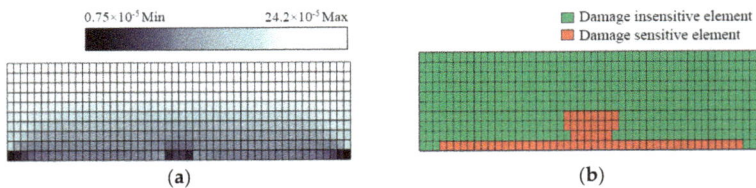

Figure 5. Simulation results for Scenario B with a stiffness reduction of 80%: (**a**) contour for strain variation; (**b**) damage-sensitive area indicated by finite elements in red.

In consideration of the simulation results for different damaged scenarios, as well as the symmetry of the FE model in Figure 1, the following four listed windows are used for comparative studies between MPCA and the proposed DWPCA:

(1). Window A involving all the sensors: $\begin{bmatrix} 1 & 2 & 3 & 4 & 5 & 6 & 7 & 8 & 9 & 10 \end{bmatrix}$;

(2). Window B involving all the sensors at the bottom: $\begin{bmatrix} 1 & 2 & 3 & 4 & 5 \end{bmatrix}$;

(3). Window C: $\begin{bmatrix} 1 & 2 & 6 & 7 \end{bmatrix}$;

(4). Window D: $\begin{bmatrix} 2 & 3 & 7 & 8 \end{bmatrix}$.

It should be noted that Window A is usually implemented in PCA and MPCA. If other windows are applied, the algorithm will belong to the DWPCA method. In the following calculations, the time window size for both MPCA and DWPCA is equal to one period of the thermal loads, namely one year (1460 measurement events). The first principal component ψ_1 is considered as the sensitive feature for damage detection because most of the variance is contained in it.

4. Results and Discussion

In this section, comparative studies between DWPCA and MPCA in previous studies for damage detection will be carried out on detection sensitivity, damage localization, and severity evaluation. Noise immunity of the proposed features will also be investigated. To begin with, the effects of the following two features on damage identification performance are investigated:

(1). CEV by MPCA: $\Delta\psi_{i,j}^{M}$;

(2). CEV by DWPCA: $\Delta\psi_{i,j}^{DW}$.

Note that Window A is usually implemented in MPCA. Other windows including Window B to Window D are implemented in the DWPCA method.

4.1. Sensitivity for Damage Detection

To begin with, the effects of different windows on the time series CEVs are investigated. In the simulations, four scenarios (A, B, C, and D) and four windows (A, B, C, and D) are considered for a comparative study. In addition, a permanent stiffness reduction of 40% was introduced for the corresponding finite elements at the beginning of the third year. Note that if Window A is used, the method belongs to MPCA because all installed sensors are taken into account. As shown in Figure 6, the time-variant CEVs of different scenarios and space windows are simulated. It can be seen that before the damage occurs, CEVs are stable for all cases. However, there exists a shift for the values after damage occurrence at all scenarios. In the unstable stage between the 2920th and 4380th measurements, strains within the moving time window involve responses from both undamaged and damaged states. As the time window moves forward, the corresponding CEVs will become stable again after 4380 measurement events. This results from the fact that all strain time series within the time window were obtained from the damaged structure after 4380 measurements. A closer look at Figure 6 indicates that a more significant change is observed for the proposed DWPCA method with Windows B, C, and D in contrast to MPCA with Window A. It is interesting to find that among the used space windows, Window B, that contains all the sensors installed at the bottom as shown in Figure 1, renders a more rapid and evident change in CEVs for the considered damaged scenarios as compared with the healthy state. The result can be explained by the fact that, as illustrated by the simulation results in Figures 4 and 5, damage has a greater influence on the responses at the bottom. From a mechanical point of view, the sensors at the bottom are near the constraint boundaries and any variations in the structure may lead to a more significant change in their responses. Additionally, one can see from Figure 6 that damage at the bottom has more influence on structural responses and CEVs than damage at the top, indicating that it is easier to detect damage at the bottom. In a word, the proposed DWPCA uses a space window to exclude sensors outside the DSA, thereby leading to an enhanced sensitivity for damage detection as compared with MPCA.

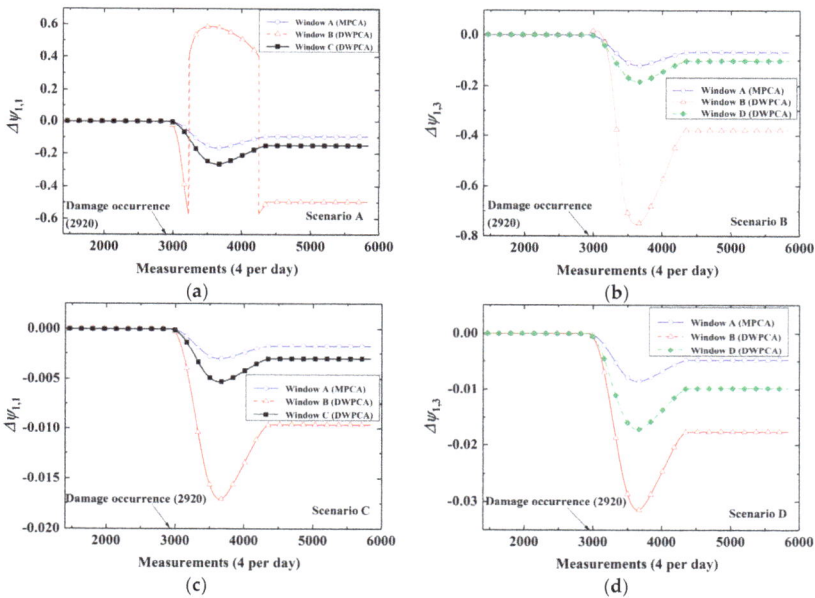

Figure 6. Evolution of CEVs for different space windows: (**a**) Scenario A; (**b**) Scenario B; (**c**) Scenario C; (**d**) Scenario D.

In order to conduct a comparative study on the detection resolution and the time to detect damage after damage occurrence between MPCA and the proposed DWPCA, a range of damage severities are considered for simulations. Detection resolution is defined as the damage level that induces a minimum detectable relative variation of CEVs in comparison with those in a healthy state. In this paper, the minimum detectable relative CEV is chosen as the value when the variation rate of the corresponding CEV with respect to time is equal to $0.67~\mu\varepsilon/\text{h}$. The time to detect damage is the period from the moment the damage occurs to that when damage is detected. Figure 7 presents the time to detect damage after damage occurrence with respect to damage levels ranging from the minimum detectable level of each method to a maximum level of 99.9% in Scenarios A, B, C and D. For DWPCA, the space window of Window B is considered in the simulations. As expected, higher damage levels at any scenario result in a shorter detection time of damage for both methods. However, as the damage level is lower than 70% for the considered scenarios, DWPCA shows a shorter time to detect damage as compared with MPCA. Furthermore, this advantage of DWPCA is increasingly evident as the damage level becomes smaller. Note that, usually, structural damage emerges is initially small and gradually evolves to larger damage. The use of a space window means that DWPCA possesses a superior capability in the early detection of damage and timely alarms in contrast to MPCA.

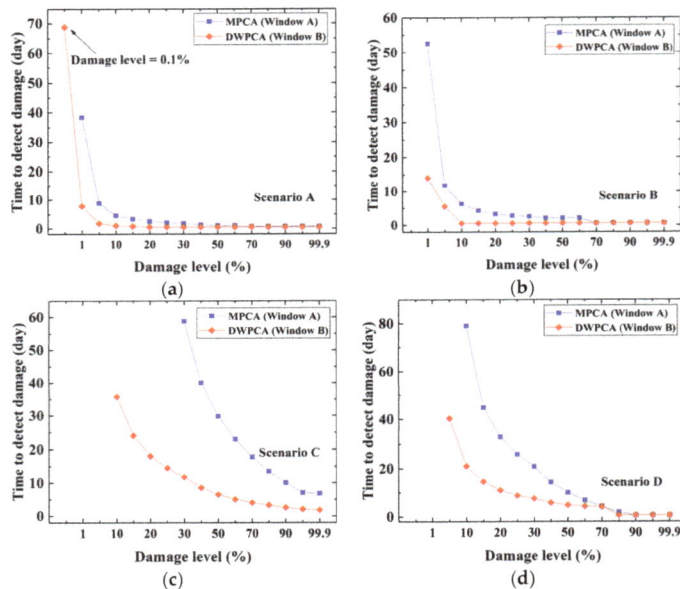

Figure 7. Time to detect damage after damage occurrence with respect to the damage level: (**a**) Scenario A; (**b**) Scenario B; (**c**) Scenario C; (**d**) Scenario D.

Table 1 presents the detection resolution of both methods according to the simulation results in Figure 7. It can be seen that DWPCA has a better detection resolution than MPCA for all scenarios, except for Scenario B, in which the same resolution is observed for both methods. In fact, the damage in Scenario B is located at the bottom of the mid-span and has significant effects on the structural responses. Hence, MPCA exhibits a comparative detection resolution even though a space window is not applied. For other cases, the detection resolution of DWPCA is commonly better than that of MPCA, owing to the fact that the space window in DWPCA excludes sensors that are not sensitive to damage. Further examination of Table 1 shows damage located at the bottom (Scenarios A and B) is easier to detect than that located at the top (Scenario C and D) because damage at the bottom, generally, has more significant effects on the structural responses. In addition, DWPCA has the best detection resolution of 0.1% in Scenario A, in which the damage is located at the bottom near the constraint boundary, while it has the worst detection resolution of 10% for Scenario C, in which the damage is located at the top near the sides of the beam. Table 2 presents the time to detect damage at a minimum common level for both MPCA and DWPCA in different scenarios. It is clear that DWPCA detects damage much earlier than MPCA. DWPCA takes about 7 to 21 days to detect damage after its occurrence, while MPCA needs about 38 to 80 days. Note that in Scenario B, DWPCA detects damage more rapidly than MPCA although MPCA has the same resolution as DWPCA. In addition, it is also apparent that damage at the bottom (Scenarios A and B) is easier and quicker to detect than that at the top (Scenarios C and D). In summary, it is demonstrated that damage detectability of the proposed DWPCA is improved as compared with MPCA due to the application of a space window which groups sensors within the DSA and excludes those insensitive to damage.

Table 1. Detection resolution of moving principal component analysis (MPCA) and double-window PCA (DWPCA) for different scenarios.

Scenario	Detection Resolution in Damage Level (%)	
	MPCA	DWPCA
A	1	0.1
B	1	1
C	30	10
D	10	5

Table 2. Time to detect damage of MPCA and DWPCA for different scenarios.

Scenario	Stiffness Reduction (%)	Time to Detect Damage (Day)	
		MPCA	DWPCA
A	1	38.25	7.75
B	1	52.5	14
C	30	58.75	11.75
D	10	79	21

4.2. Damage Localization

In Section 4.1, DWPCA has been demonstrated as a more effective tool to identify damage than MPCA. In this subsection, a methodology for damage localization will be put forward by tracking the time-variant CEVs based on the proposed DWPCA. Four Scenarios (E, F, G, and H) in Figure 3 and a stiffness reduction of 40% are considered.

At first, cases in which damage is located at the bottom (Scenarios E and F) are considered. Figure 8a–c show time-variant CEVs computed by DWPCA with different space windows for Scenario E, in which the damage is located at the bottom between Sensor 1 and Sensor 2. For Window C, as shown in Figure 8a, the CEVs corresponding to Sensor 1 and Sensor 2 show evident shifts after the occurrence of damage as compared with those corresponding to Sensor 6 and Sensor 7. For Window D, shown in Figure 8b, only the CEV corresponding to Sensor 2 has a relatively evident change owing to the fact that Sensor 2 is the sensor that is located closest to the damage in the space window. As for Window B, the CEV corresponding to Sensor 1 exhibits the most evident variation as compared with other components. For Window A, as shown as Figure 8d, the CEV corresponding to Sensor 1 by MPCA is smaller than that by DWPCA with Window C in Figure 8a or Window B in Figure 8c. It demonstrates that the CEV corresponding to Sensor 1 computed by DWPCA is larger than that obtained by MPCA in Scenario E. This proves that DWPCA is more sensitive for damage localization than MPCA. In addition, one can infer from Figure 8 that damage is located close to Sensor 1 because the variation of the corresponding CEV in various windows is the most notable. For Scenario F, in which the damage is located close to Sensor 3, similarly, the CEV related to Sensor 3 shows a significant change, as presented in Figure 9, especially for both Windows B and D as compared with Window A. Subsequently, we consider cases in which damage is located at the top (Scenarios G and H). The simulation results are shown in Figure 10 for Scenario G and in Figure 11 for Scenario H, respectively. For Scenario G, as expected, the CEV related to Sensor 6 is the most evident because the damage is in the vicinity of Sensor 6. As for Scenario H, the CEV related to Sensor 8 displays an evident shift. From Figure 8 to Figure 11, we can see that damage at the bottom has more significant effects on the corresponding CEV as compared with that at the top. Furthermore, if the damage is located at the bottom, Window B shows a better performance for damage localization because a larger variation is observed for the CEV. However, if the damage is located at the top, Window C or D is preferred. In conclusion, it is seen that DWPCA can be used to localize damage with the aid of various space windows and shows a better performance for damage localization as compared with MPCA.

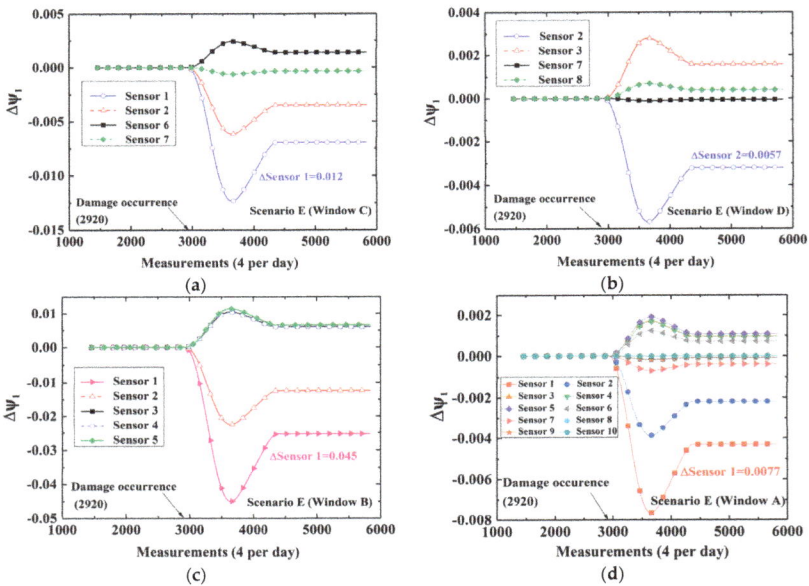

Figure 8. Time-variant CEVs with different space windows for Scenario E: (**a**) Window C; (**b**) Window D; (**c**) Window B; (**d**) Window A.

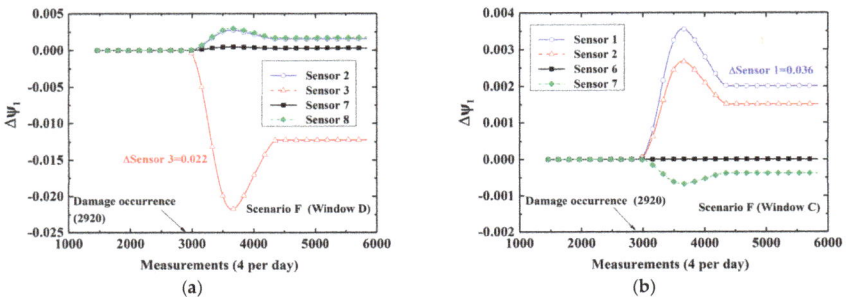

Figure 9. *Cont.*

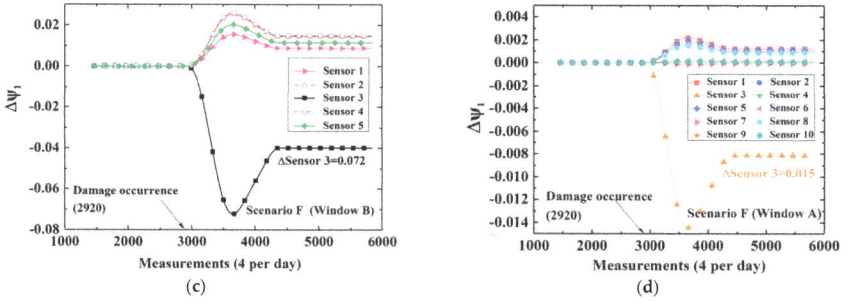

Figure 9. Time-variant CEVs with different space windows for Scenario F: (**a**) Window D; (**b**) Window C; (**c**) Window B (**d**) Window A.

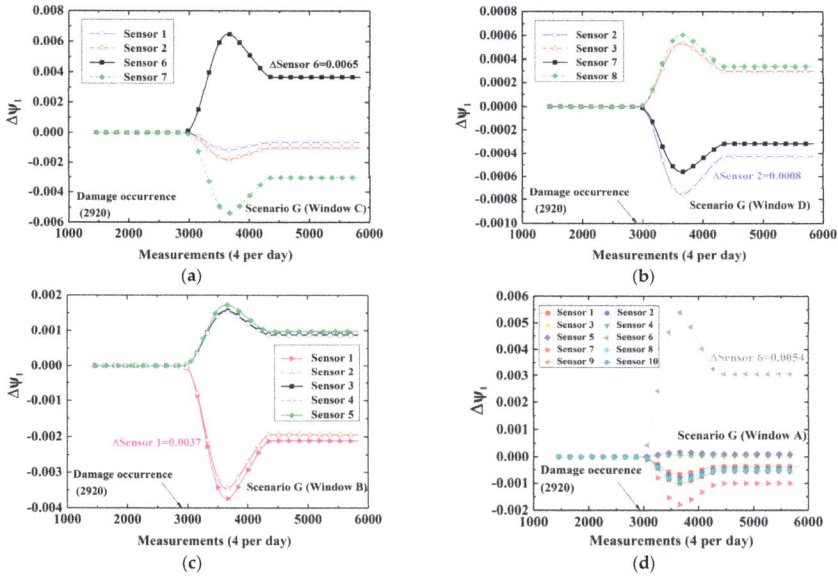

Figure 10. Time-variant CEVs with different space windows for Scenario G: (**a**) Window C; (**b**) Window D; (**c**) Window B; (**d**) Window A.

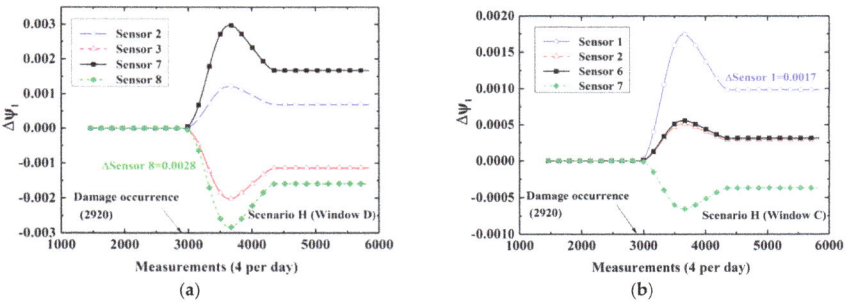

Figure 11. *Cont.*

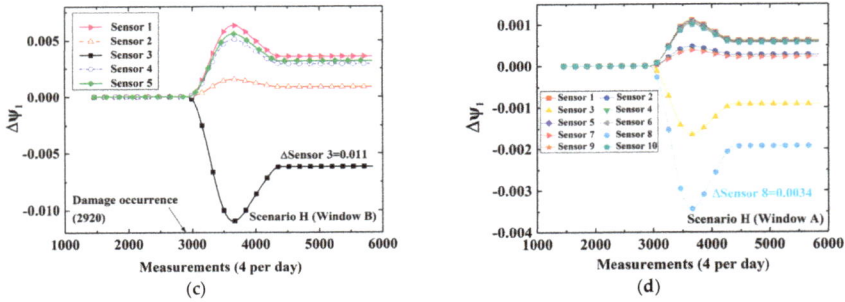

Figure 11. Time-variant CEVs with different space windows for Scenario H: (**a**) Window D; (**b**) Window C; (**c**) Window B; (**d**) Window A.

4.3. Quantitative Evaluation of Damage

Based on the discussion above, a further investigation to provide a quantitative evaluation of the damage using DWPCA is presented in this section. The relationship between the damage level and stable absolute value of CEV after damage occurrence for a range of damage severities from 0.1% to 99.9% in Scenarios A and B is presented in Figure 12. It can be seen from Figure 12a that the CEV corresponding to Sensor 1 has a monotonically ascending trend as the damage level increases for both MPCA with Window A and the proposed DWPCA with Window C. However, the corresponding CEV for DWPCA with Window C varies more dramatically as a function of the damage level than that of MPCA. A more evident discrimination between data from damaged and undamaged states is observed for the proposed method. Thus, DWPCA has a higher sensitivity for quantitative evaluation of damage as compared with MPCA. According to the simulation results of Scenario A in Figure 12a, the damage level L_D can be quantitatively evaluated in terms of $|\Delta\psi_{1.1}|$ by DWPCA with Window C, as indicated by:

$$L_D = 0.714\ln\left(\left|\Delta\psi_{1.1}^{DW}\right| + 0.190\right) + 1.175 \tag{13}$$

For Scenario B, as illustrated in Figure 12b, the related CEV also increases with an increase in damage level for MPCA with Window A or DWPCA with Window D. Similarly, the variation of CEV related to Sensor 3 by DWPCA is larger as compared with MPCA.

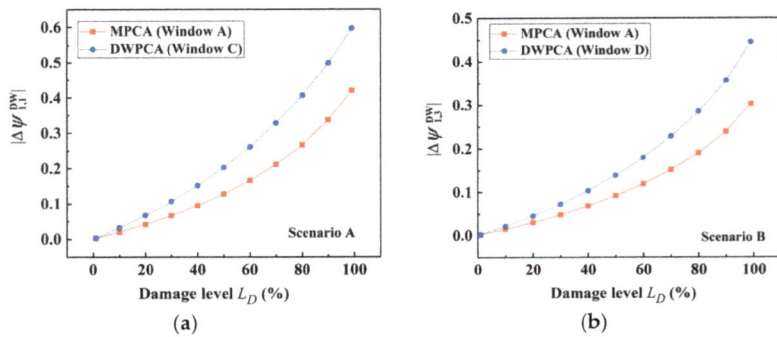

Figure 12. Variation of the absolute value of the CEVs with damage level by MPCA and DWPCA: (**a**) Scenario A; (**b**) Scenario B.

Thus, the proposed DWPCA has a higher sensitivity for damage evaluation. For DWPCA with Window D, as presented in Figure 12b, the damage level L_D can be obtained from the calculated CEV $\Delta\psi_{1.3}$ with the use of the following equation:

$$L_D = 0.588\ln\left(\left|\Delta\psi_{1.3}^{DW}\right| + 0.090\right) + 1.385 \tag{14}$$

It should be noted that the relationship between the CEV and damage level is obtained by curve fitting. This methodology for quantitative evaluation requires calibration or training with an accurate FE model.

4.4. Noise Immunity

In practice, noise caused by external environmental factors or systematic errors in SHM is inevitable. Consequently, data from SHM systems involve noise and may render damage identification methods ineffective. As a result, noise immunity of the proposed DWPCA method should be investigated. Based on the measured strain data in a large-scale bridge from the literature [39], the standard deviation of noise is considered to be from 1.25 με to 5 με. Note that strain responses in the simulations are approximately 150 με. Thus, the noise level ranges from 0.8% to 3.3%. The relationship between CEV and damage level in Scenarios A and B, in which different intensities of noise are present is presented in Figure 13. It can be seen that noise has little influence on the relationship between the CEV absolute value and the damage level in DWPCA. This is due to the favorable de-noising characteristic of PCA. In a word, the proposed DWPCA method in this study has considerably good noise immunity and shows potential for applications in practical engineering.

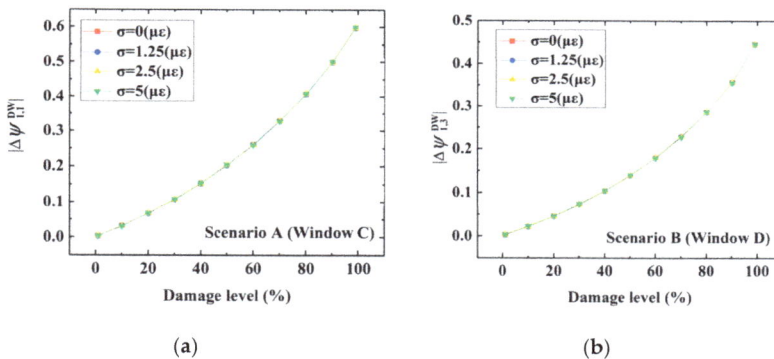

Figure 13. Variation of the absolute value of the CEVs with damage level by DWPCA in the presence of different intensities of noise: (**a**) Scenario A; (**b**) Scenario B.

5. Application to a Full-Scale Structure

Based on the validation for DWPCA with a planar beam in Sections 3 and 4, in this section, further investigation of the performance of DWPCA for a large-scale structure will be carried out to demonstrate its applicability for practical engineering purposes. The full-scale FE model will be based on the Xijiang Bridge in Zhaoqing, China. The bridge is a continuous rigid frame bridge built in 2004 in the Guangdong province, China. It consists of seven spans with a total length of 808 m. The photograph and schematic diagram of its structure are presented in Figure 14. Properties of the bridge are summarized in Table 3.

(a)

(b)

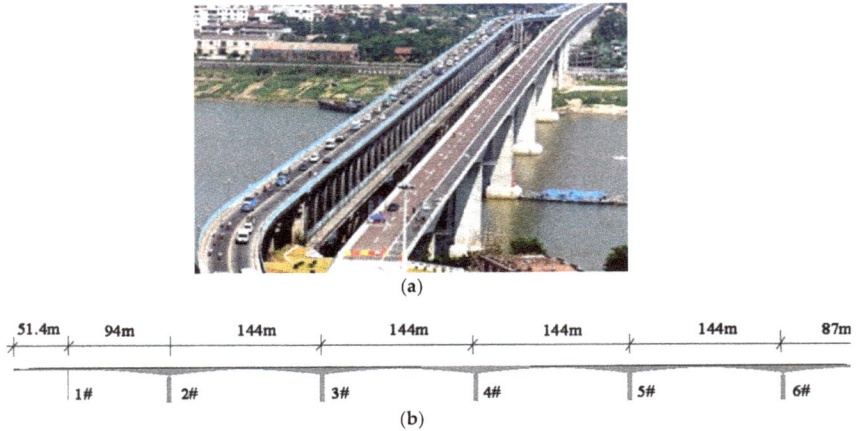

Figure 14. Xijiang Bridge: (**a**) photograph of the bridge; (**b**) schematic diagram of the structure.

Table 3. Properties of each part of the Xijiang Bridge in Zhaoqing, China.

Parts		Material	Elastic Modulus (GPa)	Poisson Ratio
Pier	1#	Concrete	34.5	0.2
	2#–6#	Concrete	30.0	0.2
Bridge deck	Box girder	Concrete	32.5	0.2
	Non-pressed and pressed steel	Steel	195.0	0.3

In order to demonstrate the sensitivity of DWPCA for damage detection of the full-scale structure, response data from different damage scenarios should be prepared. Since the bridge is in good conditions after the completion of construction stages, there are no damage events that could have generated unusual structural behavior. For the purposes of application of DWPCA on real structures, a full-scale FE model of the bridge is established. The strain responses under seasonal temperature variations presented in Figure 2 in Section 3 are obtained. Continuous structural health monitoring responses of four years at a sampling frequency of four measurements per day are collected.

Local damage is assumed to be introduced in the span between the 2# and 3# piers of the bridge, as shown in Figure 14b. Sensors are embedded every 5 m along the bridge length, as shown in Figure 15a. The arrangement of the sensor locations on the top, webs and bottom of the girder box are given in Figure 15b. Note that there are 29 monitoring sections, and each section has six sensors installed in this span. Thus, there are 174 sensors in total and these are numbered from top to bottom, from left to right (Section ① to Section ㉙) in sequence. In the FE model, damage is assumed to be at a specific element of the bridge with a permanent stiffness reduction and is introduced at the beginning of the third year. Two different damage scenarios with different damage locations marked as red are shown in Figure 16a,b:

(1). Scenario A: Damage between Section ① and Section ② in the vicinity of Sensor 8 and Sensor 10, as shown in Figure 16a;

(2). Scenario B: Damage between Section ⑭ and Section ⑮ close to Sensor 84, as shown in Figure 16b.

Space windows which are related to the DSA should be determined. During the DSA analysis in this section, the parameter η in Equation (10) is equal to 5%. The p_0 in Equation (11) is set to be 60%. After simulations for a large number of damage scenarios, it was found that the DSA is more likely to lie within two neighboring monitoring sections that are located close to the damage. Especially, when the damage is located quite close to one monitoring section, the DSA is located in the vicinity of this section. The space windows considered herein contain sensors from two neighboring monitoring sections or from one section that is closest to the damage. Thus, for brevity of demonstration, only the following spatial windows are used for comparative studies between MPCA and the proposed DWPCA:

(a) Window A involving all the sensors: $\begin{bmatrix} 1 & 2 & \ldots & 174 \end{bmatrix}$;

(b) Window B involving sensors from Section ① and Section ②: $\begin{bmatrix} 1 & 2 & \ldots & 12 \end{bmatrix}$;

(c) Window C involving sensors from Section ⑭ and Section ⑮: $\begin{bmatrix} 79 & 80 & \ldots & 90 \end{bmatrix}$;

(d) Window D involving sensors from Section ①: $\begin{bmatrix} 1 & 2 & \ldots & 6 \end{bmatrix}$;

(e) Window E involving sensors from Section ②: $\begin{bmatrix} 7 & 8 & \ldots & 12 \end{bmatrix}$;

(f) Window F involving sensors from Section ⑭: $\begin{bmatrix} 79 & 80 & \ldots & 84 \end{bmatrix}$;

(g) Window G involving sensors from Section ⑮: $\begin{bmatrix} 85 & 86 & \ldots & 90 \end{bmatrix}$.

(a)

(b)

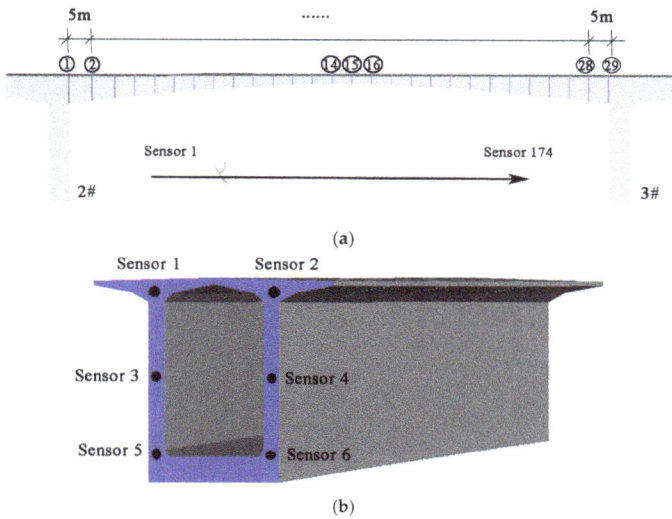

Figure 15. Spatial distribution of the sensors in the span between the 2# and 3# piers: (**a**) the arrangement of the monitoring section marked in blue along the bridge; (**b**) the arrangement of the sensors in the monitoring section.

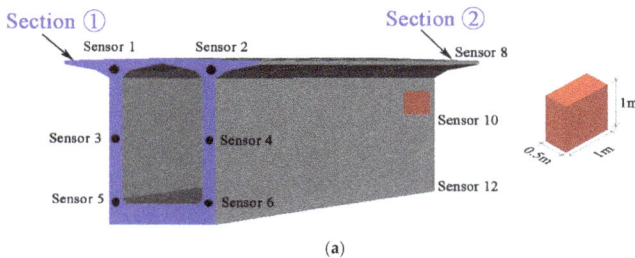

(a)

Figure 16. *Cont.*

275

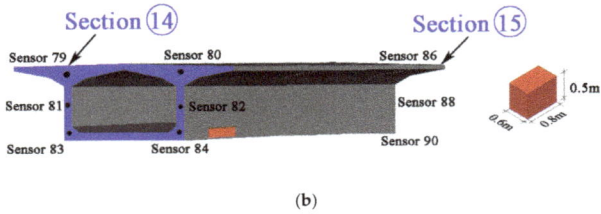

(b)

Figure 16. Damaged scenarios for evaluation of the damage detection algorithms: (**a**) damage with a stiffness reduction of 40% in the vicinity of Sensor 8 and Sensor 10; (**b**) damage with a stiffness reduction of 40% close to Sensor 84.

Comparative studies of CEVs computed by MPCA ($\Delta\psi_{ij}^{M}$) and DWPCA ($\Delta\psi_{ij}^{DW}$), respectively, on damage detection for this bridge are presented as follows. Window A is still used in MPCA. Other windows including Window B to Window G will belong to the DWPCA method in the following demonstration.

Figure 17 shows the evolution of CEVs by MPCA and DWPCA upon application in two different damage scenarios. Similarly to that of the planar beam, there are no relative variations in the corresponding CEVs in the first two years since there is no damage. In addition, there exists a shift after damage occurrence in all scenarios when the time window involves responses from both damaged and healthy states. Then, the corresponding CEVs will become stable again after 4380 measurement events when responses within the time window are obtained from the damaged structure after 4380 measurements. Note that a more significant change of corresponding CEVs by the proposed DWPCA method with Window B, C, E, or F are observed in contrast to MPCA with Window A in both scenarios. Additionally, Windows E and F, which consist of sensors from only one monitoring section, perform better than Windows B and C, which contain sensors from two neighboring monitoring sections, when damage is located quite close to one monitoring section.

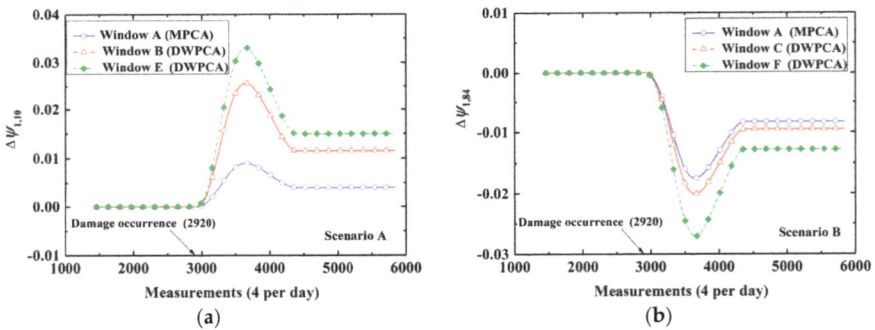

Figure 17. Evolution of the corresponding CEVs for different space windows: (**a**) Scenario A; (**b**) Scenario B.

After the investigation of DWPCA in damage identification in the bridge, a closer look at all CEVs evolutions within a spatial window will be further explored for damage localization. Figure 18a shows the evolution of CEVs computed by DWPCA with Window D and E for Scenario A, in which the damage is located between Section ① and Section ②, close to Sensor 10. For Window D and E, as shown in Figure 18a, the CEVs corresponding to Sensor 7, Sensor 8, Sensor 9 and Sensor 10 shows an evident shift after damage occurrence as compared with other sensors which are located far from damage. This is due to the fact that Sensor 7, Sensor 8, Sensor 9 or Sensor 10 is the nearest to damage in the corresponding space window. For Scenario B in Figure 18b, as expected, the CEV related to Sensor

84 is the most evident because the damage is in the vicinity of Sensor 84. Figure 18b also indicates that the CEVs corresponding to Sensor 81, Sensor 82, and Sensor 83, which are close to damage, also have a remarkable shift after damage occurrence. It is seen that DWPCA can be used to localize damage with the aid of various space windows for complex engineering structures.

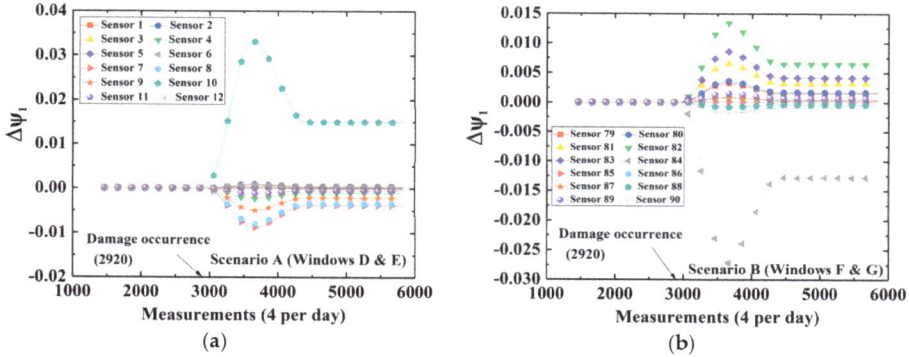

Figure 18. Time-variant CEVs with different space windows for Scenario A and B: (**a**) Windows D and E; (**b**) Windows F and G.

Based on the discussion in Section 4.3, the relationship between the damage level and stable absolute value of CEV after damage occurrence for a range of damage severities in simple beams can be quantitatively evaluated. Similarly, for Scenarios A, the damage level L_D can be quantitatively evaluated in terms of $|\Delta\psi_{1.10}|$ by DWPCA with Window E in Figure 19a, as indicated by:

$$L_D = 1.25\ln\left(|\Delta\psi_{1.10}^{DW}| + 0.037\right) + 4.121 \tag{15}$$

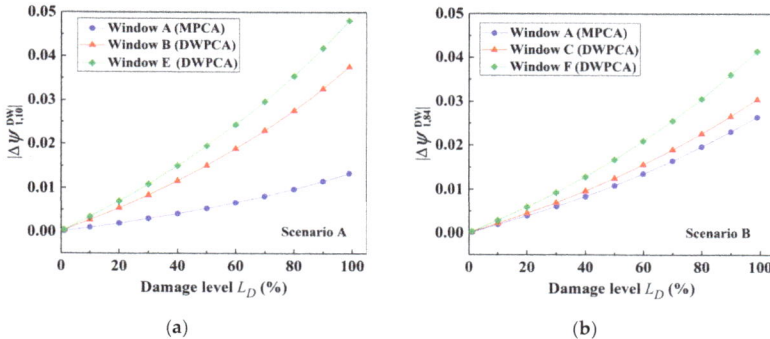

Figure 19. Variation of the absolute value of the CEVs with damage level by MPCA and DWPCA: (**a**) Scenario A; (**b**) Scenario B.

As for DWPCA with Window F for Scenario B, as presented in Figure 19b, the damage level L_D can also be obtained from the calculated $|\Delta\psi_{1.84}|$ with the use of the following equation:

$$L_D = 1.25\ln\left(|\Delta\psi_{1.84}^{DW}| + 0.032\right) + 4.303 \tag{16}$$

In summary, the proposed method DWPCA is demonstrated to be feasible for damage detection for large-scale structures. Results show that, similarly with the conclusion drawn in Section 4 for the planar beam, DWPCA has better performance in damage identification, damage localization and damage quantitative evaluation, as compared with the previous method MPCA. The is due to that the space windows used in DWPCA are capable of excluding damage-insensitive data from those sensors located far from the damage to enhance the damage detectability The proposed method is proven to have potential in applications for practical engineering.

6. Conclusions

This paper provides a novel effective method for structural damage detection by introduction of space and time windows in the traditional principal component analysis method. Due to the presence of the space window, the damage-insensitive data from those sensors located far from the damage are excluded in the analysis, and the damage detectability of the proposed method is improved in contrast to previous methods. Numerical results with a planar beam model demonstrate that the proposed method DWPCA, as compared with MPCA, improves the resolution for damage identification and is also quicker to detect damage after its occurrence. DWPCA is successful to detect minor damage with 0.1% stiffness reduction and identify damage 31 to 59 days earlier as compared with MPCA for a planar beam. With the aid of various space windows, the method is verified to have a better performance for damage localization as well. As for a quantitative evaluation of the damage severities from 0.1% to 99.9% for a planar beam, DWPCA proves to be more sensitive than previous methods. Finally, the proposed method is demonstrated to have good noise immunity and the result with a full-scale structure shows potential for applications in practical engineering. Further investigation will be focused on the feasibility of the proposed methodology to large-scale structures under more complicated loads such as real temperature variations and vehicle loads.

Author Contributions: Conceptualization, G.Z. and L.Z.; methodology, G.Z., L.Z. and Z.L.; software, L.T.; validation, L.T., Y.L., and Z.J.; formal analysis, G.Z.; investigation, G.Z., L.Z. and Z.L.; resources, L.T.; writing—original draft preparation, G.Z.; writing—review and editing, L.Z. and Z.L.; supervision, Z.J and Y.L.; funding acquisition, L.T.

Funding: This research was funded by the National Natural Science Foundation of China (Grant No. 11602087, 11472109, 11772131, 11772132, and 11772134), the Natural Science Foundation of Guangdong Province, China (Grant No. 2015A030308017, 2015A030311046, and 2015B010131009), Science and Technology Program of Guangzhou, China (Grant No. 201903010046), the State Key Lab of Subtropical Building Science, South China University of Technology (Grant No. 2014ZC17 and 2017ZD096).

Conflicts of Interest: The authors declare no conflict of interest.

Abbreviations

The following abbreviations are used in this manuscript:

CEV	Component of the eigenvector variance
DSA	Damage-sensitive area
DWPCA	Double-window principal component analysis
FE	Finite element
MPCA	Moving principal component analysis
PCA	Principal component analysis
SHM	Structural health monitoring

References

1. Gunes, B.; Gunes, O. Structural health monitoring and damage assessment Part I: A critical review of approaches and methods. *Int. J. Phys. Sci.* **2013**, *8*, 1694–1702. [CrossRef]
2. Khan, S.; Yairi, T. A review on the application of deep learning in system health management. *Mech. Syst. Signal Process.* **2018**, *107*, 241–265. [CrossRef]
3. Zhu, X.; Cao, M.; Ostachowicz, W.; Xu, W. Damage Identification in Bridges by Processing Dynamic Responses to Moving Loads: Features and Evaluation. *Sensors* **2019**, *19*, 463. [CrossRef] [PubMed]

4. Chia, C.C.; Lee, J.-R.; Bang, H.-J. TOPICAL REVIEW: Structural health monitoring for a wind turbine system: A review of damage detection methods. *Meas. Sci. Technol.* **2008**, *19*, 310–314. [CrossRef]

5. Feng, D.; Feng, M.Q. Computer vision for SHM of civil infrastructure: From dynamic response measurement to damage detection—A review. *Eng. Struct.* **2018**, *156*, 105–117. [CrossRef]

6. Adewuyi, A.P.; Wu, Z. Vibration-Based Damage Localization in Flexural Structures Using Normalized Modal Macrostrain Techniques from Limited Measurements. *Comput-Aided Civ. Infrastruct. Eng.* **2011**, *26*, 154–172. [CrossRef]

7. An, Y.; Spencer, B.F.; Ou, J. A Test Method for Damage Diagnosis of Suspension Bridge Suspender Cables. *Comput-Aided Civ. Infrastruct. Eng.* **2015**, *30*, 771–784. [CrossRef]

8. Figueiredo, E.; Figueiras, J.; Park, G.; Farrar, C.R.; Worden, K. Influence of the Autoregressive Model Order on Damage Detection. *Comput-Aided Civ. Infrastruct. Eng.* **2011**, *26*, 225–238. [CrossRef]

9. Jayawardhana, M.; Zhu, X.; Liyanapathirana, R.; Gunawardana, U. Statistical Damage Sensitive Feature for Structural Damage Detection Using AR Model Coefficients. *Adv. Struct. Eng.* **2015**, *18*, 1551–1562. [CrossRef]

10. Omenzetter, P.; Brownjohn, J.M.W. Application of time series analysis for bridge monitoring. *Smart Mater. Struct.* **2006**, *15*, 129–138. [CrossRef]

11. Bao, C.; Hao, H.; Li, Z.-X. Integrated ARMA model method for damage detection of subsea pipeline system. *Eng. Struct.* **2013**, *48*, 176–192. [CrossRef]

12. Malekzadeh, M.; Gul, M.; Kwon, I.-B.; Catbas, N. An integrated approach for structural health monitoring using an in-house built fiber optic system and non-parametric data analysis. *Smart Struct. Syst.* **2014**, *14*, 917–942. [CrossRef]

13. Posenato, D.; Kripakaran, P.; Inaudi, D.; Smith, I.F.C. Methodologies for model-free data interpretation of civil engineering structures. *Comput. Struct.* **2010**, *88*, 467–482. [CrossRef]

14. Juutilainen, I.; Röning, J. A Method for Measuring Distance From a Training Data Set. *Commun. Stat. Theory* **2007**, *36*, 2625–2639. [CrossRef]

15. Posenato, D.; Lanata, F.; Inaudi, D.; Smith, I.F.C. Model-free data interpretation for continuous monitoring of complex structures. *Adv. Eng. Inform.* **2008**, *22*, 135–144. [CrossRef]

16. Amezquita-Sanchez, J.P.; Adeli, H. Synchrosqueezed wavelet transform-fractality model for locating, detecting, and quantifying damage in smart highrise building structures. *Smart Mater. Struct.* **2015**, *24*, 065034. [CrossRef]

17. Shahsavari, V.; Chouinard, L.; Bastien, J. Wavelet-based analysis of mode shapes for statistical detection and localization of damage in beams using likelihood ratio test. *Eng. Struct.* **2017**, *132*, 494–507. [CrossRef]

18. Samaratunga, D.; Jha, R.; Gopalakrishnan, S. Wavelet spectral finite element for modeling guided wave propagation and damage detection in stiffened composite panels. *Struct. Health Monit.* **2016**, *15*, 317–334. [CrossRef]

19. Rajendran, P.; Srinivasan, S.M. Identification of Added Mass in the Composite Plate Structure Based on Wavelet Packet Transform. *Strain* **2016**, *52*, 14–25. [CrossRef]

20. Yang, C.; Oyadiji, S.O. Damage detection using modal frequency curve and squared residual wavelet coefficients-based damage indicator. *Mech. Syst. Signal Process.* **2017**, *83*, 385–405. [CrossRef]

21. Shu, J.; Zhang, Z.; Gonzalez, I.; Karoumi, R. The application of a damage detection method using Artificial Neural Network and train-induced vibrations on a simplified railway bridge model. *Eng. Struct.* **2013**, *52*, 408–421. [CrossRef]

22. Tsai, C.-H.; Hsu, D.-S. Diagnosis of Reinforced Concrete Structural Damage Base on Displacement Time History using the Back-Propagation Neural Network Technique. *J. Comput. Civ. Eng.* **2002**, *16*, 49–58. [CrossRef]

23. Story, B.A.; Fry, G.T. A Structural Impairment Detection System Using Competitive Arrays of Artificial Neural Networks. *Comput-Aided Civ. Infrastruct. Eng.* **2014**, *29*, 180–190. [CrossRef]

24. Lam, H.F.; Ng, C.T. The selection of pattern features for structural damage detection using an extended Bayesian ANN algorithm. *Eng. Struct.* **2008**, *30*, 2762–2770. [CrossRef]

25. Amiri, G.G.; Rad, A.A.; Aghajari, S.; Hazaveh, N.K. hazaveh, N.K. Generation of Near-Field Artificial Ground Motions Compatible with Median-Predicted Spectra Using PSO-Based Neural Network and Wavelet Analysis. *Comput-Aided Civ. Infrastruct. Eng.* **2012**, *27*, 711–730. [CrossRef]

26. Jajo, N.K. A Review of Robust Regression and Diagnostic Procedures in Linear Regression. *Acta Math. Appl. Sin.* **2005**, *21*, 209–224. [CrossRef]

27. Laory, I.; Trinh, T.N.; Smith, I.F.C. Evaluating two model-free data interpretation methods for measurements that are influenced by temperature. *Adv. Eng. Inform.* **2011**, *25*, 495–506. [CrossRef]

28. Kerschen, G.; Yan, A.M.; Golinval, J.C. Distortion function and clustering for local linear models. *J. Sound Vib.* **2005**, *280*, 443–448. [CrossRef]

29. Wang, C.; Guan, W.; Gou, J.; Hou, F.; Bai, J.; Yan, G. Principal component analysis based three-dimensional operational modal analysis. *Int. J. Appl. Electrom.* **2014**, *45*, 137–144. [CrossRef]

30. Liu, Y.; Li, K.; Song, S.; Sun, Y.; Huang, Y.; Wang, J. The research of spacecraft electrical characteristics identification and diagnosis using PCA feature extraction. In Proceedings of the International Conference on Signal Processing, Hangzhou, China, 19–23 October 2014; pp. 1413–1417.

31. Pei, J.; Huang, Y.; Huo, W.; Wu, J.; Yang, J.; Yang, H. SAR Imagery Feature Extraction Using 2DPCA-Based Two-Dimensional Neighborhood Virtual Points Discriminant Embedding. *IEEE J. Sel. Top. Appl. Earth Obs. Remote Sens.* **2017**, *9*, 2206–2214. [CrossRef]

32. Nguyen, V.H.; Golinval, J.-C. Fault detection based on Kernel Principal Component Analysis. *Eng. Struct.* **2010**, *32*, 3683–3691. [CrossRef]

33. Tannahill, B.K. Big Data Analytic Paradigms—From PCA to Deep Learning. In Proceedings of the Association for the Advancement of Artificial Intelligence Symposium (AAAI 2014), Quebec, QU, Canada, 27–31 July 2014; pp. 84–90.

34. Torres-Arredondo, M.A.; Tibaduiza, D.A.; Mujica, L.E.; Rodellar, J.; Fritzen, C.-P. Data-Driven Multivariate Algorithms for Damage Detection and Identification: Evaluation and Comparison. *Struct. Health Monit.* **2014**, *13*, 19–32. [CrossRef]

35. Tian, L.; Fan, C.; Ming, Y.; Jin, Y. Stacked PCA Network (SPCANet): An effective deep learning for face recognition. In Proceedings of the IEEE International Conference on Digital Signal Processing, Singapore, 21–24 July 2015; pp. 1039–1043.

36. Datteo, A.; Lucà, F.; Busca, G. Statistical pattern recognition approach for long-time monitoring of the G.Meazza stadium by means of AR models and PCA. *Eng. Struct.* **2017**, *153*, 317–333. [CrossRef]

37. Liu, F.; Wang, W.; Shen, T.; Peng, J.; Kong, W. Rapid Identification of Kudzu Powder of Different Origins Using Laser-Induced Breakdown Spectroscopy. *Sensors* **2019**, *19*, 1453. [CrossRef] [PubMed]

38. Malekzadeh, M.; Atia, G.; Catbas, F.N. Performance-based structural health monitoring through an innovative hybrid data interpretation framework. *J Civil Struct. Health Monit* **2015**, *5*, 287–305. [CrossRef]

39. Laory, I.; Trinh, T.N.; Posenato, D.; Smith, I.F.C. Combined Model-Free Data-Interpretation Methodologies for Damage Detection during Continuous Monitoring of Structures. *J. Comput. Civ. Eng.* **2013**, *27*, 657–666. [CrossRef]

40. Hubert, M.; Verboven, S. A robust PCR method for high-dimensional regressors. *J. Chemometr.* **2003**, *17*, 438–452. [CrossRef]

sensors

MDPI

Article

Length Effect on the Stress Detection of Prestressed Steel Strands Based on Electromagnetic Oscillation Method

Benniu Zhang [1,*,†], Chong Tu [1,†], Xingxing Li [2], Hongmei Cui [3] and Gang Zheng [1]

[1] School of Civil Engineering, Chongqing Jiaotong University, Chongqing 400074, China;
    tu_chong@foxmail.com (C.T.); zhenggang@cmhk.com (G.Z.)
[2] School of Information Science and Engineering, Chongqing Jiaotong University, Chongqing 400074, China;
    xingxingli331@cqjtu.edu.cn
[3] Chongqing Telecommunication Polytechnic College, Chongqing 402247, China; cui_hongmei@foxmail.com
* Correspondence: 990020030755@cqjtu.edu.cn; Tel.: +86-023-6265-2316
† These authors contributed equally to this work.

Received: 8 May 2019; Accepted: 17 June 2019; Published: 20 June 2019

check for updates

Abstract: Prestress detection of structures has been puzzling structural engineers for a long time. The inductance–capacitance (LC) electromagnetic oscillation method has shown a potential solution to this problem. It connects the two ends of a steel strand, which is simulated as an inductor, to the oscillation circuit, and the stress of the steel strand can be calculated by measuring the oscillation frequency of the circuit through a frequency meter. In the previous studies, the authors found that stress-frequency relation of 1.2 m steel strand was negatively correlated, while the stress-frequency of 10 m steel strand was positively correlated. To verify this conflict, two kinds of electrical inductance models of steel strands were established to fit the lengths. With the models, the stress-frequency relations of steel strands with different lengths were analyzed. After that, two kinds of experimental platforms were set up, and a series of stress-frequency relationship tests were carried out with 1.2 m, 5 m, 10 m and 15 m steel strands. Theoretical analysis and experimental results show that when the length is less than 2.013 m, the stress and oscillation frequencies are negatively correlated; when length is more than 2.199 m, the stress and oscillation frequencies are positively correlated; while when length is between 2.013 m and 2.199 m, the stress-frequency relationship is in transit from negative correlation to positive correlation.

Keywords: length effect; stress detection; electromagnetic oscillation; steel strand; concrete structures

1. Introduction

Compared with ordinary reinforced concrete structures, prestressed concrete structures have a lighter weight, smaller cross-section size, and better economy. The determination of the prestress loss is an important criterion for judging the bearing capacity of the prestressed structure. However, the prestress loss caused by the shrinkage and creep of the concrete and the relaxation of the prestressed steel strand directly threatens the safety of the structure. Therefore, it is particularly important to detect the effective prestressing force in the existing structure. The current prestress detection research includes the following.

1.1. Ultrasound Guided Wave Method

The ultrasonic guided wave method uses the penetration and propagation of ultrasonic waves to achieve stress detection. In recent years, ultrasonic guided waves have made some progress in the field of concrete structural stress nondestructive detection. For example, Qian et al. [1] verified the effect of ultrasonic guided wave energy entropy spectroscopy on the stress of 7-wire steel strands by numerical simulation and a series of tests. Niederleithinger et al. [2] used a network of 20 ultrasonic transducers to qualitatively detect the stress state of concrete beams. Feng et al. [3] designed an instrument with an excitation and receiving frequency greater than 100 kHz for steel strand damage detection based on the principle of the ultrasonic guided wave. The device can identify the frequency change caused by the groove with a cross-section loss of 1.13%. Xu et al. [4] used the ultrasonic guided wave below 400 kHz to realize the detection of steel strand breakage defects. Farhidzadeh et al. [5–7] achieved some success.

1.2. Embedded Fiber Sensors or Magnetoelastic Devices

Embedded fiber optic sensors use the change of optical signal to monitor the stress of structures. This type of technology has achieved some significant results in the field of civil engineering [8–10]. Typically, Huynh et al. [11] embedded optical fiber sensors in unbonded prestressed concrete beams to monitor the stress loss of prestressing tendons under temperature. Kim et al. [12] used fiber optic sensors to monitor the stress variation of the 60 m span beam during the whole construction process. Compared with the traditional Bragg grating, the advantage is that these sensors can still effectively monitor the strain after the concrete has begun cracking. In addition, Lan et al. [13] made some great progress in the research of fiber optic sensors to monitor the stress of steel strand.

The magneto-elastic devices, which are based on the magneto-elasticity method, utilize the characteristics that the ferromagnetic material changes magnetically under mechanical stress to perform stress monitoring. As a typical ferromagnetic material [14], steel strands have a corresponding relationship between stress and magnetic properties. The magneto-elastic effect is mainly used in cables, stress detection of unbonded steel strands [15,16], and prestress monitoring of new structures with a bonded steel strand [17–19]. Usually, the excitation coil emits the excitation magnetic field, and the receiving coil receives the magnetic field passing through the material. According to the relationship between the output voltage and the permeability of the material, the corresponding stress of the cable under a certain voltage is obtained. Yim at al. [20] developed a set of stress devices by using the magnetostrictive inverse effect of ferromagnetic materials, which was applied to stress monitoring of bridge cables. By comparing the numerical analysis and monitoring results, the effectiveness of the electromagnetic device in cable force monitoring was verified.

1.3. Electromagnetic Oscillation Method

Inspired by the magneto-elastic effect, the author proposed a prestressed steel strand stress detection method based on LC electromagnetic oscillation [21,22], indirectly measuring the strand stress through the electromagnetic oscillation frequency. According to the electrical theory, prestressed steel strands exhibit single conductance properties in low frequency oscillating circuits and can be divided into inductors, capacitors and resistors. When the steel strand is tensioned axially, the rate of inductance change is much larger than that of capacitance and resistance. Therefore, based on the LC electromagnetic oscillation to measure the stress of the prestressing strand, the capacitance and resistance can be neglected, and the steel strand can be modeled as an inductance. In the stress detection process, only the two ends of the strand were connected to the oscillating circuit. The axial tensile force was applied to the steel strand, and the stress signal of the prestressed steel strand was converted into a circuit oscillation frequency signal, and the stress value of the steel strand can be calculated by collecting the oscillation frequency by the frequency meter.

In the previous study [21], the theoretical model of the stress and oscillation frequencies of a steel strand was established, and the axial tensile test was carried out with six sets of 7-wire prestressed steel strands with a length of 1.2 m. This study was the first time that the LC electromagnetic oscillation method had been applied to the stress measurement of a steel strand by the authors, and the feasibility of this method was proved. The theoretical analysis and experimental results of the study show that the circuit oscillation frequency decreases with the increase of steel strand stress. In another study [22], Taylor's expansion equation was used to establish the relationship between steel strand stress and oscillation frequency, and a 7-wire prestressed steel strand with a length of 10 m was used for cyclic loading experiments. The research results show that the oscillation frequency of the circuit increases with the increase of stress.

The above studies find that the stress and oscillation frequency of the two different lengths of steel strands show opposite trends. In the case of using the same oscillating circuit and measurement method, the relationship between the stress of the 1.2 m and 10 m steel strands and the oscillation frequency is completely opposite. The results of previous studies [21,22] cannot explain the reasons for this phenomenon. Only the force–frequency variation trend of 1.2 m and 10 m steel strands were studied, and the feasibility of the inductance–capacitance (LC) electromagnetic oscillation method was verified, while the reasons for the positive (negative) correlation of force–frequency variation were not explained. According to the author's preliminary analysis, this phenomenon is directly related to the length of steel strand. Therefore, this paper has carried out related theoretical and experimental research on the length effect of the LC electromagnetic oscillation method for the stress detection of steel strands.

2. Theories and Models

According to the electrical theory, as part of the LC low-frequency electromagnetic oscillating circuit, the axial stretching of the strand is mainly represented by the change of inductance. The possible reasons for the opposite stress-frequency relationship between different lengths of steel strands can be described by the following two aspects: (a) when the length is short, the spiral characteristics of the seven steel wires are not obvious, but are similar to the segment wire, so it mainly shows the inductance characteristics of the segment wire; (b) when a certain length is reached, the spiral characteristics of seven steel wires have been formed. The steel strand approximates the spiral coil structure, and mainly exhibits the spiral coil inductance characteristics.

Different length steel strands show different inductance characteristics. Therefore, in order to further study the influence of the relationship between stress and oscillation frequency, the author establishes the mathematical models of stress-frequency under two inductance models based on the model of conductor inductance and the model of spiral coil inductance and the principle of LC electromagnetic oscillation.

2.1. Principle of LC Electromagnetic Oscillation

If a steel strand under axial tension is regarded as having the equivalent inductance of an LC oscillating circuit, its magnetic and inductance parameters will change when stretched, and then the internal stress of the steel strand can be expressed by frequency through the oscillating relationship between inductance and capacitance in the electromagnetic oscillating circuit.

As shown in Figure 1, the LC oscillating circuit, also be called the LC resonant circuit, which consists of an inductor and a capacitor connected together. A steel strand is used instead of the inductor to connect to the oscillating circuit, and a fixed capacitor component is externally connected to form an LC oscillating circuit. The basic working principle can be described as follows: (a) when the capacitor starts to discharge, the inductor is charged. When the voltage of the inductor reaches the maximum, the capacitor is discharged and the inductor is charged; (b) the inductor begins to discharge, the capacitor is charged, and when the capacitor is charged, the inductor is discharged and the capacitor is charged. In this reciprocating operation, resonance occurs.

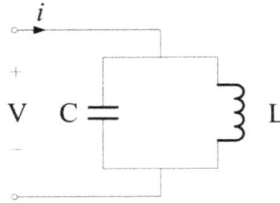

Figure 1. The inductance–capacitance (LC) oscillating circuit.

It is assumed that the circuit has no energy loss and is not affected by other external factors. When the circuit resonates, the inductive reactance X_L of the inductor L in the oscillating circuit is equal to the capacitive reactance X_C of the capacitor C.

$$X_L = X_C \tag{1}$$

The inductance X_L of the inductor can be expressed as

$$X_L = 2\pi f L \tag{2}$$

The reactance X_C of the capacitor can be expressed as

$$X_C = \frac{1}{2\pi f C} \tag{3}$$

Therefore, the relationship between the oscillation frequency of the circuit and the inductance of steel strand can be obtained by combining Equations (1)–(3).

$$2\pi f L = \frac{1}{2\pi f C} \tag{4}$$

The oscillation frequency of the circuit can be expressed as

$$f = \frac{1}{2\pi \sqrt{LC}} \tag{5}$$

In the equation, f is the oscillation frequency of the circuit, L is the inductance of the steel strand, and C is the capacitance of the circuit.

C is a constant, and the circuit oscillation frequency f can be directly measured by the frequency. Therefore, it is only necessary to establish the relationship between the inductance L and the stress σ, and the relationship model between the stress of the steel strand and the oscillation frequency of the circuit can be obtained.

2.2. Inductance Model of Segment Wire

When the steel strand is short, the spiral structure has not yet been formed, and its structural characteristics are similar to those of the line segment, so that the inductance characteristics of the segment wire are mainly exhibited. In this section, the model of stress-frequency of the steel strand is established based on the model of inductance of the segment wire.

2.2.1. Modeling

When a conductor passes a current, a magnetic field is generated inside and around the conductor. The magnetic circuit is a concentric closed magnetic field ring, and the direction is determined by the right hand rule. When the current changes, the magnetic linkage changes accordingly. For non-magnetic materials, the inductance of the conductor is the ratio of its total flux chain to the current, also known as static inductance.

$$L = \frac{\psi}{I} \tag{6}$$

In the equation, ψ is the self-inductive magnetic linkage of the conductor, I is the circuit current, and L is the inductance of the conductor.

In the previous study [21], the authors discussed the internal inductance of the wire, and the internal inductance L_{in} can be expressed as

$$L_{in} = \frac{\psi_{in}}{I} = \frac{\mu l}{8\pi} \tag{7}$$

In the equation, L_{in} is the internal inductance of the wire, μ is the permeability of the wire, and the l is the length of the wire.

As shown in Figure 2, for the external inductance produced by the wire segment, the external magnetic circuit contains the current of the whole wire. According to the Biot-Savart law, the magnetic induction intensity at any point outside the strand can be obtained.

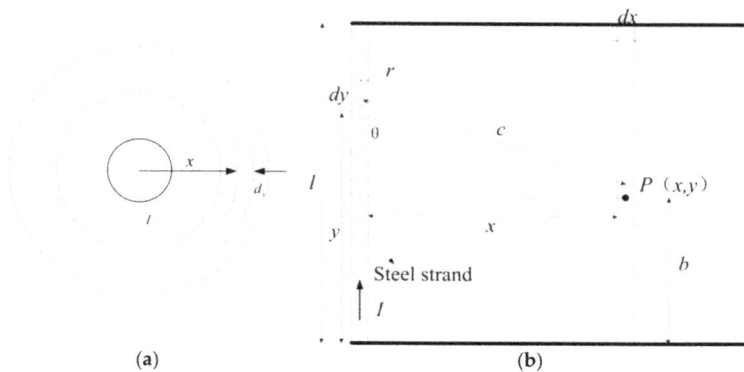

Figure 2. Calculation of external magnetic linkage of segment wire. (**a**) The distribution of external flux linkage; (**b**) Magnetic field model.

The Biot-Savart law can be defined as

$$d\vec{B} = \frac{\mu_0}{4\pi} \cdot \frac{Id\vec{l} \times \vec{r}}{r^3} \tag{8}$$

The magnetic induction intensity dB produced by current element Idy at point P can be expressed as

$$dB = \frac{\mu_0 I}{4\pi} \cdot \frac{dy}{c^2} \sin\theta = \frac{\mu_0 I}{4\pi} \cdot \frac{Ixdy}{\left[x^2 + [y-b]^2\right]^{\frac{3}{2}}} \tag{9}$$

Hence the magnitude of the magnetic induction intensity scalar produced by the segment wire at point P is as follows:

$$B = \frac{\mu_0}{4\pi} \int_0^l \frac{Ix}{\left[x^2 + [y-b]^2\right]^{\frac{3}{2}}} dy = \frac{\mu_0 I}{4\pi} \left[\frac{l-b}{x\sqrt{x^2 + [l-b]^2}} + \frac{b}{x\sqrt{x^2 + b^2}} \right] \tag{10}$$

If the wire radius is r, the total flux of the wire segment l is

$$\Phi_e = \int \vec{B} d\vec{S} = \int_r^\infty \int_0^l \frac{\mu_0 I}{4\pi} \left[\frac{l-b}{x\sqrt{x^2+[l-b]^2}} + \frac{b}{x\sqrt{x^2+b^2}} \right] dxdy = \frac{\mu_0 I}{2\pi} \left(l\cdot ln\frac{l+\sqrt{l^2+r^2}}{r} - \sqrt{l^2+r^2} + r \right) \tag{11}$$

All fluxes are only interlinked with a single wire, so the external magnetic linkage is equal to the flux. Which is

$$\Phi_e = \psi_{out} \tag{12}$$

Therefore, the external inductance of the wire can be expressed as

$$L_{out} = \frac{\psi_{out}}{I} = \frac{\mu_0}{2\pi} \left(l\cdot ln\frac{l+\sqrt{l^2+r^2}}{r} - \sqrt{l^2+r^2} + r \right) \tag{13}$$

The inductance of the wire is the sum of the internal inductance and the external inductance. Namely,

$$L = L_{in} + L_{out} = \frac{\mu l}{8\pi} + \frac{\mu_0}{2\pi} \left(l\cdot ln\frac{l+\sqrt{l^2+r^2}}{r} - \sqrt{l^2+r^2} + r \right) \tag{14}$$

The length of the wire is much larger than the radius, that is, $l \gg r$, the above formula can be simplified to

$$L = \frac{\mu l}{8\pi} + \frac{\mu_0 l}{2\pi} \left(ln\frac{2l}{r} - 1 \right) \tag{15}$$

The steel strand can be equivalent to a segment wire. Since the steel wire magnetic permeability μ is much larger than the vacuum magnetic permeability μ_0, the upper equation can be simplified to

$$L = \frac{\mu l}{8\pi} \tag{16}$$

The length of the steel strand after stretching can be expressed as

$$l = l_0 + \Delta l \tag{17}$$

In the equation, l_0 is the original length of the steel strand, and Δl is the elongation of the steel strand.

From the constitutive relation of materials,

$$\sigma = E\varepsilon = E\frac{\Delta l}{l} \tag{18}$$

The elongation Δl can be expressed as

$$\Delta l = \frac{\sigma l_0}{E} \tag{19}$$

In the equation, σ is the stress of the steel strand and E is the elastic modulus of the steel strand. And the permeability of steel strand can be expressed as

$$\mu = \mu_0 \mu_r \tag{20}$$

Combining (5), (17), (19) and (20), it can be obtained that

$$f = \frac{1}{2\pi \sqrt{\frac{\mu_0 \mu_r \left(l_0 + \frac{\sigma l_0}{E} \right)}{8\pi} \cdot C}} \tag{21}$$

2.2.2. Simulation Result

Substitute $\mu_0 = 4\pi \times 10^{-7}$ H·m^{-1}, $\mu_r = 1500$, $l_0 = 1.2$ m, $E = 1.95 \times 10^5$ MPa, C = 0.0059 uF, into Equation (21) and it can be obtained that

$$f = \frac{1000}{2\pi \sqrt{5.311 + 3.63 \times 10^{-6}\sigma}} \tag{22}$$

The simulation of the stress-frequency of the 1.2 m steel strand is shown in Figure 3.

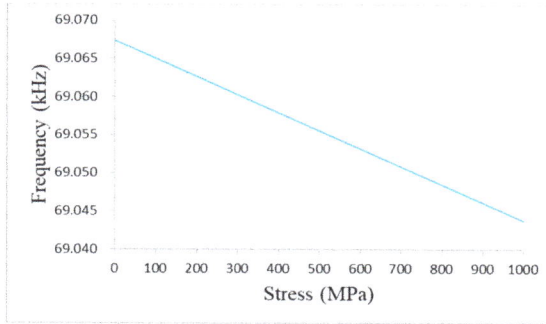

Figure 3. Stress-frequency simulation of the 1.2 m steel strand.

According to the stress-frequency relationship of the strand and Figure 3, it can be obtained that the oscillation frequency of the circuit decreases with the increase of stress. This phenomenon coincides with the author's study on the stress and frequency of a 1.2 m steel strand [21]. Therefore, it can be inferred that the 1.2 m steel strand mainly shows the inductance characteristics of the segment wire.

2.3. Inductance Model of Spiral Coil

When the steel strand reaches a certain length, the steel wire has formed a spiral structure, which is similar to the spiral coil and mainly shows the inductive characteristics of the spiral coil. It can be deduced from electrical theory that the inductance of a straight wire with a certain length is much smaller than the inductance of a spiral coil wound by a straight wire with the same length. Therefore, the inductance effect of the straight wire of the long steel strand can be neglected compared to the inductance effect of the spiral coil. Based on the spiral coil inductance model, this section establishes a relationship model between the stress and oscillation frequency of the long steel strand.

2.3.1. Modeling

As shown in Figure 4, when the steel strand is characterized by the spiral coil inductance, the peripheral steel wire can be regarded as a spiral coil, and the steel wire at the center of the cross section of the strand can be regarded as the core of the spiral coil.

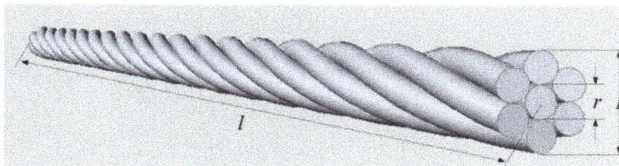

Figure 4. The structure of the steel strand.

The total magnetic linkage of the spiral coil can be expressed as

$$\psi = \mu HNS \tag{23}$$

In the equation, ψ is the total magnetic linkage of the spiral coil, μ is the magnetic permeability of the magnetic core in the spiral coil, H is the magnetic field intensity, N is the number of turns of the spiral coil, and S is the cross-sectional area of the spiral coil.

For a spiral coil, the magnetic field intensity can be defined as

$$H = \frac{NI}{l} \tag{24}$$

In the equation, I is the current of the spiral coil, and l is the length of the spiral coil.

Combining Equations (6), (23) and (24), it can be obtained that

$$L = \frac{\mu N^2 S}{l} \tag{25}$$

The cross-sectional area and length of the spiral coil will change as the steel strand is stretched. Therefore, it is necessary to calculate the amount of change of S and l after steel strand is elongated.

For the stretched steel strand, the area of coil cross-sectional can be approximately expressed as

$$S = \pi R^2 = \pi (R_0 - \Delta R) \tag{26}$$

In the equation, R_0 is the initial section radius of the steel strand, and ΔR is the amount of change of radius.

By combining Equations (17), (19), (25) and (26), and ignoring high order item ΔR^2, it can be obtained that

$$L = \frac{\mu \pi N^2 (R_0^2 + R_0 \cdot R)}{l_0 + \frac{\sigma l_0}{E}} \tag{27}$$

Since $R_0 \gg \Delta R$, Equation (27) can be simplified to

$$L = \frac{\mu \pi N^2 R_0^2}{l_0 + \frac{\sigma l_0}{E}} \tag{28}$$

By combining Equations (5), (20) and (28), it can be obtained that

$$f = \frac{1}{2\pi \sqrt{\frac{\mu_0 \mu_r \cdot \pi N^2 R_0^2}{l_0 + \frac{\sigma l_0}{E}} \cdot C}} \tag{29}$$

2.3.2. Simulation Result

Substitute $\mu_0 = 4\pi \times 10^{-7}$ H·m^{-1}, $\mu_r = 1500$, N $= 47$, $R_0 = 0.01524$ m, $l_0 = 10$ m, $E = 1.95 \times 10^5$ MPa, C $= 0.0059$ uF into Equation (29) and it can be contained that

$$f = \frac{\sqrt{1.95 \times 10^5 + \sigma}}{0.439 \pi^2} \tag{30}$$

The stress-frequency simulation result of the 10 m steel strand is shown in Figure 5.

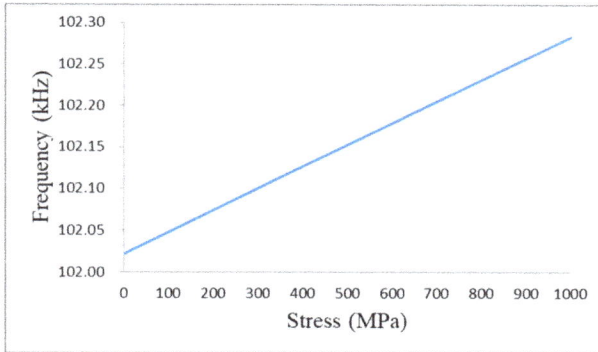

Figure 5. Stress-frequency simulation of the 10 m steel strand.

It can be seen from the relationship between the stress and frequency of the strand and Figure 5 that the oscillation frequency of the circuit increases with the increase of the stress. This phenomenon is consistent with the author's research on the stress and frequency variation of the 10 m steel strand [22]. Therefore, it can be inferred that the 10 m steel strand mainly exhibits the spiral coil inductance characteristics.

In summary, the steel strand exhibits different inductance characteristics in the LC electromagnetic oscillation circuit, and the length of steel strand is the main factor. Through the author's previous research [21,22] and the two-inductance model based on the steel strand stress-frequency model, it can be seen that different lengths of steel strands exhibit different inductance characteristics, showing different force-frequency relationships under LC electromagnetic oscillation. Namely, it can be seen that, (a) when the strand is short, it can be modeled as a inductance model of segment wire. After the axial tensile stress is applied to the steel strand, the inductance becomes large, and the relationship between the stress and the oscillation frequency is such that the stress increases and the oscillation frequency decreases; and (b) when the strand reaches a certain length, it can be modeled as a inductance model of spiral coil. After the axial tensile stress is applied to the steel strand, the inductance is reduced, and the relationship between stress and oscillation frequency is such that the stress increases and the oscillation frequency also increases.

3. Experimental Studies

3.1. Parameters of Steel Strand

In order to analyze the inductance effect model of steel strands in a practical application, steel strands with lengths of 1.2 m, 5 m, 10 m and 15 m were selected as experimental objects in combination with the test conditions, and the parameters are shown in Table 1.

Table 1. Parameters of steel strands.

Structure of Steel Strand	Length of Steel Strand l_0 m	Nominal Area of Steel Strand S mm^2	Nominal Diameter of Steel Strand D mm	Ultimate Tensile Strength R_m MPa No Less than	Maximum Tension F_m kN No Less than	Maximum Elongation A_{GT} % No Less than
1×7	1.2, 5, 10 and 15	139	15.2	1860	260	3.5

3.2. Experimental Systems

3.2.1. Experimental Devices and Procedure of Short Steel Strand (1.2 m)

The experimental system is presented in Figure 6. The steel strand, the frequency meter, and the LC oscillating circuit were connected in series to form a closed circuit, and the oscillating circuit and the frequency meter were supplied by the DC power. The hydraulic universal testing machine applied tension to the steel strand, and the stress and strain data was collected by the computer during the loading process. The accuracy of the frequency meter in the circuit is 0.0001 kHz.

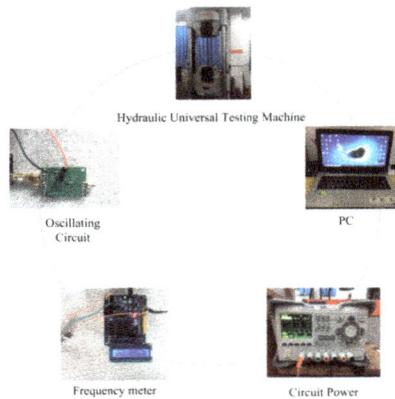

Figure 6. The experimental system of short steel strand.

The experimental process is as follows: firstly, open the material software of hydraulic universal testing machine, set parameters such as cross-sectional area of steel strand and tension rate applied in tension experiment. Then, turn on the power supply, control the universal testing machine to preload steel strand to 2 kN, and check whether the frequency meter and LC oscillation circuit are working properly. After confirming that the applied load and the strand deformation tend to be stable, uniformly apply the load to 8 kN at a loading rate of 10 mm/min.

During the loading process, it is necessary to observe the deformation of the steel strand and the accuracy of the frequency data collected by the frequency meter. The frequency meter collects frequency data once per second, and the collected data is transmitted to the mobile device via Bluetooth. Stop the loading when the load reaches 8 kN, and a set of stress and oscillation frequency test data will have been collected in one test. After the data acquisition is completed, the universal testing machine is controlled to unload the tensile force applied to the steel strand to zero while observing the trend of the frequency. A total of four sets of data were collected in the experiments.

3.2.2. Experimental Devices and Procedure of Long Steel Strand (5 m, 10 m and 15 m)

In the construction of the test system, the steel strand and hydraulic jack should be fixed on the steel reaction frame to ensure the normal follow-up work. Then, choose two areas near the middle span of the strand, grind the surface protective layer and rust with sandpaper, and paste the strain gauge. Finally, weld the wire of the circuit at the two free ends of the strand, and ensure the lead-out wire is connected to the LC oscillation circuit. Finally, connect the strain acquisition instrument, frequency acquisition instrument and oscillation circuit to the experimental system. The experimental framework is shown in Figure 7.

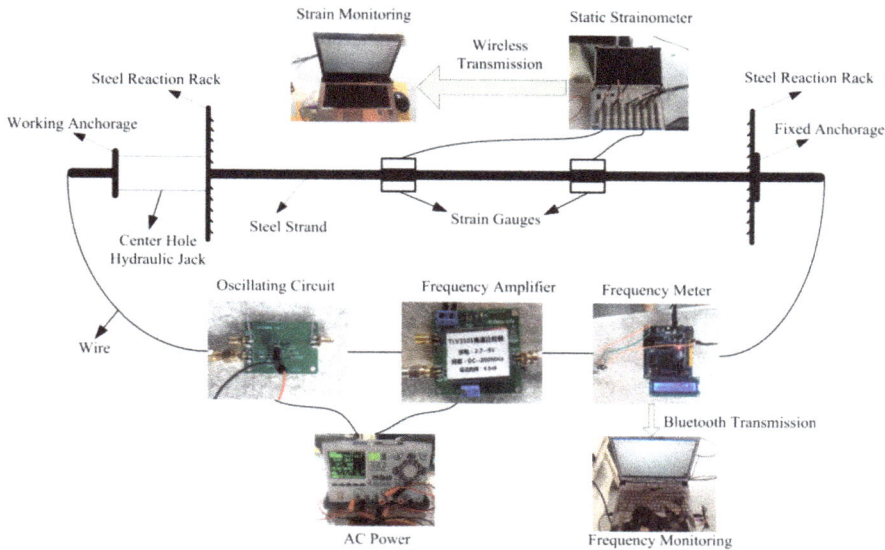

Figure 7. The experimental system of long steel strand.

Before starting the experiment, it is necessary to check the wire connection at both ends of the steel strand, and confirm that the two welded ends are wrapped with insulating tape after the welding is firm, so as to avoid separation between the two during the experiment. In addition, the data line connecting the strain gauges needs to be fixed on the steel strands, because the steel strands will rotate and drive the data lines, which will affect the connection of the data lines and the strain gauges.

After the above steps are completed, turn on the circuit to check the connection of each part and whether the circuit was oscillating. The stress and frequency test acquisition is carried out by controlling the loading and unloading of the piercing hydraulic jack to ensure the normal operation of the strain acquisition instrument and frequency acquisition device. Frequency meter and static strain acquisition instrument are set up once a second.

The stress and frequency data were collected when the stress of the steel strand reached 200 MPa, and the maximum stress applied was 900 MPa. The hydraulic jack is preloaded to 200 MPa, and the initial frequency data is collected and recorded. The stress was unloaded when loaded to 900 MPa, and the corresponding frequency data was recorded up to 200 MPa per 100 MPa of unloading. A total of four sets of data were collected in the cyclic loading experiments.

3.3. Experimental Data

In order to study the relationship between the stress and oscillation frequency of different lengths of steel strands, the median filter was used to process the frequency and stress data which were obtained under the same conditions, and the standard deviation and repeatability error of the oscillation frequency of steel strands with different lengths was calculated. Tables 2–5 show the test data of the 1.2 m, 5 m, 10 m and 15 m steel strands. Figures 8–11 show the fitting curves obtained from the test data of steel strands.

Table 2. The experimental data—1.2 m.

| Average Stress σ/MPa | Measurement Times | | | | SD | RE | MF f/kHz |
	Loading 1 f/kHz	Loading 2 f/kHz	Loading 3 f/kHz	Loading 4 f/kHz			
14.389	74.3009	74.2940	74.2748	74.2677	0.01566	0.0211%	74.2844
19.784	74.3018	74.2926	74.2746	74.2669	0.01604	0.0216%	74.2840
25.324	74.3018	74.2911	74.2740	74.2656	0.01636	0.0220%	74.2831
31.223	74.3010	74.2900	74.2733	74.2647	0.01633	0.0220%	74.2823
37.266	74.2995	74.2889	74.2720	74.2639	0.01610	0.0217%	74.2811
42.158	74.2995	74.2883	74.2720	74.2632	0.01626	0.0219%	74.2808
46.691	74.2984	74.2881	74.2712	74.2625	0.01621	0.0218%	74.2801
50.935	74.2972	74.2875	74.2705	74.2619	0.01600	0.0215%	74.2793
54.173	74.2951	74.2869	74.2696	74.2619	0.01528	0.0206%	74.2784
57.554	74.2951	74.2855	74.2687	74.2617	0.01528	0.0206%	74.2778

Table 3. The experimental data—5 m.

| Average Stress σ/MPa | Measurement Times | | | | SD | RE | MF f/kHz |
	Loading 1 f/kHz	Loading 2 f/kHz	Loading 3 f/kHz	Loading 4 f/kHz			
203	178.534	178.572	178.630	178.615	0.04345	0.0243%	178.5878
301	178.589	178.609	178.599	178.626	0.01578	0.0088%	178.6058
402	178.596	178.613	178.607	178.632	0.01508	0.0084%	178.6120
498	178.614	178.625	178.639	178.655	0.01775	0.0099%	178.6333
600	178.629	178.641	178.652	178.671	0.01784	0.0100%	178.6483
706	178.643	178.656	178.667	178.687	0.01863	0.0104%	178.6633
792	178.678	178.677	178.687	178.711	0.01582	0.0089%	178.6883
905	178.695	178.696	178.703	178.719	0.01109	0.0062%	178.7033

Table 4. The experimental data—10 m.

| Average Stress σ/MPa | Measurement Sequence | | | | | | | | SD | RE | MF f/kHz |
	Cycle 1 f/kHz Loading	Cycle 1 f/kHz Unloading	Cycle 2 f/kHz Loading	Cycle 2 f/kHz Unloading	Cycle 3 f/kHz Loading	Cycle 3 f/kHz Unloading	Cycle 4 f/kHz Loading	Cycle 4 f/kHz Unloading			
205	125.992	126.005	126.040	126.053	126.108	126.122	126.139	126.155	0.06241	0.0495%	126.0768
310	126.031	126.047	126.049	126.083	126.136	126.133	126.193	126.162	0.05999	0.0476%	126.1043
408	126.087	126.088	126.075	126.087	126.161	126.146	126.228	126.177	0.05543	0.0439%	126.1311
511	126.135	126.112	126.097	126.113	126.190	126.158	126.227	126.185	0.04571	0.0362%	126.1521
605	126.156	126.141	126.119	126.139	126.220	126.229	126.264	126.228	0.05407	0.0429%	126.1870
714	126.174	126.171	126.155	126.196	126.281	126.221	126.298	126.263	0.05474	0.0434%	126.2199
803	126.151	126.169	126.195	126.213	126.267	126.284	126.272	126.284	0.05403	0.0428%	126.2294
912	126.192	126.192	126.219	126.219	126.296	126.296	126.313	126.313	0.05427	0.0430%	126.2550

Table 5. The experimental date—15 m.

| Average Stres σ/MPa | Measurement Sequence | | | | | | | | SD | RE | MF f/kHz |
	Cycle 1 f/kHz Loading	Cycle 1 f/kHz Unloading	Cycle 2 f/kHz Loading	Cycle 2 f/kHz Unloading	Cycle 3 f/kHz Loading	Cycle 3 f/kHz Unloading	Cycle 4 f/kHz Loading	Cycle 4 f/kHz Unloading			
204	117.170	117.182	117.190	117.201	117.215	117.219	117.254	117.250	0.03049	0.0260%	117.2101
308	117.204	117.240	117.195	117.219	117.224	117.231	117.289	117.288	0.03524	0.0301%	117.2363
401	117.259	117.314	117.244	117.251	117.253	117.258	117.304	117.321	0.03172	0.0270%	117.2755
512	117.297	117.359	117.278	117.342	117.275	117.315	117.359	117.364	0.03722	0.0317%	117.3236
604	117.370	117.403	117.305	117.389	117.302	117.425	117.401	117.412	0.04747	0.0404%	117.3759
700	117.421	117.438	117.340	117.410	117.408	117.443	117.387	117.438	0.03424	0.0292%	117.4106
803	117.447	117.459	117.403	117.421	117.444	117.478	117.467	117.478	0.02681	0.0228%	117.4496
907	117.471	117.471	117.438	117.438	117.472	117.472	117.491	117.491	0.02038	0.0174%	117.4680

Notes: SD is standard deviation, RE is Repeatability error, MF is median frequency.

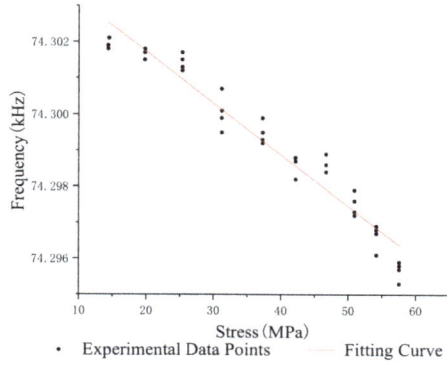

Figure 8. Stress-frequency curve—1.2 m.

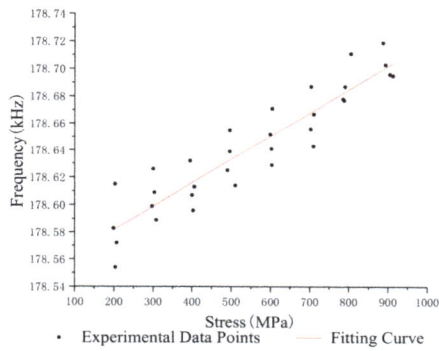

Figure 9. Stress-frequency curve—5 m.

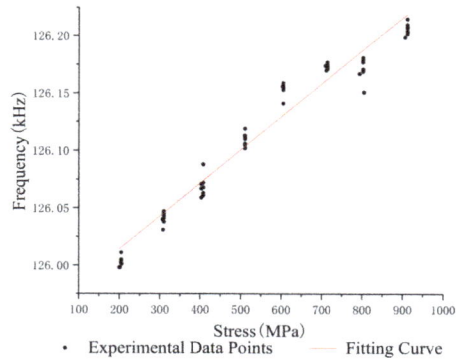

Figure 10. Stress-frequency curve—10 m.

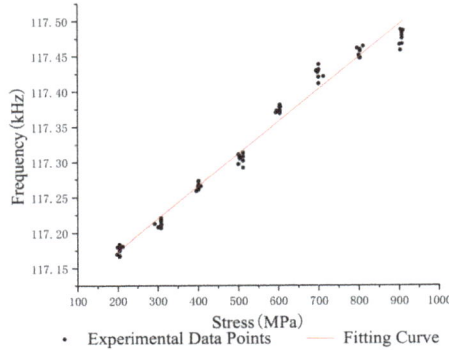

Figure 11. Curve—15 m.

The following conclusions can be drawn from the analysis:

(1) The repeatability error of the 1.2 m steel strand test data does not exceed 0.023%, and that of the 5 m, 10 m and 15 m steel strands test data are less than 0.025%, 0.05% and 0.05%.

(2) Different lengths of steel strands have different stress-frequency trends. The data analysis of the tensile test of the 1.2 m steel strand shows that the frequency decreases with the increase of stress; while the test analysis of 5 m, 10 m and 15 m steel strands shows that the frequency increases with the increase of stress. It can be seen that the 1.2 m steel strands mainly exhibit the inductance characteristics of the segment wire, while the 5 m, 10 m and 15 m steel strands exhibit the inductance characteristics of the spiral coil.

(3) The linearity of stress and frequency fitting curves of different lengths of strands is diversity. The correlation of the stress-frequency fitting curves of the 5 m, 10 m and 15 m strands are 0.8569, 0.9221 and 0.9801. With the increase in the length of the steel strand, the more concentrated the experimental data of stress-frequency is, and the better linear correlation of the curves.

(4) In summary, the steel strands exhibit different inductance characteristics in the LC electromagnetic oscillation circuit, and the length of steel strand is the main factor.

4. Results and Discussion

The theoretical derivation and experimental analysis results show that length is the critical factor affecting the stress detection of steel strand in electromagnetic resonance method. Within different length intervals, steel strands show different inductance characteristics, which make the force–frequency curve show positive and negative correlation. In order to confirm the length interval of wire inductance or spiral coil inductance of steel strand, this section is based on experimental data and the stress-frequency mathematical model to analyze the length effect of experiment and simulation.

4.1. Analysis of Length Effect

In order to obtain the critical length of the long and short steel strands determined by the experiment, four sets of experimental data of different lengths were fitted to get the force-frequency relationship curves. The experimental data were normalized and analyzed to compare the stress-frequency variation of steel strands with different lengths. The normalized stress-frequency curves are shown in Figure 12.

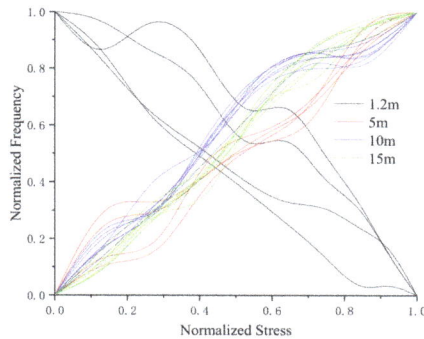

Figure 12. Stress-frequency curve of four steel strands.

It can be concluded from the figure that:

(1) The variation trend of stress-frequency curve of the four steel strands are different. The stress-frequency of the 1.2 m steel strand is negatively correlated, while the other three length are positively correlated.

(2) The stress-frequency variation trends of 1.2 m and 10 m steel strands coincided with the results of the author's previous research. What is more, the force–frequency curve of the 1.2 m steel strand shows large dispersion, and the general variation trend is that the oscillation frequency decreases with the increase of the stress. Compared with the 1.2 m steel strand, the stress-frequency curves of 5 m, 10 m and 15 m steel strands have smaller dispersion and better repeatability. The general variation trend is that the resonant frequency increases with the increase of stress.

(3) The stress-frequency variation rule of steel strands transitions from a negative correlation of 1.2 m to a positive correlation of 5 m, indicating that the critical length of the long and short steel strands is within the range of (1.2 m, 5 m). Therefore, the critical length can be obtained by fitting the relationship between f and σ in the stress-frequency fitting curve obtained from the experimental data of each length.

The critical length is related to the rate of change of the force frequency. Therefore, the length l and frequency f of the four kinds of steel strands and the slope $k_{(f/\sigma)}$ are taken as the data points of the length effect analysis. The fitting curve obtained by fitting the data points can determine the critical length. The length l of the strand has a corresponding relationship with the oscillation frequency f and the stress σ in the fitting curve equation, and the relationship can be described as a two-dimensional length effect point. Therefore, four different length steel strands correspond to four two-dimensional length effect points, and the relationship between l and $k_{(f/\sigma)}$ can be obtained by fitting. The two-dimensional length effect points obtained from the experimental data are (1.2, -0.000143), (5, 0.000174), (10, 0.000262) and (15, 0.000359). Thus, the experimental length effect curve equation can be expressed as

$$k_{(f/\sigma)} = 1.52 \times 10^{-4} \ln(l - 0.546) - 5.828 \times 10^{-5} \tag{31}$$

Similar to the experimental length effect, the simulation critical length is obtained by the relationship between f and σ in the stress-frequency fitting curve of each length. From the stress-frequency mathematical models established in the Section 2, four sets of simulated two-dimensional length effect points are (1.2, -0.000120), (5, 0.000186), (10, 0.000261) and (15, 0.000345). The equation for the simulation length effect curve can be expressed as

$$k_{(f/\sigma)} = 1.27 \times 10^{-4} \ln(l - 0.899) - 3.335 \times 10^{-5} \tag{32}$$

The experimental length effect curve and the simulation length effect curve are shown in Figure 13. In Equations (31) and (32), when $k_{(f/\sigma)}$ is zero, the critical length of the stress-frequency relationship of steel strand under the electromagnetic oscillation method can be obtained from the negative correlation to the positive correlation.

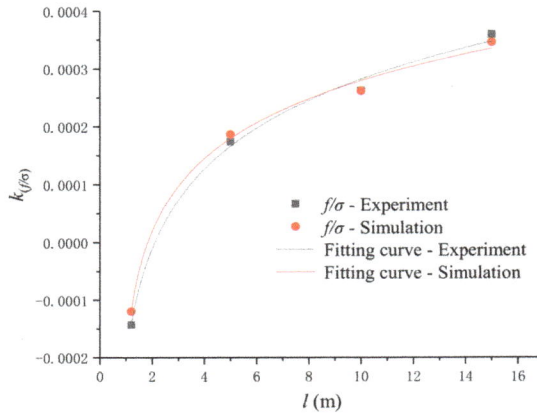

Figure 13. The length effect curves.

4.2. Analysis and Discussion

Figure 13 shows the relationship between the experimental length effect curve and the simulation length effect curve. According to length effect analysis and Figure 13, the following conclusions can be drawn from the analysis.

(1) The experimental length effect curves and Equation (31) show that the stress-frequency variation trend of steel strand is related to the length. The critical length for distinguishing long or short steel strands is $l = 2.013$ m. When $0 < l < 2.013$ m, the oscillation frequency decreases with the increase of stress; when 2.013 m $< l \leq 15$ m, the oscillation frequency increases with the increase of stress.

(2) Similar to the analysis results of experimental length effect, the stress-frequency variation trend of simulation length effect is also related to the length of steel strand. The critical length for distinguishing long or short steel strands is $l = 2.199$ m. When $0 < l < 2.199$ m, the oscillation frequency decreases with the increase of stress; when 2.199 m $< l \leq 15$ m, the oscillation frequency increases with the increase of stress.

(3) The error of critical length between experimental length effect and simulation length effect is 8.46%, which meets the requirement of application.

(4) It can be inferred from the trend of the length effect curve that when the length of steel strand is longer than 15 m, the oscillation frequency increases with the increase of stress.

It can be concluded that different length steel strands exhibit different inductance characteristics in stress detection based on LC oscillation. Both theoretical derivation and experimental analysis show that when the length of steel strand is short, the inductance characteristic of a segment wire is mainly exhibited, and the oscillation frequency decreases with the increase of stress. And when a certain length is reached, the steel strand mainly exhibits the inductance characteristics of spiral coil, and the oscillation frequency increases with the increase of stress.

To summarize, when $0 < l < 2.013$ m, the stress is negatively correlated with the oscillation frequency, that is, the oscillation frequency decreases with the increase of stress, and the main performance of steel strand is the inductance of a segment wire. When $l > 2.199$ m, the stress is positively correlated with the oscillation frequency, that is, the oscillation frequency increases with the increase of stress. At this time, the main performance of steel strand is the inductance characteristic of a spiral coil. The transition interval of length effect is 2.013 m $\leq l \leq$ 15 m, that is, the relationship between stress and oscillation frequency changes from a negative correlation to a positive correlation. The effects of length and external magnetic field are considered in this study, but if there are new influencing factors, the accuracy and consistency of the theoretical and experimental results cannot be guaranteed.

The prestress of the concrete bridge is mainly provided by the steel strand. Compared with other types of prestressed concrete bridges, the span of the simply supported girder bridges is the smallest and generally larger than 15 m. When the stress is measured by the LC electromagnetic oscillation method, the steel strand mainly shows the inductive characteristics of spiral coils. Consequently, it is appropriate to choose the stress-frequency mathematical model of steel strands based on the inductance characteristics of spiral coils in engineering applications.

5. Conclusions

This paper established the stress-frequency models of long and short steel strands. The theoretical models were validated by stress-frequency experiments of four lengths of 1.2 m, 5 m, 10 m and 15 m, and the inductance characteristics of each length of steel strand were analyzed. Both theoretical analysis and experimental research show that: (a) short steel strands mainly represent the inductance characteristics of the segment wires, the oscillation frequency decreases with the increase of stress; (b) while the long steel strands mainly represent the inductance characteristics of spiral coils, the oscillation frequency increases with stress increases.

The error of critical length obtained from the analysis of experimental length effect and simulation length effect is 8.46%, and the critical length interval for distinguishing long or short steel strands is (2.013 m, 2.199 m). When $0 < l < 2.013$ m, the oscillation frequency decreases with the increase of stress. When $l > 2.199$ m, the oscillation frequency increases with the increase of stress. When 2.013 m $\leq l \leq$ 2.199 m, the transition interval of the length effect of the stress measurement is based on the electromagnetic oscillation method, that is, the stress and the oscillation frequency are transitioned from a negative correlation to a positive correlation.

In the next stage, the influence of the length and permeability of steel strand on stress measurement will be considered comprehensively, and more frequency data will be collected to reduce the error in the experiment. In addition, the relationship between the stress and oscillation frequency of steel strands in concrete structures will also be studied as a key point in order to approach the practical engineering application.

Author Contributions: Conceptualization, B.Z.; Methodology, B.Z. and C.T.; Software, X.L. and H.C.; Validation, C.T.; Formal analysis, B.Z. and C.T.; Investigation, C.T. and X.L.; Resources, B.Z.; Data curation, C.T.; Writing—original draft preparation, C.T. and B.Z.; Writing—review and editing, C.T. and G.Z.; visualization, X.L. and H.C.; supervision, X.L. and G.Z.; Project administration, B.Z.; Funding acquisition, B.Z.

Funding: This research was funded by the National Key Research and Development Program of China, grant number 2017YFC0806007, the Program for Innovation Team Building at Institutions of Higher Education in Chongqing, grant number CXTDG201602013, the Technology Innovation Project of Chongqing Social Undertakings and People's Livelihood, grant number CSTC2016SHMSZX30026, and the Urumqi Science and Technology Plan, grant number Y161320008, the National Natural Science Foundation of China, grant number 51478072, the National Science Fund for Distinguished Young Scholars, grant number 51425801, the Joint Technical Research Project of Sichuan Tibetan Expressway Company.

Conflicts of Interest: The authors declare no conflict of interest.

Sensors **2019**, *19*, 2782

References

1. Qian, J.; Chen, X.; Sun, L.; Yao, G.; Wang, X. Numerical and experimental identification of seven-wire strand tensions using scale energy entropy spectra of ultrasonic guided waves. *Shock Vib.* **2018**, *2018*, 6905073. [CrossRef]

2. Niederleithinger, E.; Herbrand, M.; Muller, M. Monitoring of shear tests on prestressed concrete continuous beams using ultrasound and coda wave interferometry. *Bauingenieur* **2017**, *92*, 474–481.

3. Feng, H.; Liu, X.; Wu, B.; Han, Q.; He, C. Design of a miniaturised ultrasonic guided wave inspection instrument for steel strand flaw detection. *Insight* **2017**, *59*, 17–23.

4. Xu, J.; Wu, X.; Sun, P. Detecting broken-wire flaws at multiple locations in the same wire of prestressing strands using guided waves. *Ultrasonics* **2013**, *53*, 150–156. [CrossRef] [PubMed]

5. Farhidzadeh, A.; Salamone, S. Reference-free corrosion damage diagnosis in steel strands using guided ultrasonic waves. *Ultrasonics* **2015**, *57*, 198–208. [CrossRef] [PubMed]

6. Chaki, S.; Bourse, G. Guided ultrasonic waves for non-destructive monitoring of the stress levels in prestressed steel strands. *Ultrasonics* **2009**, *49*, 162–171. [CrossRef] [PubMed]

7. Rizzo, P. Ultrasonic wave propagation in progressively loaded multi-wire strands. *Exp. Mech.* **2006**, *46*, 297–306. [CrossRef]

8. Kim, J.M.; Kim, H.W.; Park, Y.H.; Yang, I.H.; Kim, Y.S. FGB sensors encapsulated into 7-wire steel strand for tension monitoring of a prestressing tendon. *Adv. Struct. Eng.* **2012**, *15*, 907–917. [CrossRef]

9. Perry, M.; Yan, Z.; Sun, Z.; Zhang, L.; Niewczas, P.; Johnston, M. High stress monitoring of prestressing tendons in nuclear concrete vessels using fibre-optic sensors. *Nucl. Eng. Des.* **2014**, *268*, 35–40. [CrossRef]

10. Leung, C.K.Y.; Wan, K.; Inaudi, D.; Bao, X.; Habel, W.; Zhou, Z.; Ou, J.; Ghandehari, M.; Wu, H.; Imai, M. Review: Optical fiber sensors for civil engineering applications. *Mater. Struct.* **2015**, *48*, 871–906. [CrossRef]

11. Huynh, T.C.; Kim, J.T. FOS-Based Prestress Force Monitoring and Temperature Effect Estimation in Unbonded Tendons of PSC Girders. *J. Aerosp. Eng.* **2017**, *30*, B4016005. [CrossRef]

12. Kim, T.M.; Kim, D.H.; Kim, M.K.; Lim, Y.M. Fiber Bragg grating-based long-gauge fiber optic sensor for monitoring of a 60 m full-scale prestressed concrete girder during lifting and loading. *Sensor. Actuators A-Phys.* **2016**, *252*, 134–145. [CrossRef]

13. Lan, C.; Zhou, Z.; Ou, J. Monitoring of structural prestress loss in RC beams by inner distributed Brillouin and fiber Bragg grating sensors on a single optical fiber. *Struct. Control Health* **2014**, *21*, 317–330. [CrossRef]

14. Militzer, M. A synchrotron look at steel. *Science* **2002**, *298*, 975–976. [CrossRef] [PubMed]

15. Duan, Y.; Zhang, R.; Zhao, Y.; Or, S.W.; Fan, K.; Tang, Z. Steel stress monitoring sensor based on elasto-magnetic effect and using magneto-electric laminated composite. *J. Appl. Phys.* **2012**, *111*, 07E516. [CrossRef]

16. Ricken, W.; Schoenekess, H.C.; Becker, W.-J. Improved multi-sensor for force measurement of pre-stressed steel cables by means of the eddy current technique. *Sens. Actuators* **2006**, *129*, 80–85. [CrossRef]

17. Guo, T.; Chen, Z.; Lu, S.; Yao, R. Monitoring and analysis of long-term prestress losses in post-tensioned concrete beams. *Measurement* **2018**, *122*, 573–581. [CrossRef]

18. Chen, Z.; Zhang, S. EM-Based monitoring and probabilistic analysis of prestress loss of bonded tendons in PSC beams. *Adv. Civ. Eng.* **2018**, 4064362. [CrossRef]

19. Kim, J.; Kim, J.W.; Lee, C.; Park, S. Development of Embedded EM Sensors for Estimating Tensile Forces of PSC Girder Bridges. *Sensors* **2017**, *17*, 1989. [CrossRef]

20. Yim, J.; Wang, M.; Shin, S. Field application of elasto-magnetic stress sensors for monitoring of cable tension force in cable-stayed bridges. *Smart Struct. Syst.* **2013**, *12*, 465–482. [CrossRef]

21. Chen, D.; Zhang, B.; Li, X.; Tu, C.; Yuan, C.; Li, W.; Zhou, Z.; Liang, Z. A stress measurement method for steel strands based on LC oscillation. *Adv. Mater. Sci. Eng.* **2018**, *2018*, 1584903. [CrossRef]

22. Li, X.; Zhang, B.; Yuan, C.; Tu, C.; Chen, D.; Chen, Z.; Li, Y. An electromagnetic oscillation method for stress measurement of steel strands. *Measurement* **2018**, *125*, 330–335. [CrossRef]

Article

Wayside Detection of Wheel Minor Defects in High-Speed Trains by a Bayesian Blind Source Separation Method

Xiao-Zhou Liu [1,2], Chi Xu [1,2] and Yi-Qing Ni [1,2,*]

[1] Hong Kong Branch of the National Rail Transit Electrification and Automation Engineering Technology Research Center, Hong Kong, China; xiaozhou.liu@connect.polyu.hk (X.-Z.L.); herbert.xu@connect.polyu.hk (C.X.)

[2] Department of Civil and Environmental Engineering, The Hong Kong Polytechnic University, Hung Hom, Kowloon, Hong Kong, China

* Correspondence: ceyqni@polyu.edu.hk; Tel.: +852-3400-8539

Received: 29 July 2019; Accepted: 12 September 2019; Published: 14 September 2019

Abstract: For high-speed trains, out-of-roundness (OOR)/defects on wheel tread with small radius deviation may suffice to give rise to severe damage on both vehicle components and track structure when they run at high speeds. It is thus highly desirable to detect the defects in a timely manner and then conduct wheel re-profiling for the defective wheels. This paper presents a wayside fiber Bragg grating (FBG)-based wheel condition monitoring system which can detect wheel tread defects online during train passage. A defect identification algorithm is developed to identify potential wheel defects with the monitoring data of rail strain response collected by the devised system. In view that minor wheel defects can only generate anomalies with low amplitude compared with the wheel load effect, advanced signal processing methods are needed to extract the defect-sensitive feature from the monitoring data. This paper explores a Bayesian blind source separation (BSS) method to decompose the rail response signal and to obtain the component that contains defect-sensitive features. After that, the potential defects are identified by analyzing anomalies in the time history based on the Chauvenet's criterion. To verify the proposed defect detection method, a blind test is conducted using a new train equipped with defective wheels. The results show that all the defects are identified and they concur well with offline wheel radius deviation measurement results. Minor defects with a radius deviation of only 0.06 mm are successfully detected.

Keywords: wheel minor defect; high-speed train; online wayside detection; Bayesian blind source separation; FBG sensor array

1. Introduction

Wheel out-of-roundness (OOR)/tread defects can impose damage to both rail tracks and vehicle components such as sleepers, wheelsets, and bearings, increasing the likelihood of derailment and undermining operational safety and ride comfort owing to high vibration amplitudes [1,2]. They can also generate ground vibration and noise that annoy residents living around the rail line [3–5]. Furthermore, while a wheel may continue to operate if it carries a small flat or polygonal shape, it is subjected to a cyclic impact load every time it rotates and the service life of key components on the vehicle-track system would be reduced [6,7]. For high-speed rail (HSR) and trains, wheel defects are the prime factor leading to faults and failures of both vehicle components and rail infrastructure in service. Due to high running speed, a wheel defect with small radius deviation within the current manufacturing/maintenance tolerance has the potential to give rise to abnormal vibration by exciting various vibration modes for the wheelsets [8,9].

To understand the causes and consequences of wheel defects, a large number of theoretical investigations and experiments have been carried out with the intention to reveal the initiation and development mechanism [2,5,7,10–13], as well as to perceive their effects on railway operation and safety through dynamic simulation [4,5,9,11,14–18]. In terms of controlling the development of wheel defects, previous studies [5,13] show that the most common and effective strategy is wheel re-profiling. In most cases, wheel defects, if caught in early stages, can be removed or machined out by re-profiling before damage becomes disastrous [19]. However, the existing mileage-based wheel re-profiling may run counter to operator's expectation by increasing the maintenance cost and reducing the service life of wheelsets. Therefore, there is a large economic incentive for adopting a condition-based maintenance (CBM) scheme which can detect and replace out-of-round wheels in time, to reduce maintenance costs for wheelsets and efficiently preventing the hazards imposed by wheel defects. Wayside wheel condition monitoring is such an efficient method under CBM scheme [20]. With the help of a wheel condition monitoring scheme, the wheelset maintenance activities can then be optimized, thereby allowing whole life costs to be reduced based on a life-cycle cost assessment.

There have been a variety of methods for wayside wheel defect detection. Included are wayside wheel load impact detectors (WILDs) [1,21–25], wayside rail acceleration detectors [26,27], wayside acoustic detectors [28,29], and wayside detectors based on laser and video camera techniques [30,31], etc. Our recent work [32] has given a brief review and comparison of wheel condition monitoring methods. It reveals that online monitoring can be more effective than offline/static inspection for wheel condition assessment and defect detection. Compared with the vehicle-borne wheel defect detection method, the wayside detectors are more suitable for massive wheel inspections. Compared with other online wayside detectors, including laser and video camera based detectors and vibration and acoustic detectors, the strain gauge based detectors confer unique benefits: (i) they are immune to train-induced vibration so they are suitable for in-service train detection, while the performance of laser and video camera based detectors can be limited by vibration during train passage; and (ii) the response signals collected by strain gauges can directly refer to wheel impact, while the effect of the excitation due to the neighboring wheels on the response features is not ignorable when the accelerometer-based and acoustic detectors are employed. The sensors in the impact detection system are usually strain gauge rosettes [33,34] or fiber Bragg grating (FBG) sensors [22]. However, most of the existing WILDs only focus on the amplitude of impact load to decide whether an impact is too great for the vehicle to remain in service [6]. If the maximum load exceeds the preset threshold, an alarm will be given. It is suitable for detection of large defects which often occur when trains run on normal-speed railways, metro lines, and freight lines. The wheel defects they investigated are deep (around 1 mm) or wide (wider than 0.1 m) flats. However, for high-speed trains, as small as 0.5 mm (radius deviation) local defect and 0.04 mm polygonal wear can be critical. As such, a more sophisticated system is needed for minor defect detection. Besides, when trains pass over the instrumented segment at low speeds, the anomalies generated by wheel defects on rail response will not be easily identified.

Therefore, in order to make a rational decision about whether a wheel should be re-profiled, a well-developed data processing procedure is demanded. This paper pursues Bayesian blind source separation (BSS) with Gaussian process (GP) model to extract the defect-sensitive feature. A defect detection procedure is then developed, which enables potential wheel minor defect identification in light of the online-monitored rail response data. The algorithms in the detection procedure are coded in MATLAB environment so that defective wheel(s) can be detected and the defect(s) can be located automatically during the passage of in-service trains.

The rest of this paper is outlined as follows. The wayside wheel condition monitoring system for rail strain data acquisition is introduced in Section 2. Section 3 presents the proposed Bayesian BSS-based wheel defect identification method, and its in-situ verification for high-speed train wheel detection is presented in Section 4. Finally, some conclusions are drawn in Section 5.

2. FBG-Based Wayside Wheel Defect Detection System

2.1. FBG Sensing System in Wayside Detection

The major challenges of the existing wayside wheel defect detection systems—including various types of WILDs, hot box detectors and laser-based systems, etc.—when applied to HSR, include the clearance required for the equipment to be installed, the need of power supply for sensors and more space claimed by deploying data acquisition system. These problems would be eliminated when using FBG sensors, which offer many advantages over conventional electrical sensors, such as immunity to electro-magnetic interference (EMI), long life-time, remote sensing and self-referencing, compact size, massive multiplexing capability, high reliability and durability, low cost and easy implementation. Specifically, in developing wayside wheel condition monitoring system, the following features of FBG sensing techniques are particularly favorable:

- Assurance of immunity to electromagnetic field: most of the conventional wheel condition monitoring systems, either resistance strain gauge- or accelerometer-based, are vulnerable to EMI induced by high voltage power supply system of modern HSR [23];
- Massive multiplexing capability: HSR always has strict requirements on clearance, which can be problematic for conventional sensing systems when considerable measuring points are needed. In contrast, FBG-based sensing system allows the use of hundreds of sensing points (FBGs) in a single fiber cable. This ability facilitates easy installation on HSR tracks with light-weight trackside equipment;
- High reliability and durability: the FBG-based sensing system can operate for more than 20 years without losses in performance even in extreme climate, such as heavy rains and snows, strong winds, or extremely hot summer days, and corrosion environment and large shocks caused by track maintenance work [22];
- Long conduction distance: the FBG-based sensing system can offer up to 100 km distant detection [23], because the optical fiber has a salient advantage in long-distance transmission with much lower signal attenuation. This allows the monitoring equipment to be installed far away from the instrumented rail section where the sensors are deployed and both the sensors and connecting fibers at the instrumented zone require no power supply.

2.2. FBG-Based Wayside Wheel Defect Detector

An FBG-based wayside rail strain response detector was developed in our recent research [35], where two FBG arrays were devised for deployment on feet of parallel rails (both left and right rails) to capture the features of potential wheel defects. The configuration of FBG array is determined based on numerical simulation presented in [36]. Through this simulation, the rail dynamic strain response subject to the excitation of defective wheel is precisely evaluated, from which the features of localized anomalies caused by wheel local defect can be revealed. This proves the feasibility of mounting strain gauges on rail foot to collect response data containing features of potential defects. The FBG sensor array deployed on rail is thereby designed which, by densely distributing the FBG sensors along a rail segment, can capture with high fidelity the localized anomaly caused by flat-defect if it exists. Figure 1 shows the deployment of FBG sensor array and configuration of the devised system. Each FBG can measure the longitudinal strain of rail foot caused by bending moment of the cross-section under the excitation of wheel dynamic load. The length of the array is slightly longer than the distance rolled over by the wheel for a complete cycle (i.e., the circumference of the wheel tread). The interval of the FBGs (denoted as *d*) along the array is around 0.15 m. This is to ensure that a few neighboring FBG sensors can concurrently detect the defect-sensitive features when a potential defect hits at any location within the instrumented rail section.

Because FBG sensors are used, power supply is not required at the railway site, the interrogator, as data logger can be installed with computer in a control room/office far away from the instrumented rail section with the use of a multi-core armored fiber optic cable to realize data transmission. As shown in Figure 1, the configuration of the devised online monitoring system consists of: (i) two FBG strain gauge arrays installed on the feet of the parallel rails; (ii) a high-speed interrogator for data collection; and (iii) a computer with data acquisition and processing software for system control and data analysis. The online detector collects rail response data at a sampling rate of 5000 Hz. The high sampling rate renders the time interval of sampling much shorter than the predicted time difference of localized anomalies, thereby the desired features can be captured.

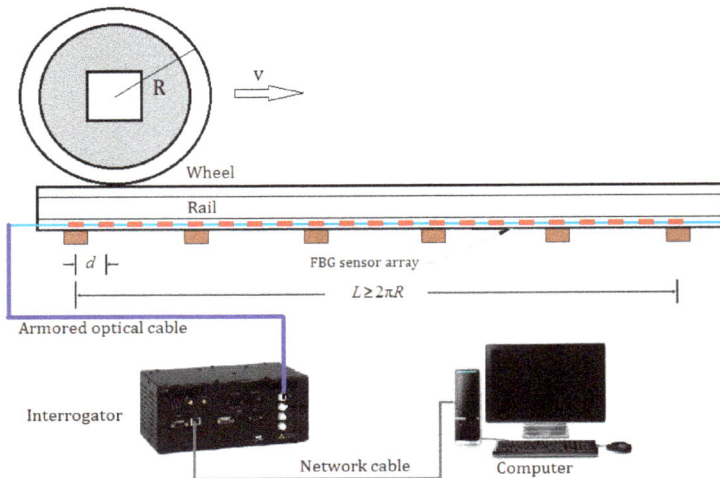

Figure 1. Deployment of FBG sensor array and configuration of the online monitoring system.

3. Wheel Defect Identification

With the online detector presented in Section 2, the rail strain responses to the excitation of passing wheels can be collected by FBG sensors deployed on the instrumented section. To detect the defective wheels as well as identify the local defects in an accurate and timely manner, a signal processing method is needed. Besides, as aforementioned, the defect detection for high-speed trains should focus on minor defects with small radius deviation, so the wheel defect detection method needs to be carefully designed in order to extract features from rail response signals that are sensitive to the defects.

The procedure of wheel defect detection in compliance with monitoring data of rail responses can be divided into three steps: firstly, the monitoring data are pre-processed using a signal extraction method, as detailed in Section 3.1. Then, the defect-sensitive feature is extracted by Bayesian BSS, as described in Section 3.2. Lastly, the potential defect is confirmed and identified by a defect confirmation scheme based on the Chauvenet's criterion, as presented in Section 3.3.

3.1. Strain Response Extraction

The requirement of real-time wheel condition monitoring and defect detection means a need for signal processing algorithm that can automatically extract response excited by each wheel from whole time history of rail strain response. To this end, the first step is to search the peak values and the corresponding time slots. The number of peaks is equal to the number of passing wheelsets. A signal processing strategy comprising four loops is developed for the response extraction, as shown in Figure 2.

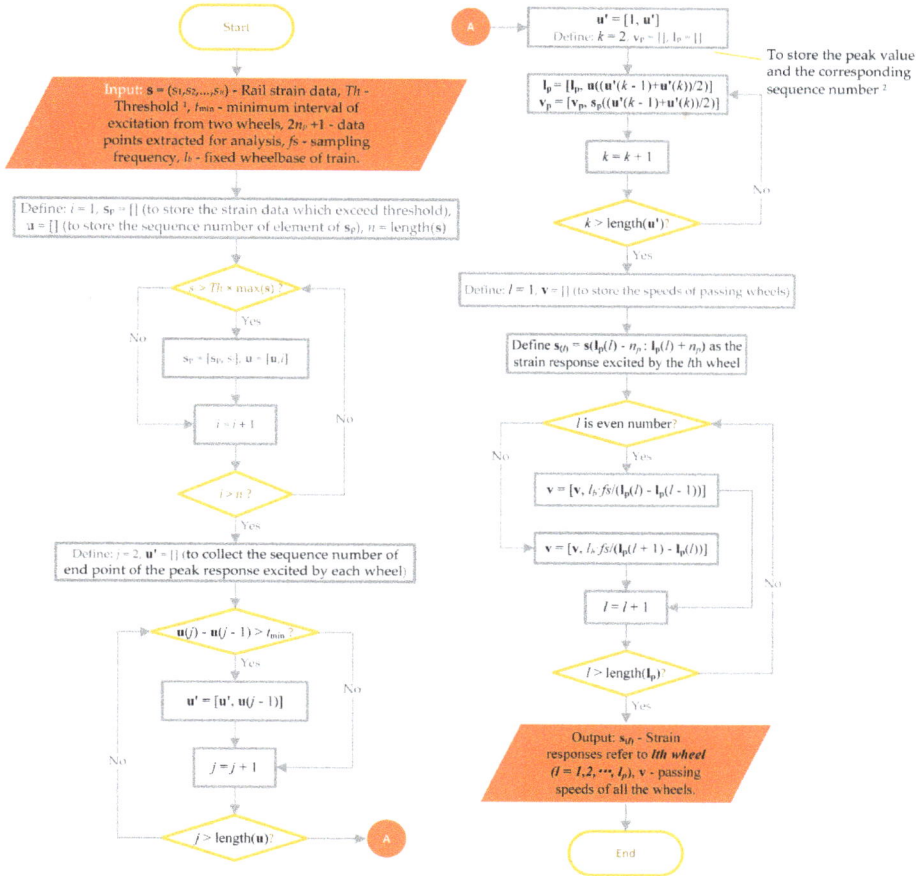

Figure 2. Procedure of rail response extraction. [1] The user-defined threshold *Th* is based on the knowledge of passing trains (e.g., heavy wagons, locomotives, metro trains, or high-speed trains). For high-speed trains concerned in this study, for example, the variance of the peak values is relatively small, the value of *Th* can be greater than 0.5. [2] In this step, the maximum strain value may not be the peak point considering that the noise in observation data may generate false peaks.

Figure 3 shows the measured strain response acquired by an FBG sensor deployed at rail foot during the passage of an eight-car train. The time history of the strain response exhibits 32 peaks in accordance with 32 wheelsets. By conducting the response extraction procedure, the strain responses corresponding to all the wheels can directly refer to the excitation of passing wheels. As such, we can extract the rail strain responses automatically with the proposed procedure, which offer a window to obtain the section of interest from the waveform of rail strain response, as indicated in Figure 3. The strain responses around their peaks obtained from different FBG sensors when a wheel passes over the instrumented rail section are illustrated in Figure 4.

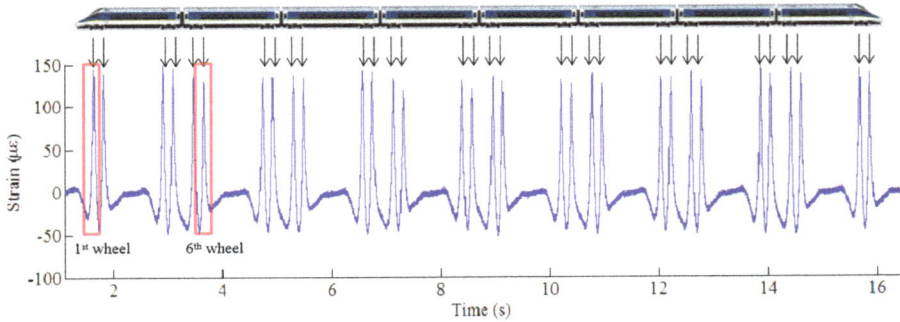

Figure 3. Measured strain response acquired by an FBG sensor deployed at rail foot.

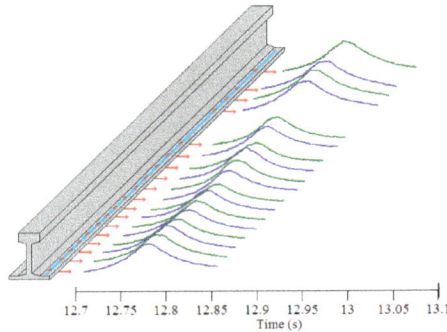

Figure 4. Strain responses acquired by different FBG sensors.

3.2. Defect-Sensitive Feature Extraction Based on Bayesian BSS

The datasets of rail response referring to all passing wheels have been obtained in Section 3.1. As shown in Figure 4, the output strain data contain a major trend that reflects the variation of rail strain response excited by the wheel passage, as well as disturbance caused by both wheel tread roughness and signal noise. In this section, the rail strain response data will be decomposed and the feature corresponding to wheel local defect will be extracted by employing Bayesian BSS, so that the effect of tread roughness of the passing wheel can be quantified.

3.2.1. Bayesian BSS

The aim of BSS is to estimate n signals (sources) and a mixing function from the m sole observations of mixtures of them. To solve the BSS problem, as an ill-posed inverse problem, the most used prior knowledge is to assume the mutual independence of each series of source signals, which leads to the development of independent component analysis (ICA) methods [37,38] and the second-order blind identification (SOBI) methods [39]. These methods cannot fully consider the temporal structure in the source signals and the difference of noise power among different channels. For rail strain response signals, the time-varying feature can be seen clearly, as shown in Figure 4. Therefore, in signal decomposition, the temporal structure in source signals should be taken into account. Besides, the FBG sensors on the array may have different performances at different locations because of uncertainties generated during the manufacturing and installation. These factors inevitably have some influence on the monitoring data. Thus, different noise power should be assigned to different channels. In recognition of this, the present study uses a Bayesian BSS framework, in which a hierarchical fully Bayesian approach for BSS problem is built.

3.2.2. Assumptions and Model Establishment

The basic form of a BSS problem at time t can be written as

$$\mathbf{X}(t) = \mathbf{Y}(t) + \mathbf{Z}(t) = \mathbf{AS}(t) + \mathbf{Z}(t) \tag{1}$$

where $\mathbf{X}(t) = [x_1(t), x_2(t), ..., x_m(t)]^T$ is a vector of size m standing for noisy observations, $\mathbf{S}(t) = [s_1(t), s_2(t), ..., s_n(t)]^T$ is a vector of size n containing the hidden sources mixing in the observation signals; $\mathbf{A}$ is called 'mixing matrix' representing the transfer function from the sources to the sensors; $\mathbf{Z}(t) = [z_1(t), z_2(t), ..., z_m(t)]^T$ is the noise vector of size m; $\mathbf{Y}(t)$ is observation signals without noise contamination. Unlike traditional BSS techniques (e.g., SOBI method) which assume that different noise sequences have a same variance, the present study considers a diagonal covariance matrix Σ_Z to model the noise sequences $\mathbf{Z}(t)$. Thus,

$$p\left(\mathbf{Z}\Big|\sum\nolimits_Z\right) = \prod_{t=1}^{L} \mathcal{N}\left(\mathbf{Z}(t); 0, \sum\nolimits_Z\right) \tag{2}$$

where L is the length of sequences $\mathbf{X}$, $\mathbf{S}$, and $\mathbf{Z}$; and $\mathcal{N}(\mathbf{Z}(t); 0, \Sigma_Z)$ represents normal distribution with the mean μ and variance σ^2. The diagonal elements in the matrix are equal to different noise power σ_i^2 of the ith observation point (i.e., $\Sigma_{Z_{ii}} = \sigma_i^2$). After modeling the noise, the likelihood function of the observation $\mathbf{X}$ can be expressed as

$$p(\mathbf{X}|\mathbf{A}, \mathbf{S}, \Sigma_Z) = \prod_{t=1}^{L} p\left(\mathbf{X}(t)\Big|\mathbf{A}, \mathbf{S}(t), \sum\nolimits_Z\right) = \prod_{t=1}^{L} \mathcal{N}\left(\mathbf{X}(t); \mathbf{AS}(t), \sum\nolimits_Z\right) \tag{3}$$

Due to the fact that each source signal of rail response is temporally correlated, GP prior is applied in the model as a prior distribution for source signals. For any finite dimensions, there always are a mean vector and a covariation matrix to describe a selected set of variables. In this model, each source signal is assumed to be a stationary GP with zero-mean, squared exponential covariation function. Thus, the source prior can be written as

$$p\left(\mathbf{S}|K_j\right) = \prod_{j=1}^{n} p\left(\mathbf{S}_j^T\right) = \prod_{j=1}^{n} \mathcal{N}\left(\mathbf{S}_j^T; 0, K_j\right) \tag{4}$$

where $\mathbf{S}_j = [s_j(t_1), s_j(t_2), ..., s_j(t_L)]$ is the jth source signal; K_j is the covariation matrix with a GP kernel expression of any two times t and t'

$$K_j(t, t') = \rho \times \exp\left(-\frac{|t - t'|^2}{2h_j}\right) \tag{5}$$

where ρ is a scale factor of the kernel that indicates the power of the generated GP, h_j is the hyperparameter of the jth source signal.

For the prior of mixing matrix $\mathbf{A}$, we consider discriminative inferences for different measuring points (FBGs) in modeling, and it is written as

$$p(\mathbf{A}|\varepsilon) = \prod_{i=1}^{m} \prod_{j=1}^{n} p\left(a_{ij}\right) = \prod_{i=1}^{m} \prod_{j=1}^{n} \mathcal{N}\left(a_{ij}; 0, \varepsilon_{ij}\right) \tag{6}$$

where a_{ij} is the element in the ith row and the jth column of the mixing matrix; ϵ_{ij} is the variance of a_{ij} and it can be considered as a hyperparameter of the mixing matrix prior.

For the distribution of the hyperparameters of noise (Σ_Z) and mixing matrix (ϵ), the conjugate prior for the variance in Gaussian likelihood is used

$$p\left(\sum{}_Z\right) = \prod_{i=1}^{m} p\left(\sum{}_{Zii}\Big|\alpha_Z, \beta_Z\right) = \prod_{i=1}^{m} IG\left(\sigma_i^2\big|\alpha_Z, \beta_Z\right) \tag{7}$$

$$p(\varepsilon) = \prod_{i=1}^{m}\prod_{j=1}^{n} p\left(\varepsilon_{ij}\big|\alpha_a, \beta_a\right) = \prod_{i=1}^{m}\prod_{j=1}^{n} IG\left(\varepsilon_{ij}\big|\alpha_a, \beta_a\right) \tag{8}$$

where α_Z, β_Z, α_a, and β_a are known parameters in inverse-gamma distribution.

Due to the non-negativity of the hyperparameter of source (h), gamma distribution is used to describe this hyperparameter's statistical feature

$$p(h) = \prod_{j=1}^{n} p(h_j) = \prod_{j=1}^{n} G\left(h_j\big|\alpha_S, \beta_S\right) \tag{9}$$

where α_S and β_S are two known parameters in the above gamma distribution.

After modeling of all the prior distributions and introducing distributions of the source and mixing matrix hyperparameters, the joint posterior distribution can be calculated by the Bayes' theorem, which is expressed as

$$\begin{aligned}
p(\mathbf{A}, \mathbf{S}, \Sigma_Z, h, \varepsilon|\mathbf{X}) &\propto p(\mathbf{X}|\mathbf{A}, \mathbf{S}, \Sigma_Z) \times p(\mathbf{A}|\varepsilon) \times p(\varepsilon) \times p(h) \times p(\Sigma_Z) \\
&= \prod_{t=1}^{L} \mathcal{N}(\mathbf{X}(t); \mathbf{AS}(t), \Sigma_Z) \times \prod_{j=1}^{n} \mathcal{N}\left(\mathbf{S}_j^T; 0, K_j\right) \times \prod_{i=1}^{m}\prod_{j=1}^{n} \mathcal{N}\left(a_{ij}\big|0, \varepsilon_{ij}\right) \\
&\times \prod_{i=1}^{m}\prod_{j=1}^{n} IG\left(\varepsilon_{ij}\big|\alpha_a, \beta_a\right) \times \prod_{j=1}^{n} G\left(h_j\big|\alpha_S, \beta_S\right) \times \prod_{i=1}^{m} IG\left(\sigma_i^2\big|\alpha_n, \beta_n\right)
\end{aligned} \tag{10}$$

To solve this joint posterior, both Gibbs sampling and Metropolis–Hastings (M-H) algorithm, as two MCMC methods, are used in Bayesian BSS model to estimate $\mathbf{A}$, $\mathbf{S}$, Σ_Z, h, and ϵ. The procedure, which was detailed in our previous research [40], consists of: (i) generating samples of the source, mixing matrix, noise covariance matrix and mixing matrix hyperparameter from the corresponding conditional posteriors $p(\mathbf{S}|\mathbf{X},\mathbf{A},\Sigma_Z,h)$, $p(\mathbf{A}|\mathbf{S},\mathbf{X},\Sigma_Z,h,\epsilon)$, $p(\Sigma_Z|\mathbf{A},\mathbf{S},\mathbf{X})$, and $p(\epsilon|\mathbf{A})$ by Gibbs sampling; (ii) deriving the expression of these conditional posteriors; and (iii) deriving the posterior of the source hyperparameter $p(h|\mathbf{S})$, which does not belong to a standard conjugate family by the M-H algorithm.

3.2.3. Defect-Sensitive Feature Extraction

The original strain response acquired by an FBG sensor situated at rail foot under the excitation of an eight-car EMU (32 wheels) is shown in Figure 4. By using the proposed Bayesian BSS method, two sources are derived. The raw response data can thereby be decomposed into two components by multiplying sources by mixing weights. Figures 5 and 6 illustrate two sets of original signals of rail response and their decomposed components. They are generated by a healthy wheel and a wheel with local defects, respectively. It is seen that in both cases, the first component is the trend of the original response signal and it reflects the rail response to an ideally rounded wheel, whereas the second component is the response excited by wheel roughness only. Comparing the two cases, the defect feature can be extracted by analyzing the second component.

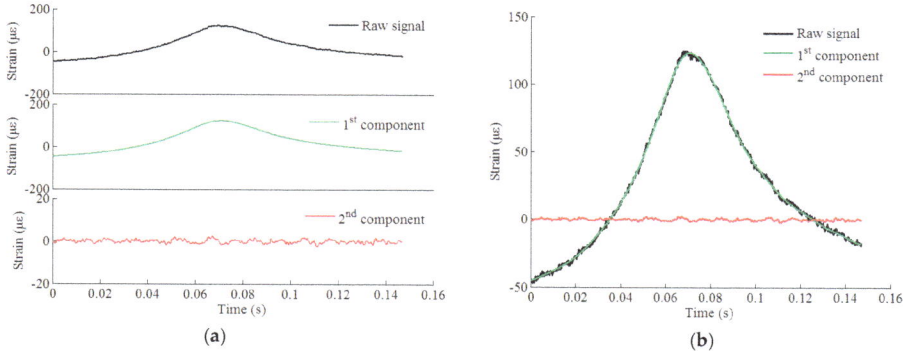

Figure 5. Raw signal of rail response to the excitation of a healthy wheel and its decomposed components: (**a**) plotted in different panels; (**b**) plotted in the same panel.

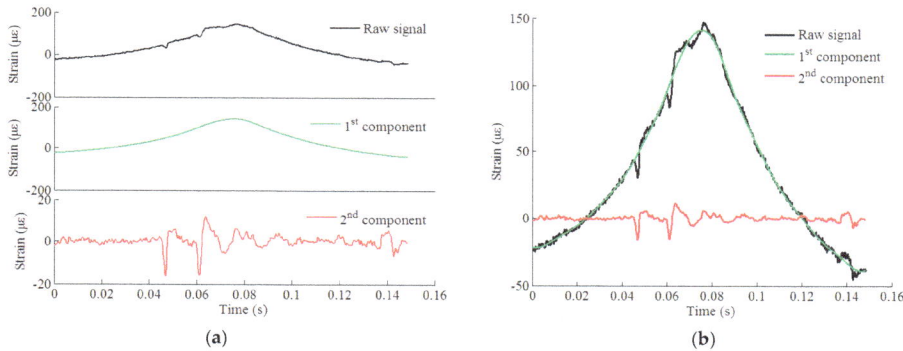

Figure 6. Raw signal of rail response to the excitation of a defective wheel and its decomposed components: (**a**) plotted in different panels; (**b**) plotted in the same panel.

3.3. Defect Identification

Pursuant to the signal decomposition and feature extraction of rail strain responses described above, this section will identify anomalies in the time history from the second signal component obtained by Bayesian BSS, targeting to detect potential defects. To the end, a threshold is to be set by a criterion for outlier detection. If a number of data points of the decomposed signal exceed the threshold, they will be recognized as localized anomalies.

Regarding the choice of criterion, it is considered that track structures and vehicle components can sustain the dynamic loads in all but the worst cases, without catastrophic failure [11]. Similarly, wheel defects that may generate such wheel–rail interaction force should be rare. Therefore, the Chauvenet's criterion is a suitable method in identifying localized anomalies that are likely to be wheel defects. A threshold (limit) for judging the anomalies from the normalized data can thereby be placed. The upper and lower limits of the probability band given by the Chauvenet's criterion are expressed in Equation (11) and Equation (12), respectively.

$$x_u = F^{-1}\left(1 - 0.25/N \middle| \mu, \sigma\right) \tag{11}$$

$$x_l = 2\mu - F^{-1}\left(1 - 0.25/N \middle| \mu, \sigma\right) \tag{12}$$

where x_u and x_l are the upper and lower limits of the probability band, F^{-1} is the normal inverse function, and N is the sample size. Given the lower and upper limits, the anomalies on the time history of normalized strain data can then be easily detected, as shown in Figure 7. Note that in this study, anomalies are the data points that are beyond lower or upper limits and the adjacent data points within a certain range in time series.

So far, the anomalies on the normalized rail strain data are obtained. However, whether these anomalies are caused by actual wheel defects still needs further investigation. Our previous studies [35,36] revealed that a strong evidence for the presence of a wheel defect is that there are more than one anomaly found on the responses collected by different FBGs and these anomalies occur at the same time period. In view of this, we can further examine the anomalies identified by the Chauvenet's criterion through a comparison of adjacent FBGs. Specifically, if an anomaly is concurrently identified from the normalized responses collected by different FBGs at the same time period, a wheel local defect can be confirmed. The features of the potential defect, including the relative response amplitude and its location on the wheel tread, can subsequently be obtained. An example of the screening mechanism was given in [35].

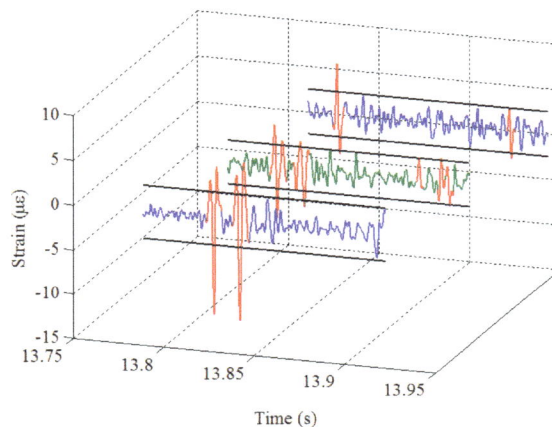

Figure 7. Detection of localized anomalies from the normalized strain data – an example of two strain response datasets: blue and green curves—normalized strain time histories from two different FBGs; black straight lines—the upper and lower thresholds specified by the Chauvenet's criterion; red curves—the anomalies identified using the Chauvenet's criterion.

The implementation procedures for online wheel defect detection in a timely manner are as follows: when a train passes over the instrumented rail section, the detector is triggered to collect data and the three-step algorithm is employed for instant data processing and evaluation. The acquired data are pre-processed first to obtain the rail response corresponding to the specific wheel load excitation (step I); afterwards, the defect-sensitive feature extraction is conducted through signal decomposition based on Bayesian BSS (step II); lastly, the localized anomalies obtained through feature extraction are used to identify potential wheel local defect by the Chauvenet's criterion (step III).

In the proposed defect detection process, there are several factors that may have effects on detection results. Among them, three issues need further exploration: the speed variation of passing trains, the temperature effect on strain measurement, and the location of FBGs with respect to sleepers. Their possible influences are discussed in the following:

- The speed variation of passing trains: The process of train passage lasts from seconds to dozens of seconds, so it is possible that the train is speeding up or slowing down during this process and the speed is not constant. However, as described in the proposed method, the condition of wheels is assessed individually, that is, the detection of each wheel is free from the interference of other wheels. Since the instrumented rail section with FBG array is only about 3 m long, the speed of each wheel is unlikely to change dramatically during its passage across the instrumented section. In addition, it has been proven that the dynamic strain monitoring data of rail obtained under different constant running speeds of a train give rise to consistent wheel defect detection results as long as the running speed is instantly measured and enough large dynamic strain of the rail is excited by the passing wheel. It is observed that when the train's running speed is lower than 20 km/h, the anomaly stemming from minor wheel defect is difficult to perceive in the measured rail dynamic strain response.

• The temperature effect on strain measurement: For strain measurement using FBG sensors, the temperature effect usually should not be ignored, since the output wavelength of FBG sensors can shift with temperature variation. However, temperature-induced wavelength change would not influence the performance of the proposed method. This is because the wavelength change caused by temperature variation mainly results in the change of baseline of the output signal. The influence of temperature can be easily eliminated by deducting the mean value of wavelength before or after train passage. Particularly in the proposed method, after pursuing BSS, the change of temperature will be reflected in the first component (source) rather than the second component (source), the latter being used for wheel defect detection. Also, the temperature variation during the short time of the wheel's passage across the instrumented section is ignorable.

• Different locations of FBGs with respect to sleepers: In this study, the FBG sensors on the array have different locations with respect to sleepers. These FBGs measure the rail strain due to bending, and the measurement result may be influenced by the distance of the sensor from the sleeper. Therefore, it is necessary to compare the signals generated by different FBGs on the array. As shown in Figure 4, under the excitation of the same wheel, the waveforms of the rail strain responses at different locations are similar. Even if there are slight differences in the amplitude of response peak, this kind of difference is mainly reflected in the first component after signal processing using BSS, rather than in the second component. Therefore, the detection results would not be affected by this issue.

4. In-Situ Verification

In this section, the proposed online detector is deployed on a rail line to verify its capability to collect singles and detect multiple wheel local defects through a blind test. Based on the test results, the performance of the devised system in local defect detection will be assessed.

4.1. Implementation of Online Detector

The devised online wheel defect detector presented in Section 2 has been implemented on a rail line, as shown in Figure 8. The devised system has a trigger module which allows the interrogator to collect wavelength data from the FBG array automatically when there is a train passing over the instrumented rail section. The monitoring data of rail strain response will be stored in a hard disk and sent to the data processing and analysis module, which integrates the wheel defect identification algorithms presented in Section 3. In this way, the condition of the wheel tread and defect information (if any) can be obtained and displayed in real time.

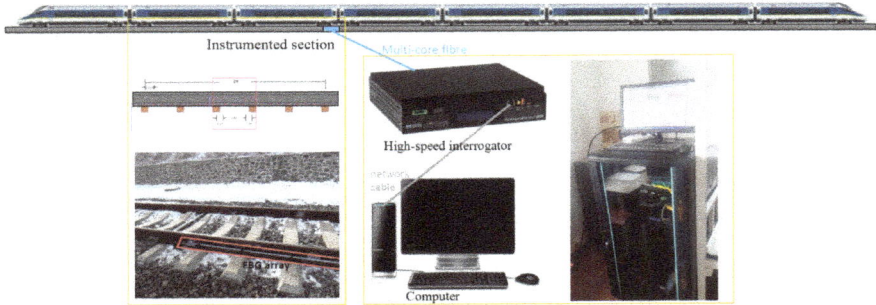

Figure 8. Configuration of the online wheel condition monitoring system.

4.2. Blind Test

To verify the proposed defect detection method, a blind test was conducted by operating a train with potential wheel local defects on the instrumented rail. The test train is a new high-speed EMU equipped with several defective wheels, as shown in Figure 9a. The distance between bogie pivot centers of the test train is 18 m and bogie wheelbase is 2.5 m. The train passes over the instrumented rail several times and the running speed ranges from 20 km/h to 50 km/h. The defective wheels have single or multiple wheel defects on their treads, but the defects were unknown before the test. The proposed defect identification method is applied to process and analyze the monitoring data collected by the online detector. By comparing the online detection results with the results of offline wheel inspection (wheel radius deviation measurement) conducted later in a depot, as shown in Figure 9b, the performance of the detector is evaluated. The detection results and performance analysis of the proposed wheel local defect detection method will be detailed in Section 4.3.

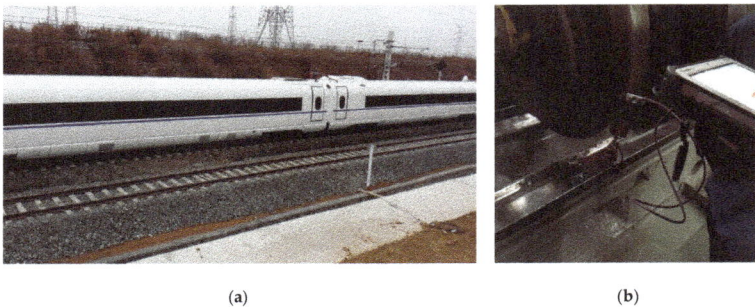

(a)   (b)

Figure 9. (**a**) The test train of an eight-car high-speed EMU; (**b**) In-depot offline wheel inspection by radius deviation measurement.

4.3. Test Results and Validation

After conducting data pre-processing (step I) and feature extraction (step II), there are 21 × 64 (21 FBGs on each array, 64 wheels) datasets of the second component of rail response signals. The defect identification algorithm (step III) is then applied to confirm the existence of potential defects. Among the rail strain responses corresponding to all 64 wheels, the right wheels of wheelsets no. 1, 6, and 24, and the left wheel of wheelset no. 27 are identified to have local anomalies. Figures 10a, 11a, 12a and 13a illustrate the second component of raw signals collected by different FBGs on the array (blue and green curves) and localized anomalies (highlighted in red) corresponding to these wheels. It is found that the defect detection results match the radius deviation measurement results (shown in Figures 10b, 11b, 12b and 13b) well in most cases, even in the multiple defect cases. Furthermore, the proposed method has excellent performance in detecting minor defects, whose depth is as low as around 0.06 mm (amplitude is −0.08 mm and baseline is around −0.02 mm), as seen in Figure 11b. It is noteworthy that in multiple defect cases, the signature of a same local defect may occur twice on the response signals collected by the FBG array because it may hit the rail twice and both hits can generate localized anomalies if the contact point is near one end of the FBG array. The online detection result for the left wheel of wheelset no. 27 is such an example, as shown in Figure 13a.

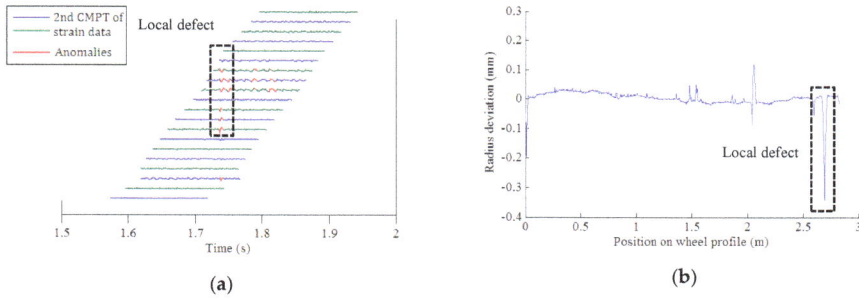

Figure 10. Defect detection results of the right wheel of wheelset no. 1: (**a**) online detection result; (**b**) offline wheel radius deviation measurement.

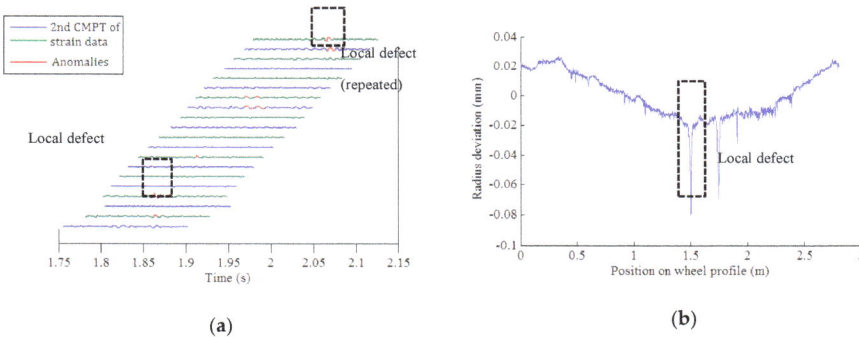

Figure 11. Defect detection results of the right wheel of wheelset no. 6: (**a**) online detection result; (**b**) offline wheel radius deviation measurement.

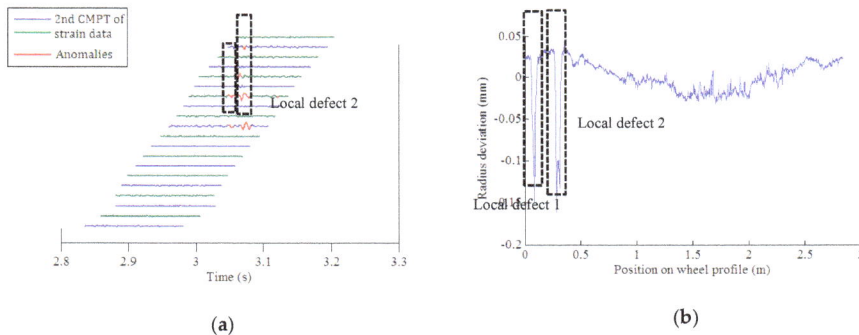

Figure 12. Defect detection results of the right wheel of wheelset no. 24: (**a**) online detection result; (**b**) offline wheel radius deviation measurement.

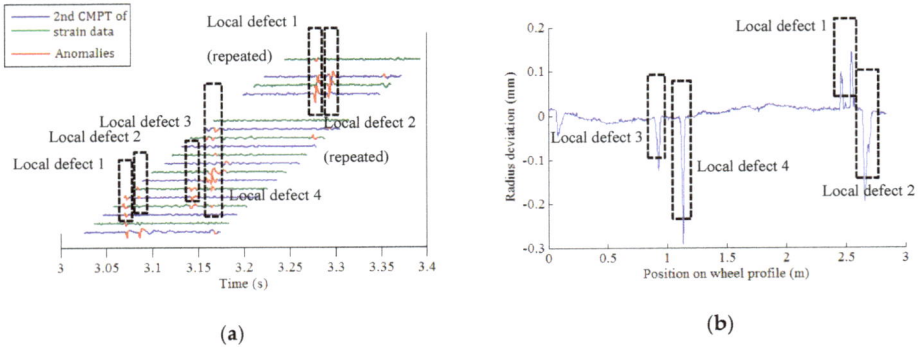

Figure 13. Defect detection results of the left wheel of wheelset no. 27: (**a**) online detection result; (**b**) offline wheel radius deviation measurement.

5. Conclusions

The demand for ensuring operation safety and reducing maintenance cost for high-speed rail calls for a well-organized CBM scheme which can provide timely and necessary information about the condition of wheels and other key components to the railway operators. To facilitate the application of online wheel defect detection methods, as a critical part of CBM scheme for rolling stocks on HSR, more efforts should be devoted to improving the performance of the existing detection methods, from the perspective of the system configuration design, performance of sensors, data acquisition system, and defect identification algorithms in line with online monitoring data. The work presented in this paper is among these efforts. It is recognized that the FBG-based wayside wheel impact detectors can be more effective for HSR wheel condition assessment and defect detection than conventional detectors. In this study, a new defect identification method for wheel minor defects which are commonly reported to cause abnormal vibration on HSR vehicle-track system is proposed, where an online monitoring system using FBG sensor arrays is employed to collect rail strain responses at multiple locations at rail feet.

In order to automatically detect potential wheel defects, this study proposes a three-step defect identification algorithm to identify local defects in light of the online monitoring data of rail strain responses. The algorithm is carefully designed to reduce both false alarms and missed hits which may induce considerable cost in HSR. Because Bayesian BSS outperforms the conventional BSS techniques in processing signals with temporal structure and quantifying measurement error/uncertainty, the proposed algorithm uses Bayesian BSS to decompose the raw signals and obtain useful features that are sensitive to wheel defects. Through data pre-processing, defect-sensitive feature extraction, and defect confirmation procedure, not only can we identify the defective wheels from all passing wheelsets, we are also able to predict the location of wheel local defects in multiple defect cases.

Blind tests were implemented to verify the performance of the proposed method. Test results indicate that the local defects can be identified with high fidelity, which are in good agreement with the offline measurements of wheel radius deviation taken in a depot. It is found that wheel defects with depth (radius deviation) as low as 0.06 mm can be successfully detected by the proposed method.

Author Contributions: Conceptualization, Y.-Q.N. and X.-Z.L.; methodology, C.X. and X.-Z.L.; doftware, C.X.; validation, X.-Z.L. and Y.-Q.N.; formal analysis, C.X.; investigation, C.X. and X.-Z.L.; resources, X.-Z.L. and Y.-Q.N.; data curation, X.-Z.L.; writing—original draft preparation, X.-Z.L. and C.X.; writing—review and editing, Y.-Q.N.; visualization, C.X. and X.-Z.L.; supervision, Y.-Q.N.; project administration, Y.-Q.N.; funding acquisition, Y.-Q.N.

Funding: This research was funded by the Research Grants Council of the Hong Kong Special Administrative Region, China (grant no. PolyU 152767/16E), the Ministry of Science and Technology of China (grant no. 2018YFE0190100), and the Innovation and Technology Commission of Hong Kong SAR Government to the Hong

Kong Branch of National Rail Transit Electrification and Automation Engineering Technology Research Center (grant no. K-BBY1).

Acknowledgments: The authors are thankful to the colleagues from Department of Electrical Engineering, The Hong Kong Polytechnic University, and CRRC Changchun Railway Vehicles Co., Ltd. who provided support in the in-situ test.

Conflicts of Interest: The authors declare no conflict of interest.

References

1. Johansson, A.; Nielsen, J.C. Out-of-round railway wheels—Wheel-rail contact forces and track response derived from field tests and numerical simulations. *Proc. Inst. Mech. Eng. Part F J. Rail Rapid Transit* **2003**, *217*, 135–146. [CrossRef]

2. Jin, X.; Wu, L.; Fang, J.; Zhong, S.; Ling, L. An investigation into the mechanism of the polygonal wear of metro train wheels and its effect on the dynamic behaviour of a wheel/rail system. *Veh. Syst. Dyn.* **2012**, *50*, 1817–1834. [CrossRef]

3. Kouroussis, G.; Connolly, D.P.; Verlinden, O. Railway-induced ground vibrations—A review of vehicle effects. *Int. J. Rail Transp.* **2014**, *2*, 69–110. [CrossRef]

4. Wu, T.X.; Thompson, D.J. A hybrid model for the noise generation due to railway wheel flats. *J. Sound Vib.* **2002**, *251*, 115–139. [CrossRef]

5. Nielsen, J. Out-of-round railway wheels. In *Wheel–Rail Interface Handbook*; Lewis, R., Olofsson, U., Eds.; Woodhead Publishing: Sawston, UK, 2009; pp. 245–279.

6. Barke, D.W.; Chiu, W.K. A review of the effects of out-of-round wheels on track and vehicle components. *Proc. Inst. Mech. Eng. Part F: J. Rail Rapid Transit* **2005**, *219*, 151–175. [CrossRef]

7. Dukkipati, R.V.; Dong, R. Impact loads due to wheel flats and shells. *Veh. Syst. Dyn.* **1999**, *31*, 1–22. [CrossRef]

8. Morys, B. Enlargement of out-of-round wheel profiles on high speed trains. *J. Sound Vib.* **1999**, *227*, 965–978. [CrossRef]

9. Wu, X.; Chi, M. Study on stress states of a wheelset axle due to a defective wheel. *J. Mech. Sci. Technol.* **2016**, *30*, 4845–4857. [CrossRef]

10. Braghin, F.; Lewis, R.; Dwyer-Joyce, R.S.; Bruni, S. A mathematical model to predict railway wheel profile evolution due to wear. *Wear* **2006**, *261*, 1253–1264. [CrossRef]

11. Liu, Y.; Stratman, B.; Mahadevan, S. Fatigue crack initiation life prediction of railroad wheels. *Int. J. Fatigue* **2006**, *28*, 747–756. [CrossRef]

12. Enblom, R. Deterioration mechanisms in the wheel–rail interface with focus on wear prediction: A literature review. *Veh. Syst. Dyn.* **2009**, *47*, 661–700. [CrossRef]

13. Wu, Y.; Du, X.; Zhang, H.J.; Wen, Z.F.; Jin, X.S. Experimental analysis of the mechanism of high-order polygonal wear of wheels of a high-speed train. *J. Zhejiang Univ. Sci. A* **2017**, *18*, 579–592. [CrossRef]

14. Baeza, L.; Fayos, J.; Roda, A.; Insa, R. High frequency railway vehicle-track dynamics through flexible rotating wheelsets. *Veh. Syst. Dyn.* **2008**, *46*, 647–659. [CrossRef]

15. Nielsen, J.C.; Lombaert, G.; François, S. A hybrid model for prediction of ground-borne vibration due to discrete wheel/rail irregularities. *J. Sound Vib.* **2015**, *345*, 103–120. [CrossRef]

16. Zhao, X.; Li, Z.; Liu, J. Wheel-rail impact and the dynamic forces at discrete supports of rails in the presence of singular rail surface defects. *Proc. Inst. Mech. Eng. Part F J. Rail Rapid Transit* **2012**, *226*, 124–139. [CrossRef]

17. Bian, J.; Gu, Y.; Murray, M.H. A dynamic wheel–rail impact analysis of railway track under wheel flat by finite element analysis. *Veh. Syst. Dyn.* **2013**, *51*, 784–797. [CrossRef]

18. Liu, X.; Zhai, W. Analysis of vertical dynamic wheel/rail interaction caused by polygonal wheels on high-speed trains. *Wear* **2014**, *314*, 282–290. [CrossRef]

19. Smith, R.A. Hatfield memorial lecture 2007 railways and materials: Synergetic progress. *Ironmak. Steelmak.* **2008**, *35*, 505–513. [CrossRef]

20. Papaelias, M.; Amini, A.; Huang, Z.; Vallely, P.; Dias, D.C.; Kerkyras, S. Online condition monitoring of rolling stock wheels and axle bearings. *Proc. Inst. Mech. Eng. Part F J. Rail Rapid Transit* **2016**, *230*, 709–723. [CrossRef]

21. Stratman, B.; Liu, Y.; Mahadevan, S. Structural health monitoring of railroad wheels using wheel impact load detectors. *J. Fail. Anal. Prev.* **2007**, *7*, 218–225. [CrossRef]

22. Filograno, M.L.; Corredera, P.; Rodríguez-Plaza, M.; Andrés-Alguacil, A.; González-Herráez, M. Wheel flat detection in high-speed railway systems using fiber Bragg gratings. *IEEE Sens. J.* **2013**, *13*, 4808–4816. [CrossRef]

23. Tam, H.Y.; Lee, T.; Ho, S.L.; Haber, T.; Graver, T.; Méndez, A. Utilization of fiber optic bragg grating sensing systems for health monitoring in railway applications. *Struct. Health Monit. Quantif. Valid. Implement.* **2007**, *1*, 1824–1831.

24. Wei, C.; Xin, Q.; Chung, W.H.; Liu, S.Y.; Tam, H.Y.; Ho, S.L. Real-time train wheel condition monitoring by fiber Bragg grating sensors. *Int. J. Distrib. Sens. Netw.* **2011**, *8*, 409048. [CrossRef]

25. Kouroussis, G.; Connolly, D.P.; Alexandrou, G.; Vogiatzis, K. Railway ground vibrations induced by wheel and rail singular defects. *Veh. Syst. Dyn.* **2015**, *53*, 1500–1519. [CrossRef]

26. Skarlatos, D.; Karakasis, K.; Trochidis, A. Railway wheel fault diagnosis using a fuzzy-logic method. *Appl. Acoust.* **2004**, *65*, 951–966. [CrossRef]

27. Belotti, V.; Crenna, F.; Michelini, R.C.; Rossi, G.B. Wheel-flat diagnostic tool via wavelet transform. *Mech. Syst. Signal Process.* **2006**, *20*, 1953–1966. [CrossRef]

28. Dybała, J.; Radkowski, S. Reduction of doppler effect for the needs of wayside condition monitoring system of railway vehicles. *Mech. Syst. Signal Process.* **2013**, *38*, 125–136. [CrossRef]

29. Zhang, D.; Entezami, M.; Stewart, E.; Roberts, C.; Yu, D. A novel doppler effect reduction method for wayside acoustic train bearing fault detection systems. *Appl. Acoust.* **2019**, *145*, 112–124. [CrossRef]

30. Asplund, M.; Lin, J. Evaluating the measurement capability of a wheel profile measurement system by using GR&R. *Measurement* **2016**, *92*, 19–27.

31. Zhang, Z.F.; Gao, Z.; Liu, Y.Y.; Jiang, F.C.; Yang, Y.L.; Ren, Y.F.; Yang, H.J.; Yang, K.; Zhang, X.D. Computer vision based method and system for online measurement of geometric parameters of train wheel sets. *Sensors* **2011**, *12*, 334–346. [CrossRef]

32. Liu, X.Z.; Ni, Y.Q.; Zhou, L. Condition-based maintenance of high-speed railway vehicle wheels through trackside monitoring. In Proceedings of the second International Workshop on Structural Health Monitoring for Railway System, Qingdao, China, 17–19 October 2018.

33. Milković, D.; Simić, G.; Jakovljević, Ž.; Tanasković, J.; Lučanin, V. Wayside system for wheel–rail contact forces measurements. *Measurement* **2013**, *46*, 3308–3318. [CrossRef]

34. Asplund, M.; Famurewa, S.; Rantatalo, M. Condition monitoring and e-maintenance solution of railway wheels. *J. Qual. Maint. Eng.* **2014**, *20*, 216–232. [CrossRef]

35. Liu, X.Z.; Ni, Y.Q. Wheel tread defect detection for high-speed trains using FBG-based online monitoring techniques. *Smart Struct. Syst.* **2018**, *21*, 687–694.

36. Ni, Y.Q.; Ying, Z.G.; Liu, X.Z. Online detection of wheel defect by extracting anomaly response features on rail: Analytical modelling, monitoring system design, and in-situ verification. In Proceedings of the 14th International Conference on Railway Engineering, Edinburgh, UK, 21–22 June 2017.

37. Hyvarinen, A. Fast and robust fixed-point algorithms for independent component analysis. *IEEE Trans. Neural Netw.* **1999**, *10*, 626–634. [CrossRef] [PubMed]

38. Delorme, A.; Makeig, S. EEGLAB: An open source toolbox for analysis of single-trial EEG dynamics including independent component analysis. *J. Neurosci. Methods* **2004**, *134*, 9–21. [CrossRef] [PubMed]

39. Belouchrani, A.; Abed-Meraim, K.; Cardoso, J.F.; Moulines, E. A blind source separation technique using second-order statistics. *IEEE Trans. Signal Process.* **1997**, *45*, 434–444. [CrossRef]

40. Xu, C.; Ni, Y.Q. A Bayesian source separation method for noisy observations by embedding Gaussian process prior. In Proceedings of the 7th World Conference on Structural Control and Monitoring, Qingdao, China, 22–25 July 2018.

sensors

MDPI

Article

Stretching Method-Based Operational Modal Analysis of An Old Masonry Lighthouse

Emmanouil Daskalakis [1,*,†], Christos G. Panagiotopoulos [2,*,†], Chrysoula Tsogka [3],
Nikolaos S. Melis [4] and Ioannis Kalogeras [4]

1   Vancouver Community College, 1155 E Broadway, Vancouver, BC V5T 4V5, Canada
2   Institute of Applied and Computational Mathematics, Foundation for Research and Technology,
    Hellas, N. Plastira 100, Vassilika Vouton, GR-700 13 Heraklion, Crete, Greece
3   Department of Applied Mathematics, University of California, Merced, 5200 North Lake Road,
    Merced, CA 95343, USA
4   Institute of Geodynamics, National Observatory of Athens, P.O. Box 20048, GR-118 10 Athens, Greece
*   Correspondence: edaskalakis@vcc.ca (E.D.); pchr@iacm.forth.gr (C.G.P.)
†   These authors contributed equally to this work.

Received: 26 July 2019; Accepted: 16 August 2019; Published: 19 August 2019

check for
updates

Abstract: We present in this paper a structural health monitoring study of the Egyptian lighthouse of Rethymnon in Crete, Greece. Using structural vibration data collected on a limited number of sensors during a 3-month period, we illustrate the potential of the stretching method for monitoring variations in the natural frequencies of the structure. The stretching method compares two signals, the current that refers to the actual state of the structure, with the reference one that characterizes the structure at a reference healthy condition. For the structure under study, an 8-day time interval is used for the reference quantity while the current quantity is computed using a time window of 24 h. Our results indicate that frequency shifts of 1% can be detected with high accuracy allowing for early damage assessment. We also provide a simple numerical model that is calibrated to match the natural frequencies estimated using the stretching method. The model is used to produce possible damage scenarios that correspond to 1% shift in the first natural frequencies. Although simple in nature, this model seems to deliver a realistic response of the structure. This is shown by comparing the response at the top of the structure to the actual measurement during a small earthquake. This is a preliminary study indicating the potential of the stretching method for structural health monitoring of historical monuments. The results are very promising. Further analysis is necessary requiring the deployment of the instrumentation (possibly with additional instruments) for a longer period of time.

Keywords: SHM; stretching method; model updating

---

1. Introduction

This work takes place in the general framework of structural health monitoring (SHM) of historic monuments [1–4]. Ideally, the monitoring system should provide a quick assessment of the state of the structure without requiring a long training period while using only a limited number of instruments and computational resources. Several vibration-based output only modal identification techniques are available in the literature for this purpose (see for example [5–7] and references therein). Since the estimated modal frequencies depend highly on changing environmental conditions (such as the temperature, and the wind direction/strength), the use of statistical methods is necessary so as to remove this dependance and provide reliable indicators of structural damage [8–15]. We also refer to [16] for a review on data normalization procedures for addressing the issue of separating structural changes of interest from operational and environmental variations. These statistical techniques, however, often require heavy computations and a long training period, of the order of one year, before the monitoring system can be operational [11]. To address this problem, we have introduced in [17] the Stretching Method (SM) which, in the context of SHM, allows the estimation of small shifts in the natural frequencies of a structure in a simple and efficient way without requiring a long training period.

The stretching method has been successfully used to detect and monitor changes in the substratum of the earth [18]. To do so, SM compares two waveforms, the current and the reference one, both obtained by cross-correlating ambient noise recordings. These two waveforms characterize the current and reference state of the substratum (e.g., the velocity with which waves propagate in the Earth) [19,20]. The difference between the two waveforms is the time-window over which the cross-correlations are averaged. A large time window is required to build the reference waveform, while a smaller one is used to obtain the current waveform. SM determines the amount of stretching that the current waveform needs so as to maximize its correlation with the reference waveform. This stretching corresponds to a relative change in the velocity with which waves propagate in the substratum.

The application of the stretching method in the structural health monitoring context was considered in [17]. There, SM was adequately modified so as to detect and monitor changes in the natural frequencies of a structure. In contrast to seismology, the cross-correlations instead of being compared in the time domain, they are processed in the frequency domain using the Fourier transform. In the SHM context, we can interpret the time-window over which the reference cross-correlation is computed as the training period of the method which is typically 7–9 days while the time needed to compute the current waveform determines the damage warning time shown to be 12–24 h [17].

In this paper we apply this SM-based methodology on output only vibration data collected on the Egyptian lighthouse located at the port of Rethymnon on the island of Crete in Greece. SM is successfully used to identify the natural frequencies of the structure related to flexural bending modes. However, due to lack of sufficient data, other modes such as axial or torsional, while shown to be present in the signals, were not vigorously identified. Based on a 2-hour time averaging period for the current quantity, we carry out an analysis of the fluctuations for the first two natural frequencies of the structure. Our analysis indicates strong dependence on temperature and also on wind direction. The fluctuations in the natural frequencies are reduced by increasing the time-averaging window over which the current waveform is computed. A time window of 24-hours seems to provide sufficiently accurate estimates of artificially imposed frequency shifts. This is determined with a sensitivity analysis that allows us to estimate with what accuracy different levels of frequency shifts in the natural frequencies can be detected. This analysis shows in particular that our methodology can be safely used for recovering frequency shifts as small as 1% when using 1-day intervals for averaging cross-correlations to compute the current quantity.

We also propose a simple model that is calibrated to match the estimated natural frequencies using model updating techniques. Generally speaking, a mathematical or numerical model may be used to simulate the behaviour and response under certain excitation of a structure. This model is often built using the finite element method. In practice, however, several properties of the real system are unknown or uncertain (i.e., material and geometric properties, boundary conditions, excitation), for which, inevitably, conjectures have to be made. Moreover, due to lack of knowledge or other restrictions, often simplifying modeling assumptions regarding the structure are required or implicitly made [21]. Due to these uncertainties, it is often better to define a simple initial model amenable to easy and efficient modifications. Furthermore, a simple surrogate model with a limited number of degrees of freedom is more appropriate to be used for possible near real time results as an output of some simulation that could be executed to embedded systems (e.g., microprocessor).

Here, we use a simple model that consists of two independent cantilever beam systems. The first refers to the east–west and the second to the south–north directions. Using model updating techniques [22] we seek to estimate elasticity (Young' modulus) and inertia (mass density) distribution over the length of the beams so as to match the model's eigenfrequencies to the estimated natural frequencies of the structure. The calibrated model, although very simple, matches (relatively well) the response of the structure to a small earthquake. This calibrated model is also used to provide possible damage scenarios associated with specific measured shifts.

The remainder of the paper is structured as follows. In Section 2, we present a brief review of the stretching method and explain how it can be used for operational modal analysis. Then, the application of the method and the analysis of the data collected on the Rethymnon lighthouse is carried out in Section 3. Section 4 is concerned with the model updating technique and Section 5 contains our conclusions.

2. Methodology Review: the Stretching Method

The Stretching Method (SM) is widely used in passive imaging applications mostly related to Geophysics [23–25]. The key idea of the method is to estimate how much we need to stretch a signal, referred to as current, so as to maximize its correlation coefficient with another signal, the reference. In Geophysics, the signals on which SM is applied are time domain cross correlations of seismic noise recordings. The reference signal characterizes the medium with no variations while the current characterizes the medium in its present condition. The difference between the two signals is the time interval over which the cross-correlations are averaged. Longer time intervals are required to obtain the reference quantity while for the current signal, the cross-correlations are averaged over a small time window. The resulting stretching parameter corresponds to the relative velocity change dv/v of the medium. Recently, SM was successfully used in the context of Structural Health Monitoring (SHM). The methodology is briefly explained below, for more details we refer the interested reader to [17].

Let $u_k(t, x_l)$ and $u_m(t, x_n)$ be the recordings of the acceleration due to random excitations in the k-th and m-th directions at locations x_l and x_n, respectively. Combination of direction and location defines a specific degree of freedom, $u^i(t) = u_k(t, x_l)$ and $u^j(t) = u_m(t, x_n)$. We use these recordings to compute the empirical cross-correlation function over the time interval $[0, T]$,

$$C_T^{i,j}(\tau) = \frac{1}{T} \int_0^T u^i(t) u^j(t + \tau) dt, \tag{1}$$

with τ denoting the lag-time. The cross-correlation function C_T is a self averaging quantity, which means that it is independent of the random excitation, provided that the averaging time window is large enough. More precisely, it can be shown that C_T converges to the average quantity C^{av},

$$C_T^{i,j}(\tau) \xrightarrow{T \to \infty} C_{av}^{i,j}(\tau), \tag{2}$$

where the average is defined with respect to the realizations of the random excitations

$$C_{av}^{i,j}(\tau) = \left\langle C_T^{i,j}(\tau) \right\rangle. \tag{3}$$

In the SHM context, SM is not applied directly to the time domain cross-correlations but rather to their Fourier transforms. Assuming that the excitations behave like random white noise and the structure is a linear and time invariant system, the elements of the cross-correlation matrix in the frequency domain correspond to power spectral densities. In this case, the resulting stretching parameter corresponds to the relative change dv/v in the natural frequencies of the structure. More precisely, we form the matrix

$$[A^{\#}(v)]_{i,j} = A_{i,j}^{\#}(v), \ \# = r, c$$

where $A_{i,j}^{\#}(v)$ is the Fourier transform of the reference ($\# = r$) or current ($\# = c$) empirical cross-correlation (1) computed for sensors i and j, $1 \leq i, j \leq n$, where n denotes the number of the available (measured) degrees of freedom,

$$A_{i,j}^{\#}(v) = \int C_T^{i,j}(\tau) e^{i 2\pi \tau v} d\tau. \tag{4}$$

The matrix $A^{\#}(\nu)$ is symmetric so it admits a symmetric singular value decomposition (SVD) of the form,

$$A^{\#}(\nu) = U^{\#}(\nu)\Sigma^{\#}(\nu)U^{\#,T}(\nu). \tag{5}$$

The matrix $\Sigma^{\#}(\nu)$ is a real diagonal matrix with the singular values $\sigma_1^{\#}(\nu), \ldots \sigma_n^{\#}(\nu)$ placed on the diagonal while the columns $\mathbf{u}_1^{\#}(\nu), \ldots, \mathbf{u}_n^{\#}(\nu)$ of the matrix $U^{\#}(\nu)$ are the corresponding singular vectors. Here $U^{\#,T}$ denotes the transpose of the matrix $U^{\#}$. The frequencies at which the first singular value admits peaks correspond to the natural frequencies of the structure while the first singular vector at the corresponding frequency is identified as the modal shape associated to this natural frequency (see Section 10.3.3 of [26]).

To monitor variations in the natural frequencies with the SM method we maximize the correlation coefficient

$$C_s(\Delta\nu) = \frac{\int_{\nu_1}^{\nu_2} \sigma_1^c(\nu + \Delta\nu)\sigma_1^r(\nu)\mathrm{d}\nu}{\sqrt{\int_{\nu_1}^{\nu_2}(\sigma_1^c(\nu + \Delta\nu))^2\mathrm{d}\nu \int_{\nu_1}^{\nu_2}(\sigma_1^r(\nu))^2\mathrm{d}\nu}}, \tag{6}$$

where σ_1^r and σ_1^c are the reference and current largest singular values of the matrices A^r and A^c respectively. The interval over which the optimization is performed is centered around each of the detected natural frequencies of the structure. The discretization is adaptively refined during the optimization process so as to provide an efficient solution. The result of this procedure provides a value for $\Delta\nu$ for each current quantity. Tracking the changes $\Delta\nu$ over time could help us detect potential structural damage [26].

3. Application to the Rethymnon Egyptian Lighthouse

In this section we apply the stretching method to structural vibration data of the Rethymnon lighthouse.

3.1. Structural Elements of the Lighthouse

The Egyptian lighthouse of Rethymnon is a masonry building constructed in 1838 while still Crete (Greece) was an Ottoman province under Egyptian control. The lighthouse came under the supervision of the French Lighthouse Company in 1864. It was operating until 1962. In 1930, for strengthening purposes, concrete reinforcement was used, as it may also be seen in the photo of Figure 1.

Figure 1. An older shot of the Egyptian lighthouse of Rethymno.

Elements of the geometric properties of the lighthouse have been adopted from data of an older rehabilitation study (1973, Lampakis, Civil Engineer), shown in Figure 2.

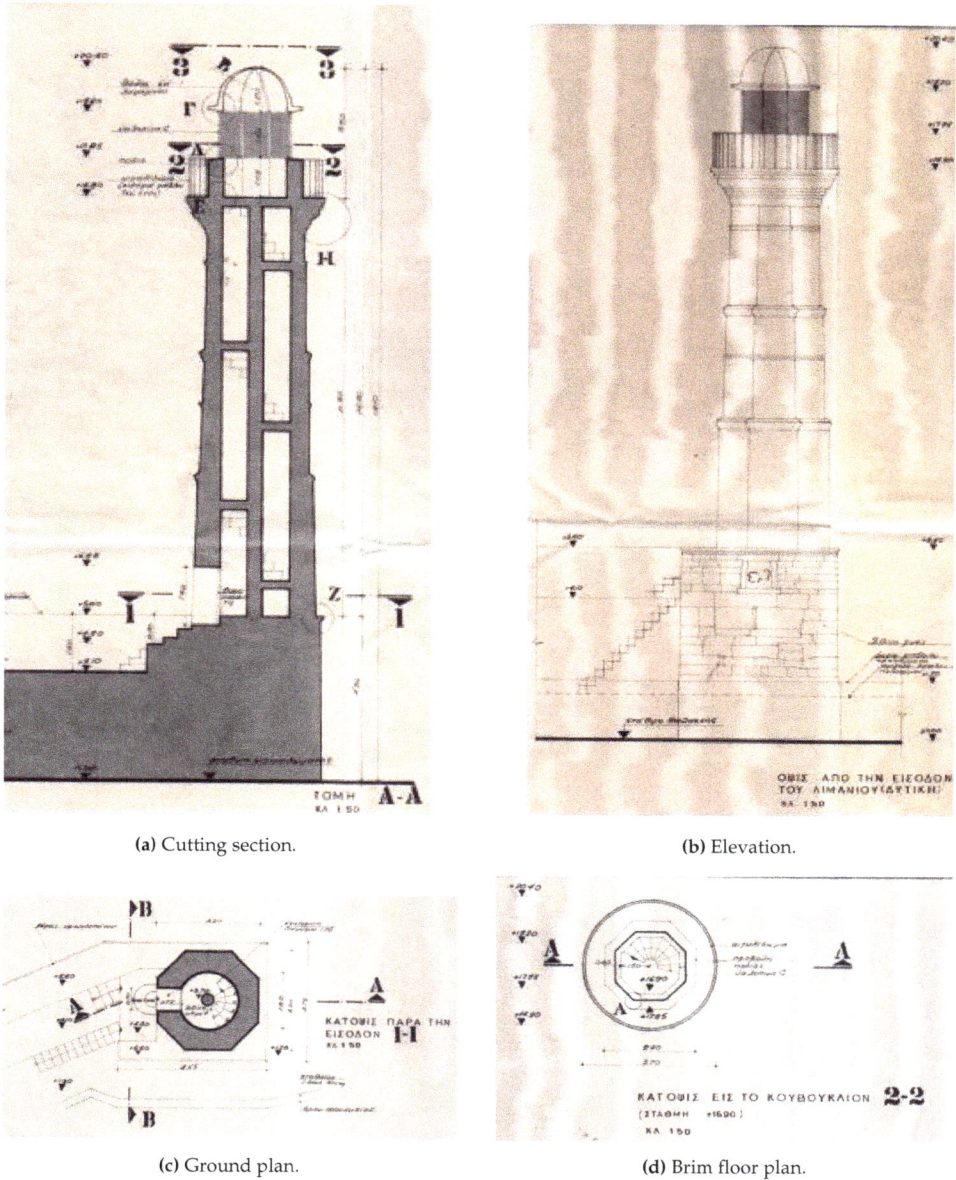

(a) Cutting section.

(b) Elevation.

(c) Ground plan.

(d) Brim floor plan.

Figure 2. Sketches of an older rehabilitation study.

Based on this study, and using a preliminary simplified model, we initially assume a length $L = 15.6$ m and consider the lighthouse as a hollow cylinder of constant cross section having internal radius $r_i = 1.2$ m and external $r_e = 1.8$ m. The material is assumed homogeneous with a modulus of Elasticity $E = 5$ GPa and mass density $\rho = 2.5$ tn/m³. These values belong in a range typical for masonry of historical buildings and give a good agreement between the model predicted eigenfrequencies (Section 3.3.1) and those estimated from the measurements (see Section 3.3.2). In Section 4.1.1, we use a heterogeneous cantilever beam model and estimate the distribution of its elasticity modulus and mass density by minimizing the misfit between the model predicted and the measured eigenfrequencies.

3.2. Instrument Placement

The deployment of the accelerographic instruments started in May 2017, after the Rethymno Ephorate of Antiquities expressed interest within the frame of the EU H-2020 STORM project. NOAIG installed two digital three-component accelerographs of CMG-5TDE type (Guralp Systems Ltd, 24 bit at 200 s/s, continuous local recording, time-stamped via GPS) at the top and the basement of the lighthouse. Care was taken for the installations to have the same orientation and to be at the same vertical axis of the lighthouse, in order to minimize possible effects from the directivity of the ground motion energy propagation and from the rotation of the body of the lighthouse.

The two instruments were recording acceleration in three directions (North, East and vertical) for a period of three months in separate 2-hour long files. We have a total of $n = 6$ degrees of freedom recorded corresponding to three directions at the top and three directions at the bottom of the lighthouse. Although the two instruments were vertically aligned, since the diameter of the lighthouse is changing with height, the alignment is not perfect.

In addition, we included in our study meteorological data (temperature, wind speed and wind direction), which were kindly offered by the Institute for Environmental Research and Sustainable Development of the National Observatory of Athens, that has developed and operates a network of automated weather stations to monitor weather conditions in Greece. Data were acquired from the weather station permanently deployed at the Municipal Enterprise of Water and Sewerage Supplies of Rethymno. The station is situated close to the coast and the Lighthouse under investigation. The meteorological station includes a Davis Instruments, type: Vantage Pro2 Wired Fan-Aspirated, weather monitoring instrument. Further detailed information on the station configuration and the produced data, as well as the NOA meteo network can be viewed in [27,28].

3.3. Modal Recognition

Using the limited number of sensors, that is two tri-axial (3D) sensors, it is possible to estimate the natural frequencies (eigenfrequencies) of the structure while for the corresponding eigenfunctions we only get a very crude estimate (i.e., their values at the two measurement locations).

3.3.1. Modal Recognition—Theoretical Expectations

Approximating the lighthouse tower as a cantilever beam of constant cross section and adopting the Euler–Bernoulli assumptions, we may analytically express the respective eigensystem and obtain an estimation for the eigenfrequencies and eigenfunctions pairs. We believe that according to the instruments' placement as described in Section 3.2, we could, with certainty, measure only flexural due to bending modes. Other modes such as axial or torsional may also exist, however, due to lack of sufficient data, we cannot safely confirm that.

The moment of inertia is taken to be $I = \pi(r_e^4 - r_i^4)/4 = 6.616 \, \text{m}^4$. The eigenfunctions corresponding to flexural vibration of the cantilever beam are (see for example in [29]),

$$\phi_n(x) = \cosh \lambda_n x - \cos(\lambda_n x) + \frac{\sinh(\lambda_n L) - \sin(\lambda_n L)}{\cosh(\lambda_n L) - \cos(\lambda_n L)} (\sinh(\lambda_n x) - \sin(\lambda_n x)) \qquad (7)$$

while, eigenfrequencies ω_n are given as,

$$\omega_n = \lambda_n^2 \sqrt{\frac{EI}{mL^4}} \qquad (8)$$

where, the first two coefficients $\lambda_n L$ are given as 1.8751 and 4.6941, respectively. Furthermore, $m=\rho A = 14.137$ tn/m, the beam's mass per length, and $A=\pi(r_e^2 - r_i^2) = 5.655$ m^2, the cross section area. The coefficient of flexural rigidity is $EI = 33080970.64$ kNm^2. The first two flexural (angular) eigenfrequencies ω_n, corresponding eigenperiods $T_n = 2\pi/\omega_n$, and ordinary eigenfrequencies $\nu_n = 1/T_n$, are calculated as:

$$\omega_1 = 22.101 \text{rad/sec}, \qquad T_1 = 0.284 \text{sec}, \qquad \nu_1 = 3.517 \text{Hz},$$
$$\omega_2 = 138.504 \text{rad/sec}, \qquad T_2 = 0.045 \text{sec}, \qquad \nu_2 = 22.043 \text{Hz}.$$

3.3.2. Modal Recognition—Signal Processing

To apply the SM method, we first compute the empirical cross-correlations according to Equation (1) for all different pairs of measurements. The reference quantity is computed by averaging the recordings over the first 8 days of acquisition. The current quantity will be computed either using a 2-hour interval or a 24-hour interval. The idea is to use for the current quantity the smallest possible window that guarantees detection of small shifts in $\Delta\nu$ with high accuracy. This allows for assessing any structural damage of the structure as soon as possible. As illustrated in Figure 4, $\Delta\nu$ measurement has a large variance due to temperature variations when a time-window of 2-hours for the current quantity is used. To chose the adequate size of the time window for the current quantity, a sensitivity analysis is carried out (results shown in Figures 7 and 8). This consists of artificially inserting a frequency shift in the monitoring data and then trying to recover it using SM. Our analysis suggests that the 24-hour interval is a good time window for detecting a frequency shift of 1% with high accuracy.

Then, the Fourier transform of the current and reference cross-correlations is computed and the corresponding 6×6 matrices are formed. Since the values of these matrices depend on the frequency, their singular value decomposition, performed frequency by frequency, provides singular values that are frequency dependent as well. Our analysis uses the first singular value of the corresponding matrix as reference and current quantity in (6).

To estimate the modes of the structure under examination, we look at the first singular value of the reference cross-correlation matrix as a function of frequency. The frequencies at which this function admits peaks are identified as structural modes. By looking at the plot in Figure 3 (i.e., the first singular value of the reference cross-correlation matrix as a function of frequency), we have identified the eigenfrequencies given in Table 1. We do not have an interpretation for the modes 3–5. We believe that they correspond to torsional and axial modes, however, not appropriately measured due to the instruments' placement.

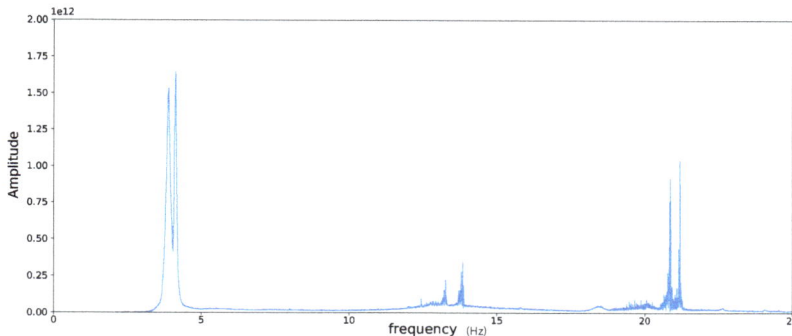

Figure 3. The first singular value of the reference cross-correlation matrix (computed using the first 8 days of acquisition) as a function of frequency (in Hz), $\sigma_1^r(\nu)$. The estimated values for the first seven natural frequencies of the structure are given in Table 1.

Table 1. The first seven eigenfrequencies estimated from the peaks of σ_1^r (see Figure 3).

Global Index n	ω_n (rad/sec)	ν_n (Hz)	Type
1	24.453	3.860	flexural, east-west
2	25.792	4.105	flexural, north-south
3	78.163	12.440	(?)
4	83.290	13.256	(?)
5	86.953	13.839	(?)
6	131.061	20.859	flexural, east-west
7	133.141	21.190	flexural, north-south

3.3.3. Modal Recognition—Theoretical Modal Values vs Recovered Modes and Conclusions

By comparing the results of Sections 3.3.1 and 3.3.2, we observe that quite good agreement has been obtained between theoretically expected, under certain rough approximations, and measured values for flexural modes of the structure. There are three estimated eigenfrequencies in the neighbourhood of 13 Hz, which due to lack of sufficient data, could not be safely recognized and attributed with specific physical meaning, while they seem to be related with torsional and axial response or other secondary motion of non-structural components on which instruments are located.

3.4. $\Delta\nu$ Measurements

We focus our attention now on the first two identified modes and look at their variations as a function of time over the acquisition period. It is well known that the variations of $\Delta\nu$ over time are mainly driven by variations in temperature [30]. To examine if the same holds here, we compare in Figure 4 the variations $\Delta\nu$ for the first two modes with the temperature variations over the same time period.

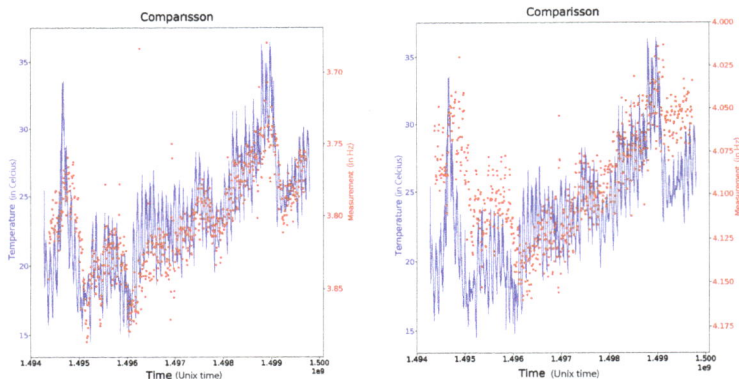

Figure 4. Comparison between the $\Delta\nu$ measurement and the temperature for the first (**left**) and second (**right**) mode.

With a first look at the results (see Figure 4), we observe that the measurement that corresponds to the first mode seems to be highly correlated with the temperature while the measurement of the second mode is less correlated with the temperature. More precisely, the measurement seems to be differently correlated at the very beginning and the very end of the measurement period in comparison with the period in between.

The explanation of this behaviour relies on the source of the excitation for these two modes. The two modes correspond to different flexural vibrations and, thus, each mode excitation is expected to depend on the wind direction.

In Figure 5, we plot the direction of the wind for two different time segments in the three month acquisition period. On the left we have a few days with wind that varies widely in direction and thus guarantees an excitation source for both modes measured in Figure 4. On the right we have a time period with winds that barely vary in direction (mostly East–West winds) and thus do not provide sufficient excitation for the second mode. Although the natural frequencies of the structure do not depend on the wind direction our ability to measure them using vibration data does depend on it. Indeed the SM assumes measurements due to white isotropic noise. If the noise is not isotropic, it affects the ability of SM to recover the measurement for $\Delta\nu$. This explains why the second mode does not correlate with the temperature in the same way as the first mode does during the periods that the wind is mostly along a single direction.

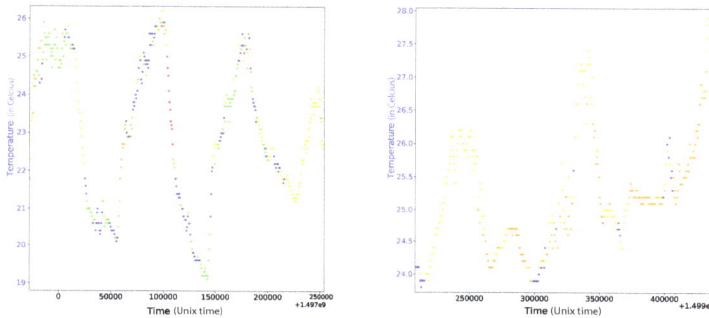

Figure 5. The temperatures for two different days and the correlation with the wind direction. The direction of the wind is represented by the different dot colors with blue to be a North–South direction and with red a West–East direction. On the left we have a couple of days at the middle of the measurement period and on the right a couple of days at the end of the measurement period.

To decrease the variations of $\Delta\nu$ so as to be able to asses damage associated with frequency shifts in the structural modes we propose to compute the current quantity using a time-window of 24-hours. This significantly reduces the variations of $\Delta\nu$ as can be seem in Figure 6.

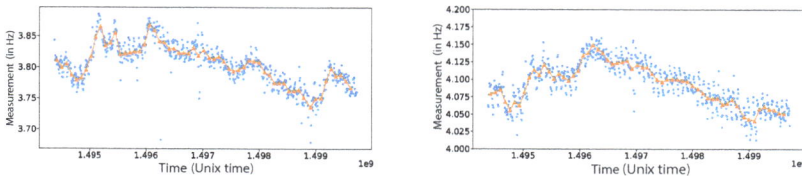

Figure 6. The estimated values for the first (right) and the second (left) eigenmode using for the current quantity a time window of 2 h (blue dots) and 24 h (orange triangles). The reference quantity is the same for both plots and is computed using the first 8 days of data.

To illustrate that SM can effectively allow us to detect small permanent shifts in the structural modes that could be associated with a damage, we artificially insert a frequency shift in the monitoring data and then try to recover it. Over the assumed period over which there is damage, we define the modified singular values $\tilde{\sigma}_i(\nu)$ as follows,

$$\tilde{\sigma}_i(\nu) = \sigma_i(\nu + \Delta\nu), \quad i = 1, \ldots, 7. \tag{9}$$

In Equation (9) we introduced a shift δv that corresponds to the small permanent shift associated to the damaged state. Then we form the correlation matrix using $\tilde{\Sigma}(v)$ instead of $\Sigma(v)$ in Equation (5) and apply the SM to this modified correlation matrix.

To determine with what accuracy we can detect different levels of frequency shifts in the natural frequencies, we carry out a sensitivity analysis. This consists in manually imposing a frequency shift as described above and then calculate the error of the SM method on recovering the imposed frequency shift. We perform this analysis for frequency shifts from 0.1% to 1% using both 1-day and 1-week measurements to define the current quantity. The results are illustrated in Figure 7, where we observe that the estimation is good for both cases with relative errors below 10% . As expected, the accuracy increases by increasing the time window used for the current quantity. Also, larger frequency shifts can be determined with higher accuracy. Noting that a frequency shift of 1% can be recovered with a relative error less than 4%, we deduce that we can trust our methodology for recovering frequency shifts as small as 1% when using 1-day measurements as current quantity.

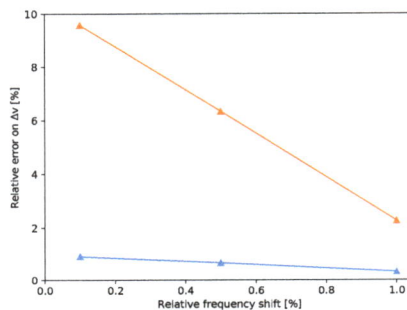

Figure 7. Relative error on Δv when recovering frequency shifts from 0.1% to 1% using either 1 day (orange) or 1 week (blue) as current quantity.

To visualize the performance of SM, we insert a 1% permanent frequency shift at the measurements shown in Figure 6 and try to recover it. The results for the first mode are illustrated in Figure 8. We observe a very good agreement between the estimated and the exact frequency shift indicating that indeed SM can be used in SHM.

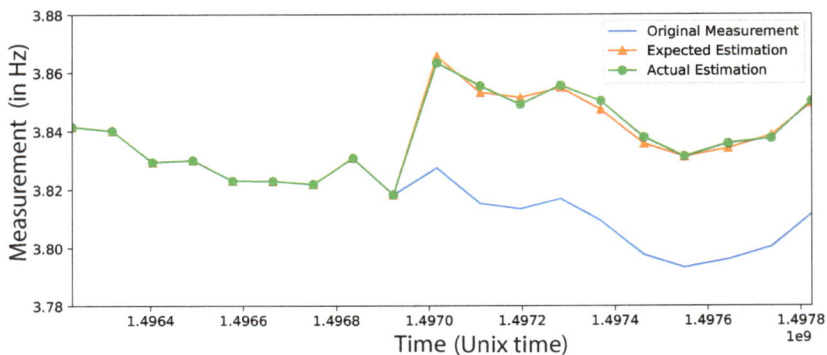

Figure 8. The estimation of first mode using a 24-hour time window for the current quantity and imposing shift of 1%. With orange we plot the exact imposed shift and with green we plot the estimation provided by the SM. We observe very good agreement.

Using a simplified numerical model we will provide in the next section possible damage scenarios that correspond to a frequency shift of 1%.

3.5. The Estimated Mode Shapes

Studying the modal values and their variations over time is not always enough for a complete understanding of what the recovered modes actually represent. The whole picture comes from studying the mode's shape as well. Using the SVD (5) we compute the singular vector corresponding to the largest singular value for the reference quantity at the identified structural frequencies reported on Table 1. The results for the first four modal shapes are reported in Table 2 while the respective degrees of freedom, are shown in Figure 9 together with a schematic illustration of the first two eigenmodes.

Table 2. Modal Shapes.

Direction/Position	First Mode	Second Mode	Third Mode	Forth Mode
East–West (EW)/bottom	−0.0801644	−0.0248191	−0.352925	−0.334469
North–South (NS)/bottom	−0.0566074	−0.0120627	−0.314454	−0.29505
Vertical (V)/bottom	−0.00216538	−0.00592483	−0.0370501	−0.121381
East–West (EW)/top	−0.887227	−0.473897	−0.788803	−0.771983
North–South (NS)/top	−0.450108	−0.87642	−0.364553	−0.385618
Vertical (V)/top	−0.024408	−0.0807035	−0.141706	−0.204166

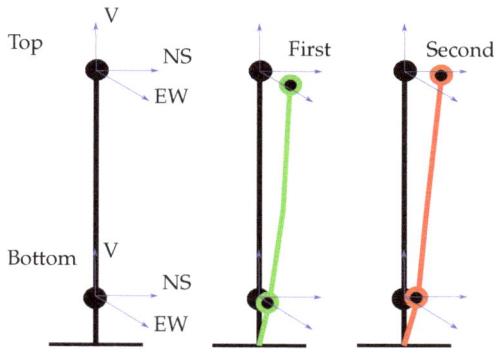

Figure 9. Measured degrees of freedom and schematic illustration of the first two eigenmodes.

4. Model Recalibration

In this section we describe the approach we have adopted in order to define a simple yet satisfactory model of the structure under consideration. Finite element simulation of a very similar lighthouse structure in Chania, an almost 80 km neighbor city west of Rethymno, has been accomplished in the past and presented in the literature [31]. The models used there were quite detailed ones with 3D simulations taking place. In the current work we have been experimenting with simpler and more adaptive 1D models with which parametric studies could be easily accomplished, and advanced techniques (minimization, transient analyses) could be more efficiently implemented. The objective here is to construct such simple models having the necessary dynamic properties and still being able to give results in near real time by simulations executed in embedded electronic devises. For that a Java finite element library together with other necessary units (e.g., Apache Maths Common) have been utilized in the development environment of SDE [32]. The simplest model that might be used for such structures is a cantilever beam, dynamic properties of which, in terms of the eigensystem, have been considered in Section 3.3.1. Here we use such a model in combination with the finite element method and appropriate calibration.

4.1. Modal Characteristics Consideration

Estimated mode shapes using the acquired data of lighthouse response (see Table 2) indicate the possible coupling of bending to torsional modes. That was expected since cross section asymmetric geometry and/or non-homogeneity of material distribution on the cross section imply coupling of bending and torsional vibrations, since shear and mass center do not coincide over the height. More generally, when geometric, shear and mass centers of the cross sections of a uniform beam do not coincide then coupling of flexural, torsional (including possible warping) and tensional vibration modes occurs [33]. Yet, that coupling might have mostly appeared because of sensors' positioning. Furthermore, as mentioned before in this work, due to lack of sufficient data, we do not consider any possible torsional and/or tensional motion and we confine ourselves to the study of structure's bending as a cantilever beam system.

Therefore, we proceed by defining two independent cantilever beam systems, avoiding in such a way the necessity of some 3D model of the beam. The first refers to the East–West and the second to the South-North direction.

4.1.1. Model Calibration as A Minimization Problem

Here we present the model updating procedure. The parameters we search for are the elasticity and inertia distribution over the length of the structure as represented by the elasticity modulus and mass density. We have admitted the quite reasonable assumption of common distribution for both elasticity and inertia.

The methodology we setup is a sensitivity-based element-by-element updating procedure [34]. We use some piecewise linear distribution $\psi_s(x)$ as:

$$\psi_s(x) = (H[x - x_{s-1}] - H[x - x_s]) \frac{x - x_{s-1}}{\Delta x} + (H[x - x_s] - H[x - x_{s+1}]) \frac{\Delta x - x - x_s}{\Delta x}, \tag{10}$$

where $x \in [0, L]$, $\Delta x = x_s - x_{s-1}$ the length of each section and $H(x)$ the Heaviside step function. Then, both elasticity modulus and mass density distributions might be given as:

$$E(x) = E_0 \sum_{s=1}^{S} a_s \psi_s(x) \text{ and } \rho(x) = \rho_0 \sum_{s=1}^{S} a_s \psi_s(x). \tag{11}$$

Here, we assume ten sections of linear distribution of material properties defined by the coefficients a_s which are actually the unknown variables of the minimization problem. Having defined these coefficients a_s we also establish the material elasticity and inertia properties distribution. The objective function f to be minimized over the S design variables a_s might be given as,

$$f(a_s) = (v_1(a_s) - \tilde{v}_1)^2 + \left(\frac{v_2(a_s)}{v_1(a_s)} - \tilde{r}_{21} \right)^2 \tag{12}$$

accompanied by simple bound constraints on a_s. We take advantage of the powerful limited-memory BFGS gradient algorithm [35]. The necessary derivative of the objective function f with respect to a_s is given by,

$$\frac{\partial f}{\partial a_s} = 2(v_1 - \tilde{v}_1) \frac{\partial v_1}{\partial a_s} + \frac{2}{v_1^3} (\tilde{r}_{21} v_1 - v_2)(v_2 \frac{\partial v_1}{\partial a_s} - v_1 \frac{\partial v_2}{\partial a_s}), \tag{13}$$

where $\frac{\partial v_i}{\partial a_s}$ denote the sensitivity indexes for the eigenfrequencies. For the dependence of a given eigenfrequency $\omega_i = 2\pi v_i$ on design variable a_s we get the explicit form [26,36],

$$\frac{\partial \omega_i}{\partial a_s} = \frac{1}{2\omega_i m_i^*} \phi_i^T \left(\frac{\partial K}{\partial a_s} - \omega_i^2 \frac{\partial M}{\partial a_s} \right) \phi_i \tag{14}$$

where ϕ_i the *i*-th and m_i^* the respective generalized mass for that mode given by the product $\phi_i^T M \phi_i$. Sensitivities of stiffness and mass matrices can be explicitly expressed on parameters such as the cross section area A, moment of inertia J, elasticity modulus E and mass density ρ among others. Sensitivity matrices with respect to such parameters are given in the literature (e.g., for the case of Euler-Bernoulli beam [34]). However, as it can be seen in Equation (14), we need to compute the sensitivities $\frac{\partial K}{\partial a_s}$ and $\frac{\partial M}{\partial a_s}$, of stiffness and mass matrix, respectively. Therefore, considering the chain rule, we get

$$
\begin{aligned}
\frac{\partial \omega_i}{\partial \alpha_s} &= \frac{\partial \omega_i}{\partial E}\frac{\partial E}{\partial \alpha_s} + \frac{\partial \omega_i}{\partial \rho}\frac{\partial \rho}{\partial \alpha_s} \\
&= \frac{1}{2\omega_i m_i^*}\phi_i^T\left(\frac{\partial K}{\partial E} - \omega_i^2\frac{\partial M}{\partial E}\right)\phi_i\frac{\partial E}{\partial a_s} + \frac{1}{2\omega_i m_i^*}\phi_i^T\left(\frac{\partial K}{\partial \rho} - \omega_i^2\frac{\partial M}{\partial \rho}\right)\phi_i\frac{\partial \rho}{\partial a_s}
\end{aligned}
\tag{15}
$$

with the sensitivities of stiffness and mass matrices over elasticity modulus and mass density parameters computed using the finite element method and $\frac{\partial E}{\partial a_s}$, $\frac{\partial \rho}{\partial a_s}$ depending only on geometrical features and being easily obtained from Equation (11).

We consider the two independent cantilever beams for the EW and SN directions respectively defining the finite element model using two hundred elements of equal length, and we set $E_0 = 5.31$ GPa and $\rho_0 = 2.5$ tn/m^3. Using Table 1, the target values for the EW model are $\tilde{v}_1 = 3.860$ Hz and $\tilde{r}_{21} = 20.859/3.860$. For the NS model these values are asked to be $\tilde{v}_1 = 4.105$ Hz and $\tilde{r}_{21} = 21.190/4.105$, respectively. Results for the distribution of elasticity modulus and mass density for both EW and SN models are shown in Figure 10, where we plot the coefficient over the length of the structure with which we should multiply E_0 and ρ_0 in order to obtain the target values for each model.

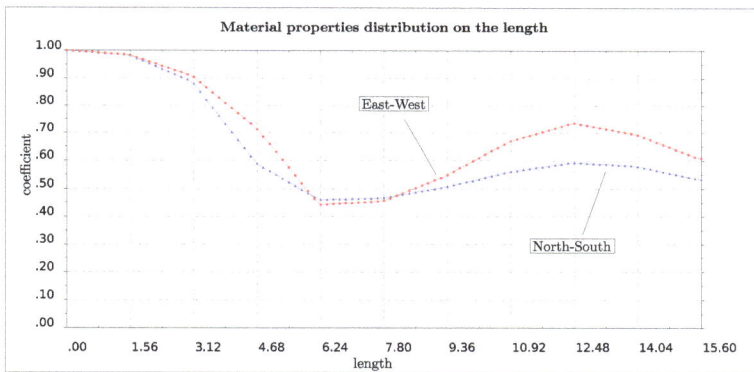

Figure 10. Material properties (elasticity modulus, mass density) nondimensional distribution (vertical axis) with the height (horizontal axis) of the lighthouse.

4.1.2. Possible Damage Profiles

As it has been shown in Section 3.4, the SM methodology is able to detect one percent variation of the eigenfrequencies. Here we present some possible damage profiles corresponding to that level of variation of eigenfrequencies. In order to do so we assume a variable damage profile of some damage starting from the base of the lighthouse and possibly spreading towards the top. It is worth mentioning that the case of localized severe damage to some restricted area near the base could occur because of an earthquake, while a distributed over the length moderate damage could be the result of some corrosion process (e.g., alveolization, the mechanical action of salt crystallization and action of wind) [37]. To define the damage height we use the coefficient h_d taking values from zero up to one and corresponding to zero height, measuring from the base of the lighthouse, up to the total height of it. For this damage area we define a second coefficient z_d with which the moment of inertia for cross sections belonging to that damaged area is multiplied. This coefficient z_d takes values from one which stands for fully operating cross section up to zero which means totally damaged cross section. The use of this damage variable is in accordance with instructions for moment of inertia reduction because of structural strength deterioration due to cracking. We again solve a minimization problem where, for some specific damage area height h_d, we ask for the value of the damage variable z_d such that the variation of the first or second eigenfrequency is of a specific given level (e.g., 1%–2%). The results are shown in Figure 11, where we present curves of constant frequency variance for one and two percent, respectively. The horizontal axis stands for the length of the damaged area measured from the bottom up to the total height. The vertical axis is the damage level of this area, represented by the coefficient $1 - z_d$ that takes values from 0 (fully operative) to one (totally damaged). We observe in Figure 11 that in order to have a frequency reduction of 1%, a damage level of 0.6 (corresponding to $z_d = 0.4$) is localized in a small area from the bottom, at about 0.5% of the total length. We also observe that to get a frequency reduction of 2% for the same length of damaged area the respective damage level should be higher, having a value at about 0.75, corresponding to damage variable $z_d = 0.25$.

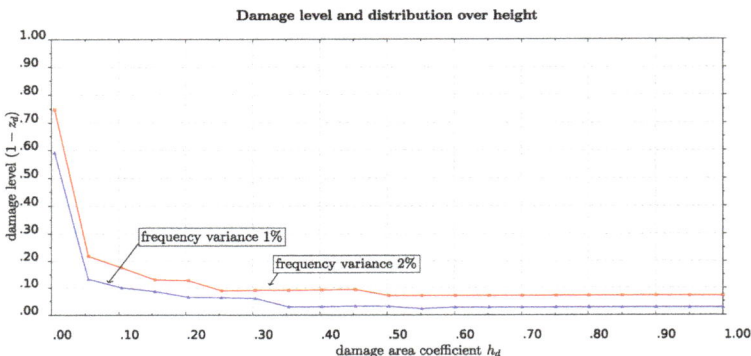

Figure 11. Combination of damage level z with damaged area length $h_d L$, for variation of eigenfreqency $\Delta v = 1\%$ and 2%.

4.2. Time History Consideration

In this last section we try to further validate our model in the time domain taking advantage of a small earthquake that occurred and recorded on June 21st, 2017. We plot in Figure 12 the recorded acceleration integrated in time to get velocities and displacements. The East–West components are shown on the left column while on the right are the North–South ones. It is observed in these plots that spurious drifts have appeared in the integrated velocities and displacements time histories. Since they have appeared in both the lower and upper levels of the structure's measured response, we keep them as they are for further analysis.

In the model, we impose the motion recorded at the lower level of the lighthouse shown in Figure 12 and compute the response at the top level which we further compare with the recorded motion on that level of the lighthouse (see Figure 13). We consider both EW and SN models for which we impose the recorded motion on the lower level. This position is at a height about 10% of the total length while we refer to that as the bottom since it is assumed to be the level on the ground floor of the lighthouse. We time integrate the equation of motion of the finite element system using the β-Newmark method. The critical damping ratio ζ has been estimated to be around 3% using the half power method [29]. We impose motion at the bottom as constraints through the consistent penalty method [38], for which both acceleration and displacement time histories are needed to be given as input.

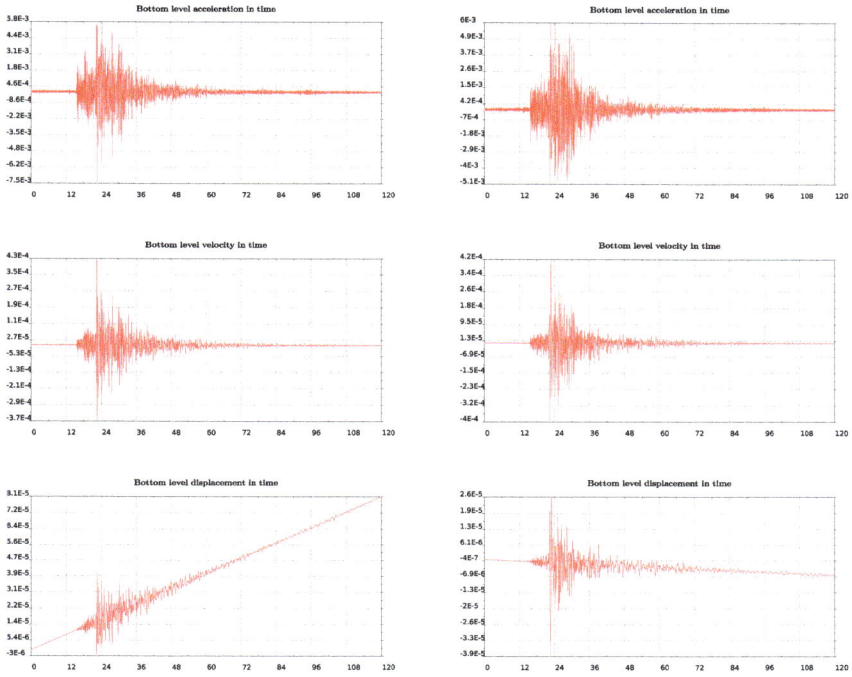

Figure 12. From top to bottom, recorded acceleration integrated in time to give velocities and displacements. On the left column East–West while on the right North–South components are plotted.

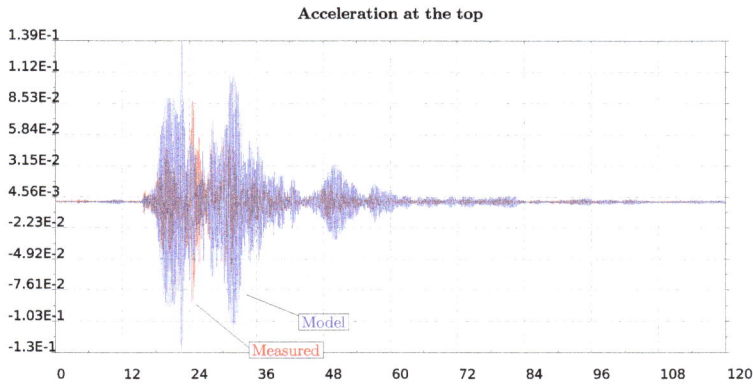

Figure 13. *Cont.*

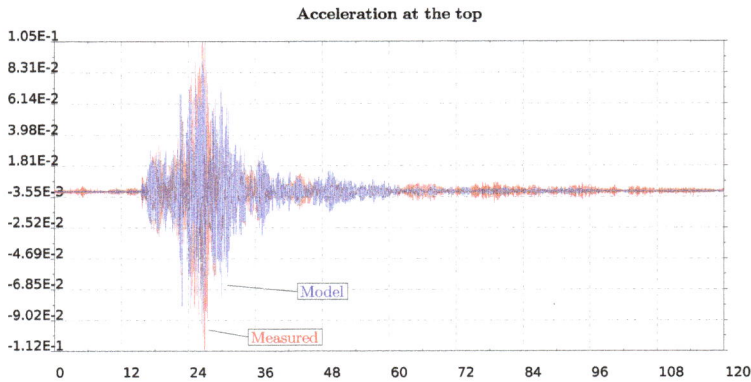

Figure 13. Response of top level acceleration for the East–West model (top plot) and for the North–South (bottom plot), respectively. Both recorded motion (red) and computed by the model (blue) are presented.

In order to compare measured and computed (by the model) results of velocities and displacements at the top of the structure, we proceed by eliminating the drifts. For that, we use a polynomial detrending of the first and second order, for velocities and displacements, respectively. The response at the top of the lighthouse, as computed by the simple (but calibrated) numerical model, is in quite good agreement with the measured response of the structure (see Figures 13 and 14) especially for the North–South component. Note, in particular, that there is an amplification factor of 10 between the bottom and the top acceleration in the measurements, which is well captured by the model.

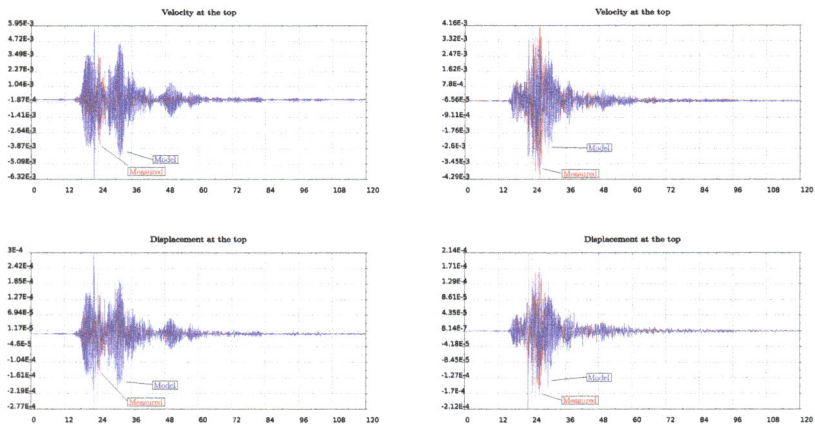

Figure 14. Response of top level velocities (top) and displacements (below) for the East–West model (left) and for the North–South (right), respectively. Both recorded motion (red) and computed by the model (blue) are presented.

5. Conclusions

The stretching method is a computationally efficient vibration-based technique that uses a very limited number of sensors permanently installed on site to measure operational structural responses for the purpose of damage detection. The preliminary study carried out in this paper on data collected on the Egyptian lighthouse of Rethymnon illustrates the potential of using SM for structural health monitoring of historic monuments. Although our results are quite promising, further effort is needed as regards to both the data acquisition and the model setup. In this study, our model is kept very simple since we do not have so many data to exploit. Note that only two instruments were placed on the lighthouse for a short period of 3 months. A more careful deployment of the instruments (so as to be able to measure axial or torsional modes) on a permanent basis would allow us to carry a more in depth study and illustrate effectively that SM allows for early warning of structural damage.

Author Contributions: Methodology, C.T., E.D and C.G.P; software, E.D. and C.G.P; validation, E.D. and C.G.P.; data curation, N.S.M and I.K.; writing–original draft preparation, C.T., E.D and C.G.P; writing–review and editing, C.T., E.D and C.G.P; supervision, C.T.

Funding: CGP acknowledges partly financial support of the Stavros Niarchos Foundation within the framework of the project ARCHERS ("Advancing Young Researchers' Human Capital in Cutting Edge Technologies in the Preservation of Cultural Heritage and the Tackling of Societal Challenges").

Acknowledgments: Images of Figures 1 and 2 courtesy of Aristodimos Chatzidakis (president of European Council of Civil Engineers, deputy president of Greek Earthquake Planning and Protection Organization).

Conflicts of Interest: The authors declare no conflict of interest.

References

1. Brencich, A.; Sabia, D. Experimental identification of a multi-span masonry bridge: The Tanaro Bridge. *Constr. Build. Mater.* **2008**, *22*, 2087–2099. [CrossRef]
2. Casarin, F.; Modena, C. Seismic Assessment of Complex Historical Buildings: Application to Reggio Emilia Cathedral, Italy. *Int. J. Archit. Herit.* **2008**, *2*, 304–327. [CrossRef]
3. Ivorra, S.; Pallarés, F.J. Dynamic investigations on a masonry bell tower. *Eng. Struct.* **2006**, *28*, 660 – 667. [CrossRef]
4. Gentile, C.; Saisi, A.; Cabboi, A. Structural Identification of a Masonry Tower Based on Operational Modal Analysis. *Int. J. Archit. Herit.* **2015**, *9*, 98–110. [CrossRef]
5. Magalhães, F.; Cunha, A.; Caetano, E. Online automatic identification of the modal parameters of a long span arch bridge. *Mech. Syst. Signal Process.* **2009**, *23*, 316 – 329. [CrossRef]
6. Reynders, E.; Houbrechts, J.; Roeck, G.D. Fully automated (operational) modal analysis. *Mech. Syst. Signal Process.* **2012**, *29*, 228 – 250. [CrossRef]
7. Ubertini, F.; Gentile, C.; Materazzi, A. Automated modal identification in operational conditions and its application to bridges. *Eng. Struct.* **2013**, *46*, 264–278. [CrossRef]
8. Worden, K.; Sohn, H.; Farrar, C. Novelty detection in a changing environment: Regression and interpolation approaches. *J. Sound Vibr.* **2002**, *258*, 741–761. [CrossRef]
9. Yan, A.; Kerschen, G.; De Boe, P.; Golinval, J. Structural damage diagnosis under varying environmental conditions part II: local PCA for non-linear cases. *Mech. Syst. Signal Process.* **2005**, *19*, 865–880. [CrossRef]
10. Farrar, C.R.; Worden, K. *Structural Health Monitoring: A Machine Learning Perspective*; John Wiley & Sons, Ltd.: Hoboken, NJ, USA, 2012.
11. Magalhães, F.; Cunha, A.; Caetano, E. Vibration based structural health monitoring of an arch bridge: From automated OMA to damage detection. *Mech. Syst. Signal Process.* **2012**, *28*, 212–228. [CrossRef]
12. Mosavi, A.; Dickey, D.; Seracino, R.; Rizkalla, S. Identifying damage locations under ambient vibrations utilizing vector autoregressive models and Mahalanobis distances. *Mech. Syst. Signal Process.* **2012**, *26*, 254–267. [CrossRef]
13. Dackermann, U.; Smith, W.; Randall, R. Damage identification based on response-only measurements using cepstrum analysis and artificial neural networks. *Struct. Health Monit.* **2014**, *13*, 430–444. [CrossRef]
14. Dervilis, N.; Worden, K.; Cross, E. On robust regression analysis as a means of exploring environmental and operational conditions for SHM data. *J. Sound Vibr.* **2015**, *347*, 279 – 296. [CrossRef]
15. Comanducci, G.; Ubertini, F.; Materazzi, A. Structural health monitoring of suspension bridges with features affected by changing wind speed. *J. Wind Eng. Ind. Aerodyn.* **2015**, *141*, 12–26. [CrossRef]
16. Sohn, H. Effects of environmental and operational variability on structural health monitoring. *Philos. Trans. R. Soc. A: Math. Phy. Eng. Sci.* **2006**, *365*, 539–560. [CrossRef]

17. Tsogka, C.; Daskalakis, E.; Comanducci, G.; Ubertini, F. The Stretching Method for Vibration-Based Structural Health Monitoring of Civil Structures. *Comput.-Aided Civil Infrastruct. Eng.* **2017**, *32*, 288–303. [CrossRef]

18. Campillo, M.; Roux, P. 1.12—Crust and Lithospheric Structure—Seismic Imaging and Monitoring with Ambient Noise Correlations. In *Treatise on Geophysics*, 2nd ed.; Schubert, G., Ed.; Elsevier: Oxford, UK, 2015; pp. 391–417.

19. Snieder, R.; Wapenar, K.; Wegler, U. Unified Green's function retrieval by cross-correlation; connection with energy principles. *Phys. Rev. E* **2007**, *75*, 036103. [CrossRef]

20. Garnier, J.; Papanicoloau, G. *Passive Imaging with Ambient Noise*; Cambridge University Press: Cambridge, UK, 2016.

21. Simoen, E.; Roeck, G.D.; Lombaert, G. Dealing with uncertainty in model updating for damage assessment: A review. *Mech. Syst. Signal Process.* **2015**, *56*, 123–149. [CrossRef]

22. Mottershead, J.E.; Friswell, M.I. *Finite Element Model Updating in Structural Dynamics*; Kluwer Academic Publishers Dordrecht: Dordrecht, The Netherlands, 1995.

23. Hadziioannou, C.; Larose, E.; Coutant, O.; Roux, P.; Campillo, M. Stability of monitoring weak changes in multiply scattering media with ambient noise correlation: Laboratory experiments. *J. Acoust. Soc. Am.* **2009**, *125*, 2654–2654. [CrossRef]

24. Hadziioannou, C.; Larose, E.; Baig, A.; Roux, P.; Campillo, M. Improving temporal resolution in ambient noise monitoring of seismic wave speed. *J. Geophys. Res.* **2011**, *116*, 2156–2202. [CrossRef]

25. Weaver, R.L.; Hadziioannou, C.; Larose, E.; Campillo, M. On the precision of noise correlation interferometry. *Geophys. J. Int.* **2011**, *185*, 1384–1392. [CrossRef]

26. Brincker, R.; Ventura, C. *Introduction to Operational Modal Analysis*; Wiley: Hoboken, NJ, USA, 2015.

27. NOA Meteo Network. Available online: http://penteli.meteo.gr/stations/rethymno/ (accessed on 10 March 2018).

28. Lagouvardos, K.; Kotroni, V.; Bezes, A.; Koletsis, I.; Kopania, T.; Lykoudis, S.; Mazarakis, N.; Papagiannaki, K.; Vougioukas, S. The automatic weather stations NOANN network of the National Observatory of Athens: operation and database. *Geosci. Data J.* **2017**, *4*, 4–16. [CrossRef]

29. Clough, R.W.; Penzien, J. *Dynamics of structures*; McGraw-Hill: Singapore, 1993.

30. Ubertini, F.; Comanducci, G.; Cavalagli, N.; Laura Pisello, A.; Luigi Materazzi, A.; Cotana, F. Environmental effects on natural frequencies of the San Pietro bell tower in Perugia, Italy, and their removal for structural performance assessment. *Mech. Syst. Signal Process.* **2017**, *82*, 307–322. [CrossRef]

31. Stavroulaki, M.; Sapounaki, A.; Leftheris, B.; Tzanaki, E.; Stavroulakis, G. Materials Modelling and Modal Analysis of the Lighthouse in the Venetian Harbour of Chania. *Tech. Mech.* **1998**, *18*, 251–259.

32. Panagiotopoulos, C.G. Symplegma, a JVM Implementation for Numerical Methods In Computational Mechanics. In Proceedings of the Free & Open-Source Software Communities Meeting, Heraklion, Greece, 13–14 October 2018. Available online: http://symplegma.org/ (accessed on 6 June 2019).

33. Friberg, P. Coupled vibrations of beams—an exact dynamic element stiffness matrix. *Int. J. Numer. Meth. Eng.* **1983**, *19*, 479–493. [CrossRef]

34. Farhat, C.; Hemez, F.M. Updating finite element dynamic models using an element-by-element sensitivity methodology. *AIAA J.* **1993**, *31*, 1702–1711. [CrossRef]

35. Zhu, C.; Byrd, R.H.; Lu, P.; Nocedal, J. Algorithm 778: L-BFGS-B: Fortran subroutines for large-scale bound-constrained optimization. *ACM Trans. Math. Softw.* **1997**, *23*, 550–560. [CrossRef]

36. Manolis, G.; Panagiotopoulos, C.; Paraskevopoulos, E.; Karaoulanis, F.; Vadaloukas, G.; Papachristidis, A. Retrofit strategy issues for structures under earthquake loading using sensitivity-optimization procedures. *Earthq. Struct.* **2010**, *1*, 109–127. [CrossRef]

37. Papayianni, I.; Pachta, V. Damages of old Lighthouses and their repair. In Proceedings of the 1st International Conference on Construction Heritage in Coastal and Marine Environments, Damage, Diagnostics, Maintenance and Rehabilitation, Lisbon, Portugal, 28–30 January 2008.

38. Paraskevopoulos, E.A.; Panagiotopoulos, C.G.; Manolis, G.D. Imposition of time-dependent boundary conditions in FEM formulations for elastodynamics: critical assessment of penalty-type methods. *Comput. Mech.* **2010**, *45*, 157. [CrossRef]

MDPI

St. Alban-Anlage 66

4052 Basel

Switzerland

Tel. +41 61 683 77 34

Fax +41 61 302 89 18

www.mdpi.com

Sensors Editorial Office

E-mail: sensors@mdpi.com

www.mdpi.com/journal/sensors

www.ingramcontent.com/pod-product-compliance
Lightning Source LLC
Chambersburg PA
CBHW051712210326
41597CB00032B/5455